Thermophiles: The Keys to Molecular Evolution and the Origin of Life?

Thermophiles: The Keys to Molecular Evolution and the Origin of Life?

JUERGEN WIEGEL and MICHAEL W. W. ADAMS

University of Georgia, Athens, USA

UK Taylor & Francis Ltd, 1 Gunpowder Square, London EC4A 3DE
USA Taylor & Francis Inc., 325 Chestnut Street, Philadelphia, PA 19106

British Library Cataloguing-in-Publication Data

A catalogue record for this book is available from the British Library.

ISBN 0–7484–0747–2 (HB)

Library of Congress Cataloging-in-Publication Data are available

Cover design by Jim Wilkie

Typeset in Times 10/12pt by Graphicraft Limited

Printed by T.J. International Ltd, Padstow, UK

Contents

Preface

This book presents updated chapters written by selected speakers at an international workshop entitled 'Thermophiles: the Keys to Molecular Evolution and the Origin of Life?', held at the University of Georgia, Athens, USA in September 1996. The Workshop was preceded by an International Conference on 'The Biology, Ecology and Biotechnology of Thermophilic Organisms' at the same venue. This was part of a series of meetings held every two years or so on thermophilic organisms, with future meetings to be held in France (1998) and India (2001).

Why did we organise a Workshop dealing with thermophiles and the controversial topics of evolution and the origin of life? The answer lies in the recent developments that have occurred in the field of thermophilic microorganisms. First, in the last decade or so, a new breed of organisms, termed 'hyperthermophiles', has been discovered, largely through the efforts of Karl Stetter at the University of Regensburg. These microorganisms have the remarkable property of growing optimally at temperatures near, and even above, 100 °C, the normal boiling point of water. Second, phylogenetic analyses based on comparisons of 16S ribosomal RNA sequences, a technique pioneered by Carl Woese of the University of Illinois, place the hyperthermophiles as the most 'slowly-evolving' of all extant life, the first to have diverged from the last common ancestor. The obvious implication is that life may have first evolved on this planet under 'hyperthermophilic' conditions, presumably when the earth was much hotter than it is at present. Consequently, evolution would then represent the adaptation of early life forms to cooler and cooler environments, leading to the mesophilic world we see today in which most life forms thrive in the 10–40 °C temperature range.

Of course, to most (although not all) who work with thermophilic organisms, the notion of life first evolving at extreme temperatures was an extremely appealing idea. If it were true, then elucidating the physiological, metabolic and molecular properties of present-day thermophiles could offer tremendous insight not only into the evolution of all living organisms at the molecular level, but also into the origin of life itself. But what evidence is there for this somewhat self-serving assumption? The majority of researchers in the thermophile field obviously have backgrounds (as do the editors of this book) in microbiology and biochemistry, but what do evolutionists, geochemists and oceanographers think about a thermophilic origin? Is there evidence in support of a hot early earth? Or do mainstream evolutionists still think in terms of a 'lukewarm' prebiotic broth? Is the Woesian 16S rRNA-based phylogenetic tree a fact of life (all three domains of it), is it supported by other molecular analyses, or are there viable alternatives? And what can be concluded from analysing the molecules and metabolic properties of hyperthermophilic organisms? Does the nature of the organisms themselves support the idea an 'ancestral', hyperthermophilic life form?

The aim of the workshop, and of this book, was, and is, to address such questions. Opinions from some of the leading protagonists in the various fields are presented, each giving their own perspective on the likely validity of a thermophilic origin of life. As editors, we have advised only on style and that each assertion, where possible, be backed up by relevant references in the primary literature.

So, what is the answer? What do we know about how life originated and the role of thermophilic organisms? In Chapter 1, John Baross, an oceanographer and microbiologist, shows how there is considerable evidence in support of a thermophilic origin from the geological record, and reviews the alternatives to the 16S R(D)NA tree. His conclusions on phylogeny are complemented and considerably extended by microbiologist Otto Kandler in Chapter 2, who introduces the concept of the evolutionary 'bush'. That the earth was indeed thermophilic during its early history is argued from a geological perspective by geologist David Schwartzman in Chapter 3. Mechanisms for how life may well have first originated based on known chemistry at thermophilic temperatures are the topics of Chapters 4 to 6, presented by evolutionist Günter Wächtershäuser, geochemist Everett Shock and colleagues, and geologist Michael Russell and coworkers, respectively. However, in Chapter 7, chemists Stanley Miller and Antonio Lazcano argue that the chemical properties of biological molecules are not consistent with a thermophilic origin. Geneticist Patrick Forterre presents a similar thesis in Chapter 8, together with a thought-provoking analysis of the meaning and interpretation of molecular phylogeny and the alternatives to the 16S-RNA based conclusions. This is taken a step further by molecular biologist James Lake and coworkers, who present data in support of a different version of the rRNA-based evolutionary tree in Chapter 9.

Peter Gogarten and coworker in Chapter 10 show how phylogenetic analyses are complicated by the phenomenon of gene transfer but can still yield information on the properties of the earliest organisms. The theme of gene exchange is continued in the next two chapters. In Chapter 11, microbiologist Juergen Wiegel discusses this as a mechanism for organisms to extend their growth temperatures, and in Chapter 12, biochemist Lars Ljungdahl and coworkers present evidence to support the exchange of genetic information for carbohydrate-degrading enzymes between a range of different life forms.

The next three chapters deal with the analyses of amino acid sequences from a phylogenetic perspective. A comparison of the properties of universal enzymes whose function is to control DNA structure (topoisomerases) by biologist Purificación López-García in Chapter 13 provides support for a thermophilic, but neither a hyperthermophilic or mesophilic, origin. In Chapter 14, geneticist James Brown discusses how aminoacyl-tRNA synthetases, which are essential enzymes in the translation of the genetic code, are providing insights into the evolution of the translation process and on the rooting of phylogenetic trees. Molecular evolutionist Bernard Labedan and biochemist Anne Boyen examine in Chapter 15 the evolutionary implications of the properties of a ubiquitous enzyme involved in nitrogen metabolism. The evolution of different protein types is given in Chapters 16 to 18. How DNA-binding proteins that facilitate DNA packing may have evolved is discussed in Chapter 16 by microbiologist Kathleen Sandman and coworkers. Biochemist Michael Danson and colleagues review in Chapter 17 the current status of enzymes of the central metabolism of hyperthermophiles, and how their properties reflect their phylogenetic placement and the factors limiting to life at extreme temperatures. In Chapter 18, biochemists Hugh Morgan and Ron Ronimus analyse the properties of a key glycolytic enzyme, phosphofructokinase, from an evolutionary perspective.

The evolution of lipids and membranes is often overlooked in phylogenetic analyses, but some information along these lines is presented by biochemist Masateru Nishihara and coworkers in Chapter 19, who focus on a particular type of enzyme involved in lipid biosynthesis. Akihiko Yamagishi and colleagues consider in Chapter 20 how some of the membrane structures in the cells of higher organisms may have evolved. In the final section, the biochemical consequences of hyperthermophilic life are considered. How hyperthermophilic organisms might solve the general problem of cofactor instability at high temperatures is assessed by biochemist Roy Daniel in Chapter 21, and how proteins might be stabilised at extreme tempertures is discussed by biochemist Reinhard Hensel and colleagues in Chapter 22. Finally, biochemist Michael Adams discusses in Chapter 23 the evolutionary implications of the use of certain metals by microorganisms that grow at 100 °C.

From this brief synopsis the reader will immediately conclude that a hyperthermophilic origin for life on this planet is not universally accepted, nor indeed is the conventional interpretation of the phylogenetic data which suggests this notion. What one can conclude from this collection of chapters is that answering such fundamental questions as how life first originated and whether it did so under high-temperature conditions, requires insight and perspective from a range of disciplines, from genetics to geology, from chemistry to climatology, and that there are not as yet any definitive answers to such questions. Nevertheless, it is hoped that this book will have helped in some way to bridge the gap between researchers in the different disciplines, such that in the future there is more interaction between them, and a better understanding of how evolution and thermophiles are viewed.

Juergen Wiegel
Department of Microbiology
University of Georgia, Athens, USA

Michael W. W. Adams
Department of Biochemistry and Molecular Biology
University of Georgia, Athens, USA

Contributors

Michael W. W. Adams
Department of Biochemistry and Molecular Biology and Center for Metalloenzyme Studies, Life Sciences Building, University of Georgia, Athens, GA 30602-7229, USA

Anne Boyen
Microbiologie, Vrije, Universiteit Brussel, and Vlaams Interuniversitair Instituut voor Biotechnologie, 1 ave E. Gryson, B 1070 Brussells, Belgium

James R. Brown
Department of Bioinformatics, SmithKline Beecham Pharmaceuticals, 1250 S. Collegeville Road, PO Box 5089, UP1345, Collegeville, PA 19426 0989, USA

Huizhong Chen
Department of Biochemistry & Molecular Biology, Life Sciences Building, A214, University of Georgia, Athens, GA 30602-7229, USA

Dan E. Daia
Department of Geology and Applied Geology, University of Glasgow, Glasgow G12 8QQ, UK

R. M. Daniel
Thermophile Research Unit, Department of Biological Sciences, University of Waikato, Bag 3105, Hamilton, New Zealand

Michael J. Danson
Centre for Extremophile Research, Department of Biology & Biochemistry, University of Bath, Bath BA2 7AY, UK

Patrick Forterre
Institut de Génétique et Microbiologie, Université Paris-Sud, CNRS, URA 1354, Bât 409, GDR 1006, 91405 Orsay Cedex, France

J. Peter Gogarten
Department of Molecular and Cell Biology, University of Connecticut, 75 North Eagleville Rd, Storrs, CT 06269-3044, USA

Allan J. Hall
Department of Geology and Applied Geology, University of Glasgow, Glasgow G12 8QQ, UK

Reinhard Hensel
FB 9 Mikrobiologie Universität GH Essen, Universitätstrasse 5, 45117, Essen, Germany

Daniel Hess
Technische Universität München, Physik Dept, James Franck Strasse, D-85748 Garching, Germany

David W. Hough
Centre for Extremophile Research, Department of Biology & Biochemistry, University of Bath, Bath BA2 7AY, UK

Ravi Jain
Molecular Biology Institute and MCD Biology, University of California, Los Angeles, CA 90095, USA

Otto Kandler
Botanical Institute of the University of Munich, Meninger Strasse 67, D-80638, Munchen, Germany

Yosuke Koga
Department of Chemistry, University of Occupational and Environmental Health, Kitakyushu, 807 Japan

Takahide Kon
Department of Molecular Biology, Tokyo University of Pharmacy and Life Science, 1432 Horinouchi, Hachioji-shi, Tokyo, 192-03, Japan

Takayuki Kyuragi
Department of Biochemical Engineering and Science, Kyushu Institute of Technology, 680-4 Kawazu, Iizuka, 820 Japan

Bernard Labedan
Institut de Génétique et de Microbiologie, CNRS, URA 1354, Université Paris-Sud, Bâtiment 409, 91405 Orsay, France

James A. Lake
Molecular Biology Institute and MCD Biology, University of California, Los Angeles, CA 90095, USA

Antonio Lazcano
Facultad de Ciencias, UNAM, Apdo, Postal 70-407, Cd, Universitaria, 04510 Mexico, D.F., Mexico

Xin-Liang Li
Department of Biochemistry & Molecular Biology, Life Sciences Building, A214, The University of Georgia, Athens, GA 30602-7229, USA

Lars G. Ljungdahl
Department of Biochemistry & Molecular Biology, Life Sciences Building, A214, The University of Georgia, Athens, GA 30602-7229, USA

Purificación López-Garcia
Institut de Génétique et Microbiologie, bat. 409, Université Paris-Sud, 91405 Orsay, France

Tom McCollom
Woods Hole Oceanographic Institute, Woods Hole, Massachusetts, USA

Stanley L. Miller
Department of Chemistry and Biochemistry, University of California, San Diego, La Jolla, CA 92093-0506, USA

Jonathan Moore
Molecular Biology Institute and MCD Biology, University of California, Los Angeles, CA 90095, USA

Hugh W. Morgan
Thermophile Research Unit, University of Waikato, Private Bag 3105, Hamilton, New Zealand

Masateru Nishihara
Department of Chemistry, University of Occupational and Environmental Health, Kitakyushu, 807 Japan

Lorraine Olendzenski
Department of Molecular and Cell Biology, University of Connecticut, 75 North Eagleville Rd, Storrs, CT 06269-3044, USA

Tairo Oshima
Department of Molecular Biology, Tokyo University of Pharmacy and Life Science, 1432 Horinouchi, Hachioji-shi, Tokyo, 192-03, Japan

John N. Reeve
Department of Microbiology, Ohio State University, Colombus, OH 43210, USA

Maria C. Rivera
Molecular Biology Institute and MCD Biology, University of California, Los Angeles, CA 90095, USA

Ron S. Ronimus
Thermophile Research Unit, University of Waikato, Private Bag 3105, Hamilton, New Zealand

Michael J. Russell
Department of Geology and Applied Geology, University of Glasgow, Glasgow G12 8QQ, UK

Rupert J. M. Russell
Centre for Extremophile Research, Department of Biology & Biochemistry, University of Bath, Bath BA2 7AY, UK

Kathleen Sandman
Department of Microbiology, Ohio State University, Columbus, OH 43210, USA

Alexander Schramm
FB 9 Mikrobiologie Universität GH Essen, Universitätstrasse 5, 45117, Essen, Germany

Mitchell D. Schulte
NASA Ames Research Center, Moffett Field, California, USA

Everett L. Shock
GEOPIG, Department of Earth & Planetary Sciences, Washington University, St Louis, Missouri, USA

David W. Schwartzman
Department of Biology, Howard University, Washington, D.C. 20059, USA

Nobuhito Sone
Department of Biochemical Engineering and Science, Kyushu Institute of Technology, 680-4 Kawazu, Iizuka, 820 Japan

Gen Takahashi
School of Medicine, Hirosaki Univeristy, 5 Zaifucho, Hirosaki 036, Japan

Michael F. Summers
Howard Hughes Medical Institute, University of Maryland Baltimore County, Baltimore, MD 21250, USA

Garry L. Taylor
Centre for Extremophile Research, Department of Biology & Biochemistry, University of Bath, Bath BA2 7AY, UK

Günter Wächtershäuser
Tal 29, 80331, Munich, Germany

Juergen Weigel
Departments of Microbiology and Biochemistry & Molecular Biology, Center for Biological Resource Recovery, University of Georgia, Athens, GA 30602-2605, USA

Akihiko Yamagishi
Department of Molecular Biology, Tokyo University of Pharmacy and Life Science, 1432 Horinouchi, Hachioji-shi, Tokyo, 192-03, Japan

Wenlian Zhu
Howard Hughes Medical Institute, University of Maryland Baltimore County, Baltimore, MD 21250, USA

PART ONE

The Early Earth

1

Do the Geological and Geochemical Records of the Early Earth Support the Prediction from Global Phylogenetic Models of a Thermophilic Cenancestor?

JOHN A. BAROSS

School of Oceanography, University of Washington, Seattle, Washington, USA

1.1 Introduction

The rooted global phylogenetic model of Woese and colleagues (Woese *et al.*, 1990; Woese, 1994), based on 16S rRNA sequences, predicts that the oldest of extant organisms are thermophilic and that the common ancestor to all extant organisms (cenancestor) was also thermophilic. The model further indicates that the earliest organisms were phenotypically similar to present-day hyperthermophiles isolated from volcanic and geothermal environments. Testing these predictions is difficult since experimental approaches are limited to extant organisms. The most common alternative approaches are to develop models of the early earth that incorporate thermal, tectonic and geochemical earth history and to look for organic chemical and specific isotopic signatures of distinct physiological groups of microorganisms along with isotopic signatures that reflect historical temperature. Trying to infer the evolutionary history of organisms on the basis of molecular phylogenetic trees from the biogeochemical signatures in the geological record is also problematic. For example, even though there is evidence for oxygenic photosynthesis, methanogenesis and microbial sulfate reduction 3 billion years ago, it is not possible to conclude that these ancient genetic lineages were the source of present-day cyanobacteria (sic), methanogens and sulfate reducers. In fact, the 16S rRNA-based phylogenetic tree does not support an ancient lineage for cyanobacteria or, for that matter, oxygenic photosynthesis, but it does indicate that archaeal sulfur reduction and methanogenesis are ancient physiologies. These findings, however, do not negate the possibilities for multiple origins of specific physiologies, for present-day organisms having evolved from one of these lineages, or for similar physiologies having evolved more than once.

There is other evidence besides that inferred from the Woese global tree of life to support the arguments for a warm to hot earth at the time of the origin of life and emergence of microbial communities. How warm or hot the earth was 4.3–3.5 billion years ago (giga-years, Ga) is a contentious issue and credible arguments can be made to support almost any temperature during the early Archaean (4.3–3.5 Ga). An important consideration is that throughout all of earth's history there have been hot aquatic environments associated with volcanic activity, regardless of the temperature of the atmosphere or the global ocean. Hydrothermal activity would have been more pronounced during the early Archaean (4.3–3.5 Ga) than at present while providing the same suite of electron acceptors and donors and sources of carbon and energy utilised by present-day microorganisms at hot spring environments (Abbott and Hoffman, 1984; Baross and Hoffman, 1985).

1.2 Was the early earth hot, warm or cold?

Answering this question may ultimately require a refinement of scale, since there have always been hot environments on earth associated with volcanic activity. Establishing that earth's ocean, atmosphere and protocontinents were warm to hot during the early Archaean (4.3–3.7 Ga) requires models that incorporate other sources of heat besides volcanic sources. These must include a greenhouse atmosphere, lack of continental mass to contribute to weathering (mechanism for removal of CO_2 and reduction of the greenhouse effect), increased radiogenic heating, and extensive bolide (extraterrestrial body) impacts. This period of earth's history is important not only because life arose then, but also because by 3.5–3.8 Ga there is evidence for complex microbial communities in the fossil record (Table 1.1).

It is generally accepted that liquid water existed on earth by 4.3 Ga (Chang, 1994; Lowe, 1994). The early Archaean ocean would have been reducing, with high concentrations of volcanic-derived volatiles including methane, carbon dioxide and hydrogen sulfide, and of reduced heavy metals, particularly iron. There is evidence for banded iron formations and red-beds during the early Archaean, indicating that there may have been some

Table 1.1 Markers in early earth history

Time (Ga)	Markers	
4.6–4.5	Formation of earth and moon	PERIOD OF INTENSE BOLIDE IMPACTS
4.4	Mean age of the atmosphere	
4.4–4.3	Oceans exist	
3.85	Biogenic signature in metasediments from Akilia Island, West Greenland (Mojzsis *et al.*, 1996; Nutman *et al.*, 1997)	
3.8	Biogenic signature in Isua (Greenland) metasediments (Moorbath *et al.*, 1973)	
3.5–3.4	Complex microfossil communities with evidence of CO_2 fixation (Apex Chert and Warrowoona, Western Australia; Onverwacht, Africa) (Schopf and Packer, 1987; Schopf, 1994)	
2.1	Accumulation of oxygen (Towe, 1994)	
1.5	Formation of ozone layer (Dott and Prothero, 1993)	
0.57	Cambrian explosion	

microbial oxygen production by as early at 3.8 Ga (Towe, 1994). The microbial fossil record and the stable carbon isotope record of kerogens to at least 3.5 Ga ago has been interpreted as indicative of oxygenic photosynthesisers (see Table 1.1 for references). It is more difficult to interpret the isotopic record in metasediments dated at 3.8 Ga and determine whether the CO_2-fixing communities at this early time were anaerobic photosynthesisers, chemolithotrophs or something else. Similarly, there is also evidence for sulfate accumulation and microbial sulfate reduction during this period (Schidlowski, 1993). The sources of sulfate and other oxidised forms of sulfur could be geochemical or microbial, although the latter requires a readily available source of oxygen (Towe, 1994).

There are three temperature models of the early earth: a hot to warm earth that cooled quickly; a hot earth that cooled slowly; and an ice-covered earth that warmed slowly or showed periods of freeze and thaw owing to bolide impacts. All of the models that incorporate multiple environmental variables depict the early earth as hot and cooling relatively quickly, with life arising and evolving during the cooling phase. A cold earth model considers only the reduced solar luminosity and indicates that the early earth would have been covered by a frozen ocean (Bada *et al.*, 1994). Periodic thawing would have taken place owing to bolide impacts and eventually the ice would have melted as the sun's luminosity increased. All other early-earth models consider one or more other sources of heat besides a solar source.

The sources of heat to the early earth were numerous (Table 1.2). Heat production from radioactive decay is estimated to have been at least five times the present value of approximately 8.38×10^7 J (2×10^7 calories) per year. Hydrothermal production of heat would have been approximately three times higher than present given the evidence for increased ridge lengths (evidence for seafloor spreading). During this early period there would have been rapid and recurring formation and reassimilation of crust at least to approximately 3.8–4.0 Ga. The oldest rocks from the earth's surface are dated at 3.8–3.96 Ga, so that the first protocontinents had probably formed by 3.8 Ga (Dott and Prothero, 1993). These continental masses are estimated to be approximately 5% of the total continental mass during late Precambrian at about 1.5 Ga (Figure 1.1). Early earth models also predict high concentrations of atmospheric CO_2, perhaps as high as 10–20 bars of pressure, four billion years ago which would have resulted in a high level of greenhouse heating. The actual temperature of the earth is difficult to know with certainty and would have depended on the rate of mantle degassing and duration of the early Archaean greenhouse atmosphere. However, it is estimated that 10 bars of CO_2 in the atmosphere could have caused a rise in the surface temperatures of 85 °C over the below-freezing conditions predicted by the lower Archaean solar luminosity, while the combination of the earth's greater rotation speed at that time and lack of significant continental mass would also have added to the surface temperature (Kasting and Ackerman, 1986; Kasting, 1993, 1997). Greenhouse gases other than CO_2, such as methane and ammonia, may have contributed significantly to maintaining high atmospheric temperatures during the early Archaean (Caldeira and Kasting, 1992; Sagan and Chyba, 1997). Sagan and Chyba (1997) argue that concentrations of ammonia alone on the early earth may have been high enough to maintain above-freezing surface temperatures despite the faint early sun. The ammonia would have been derived from the alteration of tholeiitic basalts, after being subducted to high-temperature crustal depths, or from other sources such as electric discharge. Their model predicts that the production of organic solids from the photolysis of methane at high altitude would have screened ammonia from dissociation by ultraviolet light. There is also a recent report proposing that carbon dioxide ice clouds, which reflect rather than emit infrared radiation, could maintain liquid water on the surface of Mars, and by

Table 1.2 Heat sources to the early earth (4.2–3.7 Ga)

Heat source	Comments	References
Sun	20–30% reduced luminosity compared to present day	Newman and Rood (1977); Sagan and Chyba (1997)
Radioisotope decay	Radiogenic heat 5 times greater than present	Dott and Prothero (1993)
Hydrothermal activity	Heat flow from tectonism was about 3 times present; total oceanic ridge length between 3 and 5 times present; atmospheric CO_2 estimated to be 80–600 times present levels	Baross and Hoffman (1985); Bickle (1978); Glikson (1993); Kadko *et al.* (1995); Reymer and Schubert (1986)
Atmosphere	Estimated to be up to 10 bars CO_2; possibly high concentrations of NH_3 and CH_4 remaining stable in the atmosphere resulting in greenhouse conditions	Chang (1994); Holland (1984); Hunten (1993); Sagan and Chyba (1977); Walker (1977)
Continental crust	Early evidence of crust and banded iron formations at 3.85 Ga. Crust accumulation in three stages with little stable crust before 3.8 Ga (see Figure 1.3); Weathering (CO_2 sink) is insignificant before crust accumulation	Abbott and Hoffman (1984); Lowe (1994); Nutman *et al.* (1997); Taylor and McLennan (1995, 1996)
Bolide impacts	Source of heat and water; frequent impacts by large bolides (>10 km diameter) before 3.9–4.0 Ga with potential to evaporate a 3 km global ocean; smaller but biologically significant impacts occurred periodically throughout earth's history	Glikson (1993); Maher and Stevenson (1988); Sleep *et al.* (1989); Oberbeck and Mancinelli (1994)
Glaciers[a]	Absence of glaciers before 2.7 Ga and geological evidence of glaciers (presence of tillites) before 2.0 Ga could be due to bolide impacts	Kasting (1984); Oberbeck *et al.* (1993), Oberbeck and Mancinelli (1994)

[a] While not a source of heat, the absence of glaciers in the geological record is evidence of warm to hot climactic conditions.

inference on the early earth as well, in the absence of greenhouse conditions (Forget and Pierrehumbert, 1997). Again, the Archaean atmospheric models that consider more than one source of heat indicate liquid rather than ice oceans, even if they are not good predictors of the actual surface temperatures or their fluctuations during the first 2 billion years of earth history.

The isotopic ratios of oxygen and sulfur in sediments are frequently used as geothermometers. The oxygen isotope ratios estimate early Archaean temperatures to be tens of degrees (Knauth and Epstein, 1976) to as high as 100 °C at 3.8 Ga (Karhu and Epstein, 1986), while sulfur isotope studies indicate that ocean temperatures at 3.5–2.6 Ga to be 40±10 °C (Ohmoto and Felder, 1987). Questions have been raised about the validity of

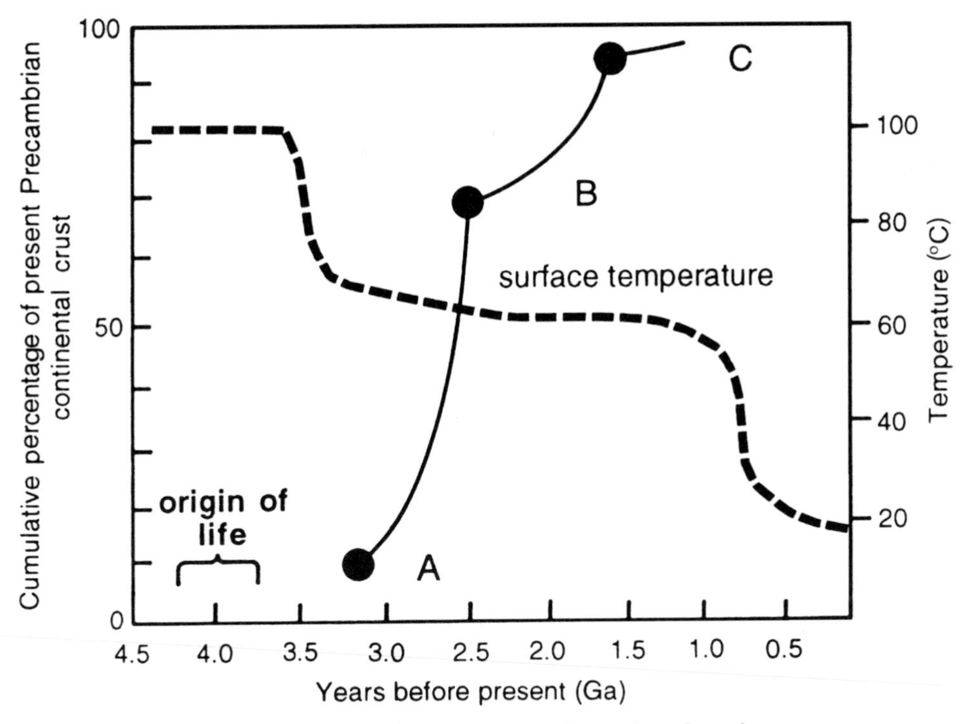

Figure 1.1 Growth curve of Precambrian continental crust based on the present age distribution of Precambrian crust. Present crustal age distribution suggests three main episodes of continental growth: (A) 3.3–3.1 Ga, when about 5% of the present crust formed; (B) 2.7–2.5 Ga, 58% of the present crust formed; (C) 2.1–1.6 Ga, 33% of the present crust formed (from Lowe, 1994); surface temperature model based on extensive bolide impacts to approximately 3700 Ma and the possibility that geological evidence of Archaean glaciers between 2.5 and 2.8 Ga may be due to bolide impact (from Oberbeck and Mancinelli, 1994).

these estimates, since the presence of tillites (deposits of angular rocks thought to be associated with glaciers) in the Archaean is evidence of glacial deposits and thus much cooler temperatures (Perry *et al.*, 1978). The glacial origin of these deposits has also been questioned, however, since there is evidence of deposits resembling tills in bolide impact craters (Oberbeck and Mancinelli, 1994; Oberbeck *et al.*, 1993) and other similar deposits that could have resulted from tectonism (Schermerhorn, 1974). Oberbeck and Mancinelli (1994) argue that bolide impacts would have been frequent enough between the time the earth formed and 3.75 Ga that earth's surface temperatures could have been as high as 100 °C during this period. This bolide argument, coupled with the argument that there were no glaciers during the early Archaean, would result in a slowly cooling earth and slowly dropping surface temperatures between 3.5 Ga and 1.0 Ga (Figure 1.2). Bolide impacts were directly involved in the slow cooling of the earth by delaying continent formation (and thus the removal of CO_2 by weathering), by maintaining a greenhouse atmosphere through the impact release of CO_2 from carbonate rocks, and by preventing the establishment of stable microbial communities that could effect climate through geochemical transformations (Oberbeck and Mancinelli, 1994). Bolides may have also been a significant source of organic carbon and water to the early earth (Chyba, 1987;

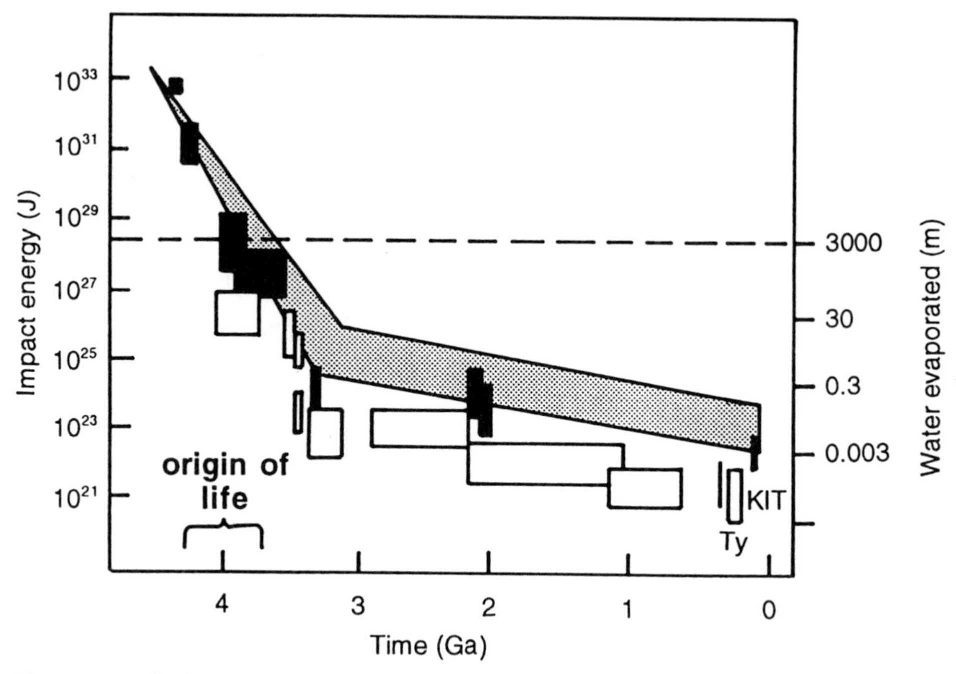

Figure 1.2 The largest bolide impacts on the Earth and the Moon. Open boxes are lunar, filled boxes terrestrial. Grey line is inferred earth impact history. Dashed line is depth of ocean vaporised by impact. K/T refers to the Cretaceous/Tertiary impact and Ty refers to the lunar crater Tycho (from Sleep *et al.*, 1989).

Chyba and Sagan, 1992; Chyba *et al.*, 1990). Based on the frequency of impacts by solar bodies and their estimated organic content (Chyba *et al.*, 1990) it is possible that there would have been enough organic material reaching the early earth from comets to maintain extensive communities of microbial heterotrophs for millions of years, thus adding credibility to the idea that heterotrophs, and not lithotrophs or phototrophs, were the earliest organisms on earth. Likewise, based on calculations that comets contain 10% water, comets could have supplied enough water to account for the current ocean volume. The implication of the Oberbeck and Mancinelli (1994) temperature model (Table 1.1) is that any microbial communities earlier than 3.75 Ga would have to have been hyperthermophilic. An alternative model could involve temperature oscillations between mesophilic and hyperthermophilic growth ranges that coincided with the frequency and magnitude of bolide impacts. Temperature oscillations also appear in a thermal evolution model presented by Chang (1994) and in the periodic ice melting model of Bada *et al.* (1994).

During the early stages in earth history (3.3–3.1 Ga) the continental mass has been estimated to be only about 5% of the present Precambrian crust and to have accumulated during three main events (Lowe, 1994; Figure 1.1). These estimates of Archaean crustal mass support the notion that weathering was not an important factor in climate change during the early Archaean. Also, this would indicate that prior to 3.8 Ga there would have been limited settings for the origin of life and habitats for the earliest microbial communities. These include the pelagic, benthic and subsurface environments associated with the early ocean, the atmosphere, and perhaps relatively short-lived protocontinents. The bolide impact models (Maher and Stevenson, 1988; Sleep *et al.*, 1989) indicate that prior to 3.8–3.9 Ga there would have been bolides of sufficient size and impact to evaporate an ocean 3 km deep and effectively sterilise the pelagic ocean, atmosphere and protocontinents (Figure 1.2). An important ramification of the bolide impact model is that the only safe havens for microbial communities prior to 3.8 Ga would have been the deep ocean and the subseafloor. The subseafloor would have had the advantages of retaining significant levels of water even after an ocean-evaporating impact while continuously producing, from hydrothermal activity, the carbon and energy sources necessary to sustain life.

1.3 Were the early organisms mesophiles, thermophiles or hyperthermophiles?

Resolving the controversy about the thermal history of the early earth is crucial to establishing the settings and conditions of chemical evolution and the origins of life. The current preponderance of geological and geochemical evidence favours a warm to hot earth, but there are other arguments to be considered against a thermophilic origin of life and hyperthermophiles being the earliest organisms on earth. These are based primarily on questions of the validity of the 16S rRNA-based phylogenetic trees, the thermal stability of specific macromolecules essential for life, and the evolution of specific macromolecules and physiological strategies used by hyperthermophiles for maintaining stability and function at high temperatures.

There are two important sources of controversy regarding predictions of evolutionary hierarchy in the microbial world from a 16S rRNA-based phylogenetic tree: Does the universal ancestor of the rooted tree represent the earliest lineage leading to extant organisms and is this the only tree of life ever on earth? Is such a rooted tree valid (Doolittle and Brown, 1994; Forterre, 1997)? The latter question has been addressed by others (see Baross and Holden, 1996; Doolittle and Brown, 1994). While it is impossible to know the answer to the former question, because molecular signatures are not available from the earliest microbial fossils, it is instructive to consider the possibilities based on extant organisms and the geochemical record. Figure 1.3 offers two of many alternative evolutionary pathways that could have led to the Woese global phylogenetic tree (Figure 1.3A). Figure 1.3B depicts a single tree in which all extant organisms evolved as one branch at some unknown period. This is similar to a tree proposed by Gogarten-Boekels *et al.* (1995) and reflects the argument proposed by Forterre *et al.* (1995) that the earliest organisms on earth could have been mesophilic. An estimation of divergence times of extant organisms based on sequences from 57 different proteins (albeit controversial) concluded that the divergence of Bacteria and Archaea dates to about 2.0 Ga, leaving about 2 billion years of microbial evolution (including the root of the Woese phylogenetic tree) that is without traces in the molecular record of extant organisms (Doolittle *et al.*, 1996). Others assume that the root of the Woese tree of life is far more ancient and perhaps represents the only tree of life to have evolved on earth (Pace, 1991). The geochemical evidence, based mostly on stable carbon ($\delta^{13}C$) and sulfur ($\delta^{34}S$) isotope data and on carbon cycle models for the emergence of different microbial physiologies, indicates that microbial CO_2 fixation occurred at 3.8 Ga and continued throughout the Archaean; oxygen was being produced as early as 3.5–3.0 Ga as evidenced by limestone, banded iron and red bed formations (Nisbet, 1987; van Andel, 1985). Moreover, evidence for methanogenesis and sulfate reduction between 2.7 and 2.5 Ga (Schidlowski, 1988) and for methane oxidation at approximately 2.7 Ga comes from depleted levels of ^{13}C in kerogens (Hayes, 1994). If the methanotrophs at 2.7 Ga were the ancestors to present-day methane oxidisers, it is possible that this physiological group could have been among the earliest of aerobic

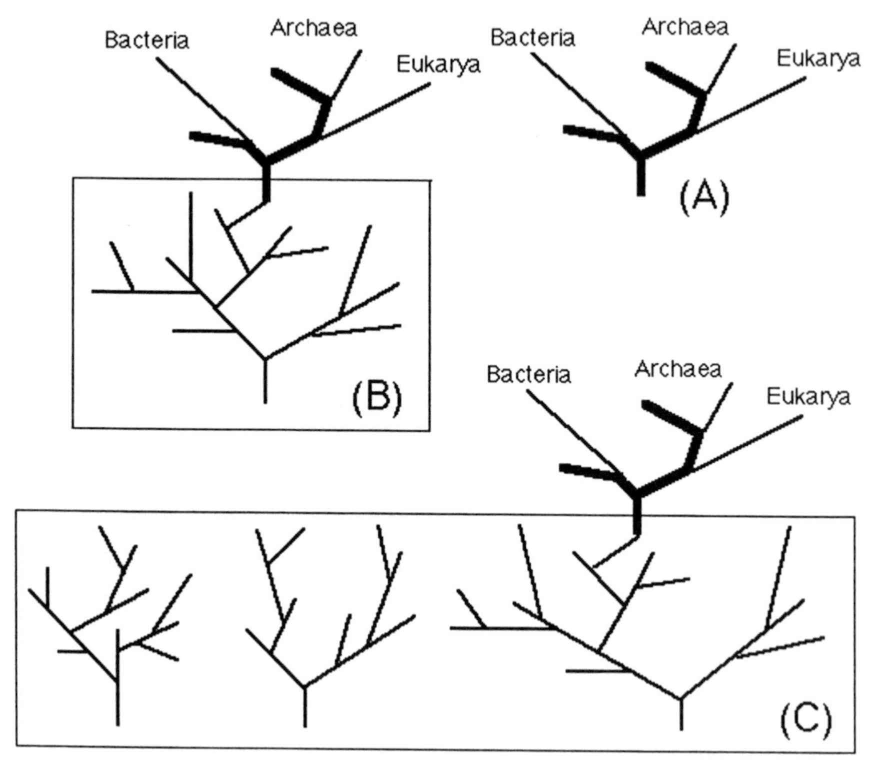

Figure 1.3 Three global phylogenetic models showing the origin of extant organisms: (A) The rooted phylogenetic tree based on 16S rRNA sequences and showing the three domains of Bacteria, Archaea, and Eukarya (Woese *et al.*, 1990). Distances derived from numbers of mutations. The root was derived from sequences of the two subunits of the F1-ATPases and the two translation elongation factors EF-1α (tu) and EF-2 (G) (Iwabe *et al.*, 1989). Bold lines lead to hyperthermophiles. (B) Model showing that life originated only once and that the earliest life forms became extinct thus implying that extant organisms have an unknown origin time in earth's geological record. (C) Same as (B) except there were multiple origins of life leading to separate phylogenetic trees and presumably different genetic codes and biochemistries.

bacteria. As indicators that other aerobes were also present, and particularly the α-proteobacteria group that include the mitochondria symbiont, they thus push back the time when aerobic eukaryotes may have evolved. By 2.1 Ga, oxygen was accumulating in the atmosphere and oceans (Towe, 1994). This indicates that the large pool of iron and other reduced metals and volatiles that had accumulated during the first 1.5–2.0 billion years must have remained mostly reduced, since they were present in considerable excess over the levels of oxygen being produced. The scenario just outlined is consistent with the Woese tree of life in that the inferred oldest branches are hyperthermophiles that have metabolic activities consistent with the geochemical signatures measured in rocks dating to at least 3.5 Ga (Schidlowski, 1993). Other lineages, including photosynthetic microorganisms, may have evolved more than once.

Figure 1.3C depicts a scenario where many different phylogenetic trees evolved simultaneously, possibly involving different genetic codes. Only one tree survived. The rationale for Figures 1.3B and 1.3C are based mostly on evolutionary history in the fossil record, which shows mass extinction of dominant animal and plant species owing to catastrophic events, climate changes, competition and other variables. That microorganisms, particularly bacteria and archaea, may have been affected by these variables or that dominant phylotypes became extinct at various times throughout earth's history is rarely assumed or even addressed. The rationale for bacteria and archaea not being subject to extinction is based on their ubiquity on the present earth. For example, hyperthermophilic archaea of several genera have been isolated from diverse environments including deep drill holes, oil wells, deep-sea and shallow hydrothermal vents and terrestrial hot springs (Huber *et al.*, 1990; L'Haridon *et al.*, 1995; Holden *et al.*, 1997; Stetter *et al.*, 1993); marine bacteria represented by closely related phylotypes, or clusters, based on 16S rRNA sequences have been documented (Giovannoni *et al.*, 1990, 1995). If the concept of microbial extinction is embraced, then a final alternative model accepts that the Woese tree of life does represent genetic lineages that date back to 3.5 Ga, but that separate lineages with similar physiologies to present-day organisms continued to evolve throughout the Precambrian along with the lineages now represented by the present tree of life, but did not survive. This may well be the case for oxygenic photosynthetic organisms.

Catastrophic events might have obliterated all of the microorganisms that live in the pelagic marine environment, but may not have effected those similar or identical species living in the deep sea. With more than 4 billion years of microbial evolution, one can assume that microbial communities have become more 'fit' to survive, grow and compete and that perhaps there is less 'evolutionary experimentation', particularly intraspecies lateral gene transfer, taking place today than at 3–4 billion years ago. Such an assumption is based on molecular and biochemical characterisation of proteins and nucleic acids that show that lateral transfer of genetic material was a dominant feature of early evolution independent of genetic relatedness of donors and recipients. There is considerable evidence for chimeric proteins in organisms from all three domains with the archaea sharing protein sequence similarities from insertions with both eukarya and Gram-positive bacteria (Baross and Holden, 1996; Gupta and Golding, 1996; Hilario and Gogarten, 1993; Sogin, 1991; Sogin *et al.*, 1996). The assumption also builds upon others; i.e. that microorganisms always possessed the ability to colonise new environments successfully and adapt to changing conditions and that early earth possessed multiple environments and combinations of physical and chemical properties that were suitable for life and the evolution of diversity as we know it today. A better understanding of conditions of the earth earlier than 3.7 Ga could provide possible answers to the latter.

The above arguments are important for inferring the physiology and temperature growth range of the earliest organisms on earth. Gogarten-Boekels *et al.* (1995) presented the case that mesophiles were the first organisms on earth and that thermophiles and hyperthermophiles evolved later. The first mesophiles became extinct, while thermophiles survived in environments not affected by the extinction events and evolved into other thermal and physiological groups of microorganisms capable of colonising the wide range of environmental conditions continuously developing on the Precambrian earth. The arguments for a mesophilic origin of life and early microbial communities are that the temperature of the early earth was cold to warm and that key macromolecules found in all present-day hyperthermophiles (necessary for growth at high temperatures) appear to have their origin in mesophiles (Forterre *et al.*, 1995). Two alternative scenarios that do not invoke macromolecules evolving first in mesophiles are: (1) that the first organisms were thermophilic, not hyperthermophilic, and thus did not require special macromolecules for growth and survival; and (2) that the first organism was indeed hyperthermophilic but environmental conditions of hydrostatic pressure and salinity aided in the stabilisation of macromolecules at hyperthermophilic temperatures (Nickerson, 1984). All three scenarios are elaborated in the following discussion.

The debate of a hot or warm origin of life has been ongoing for decades (Fox,1995; Hennet *et al.*, 1992; Marshall, 1994; Miller, 1953; Miller and Bada, 1988; Shock, 1992). The arguments can be distilled into whether or not key organic compounds for life can be synthesised and remain stable at temperatures above 150 °C. The importance of these arguments is the implication for life arising under hydrothermal vent conditions and for the earliest organisms on earth living in hydrothermal environments and being thermophilic or hyperthermophilic. Evolution from a thermophilic origin is also an assertion of the Woese rooted tree of life (Achenbach-Richter *et al.*, 1987). The origin of life and the temperature growth range of the first organisms on earth are linked by the common denominator of thermal stability of key organic compounds and macromolecules required for growth and survival at high temperatures. For example, there is considerable interest in the hypothesis that the earliest life forms used RNA and not DNA for genetic information and perhaps for enzymatic activity as well (Cech and Bass, 1986; Lazcano, 1994). Ribonucleic acids are much less stable thermally than DNA; ribose sugar in particular is believed to be very heat labile (Forterre *et al.*, 1995). It follows that an RNA world would be unlikely to develop or evolve in the hot conditions found at hydrothermal vents. There is also the compelling argument that reverse gyrase, which is required for stabilising DNA at high temperatures and found in all hyperthermophiles and some thermophiles, was formed by the fusion of a DNA helicase with a DNA topoisomerase, both of which have a mesophilic origin (Forterre *et al.*, 1995). This would imply that the ability to grow under hyperthermophilic conditions would have been dependent on an earlier evolution of mesophiles.

However, there may be environmental conditions and physiological strategies that could stabilise RNA and DNA at high temperatures, affording thermophilic independence from mesophilic precursors. Pace (1991) suggested that lower water activity could stabilise RNA. Certainly the high concentrations of the thermal protectants potassium and low-molecular-mass organic compounds, heat shock proteins, and DNA-binding proteins produced by some hyperthermophiles provide evidence of these strategies (Baross and Holden, 1996; Ciulla *et al.*, 1994; Grayling *et al.*, 1996; Martins and Santos, 1995; Ramakrishnan *et al.*, 1997; Scholz *et al.*, 1992). These are sophisticated strategies that may have taken time to evolve. Pressure is another variable known to effect the thermal stability of macromolecules. There are reports of increased thermal stability of DNA and proteins under hydrostatic pressure (Hedén, 1964; Hei and Clark, 1994) and the elevation of heat shock temperature in *Pyrococcus* ES4 (Holden and Baross, 1995). Hyperthermophilic proteins, including hydrogenases, α-glucosidases, and glyceraldehyde-3-phosphate dehydrogenase, were significantly more stable at high temperature under pressure than their mesophilic counterparts (Hei and Clark, 1994; Michels and Clark, 1992; Michels *et al.*, 1996). Similarly, DNA polymerase from hyperthermophiles and from *Thermus aquaticus* were stabilised by pressure at temperatures that were denaturing at 1 atm pressure (Summit *et al.*, 1998). There is too little information available to draw general conclusions about pressure and thermal stability in hyperthermophiles. Experiments to measure the thermal stability of RNA and DNA under pressure, and to determine whether reverse gyrase is required for DNA stability by deep-sea and particularly deep-subseafloor hyperthermophiles at *in situ* pressures, would add significantly to the dabate.

Two key points are frequently overlooked in arguments about the thermal stability of macromolecules and the thermal growth properties of the earliest organisms on earth. First, inferences are made about the molecular and physiological characteristics of organisms 4 billion years ago based on extant organisms. There is no way to know the extent to which the earliest organisms resembled present-day organisms. Did they have small genomes or multiple small genomes that were more thermally stable then the genomes in extant organisms? Did they have less efficient strategies for growing and surviving at hyperthermophilic temperatures? Did they occupy habitats with conditions that increased thermal stability of macromolecules, such as low water activity and elevated pressure? Second, the first organisms may have been moderate thermophiles rather than hyperthermophiles. Moderate thermophiles including *Thermus*, *Bacillus* and *Clostridium* species, some representatives of which have maximum growth temperatures near 85 °C, do not have reverse gyrases. The first organisms may have been thermophilic, acquiring reverse gyrase at some later stage in evolution coincident with increases in genome sizes and colonisation of higher-temperature habitats. Alternatively, the first organisms were hyperthermophiles that developed strategies to counteract the deleterious effects of high temperature on macromolecules (albeit inefficient compared to present-day hyperthermophiles) and/or to take advantage of environmental conditions such as pressure and water activity.

1.4 Conclusions

Only limited conclusions can be reached about evolutionary history based on the molecular phylogeny and physiology of extant organisms. How far back in geological history can we trace the cenancestor to all extant organisms and thus the time of emergence of hyperthermophiles? There are few data in the fossil record that can be extrapolated directly to extant organisms. The Archaean record does show that CO_2 fixation, methanogenesis and sulfate reduction are ancient physiologies and that the organisms with these metabolisms fractionated the isotopes of carbon and sulfur in the same way as present-day organisms. At the very least, it must be considered that metabolic pathways are more ancient than many of the individual genes used to construct phylogenetic trees or enable metabolic pathways, since these genes continue to evolve. The fact that there are present-day species of hyperthermophilic archaea and bacteria capable of fixing CO_2, reducing CO_2 to methane, and reducing sulfate and other oxidised forms of sulfur adds credibility to the notion that these metabolisms occurred in organisms between at 3.5 and 3.0 Ga; organisms that may be descendants of present-day hyperthermophiles expressing these same metabolic capabilities. The more than 500 million years of evolution prior to 3.5 Ga may have been a period of extensive evolutionary 'experimentation' with only one evolutionary lineage becoming the source of extant organisms.

The best evidence to support the prediction from global phylogenetic models of a hyperthermophilic (thermophilic) cenancestor is that found in the geological record. The general consensus is that the origin of life and early evolution of microbial ecosystems occurred during the earliest period of earth history (4.3–3.8 Ga). A dominant feature of this period was that the earth was relentlessly bombarded by large asteroid bodies, some of which would have been large enough to evaporate 3 km deep ocean and exterminate pelagic and terrestrial life. Moreover, these bolide impacts could have delayed the overall cooling of the earth's ocean(s) and atmosphere and delayed the formation of stable continental mass (Oberbeck and Mancinelli, 1994). The deep benthic or subseafloor zones would have been the least impacted environments on the early earth, increasing the likelihood that they were habitats for the earliest microbial communities. The subseafloor associated with hydrothermal systems offers the greatest range of conditions for diversity in evolution of microbial metabolisms and thermal growth range. Present-day subseafloor hydrothermal environments support most of the metabolisms observed to be ancient in the geological record, such as sulfur reduction and methanogenesis, in species of extant hyperthermophiles. It is unknown whether there are hyperthermophiles in the deep-hot subseafloor that possess phenotypic and genotypic characteristics that can be intrepreted as more 'ancient' than those found in hyperthermophilic species that are represented in phylogenetic trees. However, the fact that these subseafloor environments are essentially unchanged for at least 4.3 billion years (formation of the ocean) presents the possibility that there may be slowly evolving microbes that still harbour genetic information that can be used to better address questions about the physiological characteristics of the earliest organisms on earth, and the validity of the Woese rooted tree, and, particularly, the prediction that the first organisms on earth were hyperthermophiles.

Although new information could help to refine or change our understanding of the conditions on the early earth, the origin of life, and the metabolic diversity in Archaean microbial ecosystems, most of the available information from the geological and geochemical records appears consistent with the general concept, predicted by molecular phylogenies based on 16S rRNA sequences, of thermophily as an essential feature.

1.5 Summary

Global phylogenetic models indicate that the common ancestor to all extant organisms would have resembled contemporary hyperthermophiles found in volcanic environments. The implications are that the phylogenetic tree based on extant organisms is the only tree of life ever to exist on earth and thus represents 4 billion years of evolutionary history. There may be fallacies in this argument since it is impossible to determine whether there was more than one proto-phylogenetic tree during the early Archaean (4.3–3.0 Ga) and only one survived extinction events, or whether the present phylogenetic tree will have to be modified as more species are sequenced. While it is impossible to know the temperature history of the Archaean earth, there are atmosphere and bolide impact models indicating that the earth was hot (~100 °C) during the first few hundred million years after accretion (4.3–3.8 Ga) and possibly cooled slowly during the next billion years. The bolide impact models also imply that during the early stages of earth history (4.3–3.7 Ga) the deep sea and possibly the subseafloor would have been one of the most unaffected environments by catastrophic events. What is clear is that thermophily is probably an essential feature of earth history since hydrothermal vent environments are primordial, their emergence coinciding with the accumulation of liquid water on earth.

Acknowledgements

I thank Jody Deming for review of the manuscript and John Delaney, Jody Deming, Paul Johnson, Jon Kaye, Debbie Kelly, Marvin Lilley and Melanie Summit for comments and helpful discussions. The preparation of this manuscript was supported by the National Science Foundation (BCS9320070) and Washington Sea Grant (NA36RG0071).

References

Achenbach-Richter, L., Gupta, R., Stetter, K. O. and Woese, C. R. (1987) Were the original Eubacteria thermophiles? *Syst. Appl. Microbiol.*, **9**, 34–39.

Abbott, D. H. and Hoffman, S. E. (1984) Archaean plate tectonics revisited. 1. Heat flow, spreading rate, and the age of subducting oceanic lithosphere and their effects on the origin and evolution of continents. *Tectonics*, **3**, 429–448.

Bada, J. L., Bigham, C. and Miller, S. L. (1994) Impact melting of frozen oceans on the early earth: implications for the origin of life. *Proc. Natl Acad. Sci. USA*, **91**, 1248–1250.

Baross, J. A. and Hoffman, S. E. (1985) Submarine hydrothermal vents and associated gradient environments as sites for the origin and evolution of life. *Origins of Life*, **15**, 327–345.

Baross, J. A. and Holden, J. F. (1996) Overview of hyperthermophiles and their heat-shock proteins. *Adv. Protein Chem.*, **48**, 1–35.

Bickle, M. J. (1978) Heat loss from the earth: a constraint on Archaean tectonics from the relation between geothermal gradients and the rate of plate production. *Earth Planet. Sci. Lett.*, **40**, 301–315.

Caldeira, K. and Kasting, J. F. (1992) Susceptibility of the early earth to irreversible glaciation caused by carbon ice clouds. *Nature*, **359**, 226–228.

Cech, T. R. and Bass, B. L. (1986) Biological catalysis by RNA. *Annu. Rev. Biochem.*, **55**, 599–629.

Chang, S. (1994) The planetary setting of prebiotic evolution. In *Early Life on Earth*, Nobel Symposium No. 84, ed. S. Bengtson, pp. 10–23 (New York: Columbia University Press).

Chyba, C. F. (1987) The cometary contribution to the oceans of the primitive earth. *Nature*, **220**, 632–635.

Chyba, C. F. and Sagan, C. (1992) Endogenous production, exogenous delivery, and impact-shock synthesis of organic molecules: an inventory for the origins of life. *Nature*, **355**, 125–131.

Chyba, C. F., Thomas, P. J., Brookshaw, L. and Sagan, C. (1990) Cometary delivery of organic molecules to the early earth. *Science*, **249**, 366–373.

Cuilla, R. A., Burggraf, S., Stetter, K. O. and Roberts, M. F. (1994) Occurrence and role of di-*myo*-inositol-1,1′-phosphate in *Methanococcus igneus*. *Appl. Environ. Microbiol.*, **60**, 3660–3664.

Doolittle, R. F., Feng, D.-F., Tsang, S., Cho, G. and Little, E. (1996) Determining divergence times of the major kingdoms of living organisms with a protein clock. *Science*, **271**, 470–477.

Doolittle, W. F. and Brown, J. R. (1994) Tempo, mode, the progenote, and the universal root. *Proc. Natl Acad. Sci. USA*, **91**, 6721–6728.

Dott, R. H., Jr. and Prothero, D. R. (1993) *Evolution of the Earth*, 5th edn (New York, McGraw-Hill).

Forget, F. and Pierrehumbert, G. D. (1997) Warming early Mars with carbon dioxide that scatter infrared radiation. *Science*, **278**, 1273–1276.

Forterre, P. (1997) Protein versus rRNA: Problems in rooting the universal tree of life. *Am. Soci. Microbiol. News*, **63**, 89–95.

Forterre, P., Confalonieri, F., Charbonnier, F. and Duguet, M. (1995) Speculations on the origin of life and thermophily: review of available information on reverse gyrase suggests that hyperthermophilic prokaryotes are not so primitive. *Origins of Life and Evolution of the Biosphere*, **25**, 235–249.

Fox, S. W. (1995) Thermal synthesis of amino acids and the origin of life. *Geochim. Cosmochimi. Acta*, **59**, 1213–1214.

Giovannoni, S. J., Britschgi, T. B., Moyer, C. L. and Field, K. G. (1990) Genetic diversity in Sargasso Sea bacterioplankton. *Nature*, **345**, 60–63.

Giovannoni, S. J., Mullins, T. D. and Field, K. G. (1995) Microbial diversity in oceanic systems: rRNA approaches to the study of unculturable microbes. In *Molecular Ecology of Aquatic Microbes*, ed. I. Joint, pp. 217–248 (Berlin: Springer-Verlag).

GLIKSON, A. Y. (1993) Asteroids and the early Precambrian crustal evolution. *Earth-Sci. Rev.*, **35**, 285–319.

GOGARTEN-BOEKELS, M., HILARIO, E. and GOGARTEN, J. P. (1995) *Origins of Life and Evolution of the Biosphere*, **25**, 251–264.

GRAYLING, R. A., SANDMAN, K. and REEVE, J. N. (1996) DNA stability and DNA binding proteins. *Adv. Protein Chem.*, **48**, 437–467.

GU, X. (1997) The age of the common ancestor of eukaryotes and prokaryotes: statistical inferences. *Mol. Biol. Evol.*, **14**, 861–866.

GUPTA, R. S. and GOLDING, G. B. (1996) The origin of the eukaryotic cell. *Trends Bioch. Sci.*, **21**, 166–171.

HAYES, J. M. (1994) Global methanotrophy at the Archaean–Proterozoic transition. In *Early Life on Earth*, ed. S. Bengtson, pp. 220–236 (New York: Columbia University Press).

HEDÉN, C-G. (1964) Effects of hydrostatic pressure on microbial systems. *Bacteriol. Revi.*, **28**, 14–29.

HEI, D. J. and CLARK, D. S. (1994) Pressure stabilisation of proteins from extreme thermophiles. *Appl. Environ. Microbiol.*, **60**, 932–939.

HENNET, R., J.,-C., HOLM, N. G. and ENGEL, M. H. (1992) Abiotic synthesis of amino acids under hydrothermal conditions and the origin of life: a perpetual phenomenon? *Naturwissenschaften*, **79**, 361–365.

HILARIO, E. and GOGARTEN, J. P. (1993) Horizontal transfer of ATPase genes – the tree of life becomes the net of life. *BioSystems*, **31**, 111–119.

HOLDEN, J. F. and BAROSS, J. A. (1995) Enhanced thermotolerance by hydrostatic pressure In deep-sea marine hyperthermophile *Pyrococcus* strain ES4. *FEMS Microbiol. Ecol.*, **18**, 27–34.

HOLDEN, J. F., SUMMIT, M. and BAROSS, J. A. (1998) Thermophilic and hyperthermophilic microorganisms in 3–30 °C hydrothermal fluids following a deep-sea volcanic eruption. *FEMS Microbiol. Ecol.*, **25**, 33–41.

HOLLAND, H. D. (1984) *The Chemical Evolution of the Atmosphere and Oceans* (Princeton, NJ: University Press).

HUBER, R., STOFFERS, P., HOHENHAUS, S. *et al.* (1990) Hyperthermophilic archaebacteria within the crater and open-sea plume of erupting MacDonald Seamount. *Nature*, **345**, 179–182.

HUNTEN, D. M. (1993) Atmospheric evolution of the terrestrial planets. *Science*, **259**, 915–920.

KADKO, D., BAROSS, J. and ALT, J. (1995) The magnitude and global implications of hydrothermal flux. In *Physical, Chemical, Biological and Geological Interactions within Sea Floor Hydrothermal Discharge*, Geophysical Monograph 91, ed. S. Humphris, R. Zierenberg, L. Mullineaux, R. Thompson, pp. 446–466 (Washington DC: AGU Press).

KARHU, J. and EPSTEIN, S. (1986) The implication of the oxygen isotope records in coexisting cherts and phosphates. *Geochim. Cosmochim. Acta*, **50**, 1745–1756.

KASTING, J. F. (1997) Warming early earth and Mars. *Science*, **276**, 1213–1215.

KASTING, J. F. (1993) New spin on ancient climate. *Nature*, **364**, 759–760.

KASTING, J. F. (1984) Effects of high CO_2 levels on surface temperature and atmospheric oxidation state of the early earth. *J. Geophys. Res.*, **86**, 1147–1158.

KASTING, J. F. and ACKERMAN, T. P. (1986) Climatic consequences of very high carbon dioxide levels in the earth's early atmosphere. *Science*, **234**, 1383–1385.

KNAUTH, L. P. and EPSTEIN, S. (1976) Hydrogen and oxygen isotope ratios in nodular and bedded cherts. *Geochim. Cosmochim. Acta*, **40**, 1095–1108.

LAZCANO, A. (1994) The RNA world, its predecessors, and its descendants. In *Early Life on Earth*, ed. S. Bengston, pp. 70–80 (New York: Columbia University Press).

L'HARIDON, S. L., REYSENBACH, A.-L., GLÉNAT, P., PRIEUR, D. and JEANTHON, C. (1995) Hot subterranean biosphere in a continental oil reservoir. *Nature*, **377**, 223–224.

LOWE, D. R. (1994) Early environments: constraints and opportunities for early evolution. In *Early Life on Earth*, Nobel Symposium No. 84, ed. S. Bengtson, pp. 24–35 (New York: Columbia University Press).

MAHER, K. A. and STEVENSON, J. D. (1988) Impact frustration of the origin of life. *Nature*, **331**, 612–614.

MARSHALL, W. L. (1994) Hydrothermal synthesis of amino acids. *Geochim. Cosmochim. Acta*, **58**, 2099–2106.

MARTINS, L. O. and SANTOS, H. (1995) Accumulation of mannosylglycerate and di-*myo*-inositol-phosphate by *Pyrococcus furiosus* in response to salinity and temperature. *Appl. Environ. Microbiol.*, **61**, 3299–3303.

MICHELS, P. C. and CLARK, D. S. (1992) Pressure dependence of enzyme catalysis. In *Biocatalysis at Extreme Environments*, ed. M. W. W. Adams and R. Kelly, pp. 108–121 (Washington, DC: American Chemical Society Books).

MICHELS, P. C., HEI, D. and CLARK, D. S. (1996) Pressure effects on enzyme activity and stability at high temperatures. *Adv. Protein Chem.*, **48**, 341–376.

MILLER, S. L. (1953) A production of amino acids under possible primitive earth conditions. *Science*, **117**, 528–529.

MILLER, S. L. and BADA, J. L. (1988) Submarine hot springs and the origin of life. *Nature*, **334**, 609–611.

MOJZSIS, S., ARRHENIUS, G., MCKEEGAN, K. D., HARRISON, T. M., NUTMAN, A. P. and FRIEND, C. R. L. (1996) Evidence for life on earth before 3,800 million years ago. *Nature*, **385**, 55–59.

MOORBATH, S., O'NIONS, R. K. and PANKHURST, R. J. (1973) Early Archaean age of the Isua iron formation. *Nature*, **245**, 138–139.

NEWMAN, M. J. and ROOD, R. T. (1977) Implications of solar evolution for the earth's early atmosphere. *Science*, **198**, 1035–1037.

NICKERSON, K. W. (1984) An hypothesis on the role of pressure in the origin of life. *Theor. Biol.*, **110**, 487–499.

NISBET, E. G. (1987) *The Young Earth, An Introduction to Archaean Geology* (Boston: Allen and Unwin).

NUTMAN, A. P., MOJZSIS, S. J. and FRIEND, C. R. L. (1997) Recognition of ≥3850 Ma water-lain sediments in West Greenland and their significance for the early Archaean earth. *Geochim. Cosmochim. Acta*, **61**, 2475–2484.

OBERBECK, V. R. and MANCINELLI, R. L. (1994) Asteroid impacts, microbes, and the cooling of the atmosphere. *BioScience*, **44**, 173–177.

OBERBECK, V. R., MARSHALL, J. R. and AGGARWAL, H. R. (1993) Impacts, tillites, and the breakup of Gondwanaland. *J. Geol.*, **101**, 1–19.

OHMOTO, H. and FELDER, R. P. (1987) Bacterial activity in the warmer, sulfate-bearing Archaean oceans. *Nature*, **328**, 244–246.

PACE, N. (1991) Origin of life – facing up to the physical setting. *Cell*, **65**, 531–533.

PERRY, E. C. JR., AHMAD, S. N. and SWULIUS, T. M. (1978) The oxygen isotope composition of 3,800 m.y. old metamorphosed chert and iron formation from Isukasia West Greenland. *J. Geology*, **86**, 223–239.

RAMAKRISHNAN, V., VERHAGEN, M. F. J. M. and ADAMS, M. W. W. (1997) Characterisation of di-inositol-1,1′-phosphate in the hyperthermophilic bacterium *Thermotoga maritima*. *Appl. Environ. Microbiol.*, **63**, 347–350.

REYMER, A. and SCHUBERT, G. (1986) Rapid growth of some major segments of the continental crust. *Geology*, **14**, 299–302.

SAGAN, C. and CHYBA, C. (1997) The early faint sun paradox: organic shielding of ultraviolet-labile greenhouse gases. *Science*, **276**, 1217–1221.

SCHERMERHORN, L. J. G. (1974) Later Precambrian mixtites: glacial or nonglacial? *Am. J. Sci.*, **274**, 673–824.

SCHIDLOWSKI, M. (1988) A 3,800-million-year isotopic record of life from carbon in sedimentary rocks. *Nature*, **333**, 313–318.

SCHIDLOWSKI, M. (1993) The initiation of biological processes on earth: Summary of empirical evidence. In *Organic Geochemistry*, ed. M. H. Engel and S. A. Macko, pp. 639–655 (New York: Plenum Press).

SCHOLZ, S., SONNENBICHLER, J., SCHÄFER, W. and HENSEL, R. (1992) Di-*myo*-inositol-1,1-phosphate: a new inositol phosphate isolated from *Pyrococcus woesei*. *FEBS Lett.*, **306**, 239–242.

SCHOPF, J. W. (1994) The oldest known records of life: Early Archaean stromatolites, microfossils, and organic matter. In *Early Life on Earth*. Nobel Symposium No. 84, ed. S. Bengtson, pp. 193–206 (New York: Columbia University Press).

SCHOPF, J. W. and PACKER, B. M. (1987) Early archaean (3.3-billion to 3.5-billion-year-old) microorganisms from the Warrawoona Group, Australia. *Science*, **237**, 70–73.

SCHWARTZMAN, D., MCMENAMIN, M. and VOLK, T. (1993) Did surface temperatures constrain microbial evolution? *BioScience*, **43**, 390–393.

SHOCK, E. L. (1992) Chemical environments of submarine hydrothermal systems. *Origin of Life and Evolution of the Biosphere*, **22**, 67–107.

SLEEP, N. H., ZAHNLE, K. J., KASTING, J. F. and MOROWITZ, H. J. (1989) Annihilation of ecosystems by large asteroid impacts on the early earth. *Nature*, **342**, 139–142.

SOGIN, M. L. (1991) Early evolution and the origin of eukaryotes. *Curr. Opin. Genet. Dev.*, **1**, 457–463.

SOGIN, M. L., SILBERMAN, J. D., HINKLE, G. and MORRISON, H. G. (1996) Problems with molecular diversity in the eukarya. In *Society for General Microbiology Symposium: Evolution of Microbial Life*, ed. D. M. Roberts, P. Sharp, G. Alderson and M. A. Collins, pp. 167–184 (Cambridge: Cambridge University Press).

STETTER, K. O., HUBER, R., BLÖCHL, E. *et al.* (1993) Hyperthermophilic archaea are thriving in deep North Sea and Alaskan oil reservoirs. *Nature*, **365**, 743–745.

SUMMIT, M., SCOTT, B., NIELSON, K., MATHUR, E. and BAROSS, J. (1998) Pressure enhances the thermal stability of DNA polymerase from three thermophilic organisms. Extremophiles (in press).

TAYLOR, J. R. and MCLENNAN, S. M. (1996) The evolution of the continental crust. *Sci. Am.* **274**, 76–81.

TAYLOR, S. R. and MCLENNAN, S. M. (1995) The geochemical evolution of the continental crust. *Rev. Geophys.*, **33**, 241–265.

TOWE, K. M. (1994) Earth's early atmosphere: constraints and opportunities for early evolution. In *Early Life on Earth*, Nobel Symposium No. 84, ed. S. Bengtson, pp. 36–47 (New York: Columbia University Press).

VAN ANDEL, T. H. (1985) *New Views on an Old Planet* (Cambridge: Cambridge University Press).

WALKER, J. C. G. (1977) *Evolution of the Atmosphere* (London: Macmillan).

WOESE, C. R. (1994) There must be a prokaryote somewhere: microbiology's search for itself. *Microbiol. Rev.*, **58**, 1–9.

WOESE, C. R., KANDLER, O. and WHEELIS, M. L. (1990) Towards a natural system of organisms: proposals for the domains Archaea, Bacteria, and Eukarya. *Proc. Natl Acad. Sci. USA*, **87**, 4576–4579.

2

The Early Diversification of Life and the Origin of the Three Domains: A Proposal

OTTO KANDLER

Botanical Institute of the University of Munich, Munich, Germany

2.1 Introduction

Our perception of the phylogeny of life on this planet is still based on Darwin's seminal insight that 'probably all the organic beings which have ever lived on this earth have descended from one primordial form into which life was first breathed' (Darwin, 1859). The acceptance and popularisation of Darwin's message was distinctly promoted by Haeckel's (1866) universal dichotomously branched phylogenetic tree, the first figural presentation of the then supposed genealogical relationships among all groups of organisms known at that time. Although this tree has been modified in numerous details, its principles have essentially remained accepted (Whittaker, 1969) even after the prokaryote–eukaryote distinction (Chatton, 1938) and the endosymbiotic nature of the main organelles of the eukaryotic cell (comprehensively reviewed by Herrmann, 1997) became known.

Only after more than a century has the monorooted phylogenetic tree been challenged by what was called the Woesian revolution (Doolittle and Brown, 1994) and has the taxonomic dichotomy Prokaryotae/Eukaryotae been converted into the tripartite taxonomy of the three domains: Bacteria, Archaea, Eukarya (Woese *et al.*, 1990). Many biologists were sceptical, and some articulated their opposition (Mayr, 1990; Cavalier-Smith, 1992), or saw 'kingdoms in turmoil' (Margulis and Guerrero, 1991). In spite of these objections, which have been critically discussed and invalidated by Wheelis *et al.* (1992), the new concept of the three domains has found wide acceptance in recent years.

2.2 Towards a universal phylogenetic 'bush'

The tripartite split of the living world into one eukaryotic and two prokaryotic lineages was revealed by the earliest comparative studies of the ribosomal RNA sequences of organisms of most different levels of organisation (Woese and Fox, 1977). Consequently, Fox *et al.* (1980) suggested a trifurcation at an early ancestral progenotic state of life,

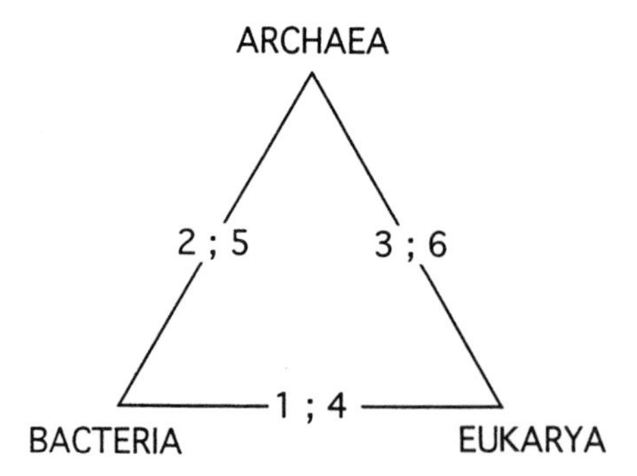

Figure 2.1 Scheme of quasi-random distribution of characteristics among the three domains. Numbers indicate characteristics common to two domains but missing in the third domain. (1) Laevorotatory acyl ester glycerolipids; (2) circular genome; (3) amino acid sequence homology of protein elongation factor EF-Tu/1α and EF-G/2; (4) ATP-phosphofructosekinase; (5) transcription units; (6) V-ATPases.

described as a genetic communion 'still in the throes of evolving the link between genotype and phenotype' (Woese, 1982).

Such a scenario, however, does not implicate the existence of a 'primordial form' *sensu* Darwin (1859) nor a 'first cell' (cf. Ponnamperuma and Chela-Flores, 1995; de Duve, 1991) from which all extant life forms descended, and thus it does not allow us to design a traditional monorooted dichotomous phylogenetic tree.

Nevertheless, there was still some hope of saving the monorooted tree when the amino acid sequences of pairs of the paralogous genes of ATPases (Gogarten *et al.*, 1989) and protein elongation factors (Iwabe *et al.*, 1989) indicated a specific relationship between archaea and eukarya. However, when the sequences of the paralogous genes were applied to rooting the universal tree (Woese *et al.*, 1990), the branching order of the three domains at the base of the tree turned out to be ambiguous. For, as examplified in Figure 2.1, each of the three statistically possible pairs among the three domains appears to be a sister group depending on the basic phenetic characteristics considered.

The quasi-random phenetic mosaicism among the three domains that this indicated was confirmed at the genomic level when the orthologous genes of the archaeon *Methanococcus jannaschii*, of the eukaryote *Saccharomyces cerevisiae*, and of several bacterial species were aligned (Clayton *et al.*, 1997). While the majority of the genes of *Methanococcus* are shared either with bacteria or/and eukaryotes, only a minority of genes seem to be confined to the domain Archaea (Bult *et al.*, 1996).

Such a quasi-random distribution of genes among the three domains cannot be explained satisfactorily by any order of dichotomous branching of a common ancestral cell or by chimerism originating from fusion or engulfment among prokaryotes at early evolutionary stages as was suggested by various authors (Zillig *et al.*, 1992; Sogin, 1994; Golding and Gupta, 1995; Doolittle, 1996; Margulis, 1996). The distribution pattern rather calls for a revival of the previous proposal of a tripartition at an early pre-cellular stage of life (Fox *et al.*, 1980), resulting in three dichotomously branched separate lineages forming a phylogenetic 'bush' rather than a tree (Figure 2.2).

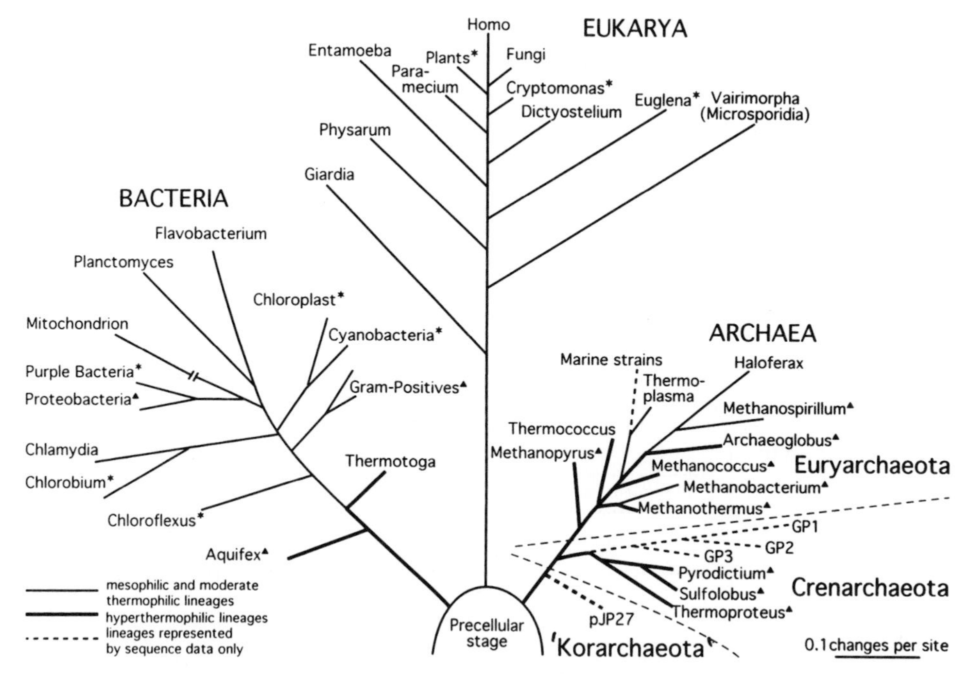

Figure 2.2 Universal phylogenetic 'bush' of organisms. Branching order and branch lengths based on SSU rRNA sequences are taken from the tree shown by Pace (1997). * = photoautotrophic; ▲ = chemoautotrophic.

2.3 Origin of the proto-cells of the three domains

As depicted in Figure 2.3 (Kandler, 1994a,b), the proto-cells, which gave rise to the lineages of the three domains, are suggested to be aboriginal products of three successive independent cellularisation events in an evolving multiphenotypical population of pre-cells. The scenario underlying Figure 2.3 suggests that life originated by reductive synthesis of organic compounds from volcanic gases in the heated hydrosphere of an 'iron–sulfur world' (Wächtershäuser, 1988a,b, 1992). This proposal is supported by the recent finding (Huber and Wächtershäuser, 1997) that an aqueous slurry of coprecipitated NiS and FeS converted CO and CH_3SH into the activated thioester of acetic acid at hydrothermal temperatures.

Although the details of the reaction mechanism leading to the formation of activated acetic acid remain to be elucidated, the abiological C–C bond-forming reaction of Huber and Wächtershäuser (1997) may be seen as a plausible preliminary step in the building up of more complex organic molecules (Crabtree, 1997) and a possible beginning of a primordial metabolism which led finally to life (Wächtershäuser, 1997).

Various redox couples, such as H_2/S^0, H_2/O_2, H_2/SO_4, H_2/CO_2, still exploited today by extant chemolithoautotrophic hyperthermophilic archaea and bacteria (Stetter *et al.*, 1990; Stetter, 1994), may have been tapped by the then nascent life and may have enabled the metabolically diversified evolving life to venture into the various habitats of the early geosphere with ample supply of such redox energy. With some delay, heterotrophic pre-cells, utilising the energy stored in the accumulating biomass by primitive fermentation may also have evolved, thus initiating carbon recycling.

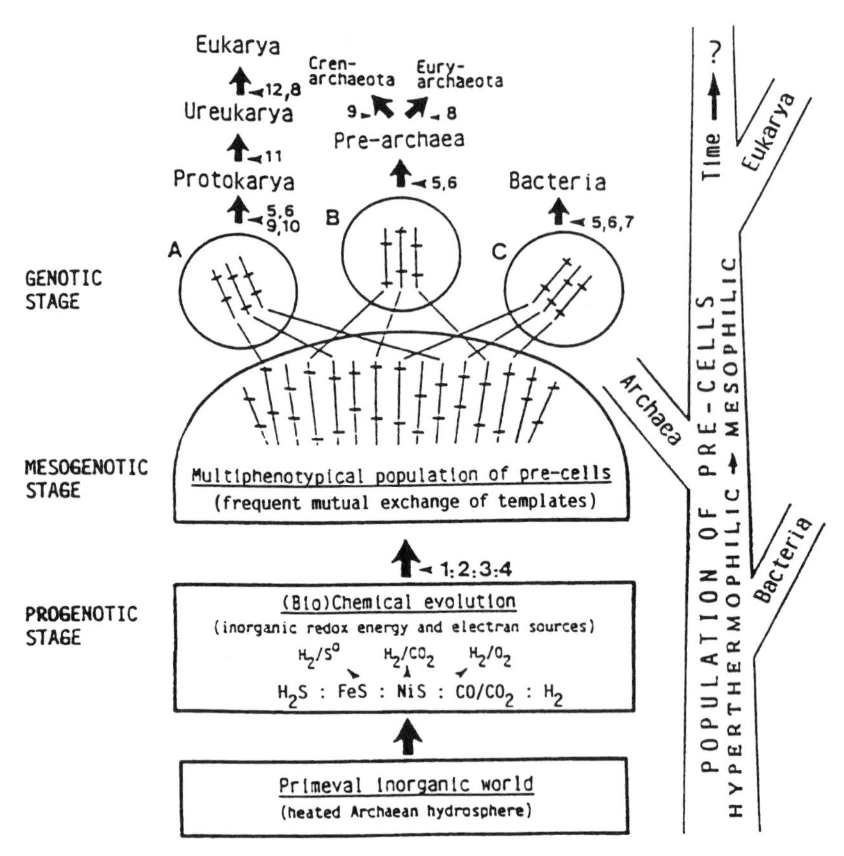

Figure 2.3 Scenario of the origin and early diversification of life and the formation of proto-cells of each of the three domains by successive cellularisation processes. A, B, C: founder groups of the three domains (Modified after Kandler 1994a.)

Important evolutionary improvements are indicated by numbers: (1) Reductive formation of organic compounds from CO or CO_2 by Me-sulfur coordinative chemistry; (2) tapping of various redox energy sources and formation of primitive enzymes and templates; (3) elements of a transcription and translation apparatus and loose associations; (4) formation of pre-cells; (5) stabilised circular or linear genomes; (6) cytoplasmic membranes; (7) rigid murein cell walls; (8) various non-murein rigid cell walls; (9) glycoproteinaceous cell envelope or glycokalyx; (10) cytoskeleton; (11) complex chromosomes and nuclear membrane; (12) cell organelles via endosymbiosis.

The ongoing extensive, but still elusive, discussions about the alleged physicochemical details leading to a complex metabolism, the evolution of templates, and the transcription and translation mechanisms are summarised in excellent reviews (cf. Shapiro, 1985; Doolittle, 1996; Deamer, 1997) and have been the subject of various symposia (Osawa and Honjo, 1991; Bengtson, 1994) and books (Morowitz, 1992; de Duve, 1991). Recently, a most comprehensive scenario of 'the process of early evolution of life that begins with chemical necessitiy and winds up in genetic chance' has been designed by Wächtershäuser (1997) by applying the principles of biochemical retrodiction.

In Woese's scenario of the evolution of life, his progenotic stage (Woese, 1987) covers a series of conditions of 'genetic communion' in evolution that run from the most primitive system with any kind of transcription probably being extremely error-prone, to a stage with a well-developed genotype–phenotype linkage prone to cellularisation which directly precedes the 'genote'. This late pre-cellular stage is denoted 'mesogenotic' in Figure 2.3 to prevent the confusion of using the term 'progenotic' for both the earliest and the most advanced stage of pre-cellular evolution.

The scenario depicted in Figure 2.3 suggests that the common ancestral mesogenotic stage, from which the proto-cells of the extant domains emerged, consisted of a multi-phenotypical population of pre-cells, i.e. metabolising, self-reproducing, loose entities exhibiting many of the basic properties of a cell but no proper cytoplasmic membrane and no stable chromosome. The genetic information was organised in modules of limited size, allowing a frequent exchange and recombination of genetic information constrained only by structural and metabolic incompatibilities.

The cellularisation events leading to the formation of the proto-cells of each of the three domains, i.e. the formation of a cytoplasmic membrane and stabilisation of the genetic information in circular or linear chromosomes, may have occurred in small spatially or ecologically separated 'founder' groups (*sensu* E. Mayr, 1954) of pre-cells containing a mixture of modules of metabolically and structurally compatible genetic information. Such a process may formally be seen as an analogue of the formation of new taxa in higher organisms, which is described as parallel polyphyly or parallelophyly (Mayr and Ashlock, 1991, pp. 256–259). However, the entities forming the founder groups of the three domains were not variants of an ancestral taxon, as in the case of parallelophyly in higher organisms, but aboriginal products of biochemical evolution (pre-cells; Figure 2.3).

The three domains may have emerged successively as indicated by the branching order of the tree at the right side in Figure 2.3. Thus, the proto-cell of the bacterial lineage arose from a much less evolved pre-cellular population than that of the eukarya, while the proto-cell of the archaea originated from an intermediate evolutionary stage of the pre-cellular population. This condition explains the distinct genetic mosaicism of the three domains, especially the close relationship of archaea and eukarya with respect to highly derived characteristics, for instance in the information-procession categories of replication, transcription and translation. On the other hand, genes for the central intermediate metabolism and amino acid biosynthesis are largely common to the three domains (Clayton *et al.*, 1997).

2.4 Origin and evolution of the domain Bacteria

The question whether at least traces of free oxygen were present in the earth's early atmosphere is decisive for our understanding of the evolution of the domain Bacteria. If we follow Towe's suggestion (Towe, 1994) that 'it is probable that free oxygen has been part of the earth's atmosphere and an important factor in the early evolution of life since the time of the oldest sedimentary rocks', we may assume that H_2/O_2 chemoautotrophic pre-cells – with respect to the energy source resembling the extant microaerophilic and hyperthermophilic genus *Aquifex* (Huber *et al.*, 1992) – were first to reach the mesogenotic stage prone to cellularisation and were a main constituent of the founder group of the domain Bacteria.

Chemoautotrophic pre-cells utilising redox couples other than H_2/O_2 as well as genetically compatible derived heterotrophic pre-cells may have amalgated into early branches of the multiplying and radiating bacterial proto-cell, resulting in the extant pattern of chemoautrophic and heterotrophic bacterial lineages, a vague analogy to the ongoing lateral gene transfer between extant organisms which leads to an increasing secondary genetic mosaicism within and among the three domains.

Because of the fragility of the cytoplasmic membrane, containing D-glycerol diacyl ester lipids, its reinforcement by a rigid cell wall sacculus was probably a prerequisite for survival, extensive radiation, and successful colonisation of virtually all habitats of the hydrosphere and geosphere.

This problem was solved by the elaboration of murein, found in numerous chemical variations (Schleifer and Kandler, 1972) in almost all branches of extant bacteria including the hyperthermophilic genera *Aquifex* (Huber *et al.*, 1992) and *Thermotoga* (Huber *et al.*, 1986) which form the two deepest branches within the domain Bacteria (Figure 2.2). The few bacterial taxa without a rigid murein cell wall, such as *Chlamydia*, *Mycoplasma*, or *Planctomyces* represent highly derived forms which are supposed to have lost their cell wall during later phases of evolution.

An ecologically most important evolutionary breakthrough reached by some of the moderately thermophilic and mesophilic branches of the domain Bacteria was the development of various modes of phototrophic growth, e.g. by purple bacteria, and of oxygenic photosynthesis by the cyanobacteria. The evolutionary interplay between H_2/O_2 chemoautotrophy and photosynthesis, as indicated by related cofactors for electron transport (Pierson, 1994) and enzymes for CO_2 fixation (Fuchs, 1989), remains to be further investigated.

2.5 Origin and evolution of the domain Archaea

The domain Archaea comprises all extant hyperthermophilic H_2/S^0 and methanogenic H_2/CO_2 chemolithoautotrophs. Since the environmental conditions for these modes of chemolithoautotrophic life have remained almost constant in the volcanic areas up to the present day, the stimulation for evolution by environmental factors was much smaller than in the case of the O_2-dependent life. Therefore, under these conditions, the evolutionary tempo in the H_2/S^0 and H_2/CO_2 chemolithoautotrophs was probably slower than that in the H_2/O_2 chemolithoautotrophs, as is indicated by the phylogenetic distances between the early branchings, which are larger in the bacterial than in the archaeal tree (Figure 2.2).

We therefore conjecture that the formation of the bacterial proto-cell preceded that of the archaeal proto-cell (Figure 2.3), although H_2/S^0 and H_2/CO_2 chemolithoautotrophies are most likely older (more archaic) than H_2/O_2 chemolithoautotrophy.

The cellularisation event leading to the proto-cell of the domain Archaea must have been more complex than that leading to that of the domain Bacteria. Presumably a metabolically complex pre-archaeal proto-cell surrounded by an isopranyl ether lipid membrane may have spawned the two lineages, comprising either H_2/S^0 or H_2/CO_2 chemolithoautotrophs corresponding to the extant kingdoms Crenarchaeota and Euryarchaeota.

Because of the greater physical strength of the archaeal cytoplasmic membranes containing L-glycerol isopranyl ether lipids (Kates, 1993), instead of the bacterial D-glycerol acyl ester lipids, the invention of a further stabilising rigid cell wall was not as essential for survival as in the domain Bacteria. Thus selection was less stringent, and radiation into various lineages preceded the invention of rigid cell walls or envelopes of different chemical nature of the particular lineage (Figure 2.4; Kandler and König, 1993; Kandler, 1994b).

For the time being, we may assume that the extant hyperthermophilic chemolithoautotrophic or sulfur-respiring organotrophic crenarchaeotes and the hyperthermophilic methanogenic euryarchaeotes represent the least modified descendants of the primordial chemolithoautotrophic phenotypes. Therefore, in spite of Doolittle's critique (Doolittle, 1995), it seems still justified to call the domain, embracing both kingdoms, Archaea (Woese *et al.*, 1990), although, as mentioned above, the speciation of the domain Bacteria presumably occurred earlier.

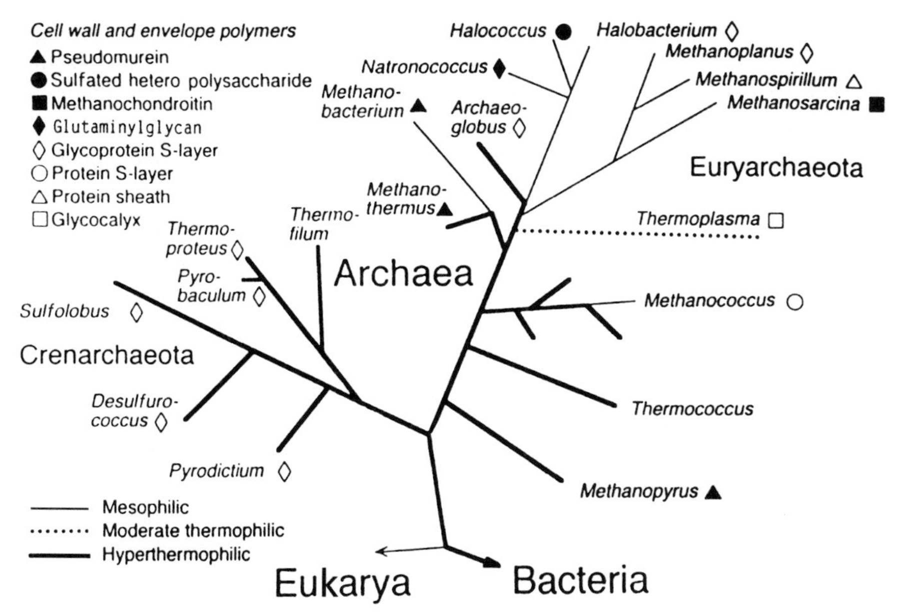

Figure 2.4 Distribution of cell wall and cell envelope polymers among archaea. Branching order and branch lengths are based on 16S rRNA sequences. (Modified after Kandler and König, 1993.)

Both archaeal kingdoms spawned several lineages adapted to mesophilic and heterotrophic life (Figure 2.2.). The recent application of molecular phylogenetic methods to the study of populations of natural microbial ecosystems without cultivation has resulted in the discovery of many more of such lineages than hitherto known (Pace, 1997).

Even a third deeply rooted kingdom, the Korarchaeota (Barns *et al.*, 1996), has been provisionally proposed. However, the morphological and metabolic characteristics of the new kingdom are still unknown since the proposal is based only on 16S rRNA sequences obtained from plankton of a heated volcanic pool.

2.6 Origin and evolution of the domain Eukarya

The domain Eukarya contains neither any genuine chemolithautotrophic nor thermophilic members growing optimally above 50 °C. The lack of thermophilic eukaryotes is most likely due to the thermolability of their intracellular membrane systems, which, as suggested by Brock (1978), eukaryotes cannot construct at high temperatures. Hence, founder groups of the domain Eukarya most likely consisted of highly derived mesophilic heterotrophic pre-cells, adapted to an almost 'modern', at least late archaean or early proterozoic, environment of a predominantly mesophilic temperature regime favourable for the evolution of intracellular membrane systems typical of eukaryotes. According to Schwartzman (1998), the time required for the cooling of the hot archaean oceanic and terrestrial environments to a mesophilic temperature range may have been responsible for the large interval of 1.5–2.0 Ga between the emergence of (thermophilic?) prokaryotic life at about 3.5 Ga BP (Schopf, 1994) and that of (mesophilic?) eukaryotic life at about 2.1 Ga BP (Runnegar, 1994; Hofman, 1994; Weiguo, 1994).

Among the many discontinuities between prokaryotes and eukaryotes with respect to their cell organisation, their gene and genome structure shows the most spectacular differences. The large size of the eukaryotic nucleus and the many linear chromosomes have often been understood as being improvements or advancements acquired after the emergence of eukaryotes from prokaryotes. However, the opposite is more likely, as pointed out by Doolittle and Brown (1994):

> Eukaryotic nuclear genomes are after all very messy structures, with vast amounts of seemingly unnecessary junk DNA, difficult-to-rationalise complexities in mechanisms of transcription and mRNA modification and processing, and needless scattering of genes that often in prokaryotes would be neatly arranged into operons. It might be easiest to see nuclear genomes as in a primitive state of organisation, which prokaryotes, by dint of vigorous selection for economy and efficiency (streamlining), have managed to outgrow.

Doolittle's view is in accordance with our argument based on the scenario depicted in Figure 2.3 and on the palaeontological evidence that eukaryotes are still in a 'juvenile' age of about 1.5–2 Ga as compared to prokaryotes. They look back to a cellular life and streamlining by Darwinian selection of about 3.8 billion years in the case of bacteria, or somewhat less in the case of archaea because of the above-mentioned delayed formation of the archaeal proto-cell (Figure 2.3).

Accordingly, derived traits such as the transcription apparatus of archaea exhibit an intermediate degree of refinement between the most streamlined and the most unrefined version of the bacteria and the eukarya, respectively (Zillig *et al.*, 1993). Correspondingly, the presence of introns in some lineages of archaea is indicative of an incomplete removal of introns which are still common in eukarya but virtually absent in bacteria.

The intermediate evolutionary position of the archaea between bacteria and eukarya is also evident when the distribution of 'modern' biosynthetic capabilities, for instance the *N*-glycosylation of proteins, is considered: glycoproteins are most common in eukarya, also frequent in archaea (cf. Kandler and König, 1993), but very rare in bacteria. Similarly, the reinforcement of the cytoplasmic membrane by a glycocalyx consisting mainly of glycoprotein and lipoglycans is restricted to eukarya with only one known exception: the archaeon *Thermoplasma* which exhibits a glycocalyx-like reinforcement of the cytoplasmic membrane (cf. Kandler and König, 1993).

Undoubtedly, the development of actin and of a cytoskeleton was a most important event in the development of the eukarya which converted the ureukaryal cell into a 'modern' eukaryotic cell, able to engulf prokaryotes, to feed on them by phagocytosis, or to convert them into endosymbionts and finally to cell organelles for energy production (mitochondria) and photoautotrophic growth (chloroplasts). Although the endosymbiotic origin of these cell organelles was proposed almost a century ago on the basis of cytological (Mereschkowsky, 1905) and genetic (Renner, 1934) studies, the endosymbiosis hypothesis was only substantiated by biochemical and fine-structural evidence accumulating during the 1960s, and was finally proved by the 16S rRNA sequence data (Woese and Fox, 1977).

Today, the elucidation of the intimate interplay between the prokaryotic genomes of the organelles and the nuclear genome of the eukaryotic host is one of the most exciting tasks of modern biology (cf. Herrmann, 1997).

2.7 The fate of the progenotic and pre-cellular life on earth

After all, what was (is) the fate of the pre-cellular life after it gave birth to the domain Eukarya? Was it destroyed by changes in the environment; was it devoured by the many hungry more 'highly' developed descendants; or can we expect to see an additional domain emerge in the distant future, and, thus, does it make sense to hunt for pre-cellular life in creepy, submarine volcanic areas?

The answer to the last question must be yes, if we believe in the correctness and the constancy of the physicochemical laws and in our present understanding of the geological history of our planet. Since volcanism still persists in the outer mantle of the earth, the respective processes should still be going on: reductive synthesis and accumulation of organic matter from volcanic gases, as examplified by Huber and Wächtershäuser's (1997) test tube experiments and the supposed further reactions leading to more complex organic compounds resulting, finally, in progenotic life. However, during the past 3 Ga the ocean water has become aerobic because of the evolution of oxygenic photosynthesis. In addition, the outer crust became densely populated by hyperthermophilic prokaryotes (Stetter *et al.*, 1993; Gold, 1992). Thus, today, nascent life would be confronted with chemical and ecological conditions very different from those on the virgin earth 4 Ga ago. Hence, if possible at all, evolution would be forced into other directions.

2.8 Extraterrestrial life?

Exobiologists hold the position that life and even civilisations may be widespread in the universe (cf. Horneck, 1995). In fact, it seems most likely that planetary bodies with similar subsurface volcanic conditions as those known to exist in our planet are common in the universe and may give rise to the evolution of life even if the surface is totally inhospitable (Gold, 1992). Thus, any progress in understanding the origin of life from inanimate matter and the early evolution of life on earth would also stimulate exobiological research and vice versa.

2.9 Conclusion

Life originated by reductive synthesis of organic compounds from volcanic gases in a heated 'iron–sulfur world' of the primeval earth and evolved via a progenotic stage into a multiphenotypical population of mesogenotic pre-cells, prone to cellularisation. The chemolithoautotrophic proto-cells of the domains Bacteria and Archaea originated first, under a thermophilic regime, while the proto-cell of the domain Eukarya arose distinctly later from a derived heterotrophic population of pre-cells under mesophilic conditions. The concept of the origin of life from inanimate matter based exclusively on physicochemical processes conforms Immanuel Kant's (1790) insight, far ahead of the ideas ruling his time and even those dominating the nineteenth and early twentieth centuries:

> This analogy of forms . . . strengthens the suspicion that they have an actual kinship . . . in the gradual approximation of one animal species to another, from . . . man, back to the polyp, and from this back even to mosses and lichens, and finally . . . to crude matter; and from this, and the forces, which it exerts in accordance with mechanical laws, resembling those by which it acts in the formation of crystals, seems to be developed the whole technic of nature.

2.10 Summary

A critical consideration of the relationship of basic phenetic characteristics of the three domains of life shows that each of the three statistically possible pairs among the three domains appears to be a sister group depending on the characters considered. The quasi-random distribution of basic characters among the three domains can be explained neither by any order of dichotomous branching of a common ancestral cell nor by chimerism between the domains at an early evolutionary stage.

It is therefore proposed that the proto-cells of each of the three domains were formed at different evolutionary stages by independent cellularisation events in founder groups of an evolving pre-cellular population. Accordingly, a 'first cell', ancestral to the three domains, has never existed. Thus, the figural representation of the genealogical relationship among all extant living beings results in a tripartite 'phylogenetic bush' rather than a traditional monorooted dichotomously branched phylogenetic tree.

A scheme depicting the proposed evolution of life beginning with spontaneous synthesis of carbon compounds from volcanic gasses in a heated primeval inorganic world and proceeding via progenotic and mesogenotic stages to the extant genotic stage is presented. Important aspects of the evolution of each of the three domains are discussed.

Acknowledgements

I am indebted to K. H. Schleifer, K. O. Stetter, G. Wächtershäuser, and C. R. Woese for communicating unpublished work, stimulating discussions and advice.

References

BARNS, S. M., DELWICHE, C. F., PALMER, J. D. and PACE, N. R. (1996) Perspectives on archaeal diversity, thermophily and monophyly from environmental rRNA sequences. *Proc. Natl. Acad. Sci. USA*, **93**, 9188–9193.

BENGTSON, S. (ed.) (1994) *Early Life on Earth*, Nobel Symposium No. 84, p. 630 (New York: Columbia University Press).

BROCK, T. D. (1978) *Thermophilic Microorganisms and Life at High Temperatures* (New York: Springer-Verlag).

BULT, C. J., WHITE, O., OLSEN, G. J. *et al.* (1996) Complete genome sequence of the methanogenic archaeon, *Methanococcus jannaschii*. *Science*, **273**, 1058–1072.

CAVALIER-SMITH, T. (1992) Bacteria and Eukaryotes. *Nature*, **356**, 570.

CHATTON, E. (1938) *Titres et travaux scientifiques (1906–1937) de Eduard Chatton* (Sottano, Sète).

CLAYTON, R. A., WHITE, O., KETCHUM, K. A. and VENTER, J. C. (1997) The first genome from the third domain of life. *Nature*, **387**, 459–462.

CRABTREE, R. H. (1997) Where smokers rule. *Science*, **276**, 222–247.

DARWIN, C. (1859) *The Origin of Species by Means of Natural Selection or the Preservation of Favoured Races in the Struggle for Life.* (London: John Murray), p. 69; London (1968): Penguin Books.

DEAMER, D. W. (1997) The first living systems: a bioenergetic perspective. *Microbiol. Mol. Biol. Rev.*, **61**, 239–261.

DE DUVE, C. (1991) *Blueprint for a Cell – The Nature and Origin of Life*, p. 275 (Burlington, NC: Neil Patterson Publishers).

DOOLITTLE, R. F. (1995) Of Archae and Eo: what's in a name?, *Proc. Natl Acad. Sci. USA*, **92**, 2421–2423.

DOOLITTLE, R. F. (1996) Some aspects of the biology of cells and their possible evolutionary significance. In *Evolution of Microbial Life*, ed. D.McL. Roberts, P. Sharp, G. Aldersond and M. Collins, pp. 1–20 (Cambridge: University Press).

DOOLITTLE, W. F. and BROWN, J. R. (1994) Tempo, mode, the progenote and the universal root, *Proc. Natl Acad. Sci. USA*, **91**, 6721–6728.

FOX, G. E., STACKEBRANDT, E., HESPELL, R. B. *et al.* (1980) The phylogeny of prokaryotes. *Science*, **209**, 457–463.

FUCHS, G. (1989) Alternative pathways of autotrophic CO_2 fixation. In *Autotrophic Bacteria*, ed. H. G. Schlegel and B. Bowien, pp. 365–382 (Madison, WI: Science Tech Publishers).

GOGARTEN, J. P., KIBAK, H., DITTRICH, P. *et al.* (1989) Evolution of the vacuolar H-ATPase: implication for the origin of eukaryotes. *Proc. Natl Acad. Sci. USA*, **86**, 6661–6665.

GOLD, T. (1992) The deep, hot biosphere. *Proc. Natl Acad. Sci. USA*, **89**, 6045–6049.

GOLDING, G. B. and GUPTA, R. S. (1995) Protein-based phylogenies support a chimeric origin of the eukaryotic genome. *Mol. Biol. Evol.*, **12**, 1–6.

HAECKEL, E. (1866) *Generelle Morphologie der Organismen* (Berlin: Verlag Georg Reimer).

HERRMANN, R. G. (1997) Eukaryotism, towards a new interpretation. In *Eukaryotism and Symbiosis*, ed. H. E. A. Schenk, R. G. Herrmann, K. W. Jeon, N. E. Müller and W. Schwemmler, pp. 73–118 (Heidelberg: Springer-Verlag).

HOFMANN, H. J. (1994) Proterozoic carbonaceous compressions ('metaphytes' and 'worms'). In *Early Life on Earth*, Nobel Symposium No. 84, ed. S. Bengtson, pp. 342–357 (New York: Columbia University Press).

HORNECK, G. (1995) Exobiology, the study of the origin, evolution and distribution of life within the context of cosmic evolution: a review. *Planet. Space Sci.*, **43**, 189–217.

HUBER, C. and WÄCHTERSHÄUSER, G. (1997) Activated acetic acid by carbon fixation on (Fe, Ni) S under primordial conditions. *Science*, **276**, 245–24.

HUBER, R., LANGWORTHY, T. A., KÖNIG, H. *et al.* (1986) *Thermotoga maritima* sp. nov. represents a new genus of unique extremely thermophilic Eubacteria growing up to 90 °C. *Arch. Microbiol.*, **144**, 324–333.

HUBER, R., WILHARM, T., HUBER, D. *et al.* (1992) *Aquifex pyrophilus* gen. nov. sp. nov., represents a novel group of marine hyperthermophilic hydrogen-oxidising bacteria, *System. Appl. Microbiol.*, **15**, 340–351.

IWABE, N., KUMA, K., HASEGAWA, M., OSAWA, S. and MIYATA, T. (1989) Evolutionary relationship of Archaebacteria, Eubacteria, and eukaryotes inferred from phylogenetic trees of duplicated genes. *Proc. Natl Acad. Sci. USA*, **86**, 9355–9359.

KANDLER, O. (1994a) The early diversification of life. In *Early Life on Earth*, Nobel Symposium No. 84, ed. S. Bengtson, pp. 152–161 (New York: Columbia University Press).

KANDLER, O. (1994b) Cell wall biochemistry and three-domain concept of life. *System. Appl. Microbiol.*, **16**, 501–509.

KANDLER, O. and KÖNIG, H. (1993) Cell envelopes of archaea: structure and chemistry. In *The Biochemistry of Archaea (Archaebacteria)*, ed. M. Kates *et al.*, vol. 26, pp. 223–259 (Amsterdam: Elsevier Science).

KANT, I. (1790) *The Critique of Judgement*, translation by Meredith, J. C. (1952), pp. 77–82 (Oxford: Clarendon Press).

KATES, M. (1993) Membrane lipids of archaea. In *The Biochemistry of Archaea (Archaebacteria)*, ed. M. Kates, pp. 261–295 (Amsterdam: Elsevier Science).

MARGULIS, L. (1996) Archaea-eubacterial mergers in the origin of Eukarya: phylogenetic classification of life. *Proc. Natl Acad. Sci. USA*, **93**, 1070–1076.

MARGULIS, L. and GUERRERO, R. (1991) Kingdoms in turmoil. *New Scientist*, 23 March, 46–50.

MAYR, E. (1954) Change of genetic environment and evolution. In *Evolution as a Process*, ed. J. Huxley, Ford E. B. Hardy, pp. 157–180 (London: Allen and Unwin).

MAYR, E. (1990) A natural system of organisms. *Nature*, **348**, 491.

MAYR, E. and ASHLOCK, P. D. (1991) *Principles of Systematic Zoology*, pp. 255–258 (New York: MacGraw-Hill).

MERESCHKOWSKY, C. (1905) Über Natur und Ursprung der Chromatophoren im Pflanzenreiche. *Biol. Centralblatt*, **25**, 593–604.

MOROWITZ, H. J. (1992) *Beginnings of cellular life.* (New Haven, CT: Yale University Press).

OSAWA, S. and HONJO, T. (eds) (1991) *Evolution of Life*, p. 460 (Tokyo: Springer-Verlag).

PACE, N. R. (1997) A molecular view of microbial diversity and the biosphere. *Science*, **276**, 734–740.

PIERSON, B. K. (1994) The emergence, diversification, and role of photosynthetic eubacteria. In *Early Life on Earth*, Nobel Symposium No. 84, ed. S. Bengtson, pp. 161–180 (New York: Columbia University Press).

PONNAMPERUMA, C. and CHELA-FLORES, J. (eds) (1995) *Chemical Evolution: Structure and Model of the First Cell*, p. 383 (Dordrecht: Kluwer Academic).

RENNER, O. (1934) Die pflanzlichen Plastiden als selbständige Elemente der genetischen Konstitution. *Ber. Verh. Sächs. Akad. Wiss. Leipzig, Math.-Phys. Kl.*, **86**, 214–266.

RUNNEGAR, B. (1994) Proterozoic eukaryotes: evidence from biology and geology. In *Early Life on earth*, Nobel Symposium No. 84, ed. S. Bengtson, pp. 287–297 (New York: Columbia University Press).

SCHLEIFER, K. H. and KANDLER, O. (1972) Peptidoglycan types of bacterial cell walls and their taxonomic implications. *Bacteriol. Rev.*, **36**, 404–477.

SCHOPF, J. W. (1994) The oldest known records of life: Early Archaean stromatolites, microfossils, and organic matter. *Early Life on earth*, Nobel Symposium No. 84, ed. S. Bengtson, pp. 193–206.

SCHWARTZMAN, D. W. (1998) Life was thermophilic for the first two-thirds of earth history (this volume).

SHAPIRO, R. (1985) *Origins: The Possibilities of Science for the Genesis of Life on Earth* (New York: Summit Books).

SOGIN, M. L. (1994) The origin of eukaryotes and evolution into major kingdoms. In *Early Life on Earth*, Nobel Symposium No. 84, ed. S. Bengtson, pp. 181–192 (New York: Columbia University Press).

STETTER, K. O. (1994) The lesson of Archaebacteria. In *Early Life on Earth*, Nobel Symposium No. 84, ed. S. Bengtson, pp. 143–151 (New York: Columbia University Press).

STETTER, K. O., FIALA, G., HUBER, G., HUBER, R. and SEGERER, A. (1990) Hyperthermophilic microorganisms. *FEMS Microbiol. Rev.*, **75**, 117–124.

STETTER, K. O., HUBER, R., BLÖCHL, E. *et al.* (1993) Hyperthermophilic archaea are thriving in deep North Sea and Alaskan oil reservoirs. *Nature*, **365**, 743–745.

TOWE, K. M. (1994) Earth's early atmosphere: constraints and opportunities for early evolution. In *Early Life on Earth*, Nobel Symposium No. 84, ed. S. Bengtson, pp. 36–47 (New York: Columbia University Press).

WÄCHTERSHÄUSER, G. (1988a) Pyrite formation, the first energy source for life: a hypothesis. *System Appl. Microbiol.*, **10**, 207–210.

WÄCHTERSHÄUSER, G. (1988b) Before enzymes and templates: theory of surface metabolism, *Microbiol. Rev.*, **52**, 452–484.

WÄCHTERSHÄUSER, G. (1992) Groundworks for an evolutionary biochemistry: the iron–sulfur world. *Prog. Biophys. Mol. Biol.*, **58**, 85–201.

WÄCHTERSHÄUSER, G. (1997) The origin of life and its methodological challenge. *J. Theor. Biol.*, **187**, 483–494.

WEIGUO, S. (1994) Early multicellular fossils. In *Early Life on Earth*, Nobel Symposium No. 84, ed. S. Bengtson, pp. 358–369 (New York: Columbia University Press).

WHEELIS, M., L., KANDLER, O. and WOESE, C. R. (1992) On the nature of global classification. *Proc. Natl Acad. Sci. USA*, **89**, 2930–2934.

WHITTAKER, R. H. (1969) New concepts of kinkdoms of organisms. *Science*, **163**, 150–160.

WOESE, C. R. (1982) Archaebacteria and cellular origins: an overview. *Zentralblatt f. Bakteriologie u. Hygiene I. Abt. Originale.*, **C3**, 1–17.

WOESE, C. R. (1987) Bacterial evolution. *Microbiol. Rev.*, **51**, 221–271.

WOESE, C. R. and FOX, G. E. (1977) Phylogenetic structure of the prokaryotic domain: the primary kinkdoms. *Proc. Natl Acad. Sci. USA*, **75**, 5088–5090.

WOESE, C. R., KANDLER, O. and WHEELIS, M. L. (1990) Towards a natural system of organisms: proposal for the domains Archaea, Bacteria and Eukarya. *Proc. Natl Acad. Sci. USA*, **87**, 4576–4579.

ZILLIG, W., PALM, P. and KLENK, H. P. (1992) A model of the early evolution of organisms: the arisal of the three domains of life from the common ancestor. In *The Origin and Evolution of Prokaryotic and Eukaryotic Cells*, ed. H. Hartman and K. Matsuno, pp. 163–182 (New Jersey: World Scientific).

ZILLIG, W., PALM, P. and KLENK, H. P. (1993) Transcription in archaea. In *The Biochemistry of Archaea (Archaebacteria)*, ed. M. Kates *et al.*, vol. 26, pp. 367–391 (Amsterdam: Elsevier Science).

3

Life was Thermophilic for the First Two-thirds of Earth History

DAVID W. SCHWARTZMAN

Department of Biology, Howard University, Washington, DC, USA

3.1 Introduction

Maynard Smith and Szathmary (1995) remark that 'Astonishingly, the time that was needed to pass from inanimate matter to life is four times shorter than needed for passing form prokaryotes to eukaryotes [this is likely a very conservative estimate] . . . it is hard to argue that they [the steps to form eukaryotes] are more difficult than to establish a genetic code'. They then go on to propose an explanation based on existing competition of prokaryotes. Further, they are also puzzled about the timing of the subsequent emergence of Metazoa, but offer the conventional explanation, the rise of atmospheric oxygen. Let us examine this critical problem of the timing of emergence more closely.

Cloud (1976) proposed that atmospheric pO_2 levels determined the timing of major events in biotic evolution, such as the emergence of eukaryotes and Metazoa (the 'animal' kingdom). It has been recently suggested that aerobic respiration emerged very early (Castresana and Saraste, 1995). If oxygenic photosynthesis occurred by no later than 3.5 billion years ago (Ga) (Schopf, 1992), the occurrence of aerobic microenvironments (e.g., in cyanobacterial mats) plausibly predated the emergence of aerobic eukaryotes and Metazoa. The rise of atmospheric oxygen by 1.9 Ga (15% of the present atmospheric level (PAL) according to Holland, 1994) predated the emergence of Metazoa, which may require less than 2% PAL; this scenario has been challenged by Ohmoto (1996, 1997), who argue for pre-1.9 Ga atmospheric oxygen levels comparable to today's value. On the other hand, the rise of atmospheric oxygen may well have constrained the emergence of megascopic eukaryotes, particularly Metazoa, as originally argued by Cloud (1976). Thus, why did aerobic environments plausibly sufficient for their metabolism predate the emergence of these organisms by at least 0.5 to 1 billion years? The following possibilities may explain this observation:

1 The earliest aerobic (mitochondrial) eukaryote and Metazoan fossils are not yet recognised or not preserved.
2 Physical constraints other than oxygen level prevented their emergence; temperature is one likely constraint.
3 Evolution is that slow (least likely explanation?).

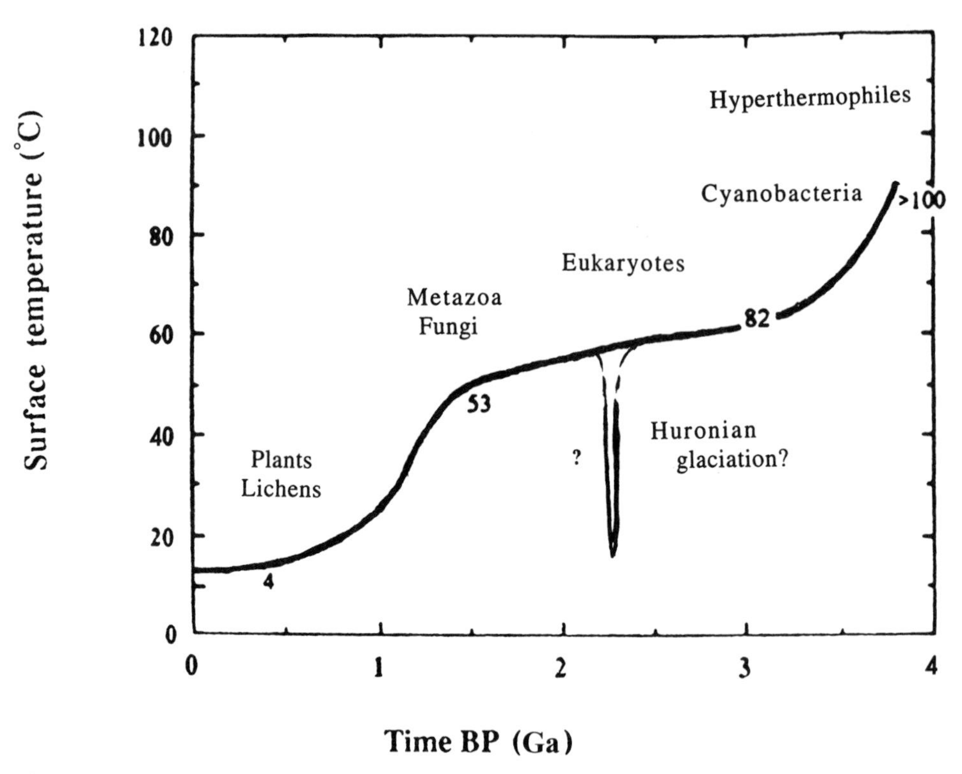

Figure 3.1 Proposed surface temperature history of the earth. Evolutionary developments are indicated. Numbers on curve are model calculations of biotic enhancement ratio (present to past); see Schwartzman and Shore (1996).

If surface temperature was the critical constraint on microbial evolution, as we have suggested (Schwartzman *et al.*, 1993), then the approximate upper temperature limit for viable growth of a microbial group should equal the actual surface temperature at time of emergence, assuming that an ancient and necessary biochemical character determines the presently determined upper temperature limit of each group. Turning to the proposed surface temperature history of the earth, cyanobacteria emerged at 3.5 Ga (70 °C), aerobic eukaryotes at about 2.6 Ga (60 °C) and Metazoa at 1.0–1.5 Ga (50 °C), as shown in Figure 3.1, Table 3.1 (Schwartzman, 1995). This temperature history is consistent with palaeotemperatures from the oxygen isotopic record of chert (marine sediments composed of chemically or biochemically precipitated silica) and carbonates (discussed below). Thus, with the possible exception of high mountains, surface environments were hot enough to prevent the emergence of mesophiles until about 1.5 Ga.

In this scenario, the genetic potentiality for rapid evolution is simply realised as soon as – relative to a geological time scale of course – external conditions allow its expression. If confirmed, this conclusion would have major implications to evolutionary biology and our understanding of the evolution of the biosphere.

The upper temperature limit for viable growth has been determined for the main organismal groups (Table 3.1). This limit is appoarently determined by the thermolability of biomolecules (e.g., nucleic acids), organellar membranes and enzyme systems (e.g., heat shock proteins?) (Brock and Madigan, 1991). For example, the mitochondrial membrane is particularly thermolabile, apparently resulting in an upper temperature limit of 60 °C for aerobic eukaryotes. Could the upper temperature limit of 50 °C for Metazoa

Table 3.1 Upper temperature limits for growth of living organisms, times of their emergence

Group	Approximate upper temperature limit (°C)	Time of emergence (Ga)
Plants	45–50	~0.5
Metazoa	50	1–1.5[a]
Aerobic eukaryotes	60	~2.6[c]
Prokaryotic microbes		
Cyanobacteria	70–73	≥3.5[b]
Methanogens	>100	≥3.8[c]
Extreme thermophiles	>100	≥3.8[c]

(Temperatures from Brock and Madigan, 1991.)
[a] Problematic fossil evidence, molecular phylogeny.
[b] Fossil evidence.
[c] Molecular phylogeny; earliest fossil evidence for eukaryotes at 2.1 Ga.

be linked to the thermolability of proteins essential to blastula formation? Anaerobic eukaryotes, thought to have preceded aerobic eukaryotes, before the endosymbiogenic event creating mitochondria (Sogin *et al.*, 1989), may have emerged at somewhat higher temperatures than the assumed limit of 60 °C for eukaryotes. A higher temperature limit for ancestral amitochondrial than mitochondrial eukaryotes would be consistent with the inferred molecular phylogeny, as well as suggestions that mitochondria are more thermolabile than nuclei. It would be interesting if any anaerobic eukaryotes are viable above the apparent limit for aerobic eukaryotes (about 60 °C). Clearly, fundamental research is needed to better understand the biochemical and biophysical basis for the upper temperature limits of the organismal groups.

3.2 A case for thermophilic surface conditions for the Archaean/Early Proterozoic (3.5–1.5 Ga)

We now proceed to consider systematically the case for the temperature scenario suggested above (Figure 3.1).

3.2.1 *Fossil record and molecular phylogeny*

Microbial fossils with ages of 3.5–1.5 Ga are consistent with thermophily, based on morphological similarities with living prokaryotes (Schopf, 1992) which have thermophilic varieties (e.g., Chroococcales, Chloroflexus, Oscillatoriales, Nostocales). For example, the presumed cyanobacterial fossils from 3.5 Ga Australian cherts are consistent with thermophily. The remarkable correspondence of the sequence of appearance of life forms (see Table 3.1) in the Precambrian to their upper temperature limits led Hoyle (1972) to suggest that hot surface conditions on earth may have held up the emergence of complex life. The thermophilic character of deeply rooted eubacteria and archaebacteria is consistent with very warm conditions on the Archaean earth surface (Woese, 1987; Spooner, 1992), but of course does not require them, since early thermophilic forms may have been restricted to similar local environments as now (i.e., hot springs, thermal vents). However, present-day thermophilic communities could be living models for a surface biota inhabiting both oceanic and terrestrial environments in the Archaean/Early Proterozoic if thermophily was required at that time.

3.2.2 Palaeotemperatures from the oxygen isotopic record of cherts and carbonates

Palaeotemperatures have been derived from the oxygen isotope record of pristine cherts (Knauth and Epstein, 1976; Karhu and Epstein, 1986; Knauth and Lowe, 1978; Knauth, 1992) and sedimentary carbonates (Burdett *et al.*, 1990; Winter and Knauth, 1992). A necessary assumption for these temperature calculations, that the Precambrian (at least from 3.5 Ga to its termination at about 0.6 Ga) oxygen isotopic composition of sea water was close to the present, received recent support from Holmden and Muehlenbacks (1993). Knauth and Clemens (1995) have recently reported a large new data set which includes unmetamorphosed cherts with microfossils. Their results are consistent with the inferred Precambrian temperature history given in Knauth (1992), essentially identical to the first-order temperature history in Figure 3.1.

3.2.3 Sedimentological evidence

Archaean (3.8–2.5 Ga) first-cycle clastic sediments are deeply weathered, with arkoses only appearing in abundance in the Proterozoic (2.5–0.6 Ga) (Lowe, 1994), consistent with high Archaean chemical weathering intensities. The latter plausibly resulted from high surface temperatures and atmospheric pCO_2 levels balancing a higher volcanic/metamorphic CO_2 source on smaller continental areas relative to more recent geological time. Arkoses are sandstones with feldspar fragments; feldspars weather more rapidly than quartz, the main component of sandstones.

3.2.4 Sedimentary gypsum is not a good temperature constraint

An upper limit on surface temperature has been commonly cited, this for 3.5 Ga, from evidence for primary evaporitic gypsum precipitation (Walker, 1982), providing a limit of 58 °C since anhydrite is stable above that temperature in fresh water; in sea water the temperature is still lower. However, this constraint appears tenuous given the metastable precipitation of gypsum in nature and in laboratory experiments, far above its stability field, in place of anhydrite, even at 80 °C (Berner, 1971; Gunatilaka, 1990). Metastable precipitation of gypsum above its stability field may also undercut the claim that evidence of possible coexisting evaporitic gypsum and halite in the Proterozoic gives an upper temperature limit of 18 °C (Walker, 1982). When anhydrite does form from a gypsum precursor it often inherits the gypsum crystal form, the very evidence (in the form of pseudomorphs or molds) that is cited for the 58 °C and 18 °C temperature limits.

3.2.5 Lack of fractionation of sulfur isotopes

There is a comparative lack of fractionation of sulfur isotopes between coexisting sulfide and sulfates in Archaean and early Proterozoic sediments relative to Phanerozoic (Ohmoto and Felder, 1987; Knoll, 1990; Bottomley *et al.*, 1992). Knoll (1990) reviews possible explanations for this lack of fractionation in sediments dated at 3.5 Ga, concluding: '. . . perhaps almost all sulfate in pore fluids was reduced biologically to sulfide in an essentially closed system with little fractionation because of high ambient temperatures (70 °C or more) – a theory for which the geological record provides little supporting evidence. A generally acceptable solution to this problem has not yet been proposed' (p. 12). We cite here and in published work just such supporting evidence.

3.2.6 Constraints on pH/pCO_2 ocean, pCO_2 atmosphere

Atmospheric levels of pCO_2 on the order of 1 bar correspond to temperatures ranging from 50 °C to 70 °C in the Archaean/early Proterozoic (Schwartzman *et al.*, 1993; Schwartzman and Shore, 1996). This level is compatible with constraints on oceanic pH/pCO_2 (Grotzinger and Kasting, 1993). A methane and/or ammonia greenhouse (e.g., Sagan, 1977; Lovelock, 1988; Kasting and Grinspoon, 1990) has been proposed for the Archaean. If post-Huronian temperatures in the early Proterozoic were high, a reduced gas (e.g., methane) greenhouse is unlikely given the rise of atmospheric oxygen during this time, thus requiring high atmospheric pCO_2 levels. A recent ingenious attempt to infer atmospheric pCO_2 levels prior to 2.2 Ga from the mineralogy of palaeosols (fossil soils) has derived an upper limit of $10^{-1.4}$ atm (Rye *et al.*, 1995). However, this result may be tenuous given the metamorphic/metasomatic alteration of the original palaeosol and questionable assumptions (such as surface temperature and relevant phase equilibria; for example, an atmospheric pCO_2 on the order of 1 bar, by lowering the pH of water involved in weathering, could actually prevent the formation of the phase, siderite, the absence of which is taken as the basis of the upper limit on pCO_2). In contrast, for Archaean and early Proterozoic banded iron formations (sediments composed largely of iron oxides), siderite is commonly a primary phase, stable at the same high pCO_2 (Klein and Bricker, 1977), consistent with sea water pH buffered above 6 (Grotzinger and Kasting, 1993). There is intriguing evidence of present-day bacterial biomineralisation of iron hydroxide and silica in hot springs, a plausible model of some Precambrian banded iron formations (Konhauser and Ferris, 1996).

3.2.7 Cyanobacterial CO_2 concentrating mechanism

The ability of cyanobacteria to increase their internal pCO_2 by up to 1000 times is consistent with an adaptation to declining CO_2/O_2 in the external environment (Strauss *et al.*, 1992; Badger and Andrews, 1987) and therefore much higher levels of CO_2 in the Archaean atmosphere.

3.2.8 What about the Huronian glaciation at about 2.3 Ga?

The Huronian glaciation is generally accepted by geologists as the earliest documented glaciation. The Neoproterozoic glaciations curiously follow more than a billion years later. However, these ancient tillites (coarse-grained sediments of glacial origin) have been reinterpreted as possible impact deposits (Oberbeck *et al.*, 1993; Rampino, 1994), possibly removing a key upper temperature constraint. The possibility of late Precambrian glaciations is consistent with, but not required by, the proposed temperature history discussed here, since temperatures drop significantly by 1.2 Ga. At least three possible explanations of the Huronian event are consistent with the above high temperature scenario:

(a) A large global temperature excursion (drop of 50 °C) occurred, perhaps triggered by an albedo increase associated with impact generated debris (Rampino *et al.*, 1996). Such an excursion was originally suggested by Knauth and Epstein (1976). An oblique impact might have triggered just this excursion by creating an orbiting ring around the earth, stable for sufficient time to reduce solar radiation reaching the atmosphere and surface (Schultz and Gault, 1990). Reflective CO_2 clouds might prolong low surface temperatures (Caldeira and Kasting, 1992; but see Forget and Pierrehumbert, 1997).

(b) Glaciation did not occur; the Huronian 'tillites' are simply normal debris flows or are really the result of impact, which can generate features commonly identified as glacial (e.g., striations, polishing of clasts) which are found in K–T and other known impact deposits (Rampino *et al.*, 1997).

(c) A methane-dominated greenhouse from 2.8 to 2.3 Ga was destroyed by the rise of atmospheric oxygen, triggering glaciation (see Hayes, 1994, for evidence of methane production). It then took some 10^8 years for volcanically derived CO_2 to build up in the atmosphere, returning surface temperatures to 50–60 °C, consistent with the oxygen isotope record.

The Huronian 'tillites' should be reinvestigated with these possibilities in mind. Evidence for impact would include shock features in quartz, nanodiamonds and other sedimentological characteristics consistent with impact but not glacial origin.

3.3 A geophysiological model of biospheric evolution

A high carbon dioxide 'pressure cooker' atmosphere, inherited from the vigorous degassing and cometary impacts on the early earth, is assumed just prior to the origin of life, with surface temperatures 70–100 °C (Kasting and Ackerman, 1986; Chyba *et al.*, 1990; Delsemme, 1992). Soon after the origin of life, thermophiles colonised the land, enhancing chemical weathering rates, sequestering carbon dioxide into limestone deposits and thereby cooling the earth's surface (Schwartzman and Volk, 1989, 1991). This marked the commencement of the biotically mediated carbonate–silicate geochemical cycle (Walker *et al.*, 1981). Surface cooling is determined by the evolution of the biosphere, constrained by abiotic boundary conditions (i.e., luminosity of the sun, continental area, outgassing rate).

Each new innovation in the microbial soil community resulted in greater biotic enhancement of weathering, culminating in the rhizosphere of higher plants (Schwartzman *et al.*, 1993). There are of course second-order pertubations arising from such factors as continental drift, episodic burial of organic carbon and pulses of intense volcanism (particularly relevant in the Phanerozoic, when atmospheric pCO_2 levels dropped) but these are ignored in the first-order temperature scenario shown in Figure 3.1. Marked cooling in the mid Proterozoic is indicated, consistent with the apparent emergence of Metazoa at 1.0–1.5 Ga (Chapman, 1992; Morris, 1993) and palaeotemperatures of 25–43 °C on 1.1–1.2 Ga cherts (Kenny and Knauth, 1992). This cooling may have been the result of the rise of atmospheric oxygen and more extensive colonisation of the land surface by algal mats (see Beeunas and Knauth, 1985; Horodyski and Knauth, 1994). This increase in terrestrial biotic productivity, along with the onset of frost wedging in mountains, likely substantially increased the global biotic enhancement of weathering (Schwartzman, 1994). While the first definitive fossil evidence for Metazoa dates at 0.65 Ga, an older problematical record does exist (e.g., Robbins *et al.*, 1985; Wray *et al.*, 1996), along with molecular evidence for their emergence at about 1 Ga (Wray *et al.*, 1996). The story of the earliest Metazoa may be analogous to that of Eukarya (Sogin *et al.*, 1989), in that the preservation of microscopic soft-bodied organisms as fossils is unlikely.

A geophysiological model for the evolution of the biosphere, summarising the above discussion, is shown in Figure 3.2. The evolution of the biosphere does not optimise conditions for existing biota, unlike the original Gaia hypothesis. At least two catastrophes for existing life apparently occurred; the well-known oxygen catastrophe and a temperature catastrophe for thermophilic bacteria, once the likely colonisers of the ocean and land, now mainly restricted to hot springs, hydrothermal vents on the ocean floor, porosity in the first few kilometres of the crust and water heaters, although some thermophiles grow in mesophilic environments.

The progressive increase in the biotic enhancement of weathering as a product of biotic and biospheric evolution intensified the carbon sink with respect to the atmosphere–ocean system leading to the transition of climate from a 'hothouse' in the early Precambrian to an 'icehouse' in the Phanerozoic. Climate and life coevolved as a tightly coupled system, constrained by abiotic factors (varying solar luminosity, and the crust's tectonic and impact history). Self-regulation of this coupled system is a property of geophysiology.

3.4 Conclusions

Both the beginning and end of life on this planet are determined by purely non-biological conditions, the beginning by the hydrothermal activity on the ocean floor, the end by the rising radiant energy flux from the sun. But between these times, predetermined by the initial conditions of our solar system, the biosphere evolves in its overall patterns deterministically, going from a hothouse with surface temperatures near 100 °C to an icehouse, with intermittent glacial periods, then in the future back into a hothouse regime before its destruction. Surface temperatures on earth were above the upper limit for thermophily (50 °C) for the first two-thirds of earth's history. The upper temperature limit for growth of organismal groups is apparently determined by the thermolability of biomolecules, organellar membranes and enzyme systems. Aerobic microenvironments plausibly predated the rise of atmospheric oxygen. In addition, atmospheric levels of oxygen sufficient for Metazoan metabolism may have preceded its emergence by at least 0.4 billion years. Thus, high Archaean and early Proterozoic surface temperatures (50–70 °C), inferred from the oxygen isotopic record in cherts and carbonates, could have held back the emergence of eukaryotes and Metazoa, with their upper temperature limits for growth corresponding to the ambient surface temperature at the time of emergence.

3.5 Summary

Surface temperatures on earth were above the upper limit for thermophily (50 °C) for the first two-thirds of earth history. The main evidence for this assertion is the palaeotemperature record of the Precambrian surface environment derived from the oxygen isotopic ratios preserved in ancient chert and carbonates. Surface temperature has been a critical constraint on the tempo of major events in biotic evolution, while itself having

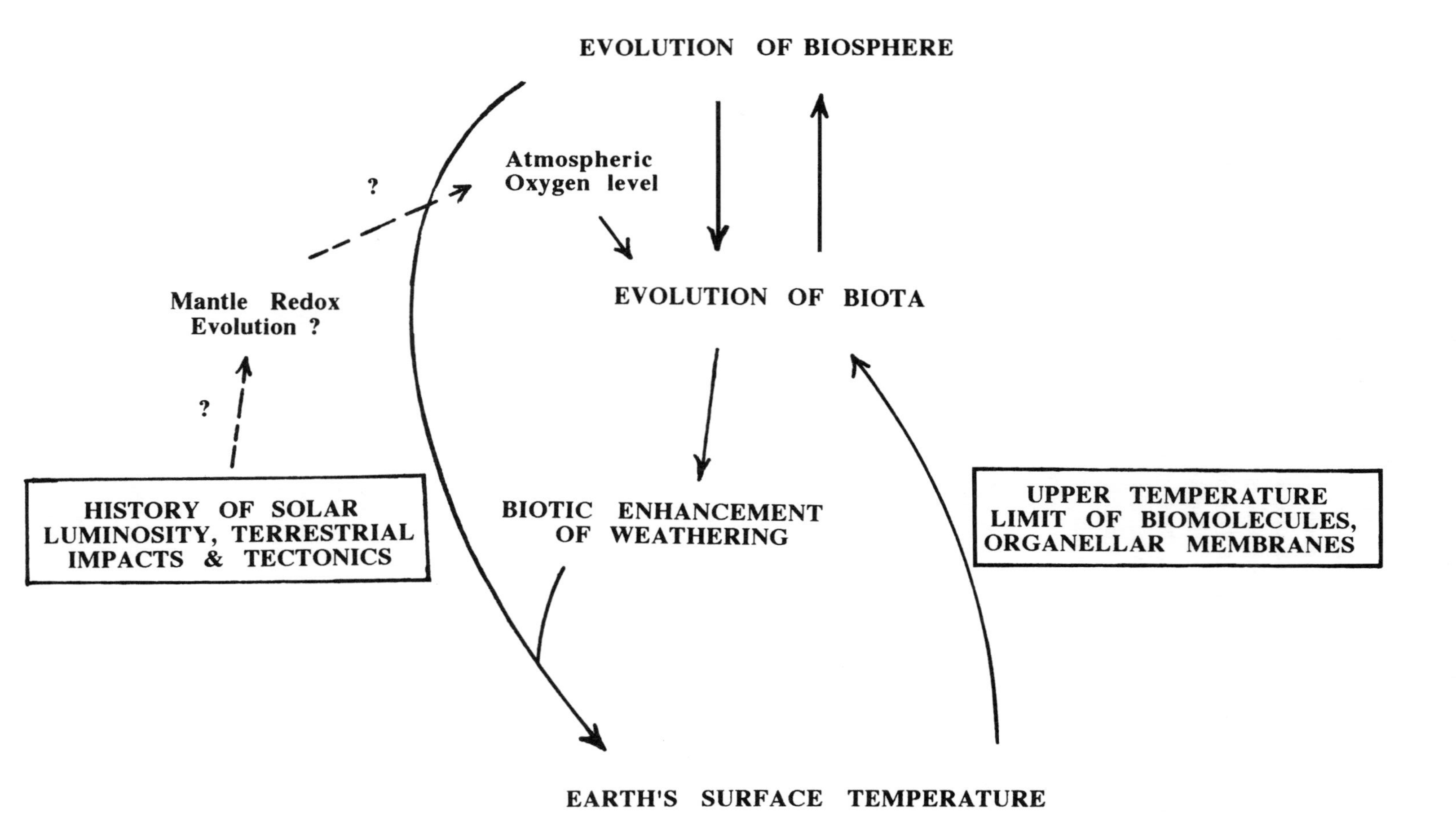

Figure 3.2 Geophysiological model of biospheric evolution. Constraints are shown in boxes. A mantle buffering of atmospheric oxygen is suggested by Kasting *et al.* (1993).

been determined by a progressively increasing role of biota in climatic change over geological time, within the context of abiotic evolution (solar and terrestrial). The temperature constraint has occurred because each major innovation in biological evolution, such as oxygenic photosynthesis (emergence of cyanobacteria), has an inherent biochemical and biophysical upper temperature limit for its metabolism. With the long-term cooling of the earth's surface, new metabolisms and cell types became possible as their upper temperature limits were reached. Cooling occurred because of the combined effects of abiotic variations (e.g., volcanic outgassing rates, solar luminosity) and the progressively powerful effect of land biota on the sequestering of carbon from the atmosphere via chemical weathering in soils.

References

BADGER, M. R. and ANDREWS, T. J. (1987) Co-evolution of Rubisco and CO_2-concentrating mechanisms. *Prog. Photosynthesis Res.*, **3**, 601–609.

BEEUNAS, M. A. and KNAUTH, L. P. (1985) Preserved stable isotopic signature of subaerial diagenesis in the 1.2-b.y. Mescal Limestone, central Arizona: implications for the timing and development of a terrestrial plant cover. *Geol. Soc. Am. Bull.*, **96**, 737–745.

BERNER, R. A. (1971) *Principles of Chemical Sedimentology* (New York: McGraw-Hill).

BOTTOMLEY, D. J., VEIZER, J., NIELSEN, H. and MOCZYDLOWSKA, J. (1992) Isotopic composition of disseminated sulfur in Precambrian sedimentary rocks. *Geochim. Cosmochim. Acta*, **56**, 3311–3322.

BROCK, T. D. and MADIGAN, M. T. (1991) *Biology of Microorganisms*, 6th edn (Englewood Cliffs, N.J.: Prentice-Hall).

BURDETT, J. W., GROTZINGER, J. P. and ARTHUR, M. A. (1990) Did major changes in the stable-isotope composition of Proterozoic sea water occur? *Geology*, **18**, 227–230.

CALDEIRA, K. and KASTING, J. F. (1992) Susceptibility of the early Earth to irreversible glaciation caused by carbon dioxide clouds. *Nature*, **359**, 226–228.

CASTRESANA, J. and SARASTE, M. (1995) Evolution of energetic metabolism: the respiration-early hypothesis. *Trends Biochem. Sci.*, **20**, 443–448.

CHAPMAN, D. J. (1992) Origin and divergence of protists. In *The Proterozoic Biosphere*, ed. J. W. Schopf and C. Klein, pp. 477–483 (New York: Cambridge University Press).

CHYBA, C. F., THOMAS, P. J., BROOKSHAW, L. and SAGAN, C. (1990) Cometary delivery of organic molecules to the early Earth. *Science*, **249**, 366–373.

CLOUD, P. (1976) Beginnings of biospheric evolution and their biogeochemical consequences. *Paleobiology*, **2**, 351–387.

DELSEMME, A. H. (1992) Cometary origin of carbon, nitrogen and water on the Earth. *Origins of Life and Evolution of the Biosphere*, **21**, 279–298.

FORGET, F. and PIERREHUMBERT, R. T. (1997) Warming early Mars with carbon dioxide clouds that scatter infrared radiation. *Science*, **278**, 1273–1276.

GROTZINGER, J. P. and KASTING, J. F. (1993) New constraints on Precambrian ocean composition. *Journal of Geology*, **101**, 235–243.

GUNATILAKA, A. (1990) Anhydrite diagenesis in a vegetated sabkha, Al-Khiran, Kuwait, Arabian Gulf. *Sediment. Geol.*, **69**, 95–116.

HAYES, J. M. (1994) Global methanotrophy at the Archean–Proterozoic transition. In *Early Life on Earth*, Nobel Symposium No. 84, ed. S. Bengtson, pp. 220–236 (New York: Columbia University Press).

HOLLAND, H. D. (1994) Early Proterozoic atmospheric change. In *Early Life on Earth*, Nobel Symposium No. 84, ed. S. Bengtson, pp. 237–244 (New York: Columbia University Press).

HOLMDEN, C. and MUEHLENBACKS, K. (1993) The $^{18}O/^{16}O$ ratio of 2-billion-year-old sea water inferred from ancient oceanic crust. *Science*, **259**, 1733–1736.

HORODYSKI, R. J. and KNAUTH, L. P. (1994) Life on land in the Precambrian. *Science*, **263**, 494–498.

HOYLE, F. (1972) The history of the Earth. *Q. J. Ray. Astron. Soc.*, **13**, 328–345.

KARHU, J. and EPSTEIN, S. (1986) The implication of the oxygen isotope records in coexisting cherts and phosphates. *Geochim. Cosmochim. Acta*, **50**, 1745–1756.

KASTING, J. F. and ACKERMAN, T. P. (1986) Climatic consequences of very high CO_2 levels in earth's early atmosphere. *Science*, **234**, 1383–1385.

KASTING, J. F. and GRINSPOON, D. H. (1990) The faint young Sun problem. In *The Sun in Time*, ed. C. P. Sonnett, M. S. Giampapa and M. S. Matthews, pp. 447–462 (Tucson, AZ: University of Arizona Press).

KASTING, J. F., EGGLER, D. H. and RAEBURN, S. P. (1993) Mantle redox evolution and the oxidation state of the Archean atmosphere. *J. Geol.*, **101**, 245–25.

KENNY, R. and KNAUTH, L. P. (1992) Continental palaeoclimates from δD and $\delta^{18}O$ of secondary silica in palaeokarst chert lags. *Geology*, **20**, 219–222.

KLEIN, C. and BRICKER, O. P. (1977) Some aspects of the sedimentary and diagenetic environment of Proterozoic banded iron formation. *Econ. Geol.*, **72**, 1457–1470.

KNAUTH, L. P. (1992) Origin and diagensis of cherts: an isotopic perspective. In *Isotopic Signatures and Sedimentary Records, Lecture Notes in Earth Sciences No. 43*, ed. N. Clauer and S. Chaudhuri, pp. 123–152 (Berlin: Springer-Verlag).

KNAUTH, L. P. and CLEMENS, P. L. (1995) Climatic history of the Earth based on isotopic analyses of cherts. *Annu. Mtg, Geol. Soc. Am., Abstr. with Programs*, **27**, no. 6, A-205.

KNAUTH, L. P. and EPSTEIN, S. (1976) Hydrogen and oxygen isotope ratios in nodular and bedded cherts. *Geochim. Cosmochim. Acta*, **40**, 1095–1108.

KNAUTH, L. P. and LOWE, D. R. (1978) Oxygen isotope geochemistry of cherts from the Onverwacht Group (3.4 billion years), Transvaal, South Africa, with implications for secular variations in the isotopic composition of cherts. *Earth Planet. Sci. Lett.*, **41**, 209–222.

KNOLL, A. H. (1990) Precambrian evolution of prokaryotes and protists. In *Palaeobiology. A Synthesis*, ed. D. E. G. Briggs and P. R. Crowther, pp. 9–16 (Oxford: Blackwell Science).

KONHAUSER, K. O. and FERRIS, F. G. (1996) Diversity of iron and silica precipitation by microbial mats in hydrothermal waters, Iceland: implications for Precambrian iron formations. *Geology*, **24**, 323–326.

LOVELOCK, J. (1988) *The Ages of Gaia* (New York: W.W. Norton).

LOWE, D. R. (1994) Early environments: constraints and opportunities for early evolution. In *Early Life on Earth*, Nobel Symposium No. 84, ed. S. Bengtson, pp. 24–35 (New York: Columbia University Press).

MAYNARD SMITH, J. and SZATHMARY, E. (1995) *The Major Transitions in Evolution* (Oxford: W. H. Freeman).

MORRIS, S. C. (1993) The fossil record and the early evolution of the Metazoa. *Nature*, **361**, 219–225.

OBERBECK, V. R., MARSHALL, J. R. and AGGARWAL, H. (1993) Impacts, tillites, the breakup of Gondwanaland. *J. Geol.*, **101**, 1–19.

OHMOTO, H. (1996) Evidence in pre-2.2 Ga palaeosols for tthe early evolution of atmospheric oxygen and terrestrial biota. *Geology*, **24**, 1135–1138.

OHMOTO, H. (1997) When did the earth's atmosphere become oxic? *Geochem. News*, no. 93, pp. 12–13, 26.

OHMOTO, H. and FELDER, R. P. (1987) Bacterial activity in the warmer, sulfate-bearing, Archean oceans. *Nature*, **328**, 244–246.

RAMPINO, M. R. (1994) Tillites, diamictites, and ballistic ejecta of large impacts. *J. Geol.*, **102**, 439–456.

RAMPINO, M. R., ERNSTSON, K. and ANGUITA, F. (1997) Striated and polished clasts in impact-ejecta and the 'tillite problem'. *Annu. Mtg, Geol. Soci. Am. Abstr. with Programs*, **29**, no. 7, A-81.

RAMPINO, M. R., SCHWARTZMAN, D. W., CALDEIRA, K. and SCHWARTZMAN, P. D. (1996) Impacts and Precambrian climate: the Huronian enigma. *5th International Conference on Bio-astronomy, IAU Colloquium No. 161. Capri, July 1–5, 1996*, abstract.

ROBBINS, E. I., PORTER, K. G. and HABERYAN, K. A. (1985) Pellet microfossils: possible evidence for metazoan life in early Proterozoic time. *Proc. Natl Acad. Sci USA*, **82**, 5809–5813.

RYE, R., KUO, P. H. and HOLLAND, H. D. (1995) Atmospheric carbon dioxide concentrations before 2.2 billion years ago. *Nature*, **378**, 603–605.

SAGAN, C. (1977) Reducing greenhouses and the temperature history of Earth and Mars. *Nature*, **269**, 224–226.

SCHOPF, J. W. (1992) Paleobiology of the Archean. In *The Proterozoic Biosphere*, ed. J. W. Schopf and C. Klein, pp. 25–39 (New York: Cambridge University Press).

SCHULTZ, P. H. and GAULT, D. E. (1990) Prolonged global catastrophes from oblique impacts. In *Global Catastrophes in Earth History; An Interdisciplinary Conference on Impacts, Volcanism, and Mass Mortality*, ed. V. L. Sharpton and P. D. Ward (Geological Society of America Special Paper 247), pp. 239–261.

SCHWARTZMAN, D. (1994) Biotic enhancement of weathering redux, *Mineral. Mag.*, **58A**, (L-Z) 815–816.

SCHWARTZMAN, D. (1995) Temperature and the evolution of the biosphere. In *Progress in the Search for Extraterrestrial Life, 1993 Bioastronomy Symposium, Santa Cruz, California*, ed. G. Seth Shostak, pp. 152–161 (San Francisco: Astronomical Society of the Pacific).

SCHWARTZMAN, D. W. and SHORE, S. N. (1996) Biotically mediated surface cooling and habitability for complex life. In *Circumstellar Habitable Zones. Proceedings of the First International Conference*, ed. L. R. Doyle, pp. 421–443 (Menlo Park, California.: Travis House Publishers).

SCHWARTZMAN, D. W. and VOLK, T. (1989) Biotic enhancement of weathering and the habitability of earth. *Nature*, **340**, 457–460.

SCHWARTZMAN, D. and VOLK, T. (1991) Biotic enhancement of weathering and surface temperatures on earth since the origin of life. *Palaeogeography, Palaeoclimatolology, Palaeoeclogy (Global Planet. Change Section)*, **90**, 357–371.

SCHWARTZMAN, D., MCMENAMIN, M. and VOLK, T. (1993) Did surface temperatures constrain microbial evolution? *BioScience*, **43**, 390–393.

SOGIN, M. L., GUNDERSON, J. H., ELWOOD, H. J., ALONSO, D. A. and PEATTIE, D. A. (1989) Phylogenetic meaning of the kingdom concept: An unusual ribosomal RNA from Guardia lamlia. *Science*, **243**, 75–77.

SPOONER, E. T. C. (1992) Similarities between environmental requirements for the deepest known branches of the universal phylogenetic tree and early Archean (~3.0–3.5 Ga) whole ocean conditions. *Annu. Mtg, Geol. Soc. Am., Abst. with Programs*, **24**(7), A137.

STRAUSS, H., DES MARAIS, D. J., HAYES, J. M. and SUMMONS, R. E. (1992) Proterozoic organic carbon – its preservation and isotopic record. In *Early Organic Evolution: Implications for Mineral and Energy Resources*, ed. M. Schidlowski, S. Golubic, M. M. Kimberly, D. M. McKirdy and P. A. Trudinger, pp. 203–211 (Berlin: Springer-Verlag).

WALKER, J. C. G. (1982) Climatic factors on the Archean earth. *Palaeogeog., Palaeoclimatol., Palaeoecol.*, **40**, 1–11.

WALKER, J. C. G., HAYS, P. B. and KASTING, J. F. (1981) A negative feedback mechanism for the long-term stabilisation of earth's surface temperature. *J. Geophys. Res.*, **86**, 9776–9782.

WINTER, B. L. and KNAUTH, L. P. (1992) Stable isotope geochemistry of early Proterozoic carbonate concretions in the Animikie Group of the Lake Superior region: evidence for anaerobic bacterial processes. *Precambrian Res.*, **54**, 131–151.

WOESE, C. R. (1987) Bacterial evolution. *Microbiol. Rev.*, **51**, 221–271.

WRAY, G. A., LEVINTON, J. S. and SHAPIRO, L. H. (1996) Molecular evidence for deep Precambrian divergences among metazoan phyla. *Science*, **274**, 568–573.

PART TWO

The Origin of Life

4

The Case for a Hyperthermophilic, Chemolithoautotrophic Origin of Life in an Iron–Sulfur World

GÜNTER WÄCHTERSHÄUSER

TAL 29, Munich, Germany

4.1 Introduction

The unobservability of the past has the inescapable consequence that the problem of the origin and early evolution of life can only be solved by theory. Alternative theories on the origin of life compete to explain the facts of the extant biosphere. Therefore, selection among such theories should be carried out on the basis of their *explanatory power* (Popper, 1963): their power to explain many observable facts with few evolutionary assumptions. In constructing a theory of early evolution, explanatory power can best be achieved by the postulation of common evolutionary precursor functions for disparate extant successor functions.

Spontaneous generation is a synonym for the origin of life. Even the simplest extant cellular organisms are too complex for spontaneous generation. It is, therefore, our task to locate a subcellular component for which we can construct an archaic functional precursor in such a fashion that all other subcellular components can be explained as arising therefrom by evolution.

Conventional theories assuming that life began with replicating empty lipid vesicles (Morowitz, 1992) or with replicating nucleic acids (RNA world) (Joyce, 1989) are unacceptable for several reasons. The assumed primordial entity of life is (1) too complex for spontaneous generation; (2) dependent on a geochemically obscure prebiotic broth; and (3) based on chemically unsound assumptions regarding the availability of activated starting materials for replication.

The theory that life began with a metabolism is free of the above difficulties (Wächtershäuser, 1988a). It implies that the primordial organism (metabolist) later evolved a cell envelope for keeping the constituents of the metabolism together and a genetic machinery for controlling the metabolism. This means that there must be a general mechanism of evolution which is independent of nucleic acid replication and copying mistakes. It should contain the nucleic acid-implemented mechanism of evolution as a special case, and show this special mechanism of evolution as emerging within a more general process of evolution.

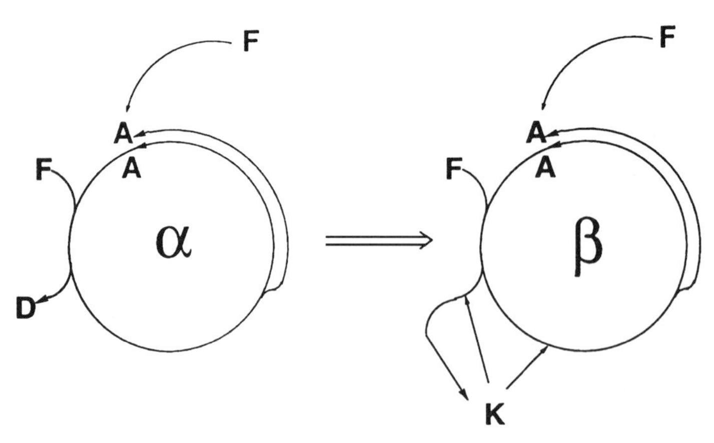

Figure 4.1 Evolution by dual feedback. The double arrow (⇒) signifies an evolutionary change from the α-cycle to the β-cycle (F, food; A, food acceptor; D, product of decay; K, vitaliser).

4.2 General mechanism of evolution

A general mechanism of evolution must be based on translation of the biological notion of reproduction into the chemical notion of chain reactions with branching. The theory of chain reactions comprises three types of reaction: (1) reactions of propagation, whereby an autocatalyst A catalyses its own formation from starting material F (e.g. $F + A \rightarrow 2A$) (reproduction cycle); (2) reactions of initiation, whereby the autocatalyst arises *de novo* from starting material ($F \rightarrow A$) (spontaneous generation); and (3) reactions of decay, whereby the autocatalyst deteriorates into products of decay D, ($A \rightarrow D$).

A general notion of mutation can be postulated within the theory of chain reactions: the appearance of special products of decay K (*vitalisers*), which show a dual positive catalytic feedback effect, 'altruistically' into the metabolism whence they derive and 'egotistically' into their own pathways of derivation (Figure 4.1) (Wächtershäuser, 1988a, 1992). Nucleic acids are a special case of vitalisers, coding for all enzymes including the polymerases. Ribosomes are vitalisers, producing all proteins including the ribosomal proteins. Coenzymes are vitalisers, catalysing a spectrum of reactions including steps in their own biosynthesis.

4.3 Thermophilic surface metabolism

Physical coherence of all constituents of the metabolism is the precondition for a feedback of vitalisers, but starting materials and products of decay should be able to enter or leave freely. A solution of this problem is mineral surface bonding. This means that the first organisms are of a composite nature, comprising a mineral material and its ligand sphere for a *surface metabolism* (Wächtershäuser, 1988a).

Mineral surface bonding is seen as a functional precursor of cellular enclosure and of bonding to enzyme surfaces. It establishes the first organism–environment dichotomy: the chemical reaction milieu near the surface (*ambiance*) can be different from the chemical milieu in the bulk aqueous phase. The ambiance can be subject to an evolvable modification by products of the metabolism itself, e.g. by surface-bonded buffers, or by lipids which eventually evolve into cellular membranes.

Temperature is an important parameter for a surface metabolism. In contrast to the chemical activity parameters, there cannot be a thermal difference between the ambiance and the bulk aqueous phase. A hot early earth is incompatible with a prebiotic broth (Miller and Lazcano, 1995), but compatible with a surface metabolism. The reactions of the latter have a smaller reaction entropy than corresponding solution reactions. Therefore, elevated temperatures are tolerated by the surface metabolism. They are even required, since surface-bonded constituents have a narrower Boltzmann distribution of kinetic energy than dissolved constituents. The surface metabolists were (hyper)thermophilic. This agrees well with the fact that the universal ancestor at the deepest node in Woese's tree of life was hyperthermophilic (Woese, 1987).

4.4 The universal pattern of metabolism

A universal pattern of metabolism is a main consequence of evolution by vitalisers. Any metabolism, at any time and anywhere in the universe, will show a pattern that comprises a core metabolism, consisting of an archaic pathway of initiation (or a successor) and of an archaic autocatalytic propagation cycle (or a successor), and biosynthetic pathways beginning at the core metabolism and generating vitalisers. The latter will show a radial pattern. This is due to the greater likelihood that a new vitaliser derives from and feeds back into the periphery of a metabolism rather than its more highly integrated centre.

A principle of evolutionary conservatism is another consequence of evolution by vitalisers. The core of a metabolism is more conserved than its periphery. Therefore, the core must be closest to the origin of life. This is of great heuristic value. It means that the remnants of the oldest phases of evolution can be detected by tracing the biosynthetic pathways to the core metabolism.

The primordial metabolism can be retrodicted by applying the consequences of the above principles to the extant metabolism (Figure 4.2) with the following conclusions: (1) The extant reductive acetyl-CoA pathway (found in certain Bacteria and Archaea; and partially and/or reversed in all other organisms) is the evolutionary successor of the primordial pathway of initiation. (2) The extant reductive citrate cycle (found in certain Bacteria and Archaea; and partially and/or reversed in all other organisms) is the evolutionary successor of the primordial autocatalytic propagation cycle (*alpha cycle*) (Wächtershäuser, 1990). (3) The biosynthetic pathways reveal the oldest vitalisers by their constituents closest to the core metabolism.

4.5 Rules of retrodiction

A chemoautotrophic origin of life (Wächtershäuser, 1988b) is the immediate consequence of the above. The primordial organism must have used CO_2 and/or CO as carbon source. Both occur in volcanic exhalations and hydrothermal vents. A prebiotic broth, if it ever existed, is not required for or relevant to the origin of life.

The method of connecting extant biochemical features with assumptions about primordial evolution by the principle of explanatory power is called *retrodiction*. The rules of retrodiction, required for reconstructing the archaic pathways, follow from the assumption of an autotrophic origin:

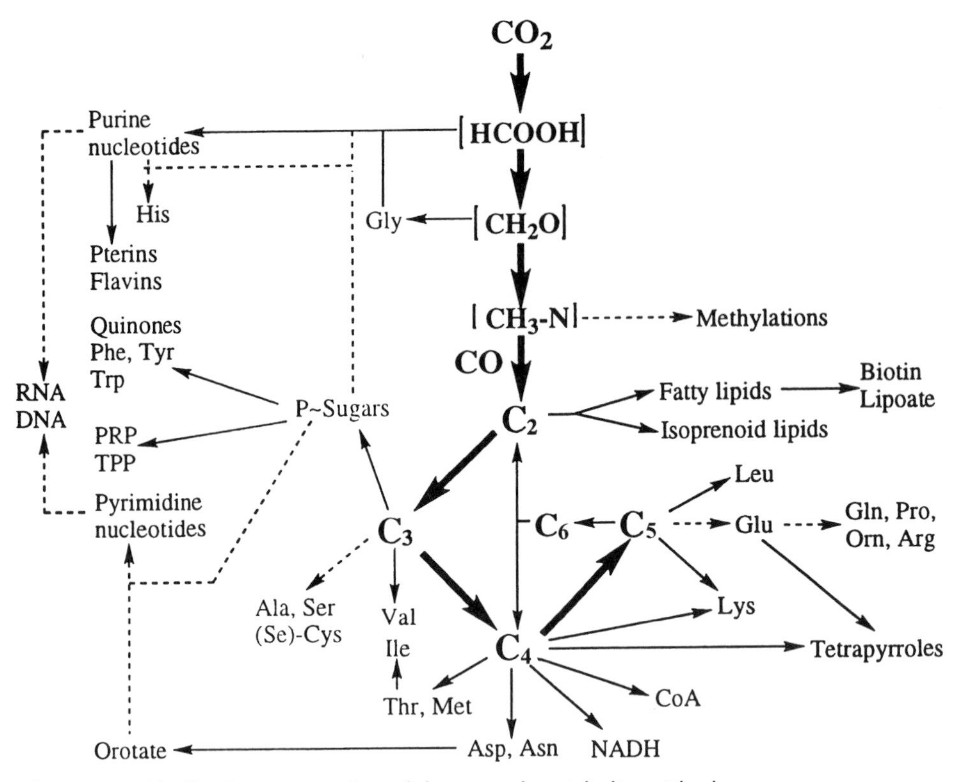

Figure 4.2 Idealised representation of the central metabolism. The heavy arrows represent the core metabolism; the light solid arrows represent C—C bond-forming reactions; the broken arrows represent other synthetic reactions.

Rule 1 Biocatalytic thiol groups are retrodicted into inorganic H_2S.

Rule 2 Extant biochemical Fe–S or Fe–Ni–S clusters are retrodicted into FeS or NiS.

Rule 3 Organic amine or alcohol groups are retrodicted into thiol groups.

Rule 4 Keto groups are retrodicted into an equilibrium with thio groups (e.g. —HC═C(SH)—).

Rule 5 Thioester groups (—COSR) are retrodicted into thioacid groups (—COSH).

Rule 6 Extant biochemical reducing agents are retrodicted into H_2 (Wächtershäuser, 1988d) or into FeS/H_2S.

It is of particular interest that the assumption of H_2S and FeS yields a primordial reducing agent without any further assumption (Wächtershäuser, 1988b).

$$FeS + H_2S \rightarrow FeS_2 + H^+ + 2e^- \tag{4.1}$$

$$2\,FeS \rightarrow FeS_2 + Fe^{2+} + 2e^- \tag{4.2}$$

The standard reducing potential of reaction (4.1) is $E^{\circ\prime} = -620$ mV, more than enough for all biochemical reducing reactions. It is due to the extreme stability of the pyrite lattice. Pyrites are the most ubiquitous iron minerals under anaerobic conditions. Moreover, transition metal sulfides and pyrites have exposed metal valences. Therefore, the anionic primary products of carbon fixation ($RCOO^-$, $RCOS^-$, RS^-). will bind *in statu nascendi* to the mineral surface. We will now apply these rules to the retrodiction of the primordial pathway of initiation, the alpha cycle and the earliest vitalisers in an iron–sulfur world.

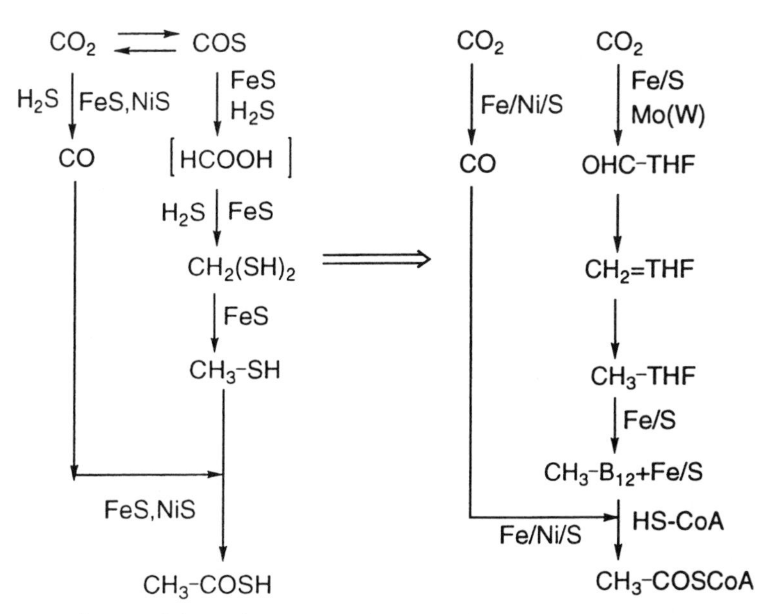

Figure 4.3 Evolution of the reductive acetyl-CoA pathway from a hypothetical archaic precursor pathway (Fe/S, iron–sulfur enzyme; Fe/Ni/S, iron–nickel–sulfur enzyme; THF, tetrahydrofolate or its methanopterin analogues; Mo(W), molybdo(tungsto)pterin).

4.6 The primordial pathway of initiation

The primordial pathway of initiation can be retrodicted by applying the above rules to the extant reductive actyl-CoA pathway (Figure 4.3). By rule 2 the Fe–Ni–S cluster is retrodicted into a mixture of FeS and NiS; and by rule 3 the methylamino group of methylpterin is retrodicted into methylmercaptane, CH_3SH. In experiments, CH_3SH and 1 bar CO were reacted in water at 100 °C and in the presence of coprecipitated NiS and FeS. Acetic acid was formed in the narrow pH range of 6–7 with a maximum of up to 40 mol%, based on CH_3SH (Huber and Wächtershäuser, 1997),

$$CH_3—SH + CO + H_2O \rightarrow CH_3—COOH + H_2S \tag{4.3}$$

This confirmed a previous prediction (Wächtershäuser, 1990). A reaction of CO and H_2S (without CH_3SH) on coprecipitated NiS–FeS yielded acetic acid, CH_3SH and COS. Traces of CH_3SH were also reported in the reaction of CO_2 and H_2S on FeS.

Activated thioacetic acid (CH_3COSH) is expected as an intermediate of the above reaction by rule 5. This was corroborated by detection of superequilibrium amounts (5 mol%) of acetanilide in the presence of aniline, and of superequilibrium ratios of CH_3COSCH_3:CH_3COOH (8:25):

$$CH_3—SH + CO \rightarrow CH_3—CO—SH \tag{4.4}$$

$$CH_3—CO—SH + Ph—NH_2 \rightarrow CH_3—CO—NH—Ph + H_2S \tag{4.5}$$

$$2\ CH_3—SH + CO \rightarrow CH_3—CO—S—CH_3 + H_2S \tag{4.6}$$

Thus, the reported carbon fixation generates *structure* and *reactivity*. The group activation is available for coupling into other surface-metabolic compounds via anhydrides (—CO—O—CO— or —CO—O—PO_2^-—O—) and may give rise to pyrophosphate. This means that the important theory (Baltscheffsky *et al.*, 1986) of pyrophosphate being the functional precursor of ATP can be readily accommodated within the iron–sulfur world.

4.7 The alpha cycle

The primordial autocatalytic cycle, the alpha cycle, can be retrodicted by applying the above rules to the extant reductive citrate cycle (Figure 4.4). This cycle is rather complex.

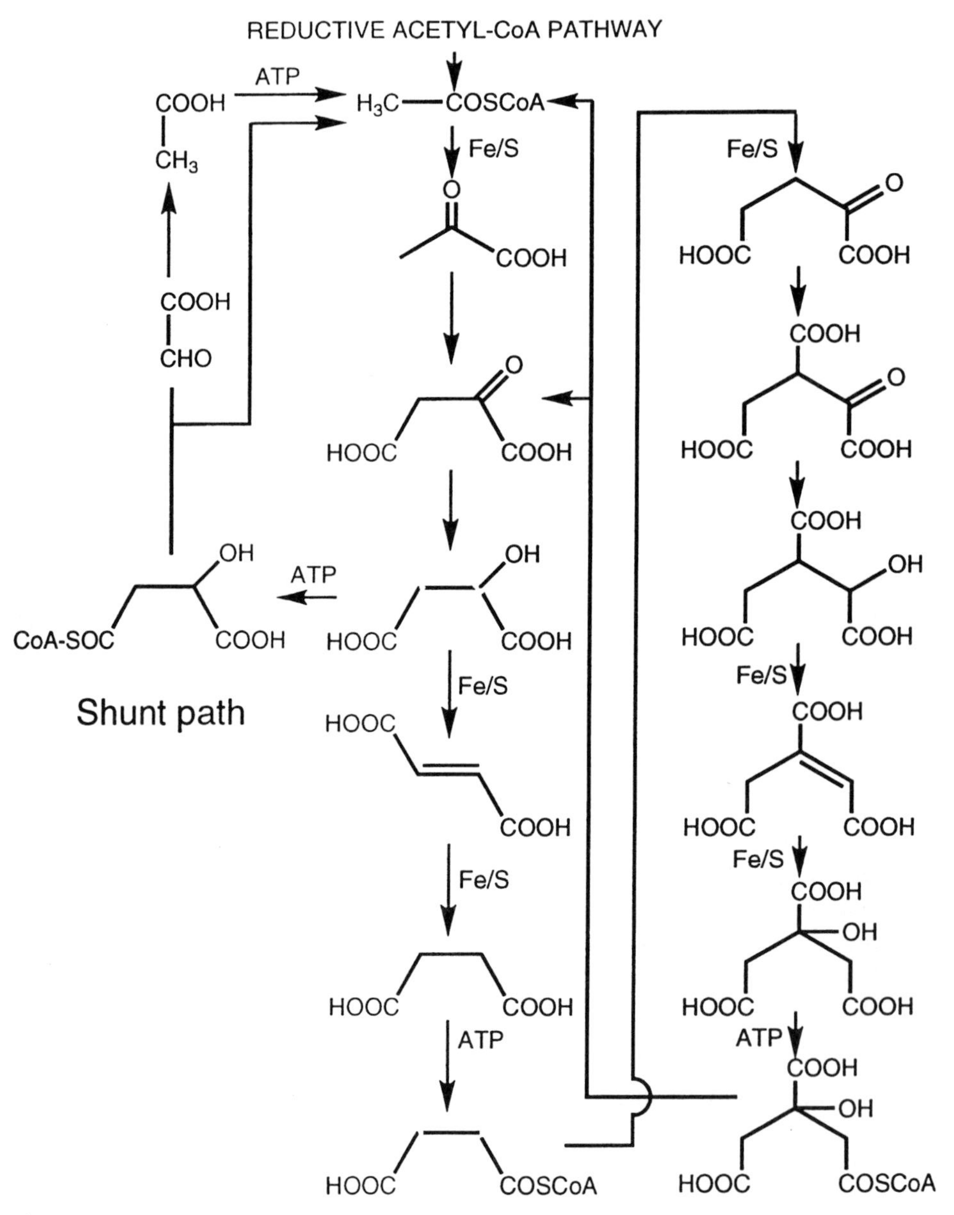

Figure 4.4 Extant reductive citrate cycle with shunt pathway (Fe/S represents iron–sulfur enzyme).

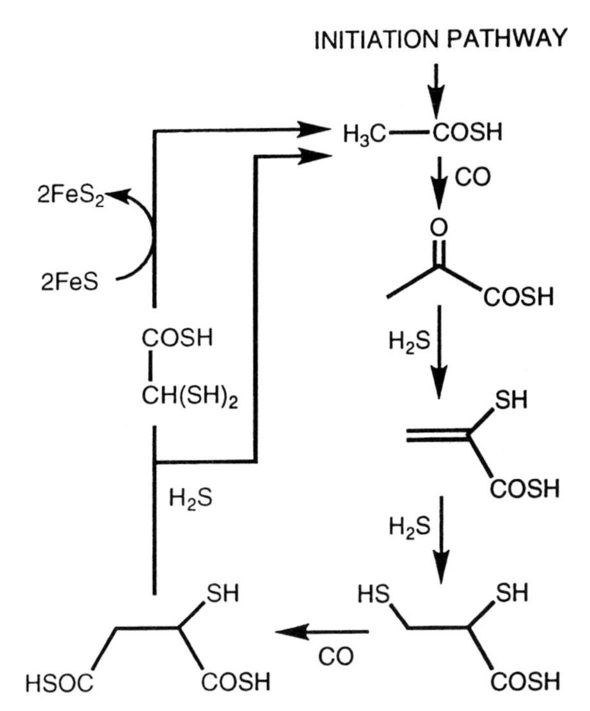

Figure 4.5 A proposal for the alpha cycle.

It comprises four CO_2-fixation steps belonging to two different reaction types, and two ATP-consuming activation steps. A hypothetical shunt path, based on existing enzmes, can be seen as part of a much simpler reductive 'malate cycle'. Based on this reductive malate cycle we can retrodict the alpha cycle (Figure 4.5). It is obtained by replacing CO_2 by CO, iron–sulfur clusters by FeS, and thioesters by thioacids. In this alpha cycle, activation is generated automatically. Departing from this alpha cycle, evolution may have proceeded through the stages of a CO_2-based, pyrite-pulled reductive malate cycle (Wächtershäuser, 1992), a CO_2-based, pyrite-pulled reductive citrate cycle (Wächtershäuser, 1990), to the extant reductive citrate cycle.

4.8 The earliest vitalisers

The earliest vitalisers can be retrodicted by recalling that rule 2 introduces transition metal sulfides as catalysts for the primordial metabolism. Therefore, the earliest feedback was due to by-products of the core metabolism operating as ligands, which rendered the catalytic metal centres more active and/or more selective. From this many consequences were drawn, for example, that the earliest ferredoxins must predate NAD(P)H. (Wächtershäuser, 1988a, 1992).

It is common to ask the chicken-or-egg question: whether proteins preceded nucleic acids or vice versa. This question is rather short-sighted. A more fundamental question appeals to the famous controversy regarding the nature of life's catalysts (cf. Willstätter, 1927). Sumner favoured proteins, while Willstätter favoured metals. The above proposal means that life began with Willstätter catalysts and invented Sumner catalysts later. Early biochemistry was largely coordination chemistry and early biochemical evolution was largely an evolution of ligands.

Amino acids are important branch products of the core metabolism and were surely among the earliest vitalisers. Glutamate synthase, an iron–sulfur enzyme, is in charge of the reductive amination of α-keto glutarate. With rules 2 and 6 we can retrodict the iron–sulfur enzyme and the reducing agent into FeS in conjunction with H_2S or H_2 and optionally NiS. The reductive amination of a variety of α-keto acids (except oxaloacetate) is facile if FeS precipitated *in situ* is used. Yields of up to 90% are obtained in 17 h at 75 °C (Hafenbradl *et al.*, 1995, Huber and Wächtershäuser, unpublished results):

$$RCOCOOH + HNR_2 + FeS + H_2S \rightarrow RCH(NR_2)COOH + FeS_2 + H_2O \quad (4.7)$$

Therefore, the archaic amino acids Gly, Cys, Ala, Glu, α-amino-adipate, etc. must have been among the earliest vitalisers. Amino acids are well known ligands in metal complexes and may have served directly as vitalisers. In the presence of FeS, NiS, H_2S, CO in water at 100 °C and pH 7 to 10 amino acids (5.10^{-2}m) become activated to form a continuously recycling library of peptides (Huber and Wächtershäuser, 1998). All biosynthetic conversions of the archaic amino acids must have produced, step by step, more and more effective ligands for modifying the transition metal centres. This makes the evolution of the lengthy pathways to derived amino acids (e.g. Arg or Lys) understandable. Intermediates in these pathways, like saccharopin or cystathionine, must have been ligands in the earliest coordination chemistry of life.

Surface-bonded or cluster-bonded peptides are a special form of ligand. For these the theory of surface metabolism has stipulated a polyanionic structure with units of anionic amino acids (e.g. Glu, Cys, phosphoserine) strongly bonded to exposed metal valences (Wächtershäuser, 1988a). This strong surface bonding has the effect that the peptides remain on the surface and participate in the surface metabolism and that they are stabilised even at elevated temperatures. This proposal has been criticised (De Duve and Miller, 1991), but it has experimental support (Ferris *et al.*, 1996).

A *source of ammonia* is found in volcanic gases (Corazza, 1986). Iron sulfide may be the transition metal catalyst for its formation by nitrogen fixation under volcanic conditions (Wächtershäuser, 1988a). An alternative ammonia source has been suggested to be nitrate produced by discharges in an anoxic atmosphere of N_2/CO_2 via NO_x in the hot primitive ocean, which is reduced to NH_3 by FeS/H_2S as demonstrated experimentally (Blöchl *et al.*, 1992).

Thio-sugars may be the earliest sugar types (Wächtershäuser, 1992) for the extant pathway from pyruvate to phosphoglycerate is too endergonic for an archaic pathway. Thio-sugars are excellent ligands and surface bonders. To this day the thio-sugar derivatives molybdopterin and tungstopterin are important ligands.

4.9 The genetic machinery

Phosphorylated sugars, the successors of the thio-sugars, are also excellent ligands and surface bonders. In a surface metabolism all archaic sugars must have been phosphorylated. Surface-bonded phosphotrioses have been proposed to undergo an interconversion between dihydroxyacetone phosphate, 3-phosphoglyceraldehyde and 2-phosphoglyceraldehyde (Wächtershäuser, 1988a). As a key step of the surface metabolism, the formation of phosphopentose by condensation of phosphoglycolaldehyde and phosphoglyceraldehyde has been postulated (Wächtershäuser, 1988a). This prediction has recently been confirmed experimentally (Pitsch *et al.*, 1995).

Purines must also be seen as early vitalisers. All intermediates in the purine pathway must have operated as ligands. Acid–base catalysis may also have been a purine function early on. Throughout the evolution of purine biosynthesis all intermediates must have been surface-bonded, mediated by a phosphorylated sugar substructure. The earliest base pairing has been suggested as occurring between 3-bonded and 9-bonded purines (Wächtershäuser, 1988c). It was at first detrimental, only later turning into a benefit by dual feedback. Pyrimidines are seen as entering later still.

The emergence of the genetic machinery was based throughout on vitalisers. A most primitive nucleic acid may be a surface-bonded homopolymer providing only one kind of base pair. As a surface-bonded proto-mRNA it would simply have the function of a surface-metabolic positioning gadget. In this form the egotistic feedback could operate by the positioning of surface-bonded nucleotides to catalyse the formation of a new proto-mRNA strand (proto-replication). The altruistic feedback could operate by the positioning of surface-bonded proto-tRNAs to catalyse amide (peptide) bond formation (proto-translation). The present theory, therefore, supports Noam Lahav's notion of coevolution of replication and translation (Lahav, 1991). Sequences and sequence specificity are seen as coming in later and in steps. This marks the changeover from the analogue information of simple autocatalytic cycles to the digital information of nucleic acid sequences. In this changeover, the ribosome (digital-to-analogue converter) is again a vitaliser, producing all proteins including ribosomal proteins. Nucleic acids having a replication function without a translation function would not be vitalisers but rather virulysts (proto-viruses). They would kill their host organism and ultimately themselves. It is ironic that within a revisionist RNA-world theory the emergence of self-replicating RNA, not within a prebiotic broth as originally assumed but within pre-evolved organisms, is considered to be driven by a selective advantage, when in fact it is a selective disadvantage for the organism. This is a perfect example for the muddles created in science if it is attempted to avoid a refutation by the redefinition of a term.

Protein folding in three dimensions replaces peptide surface bonding. The earliest folding structures are seen as being determined by the covalent bonding to metal (iron–sulfur) clusters and later by covalent disulfide bonds. With the emergence of lipophilic amino acids the formation of a lipophilic core comes additionally into the picture and with the emergence of basic amino acids the formation of stabilising salt bridges on the surface of the folding structures. With the perfection of the sequence fidelity of the genetic machinery, folding by strictly noncovalent bonds becomes possible and some of the covalent folding determinants by metal clusters can fall by the wayside.

4.10 The thermal irreversibility of evolution

The thermal irreversibility of evolution can be understood in one of its aspects as a consequence of the evolution of protein folding. At the high temperatures of the primordial organisms, protein folding is accommodated by covalent bonding to metal clusters. The first hyperthermophilic protein folding structures without covalent folding mediators require precise sequences within a small and fragmented sequence space. On the evolutionary way down toward mesophily the sequence space for protein folding grows. Organisms can evolve easily downwards to mesophily by changing their proteins one after the other to mesophily, with a subsequent random walk of their sequences within the enlarged sequence space. A reversed evolution from mesophily to hyperthermophily would require a concerted adaptation of all proteins to higher temperatures and into a smaller and smaller and more and more fragmented sequence space. This is highly improbable.

4.11 Conclusions

The organism–environment dichotomy is fundamental for biology. The theory of a chemoautotrophic origin of life has highly specific consequences for both sides of this dichotomy. Organismically speaking, it postulates an evolving primordial reductive carbon-fixation metabolism. Environmentally speaking, it stipulates for the original homestead of life a region with sulfide activity and precipitating transition metal sulfides and with the formation of pyrite; and the availability of carbon monoxide and carbon dioxide and the reductive formation of ammonia, the availability of phosphate bonded to sulfide mineral surfaces, and a temperature high enough for the thermal requirements of a surface metabolism. From the vantage point of this restrictive initial condition, we may develop the theoretical means for a coherent evolutionary explanation of the exploding set of metabolic and genomic data, and for achieving a hitherto unthinkable degree of completeness in biology. From this vantage point, the overall process of evolution can be interpreted as a process of self-liberation from the narrow chemical confines of an iron–sulfur world and from a two-dimensional existence on mineral surfaces. It is a process that begins with chemical necessity and winds up in genetic chance.

4.12 Summary

It is proposed that salient facts of biochemistry can be explained by the assumption that life began with a thermophilic, chemoautotrophic metabolism, organised and catalysed by bonding to sulfide mineral surfaces, and evolving through by-products (vitalisers) with a dual autocatalytic feedback: directly into their pathway of derivation and indirectly into the metabolism whence they derive. The homestead for such an origin of life is a setting with a flow of magmatic exhalations.

References

Baltscheffsky, H., Lundin, M., Luxemburg, C., Nyren, P. and Baltscheffsky, M. (1986) Inorganic pyrophosphate and the molecular evolution of biological energy coupling. *Chem. Scr.*, **26B**, 259–262.

Blöchl, E., Keller, M., Wächtershäuser, G. and Stetter, K. O. (1992) Reactions depending on iron sulfide and linking geochemistry with biochemistry. *Proc. Natl Acad. Sci. USA*, **89**, 8117–8120.

Corazza, R. (1986) Field workshop on volcanic gases, Vulcano (Italy), 1982, General Report. *Geothermics*, **15**, 197–200.

De Duve, C. and Miller, S. L. (1991) Life in two dimensions? *Proc. Natl Acad. Sci. USA*, **88**, 10014–10017.

Ferris, J. P., Hill Jr., A. R., Liu, R. and Orgel, L. E. (1996) Synthesis of long prebiotic oligomers on mineral surfaces. *Nature*, **381**, 59–61.

Hafenbradl, D., Keller, M., Wächtershäuser, G. and Stetter, K. O. (1995) Primordial amino acids by reductive amination of α-oxo acids in conjunction with the oxidative formation of pyrite. *Tetrahedron Lett.*, **36**, 5179–5182.

Heinen, W. and Lauwers, A. M. (1996) Organic sulfur compounds resulting from the interaction of iron sulfide, hydrogen sulfide and carbon dioxide in an anaerobic aqueous environment. *Origins of Life*, **26**, 131–150.

Huber, C. and Wächtershäuser, G. (1997) Activated acetic acid by carbon fixation on (Fe, Ni)S under primordial conditions. *Science*, **276**, 245–247.

HUBER, C. and WÄCHTERSHÄUSER, G. (1998) Peptides by activation of amino acids with CO on (Ni,Fe)S surfaces and implications for the origin of life. *Science*, in print.

JOYCE, G. F. (1989) RNA evolution and the origin of life. *Nature*, **338**, 217–224.

LAHAV, N. (1991) Prebiotic co-evolution of self-replication and translation or RNA world? *J. Theor. Biol.*, **151**, 531–539.

MILLER, S. L. and LAZCANO, A. (1995) *J. Mol. Evol.*, **41**, 689–692.

MOROWITZ, H. J. (1992) *Beginnings of Cellular Life: Metabolism Recapitulates Biogenesis* (New Haven, CT: Yale University Press).

PITSCH, S., ESCHENMOSER, A., GEDULIN, B., HUI, S. and ARRHENIUS, G. (1995) Mineral-induced formation of sugar phosphates. *Origins of Life*, **25**, 297–334.

POPPER, K. R. (1963) *Conjectures and Refutations* (London: Routledge).

WÄCHTERSHÄUSER, G. (1988a) Before enzymes and templates: theory of surface metabolism. *Microbiol. Rev.*, **52**, 452–484.

WÄCHTERSHÄUSER, G. (1988b) Pyrite formation, the first energy source for life: a hypothesis. *System Appl. Microbiol.*, **10**, 207–210.

WÄCHTERSHÄUSER, G. (1988c) An all-purine precursor of nucleic acids. *Proc. Natl Acad. Sci. USA*, **85**, 1134–1135.

WÄCHTERSHÄUSER, G. (1990) The evolution of the first metabolic cycles. *Proc. Natl Acad. Sci. USA*, **87**, 200–204.

WÄCHTERSHÄUSER, G. (1992) Groundworks for an evolutionary biochemistry: the iron–sulfur world. *Prog. Biophys. Mol. Biol.*, **58**, 85–201.

WILLSTÄTTER, R. (1927) *Problems and Methods in Enzyme Research* (Ithaca, NY: Cornell University Press).

WOESE, C. R. (1987) Bacterial evolution. *Microbiol. Rev.*, **51**, 221–271.

5

The Emergence of Metabolism from Within Hydrothermal Systems

EVERETT L. SHOCK[1], TOM McCOLLOM[1,2] AND MITCHELL D. SCHULTE[1,3]

[1] *GEOPIG, Department of Earth & Planetary Sciences, Washington University, St Louis, Missouri, USA*
[2] *Woods Hole Oceanographic Institute, Woods Hole, Massachusetts, USA*
[3] *NASA Ames Research Center, Moffett Field, California, USA*

5.1 Introduction

According to the phylogenetic tree based on 16S rRNA (Woese *et al.*, 1990), the organisms that branch closest to a common ancestor gain their metabolic energy from inorganic chemical reactions. This is permissive evidence for models of early ecosystems in which geochemical energy sources were tapped to fuel primary productivity rather than sunlight. It also suggests that metabolic systems emerged via energy transfer in coupled organic and inorganic redox reactions. Furthermore, we can infer from the central position of hyperthermophiles at the trunk of the 16S rRNA tree that early ecosystems were hydrothermal and that the emergence of life may have occurred in hydrothermal systems. If so, examining the geochemistry of hydrothermal systems should reveal the energy sources that supported early ecosystems, as well as the required thermodynamic drive for a metabolic system to emerge from a network of interdependent reactions.

An underlying premise of geology is that the present is the key to the past. Processes that can be observed to occur are those most likely to be invoked successfully to explain the geological record. Details can differ, but currently active processes reveal how ancient systems functioned. Hydrothermal systems are no exception. We can study present-day submarine hydrothermal systems to gain insight into what may have been the earliest ecosystems on earth. At the same time, we should keep in mind that the composition of sea water and oceanic crustal rocks may have changed, implying that the composition of hydrothermal fluids may also have differed. However, as described below, the most likely conditions in the past make ancient hydrothermal systems even more attractive than present systems as sites for early ecosystems and the emergence of life. And, as this chapter describes, present systems are extremely capable of supplying ample geochemical energy for microbial metabolism, organic synthesis and transformation of compounds.

5.2 The energetics of methanogenesis in submarine hydrothermal ecosystems

In submarine hydrothermal ecosystems, the lowest-branching hyperthermophilic Archaea are methanogens, which gain energy through CO_2 reduction via the overall reaction

$$CO_2 + 4\,H_2 = CH_4 + 2\,H_2O \tag{5.1}$$

If energy is to be gained from this reaction, then CH_4 concentrations must be lower than values in equilibrium with the concentrations of CO_2 and H_2 present where the methanogens live. This energy can be quantified by calculating the overall Gibbs free energy of this reaction (ΔG_r) from the familiar expression

$$\Delta G_r = \Delta G_r^\circ + RT \ln Q_r \tag{5.2}$$

where ΔG_r° represents the standard-state Gibbs free energy of reaction, R is the gas constant, T is temperature in kelvins, and Q_r stands for the activity product for the natural conditions. In the case of reaction (5.1), Q_r becomes

$$Q_r = \frac{a(CH_4)(a(H_2O))^2}{a(CO_2)(a(H_2))^4} \tag{5.3}$$

where a stands for activity.

If CH_4, H_2 and CO_2 are aqueous species, they are not likely be at standard-state conditions (unit activities) in the solution, because the standard state for aqueous species in unit activity is a *hypothetical* one molal solution referenced to infinite dilution at any temperature and pressure. Given that the standard state is hypothetical, it is not common that real solutes have unit activities; in fact, activities are often many orders of magnitude different from 1. As a result, the magnitude of Q_r can be enormous. This is why the sign of ΔG_r° is often useless for determining how a reaction involving aqueous species will proceed (Shock *et al.*, 1995; Anderson, 1996). In contrast, the activity of water in the solution, which has a standard state of unit activity for the pure solvent at any temperature and pressure, is not far from unity in many geological fluids, and we can set $a(H_2O) = 1$ for sea water and submarine hydrothermal fluids without introducing significant error.

Consider an autotrophic methanogen living at 100 °C and 300 bars in a submarine hydrothermal system. How much energy can it obtain from reaction (5.1)? Answering this question is facilitated by evaluating ΔG_r at this elevated temperature and pressure. The standard-state term, $\Delta G^\circ_{r,P,T}$, can be calculated by summing the apparent standard state Gibbs free energies $G^\circ_{i,P,T}$ of the species in the reaction via

$$\Delta G^\circ_{r,P,T} = \sum \nu_{i,r} G^\circ_{i,P,T} \tag{5.4}$$

where $\nu_{i,r}$ stands for the stoichiometric reaction coefficient of the ith species in the reaction (positive for products and negative for reactants), and values of $G^\circ_{i,P,T}$ for these dissolved gases are readily calculated using data and parameters for the revised Helgeson–Kirkham–Flowers (HKF) equation of state (Shock *et al.*, 1989, 1992; Shock and Helgeson, 1990), together with the SUPCRT92 computer code (Johnson *et al.*, 1992). Values of $\Delta G^\circ_{r,P,T}$ for reaction (5.1) calculated in this manner are depicted in Figure 5.1. At 100 °C and 300 bars, $\Delta G^\circ_{r,P,T}$ is −182.72 kJ/mol (−43.67 kcal/mol).

The second step in evaluating the energy available to an autotrophic methanogen from reaction (5.1) is to quantify Q_r for the natural setting of interest (McCollom and Shock, 1997). One way of reaching 100 °C in a submarine hydrothermal system is by mixing cold

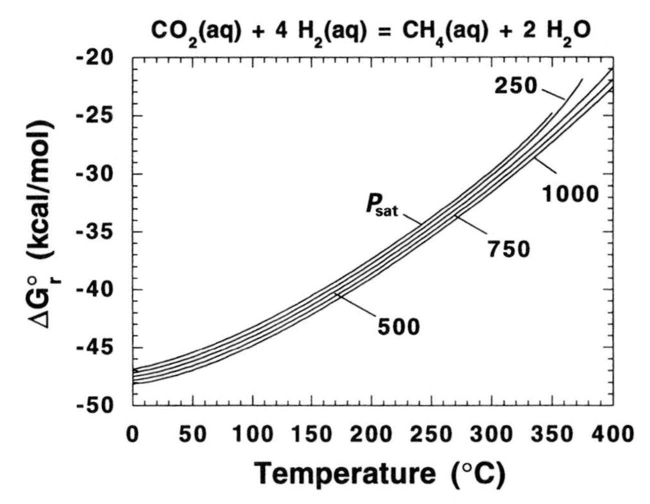

Figure 5.1 Standard-state Gibbs free energy of the methanogenic reaction (5.1) in aqueous solution as a function of temperature at various pressures (bars). P_{sat} refers to pressures consistent with vapour–liquid saturation for H_2O (boiling curve). Other curves are isobars from 250 to 1000 bars. Note that the temperature dependence is considerably greater than the pressure dependence.

sea water and hot hydrothermal vent fluid. This permits a relatively simple way to evaluate the activities of $H_2(aq)$, $CO_2(aq)$ and $CH_4(aq)$, because their concentrations are known for sea water and vent fluids. It takes about 2.55 kg of 2 °C sea water mixed with 1 kg of 350 °C vent fluid to generate a solution at 100 °C. Sea water and hydrothermal fluid compositions used by McCollom and Shock (1997), yield activities of $CO_2(aq)$, $H_2(aq)$ and $CH_4(aq)$ of $10^{-2.49}$, $10^{-3.32}$ and $10^{-4.71}$, respectively, in the 100 °C solution, if no reactions involving these aqueous species occur. Using these values, Q_r equals $10^{11.06}$, and, transformed to energy units via $RT \ln Q_r$, these geochemical constraints on fluid composition contribute 79.1 kJ/mol (18.9 kcal/mol) to the overall Gibbs free energy, which becomes −103.7 kJ/mol (−24.8 kcal/mol) for the overall methanogenesis reaction (5.1) at 100 °C.

Of the two reactants, H_2 is at the lower concentration and is likely to be the limiting factor for microbial growth via methanogenesis (see McCollom and Shock, 1997). Combination of all the H_2 with CO_2 to yield CH_4 will supply about 29.3 kJ (7 cal) per kg of 100 °C solution. Because all of the H_2 comes from the vent fluid, this can be recast as a metabolic yield of 29.3 kJ (7 cal) per kg vent fluid at 100 °C. (It should be emphasised that this value refers only to 100 °C, and that values at other temperatures will differ because the composition of the fluid mixtures to reach these temperatures will differ.) Consistent with discussions by McCollom and Shock (1997), this amount of energy is enough to support about 4 mg of cells per kg vent fluid if efficiencies are about 10%. Assuming a global flux of hydrothermal fluid of 3×10^{13} kg/year (Elderfield and Schultz, 1996) implies an annual biomass productivity from methanogenesis of about 1.5×10^{11} kg.

Hyperthermophiles in vent systems pursue a variety of reductive strategies to gain energy from thermodynamically favoured reactions. Energies from many inorganic reactions used by chemolithoautotrophs are quantified by McCollom and Shock (1997), who studied reductive strategies like methanogenesis, sulfate reduction and sulfur reduction, as well as oxidative strategies like methanotrophy, sulfur oxidation and sulfide oxidation. Results as metabolic yield per kg of vent fluid are summarised in Figure 5.2, where it can

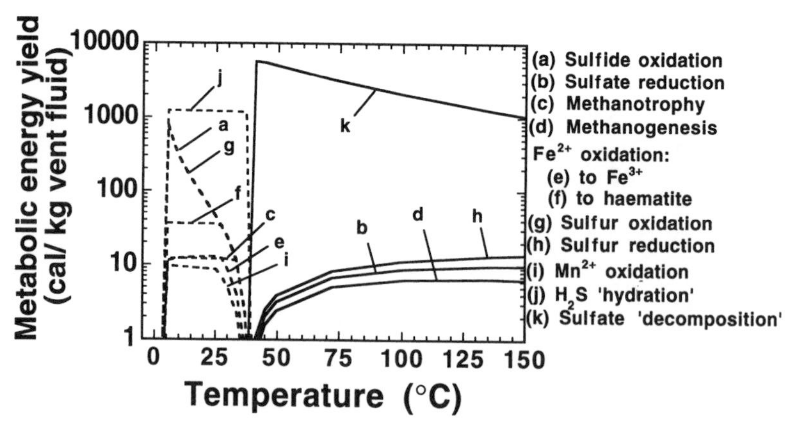

Figure 5.2 Metabolic yield from several inorganic reactions capable of supporting chemolithoautotrophic organisms in present-day submarine hydrothermal systems in which temperature is controlled by fluid mixing (McCollom and Shock, 1997). Oxidative reactions (a, c, e, f, g, i, and j) can provide energy at low temperatures, but are not energy sources above about 45 °C. Reductive reactions (b, d, h, and k) are the plausible energy sources at high temperature, but, with the exception of sulfate decomposition to sulfide and O_2 (k), do not yield as much energy as some of the oxidative reactions.

be seen that the availability of energy from oxidative reactions is limited to temperatures ≤ 45 °C with reductive reactions yielding energy at higher temperatures. This figure reveals how the foundation of hydrothermal ecosystem structure is provided by geochemical processes. Specifically, these results show that at 100 °C there is about 43.9 kJ/kg vent fluid available from sulfur reduction and 41.8 kJ/kg vent fluid from sulfate reduction. In addition, sulfate disproportionation to $H_2S(aq)$ and $O_2(aq)$ ('decomposition' in Figure 5.2) can yield an enormous amount of energy, but no organisms have yet been shown to use this highly exergonic pathway. Although reactions among inorganic compounds can yield large amounts of energy in these systems, other reactions in which organic compounds are formed can also release energy.

5.3 The energetics of acetogenesis in submarine hydrothermal ecosystems

Production of acetic acid from CO_2 and H_2 is an example of a thermodynamically favoured organic synthesis reaction that can supply energy in submarine hydrothermal ecosystems. The overall reaction of interest is given by

$$2\,CO_2(aq) + 4\,H_2(aq) = CH_3COOH(aq) + 2\,H_2O \qquad (5.5)$$

The 2:1 ratio of H_2 to CO_2 in this reaction reflects the overall oxidation state of carbon in acetic acid (zero), and can be compared to the 4:1 ratio required for methanogenesis, where the oxidation state of carbon in the product (methane) is –4. Accordingly, the energy available from acetogenesis is less than that from methanogenesis, all else being equal. A typical concentration for acetic acid in sea water is 50 μg/l, consistent with an activity of 10^{-6} (Thurman, 1985; Shock, 1995). Assuming the concentration of acetic acid

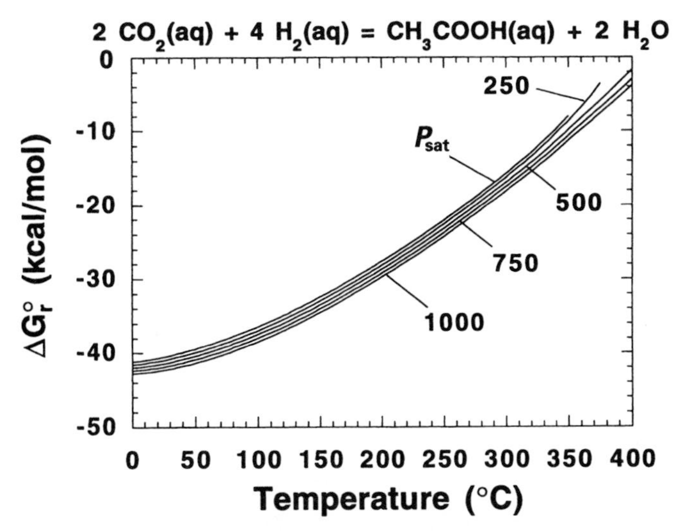

Figure 5.3 Standard-state Gibbs free energy of the acetogenesis reaction (5.5) in aqueous solution as a function of temperature at various pressures (bars). As in the case of the methanogenesis reaction in Figure 5.1, the temperature dependence is considerably greater than the pressure dependence. Comparison with Figure 5.1 shows that the temperature dependence of ΔG°_r for acetogenesis is greater than that for methanogenesis.

in the vent fluid is very much lower (consistent with measurements made by D. Butterfield, personal communication), a plausible activity for acetic acid is ~$10^{-6.17}$ at 100 °C in a mixed fluid. Using values of $a(CO_2(aq))$ and $a(H_2(aq))$ from the methanogenesis reaction discussed above, $Q_r = 10^{12.11}$ for reaction (5.5). Values of $\Delta G^\circ_{r,P,T}$ for reaction (5.5) can be calculated with the revised HKF equations using data and parameters from Shock *et al.* (1989), Shock and Helgeson (1990) and Shock (1995), as shown in Figure 5.3. Combining ΔG°_r and $RT \ln Q_r$ for a mixed solution in a hydrothermal system at 100 °C yields $\Delta G_r = -23.5$ kJ/mol (−15.8 kcal/mol), indicating that energy is released as acetic acid is made. In this case, H_2 would again be the limiting factor, with the result that a maximum of about 4.5 cal/kg vent fluid is available for acetogenesis in the 100 °C fluid. This energy is capable of supporting about 3 mg cells per kg vent fluid or an annual biomass productivity of 0.9×10^{11} kg. This is somewhat less than for methanogenesis, where H_2 is also the limiting factor.

The energetic relations between methanogenesis and acetogenesis can be summarised as shown in Figure 5.4. In this plot of energy vs reaction progress, the path from the unstable mixture of CO_2 and H_2 leads over a small bump to acetic acid + H_2O, which together occupy a lower energy state, and a larger bump before dropping to the lowest energy state occupied by CH_4 and H_2O. The bumps represent kinetic barriers that prevent these thermodynamically favoured reduction reactions from proceeding on their own. Because these reactions are kinetically inhibited yet thermodynamically favoured, the energy is stored as a disequilibrium state in the mixture. Organisms use enzymes to catalyse these reactions by lowering the activation energies of the kinetic barriers and gain the free energy liberated as the carbon and hydrogen are transformed to either the metastable assemblage of acetic acid and water or the stable assemblage of methane and water.

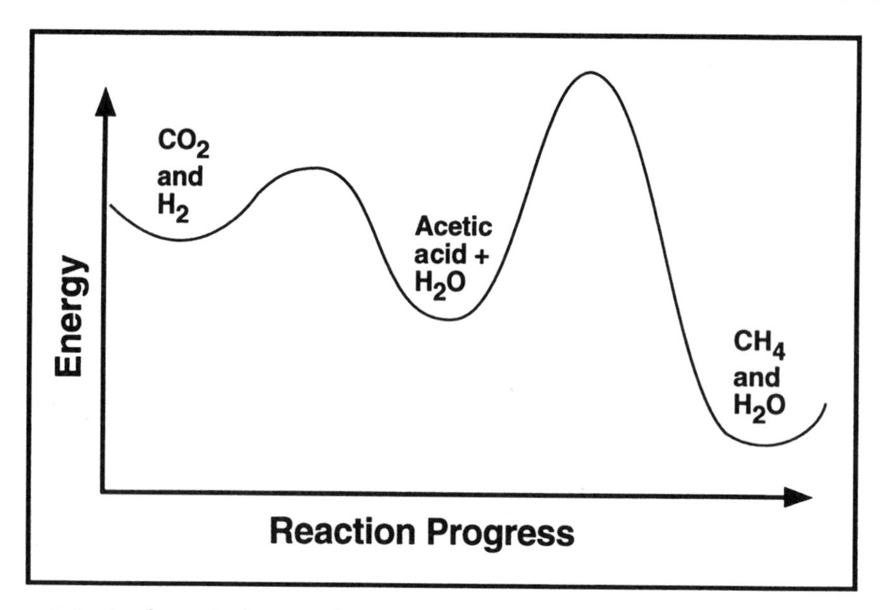

Figure 5.4 A schematic diagram showing changes in energy associated with reductive reactions involving CO_2 and H_2 at 100 °C in present-day submarine hydrothermal systems. The conversion of CO_2 and H_2 to acetic acid and H_2O is exergonic, as indicated by the lower position of the latter set of compounds. Therefore, acetogenesis can provide energy. Methanogenesis, the conversion of CO_2 and H_2 to methane and H_2O, releases even more energy than acetogenesis. There are kinetic barriers to both of these thermodynamically favoured reactions, as indicated by the bumps in the path of reaction progress. Microbes use enzymes to lower the kinetic barriers and allow the energy-releasing reactions.

5.4 Amino acid synthesis in hydrothermal ecosystems

Acetogenesis is by no means the only organic synthesis reaction that is thermodynamically favoured in hydrothermal systems. As shown by the following examples, amino acid synthesis can also release energy. Overall reactions that yield alanine, aspartic acid, glutamic acid, glycine, leucine, serine and valine from the inorganic constituents of hydrothermal fluids are given in reactions (5.6) to (5.12) respectively.

$$3\,CO_2(aq) + NH_4^+ + 6\,H_2(aq) = CH_3CHNH_2COOH(aq) + H^+ + 4\,H_2O \tag{5.6}$$

$$4\,CO_2(aq) + NH_4^+ + 6\,H_2(aq) = HOOCCH_2CHNH_2COOH(aq) + H^+ + 4\,H_2O \tag{5.7}$$

$$5\,CO_2(aq) + NH_4^+ + 9\,H_2(aq) = HOOC(CH_2)_2CHNH_2COOH(aq) + H^+ + 6\,H_2O \tag{5.8}$$

$$2\,CO_2(aq) + NH_4^+ + 3\,H_2(aq) = CH_2NH_2COOH(aq) + H^+ + 2\,H_2O \tag{5.9}$$

$$6\,CO_2(aq) + NH_4^+ + 15\,H_2(aq) = (CH_3)_2CHCH_2CHNH_2COOH(aq) + H^+ + 10\,H_2O \tag{5.10}$$

$$3\,CO_2(aq) + NH_4^+ + 5\,H_2(aq) = CH_2OHCHNH_2COOH(aq) + H^+ + 3\,H_2O \tag{5.11}$$

$$5\,CO_2(aq) + NH_4^+ + 12\,H_2(aq) = (CH_3)_2CHCHNH_2COOH(aq) + H^+ + 8\,H_2O \tag{5.12}$$

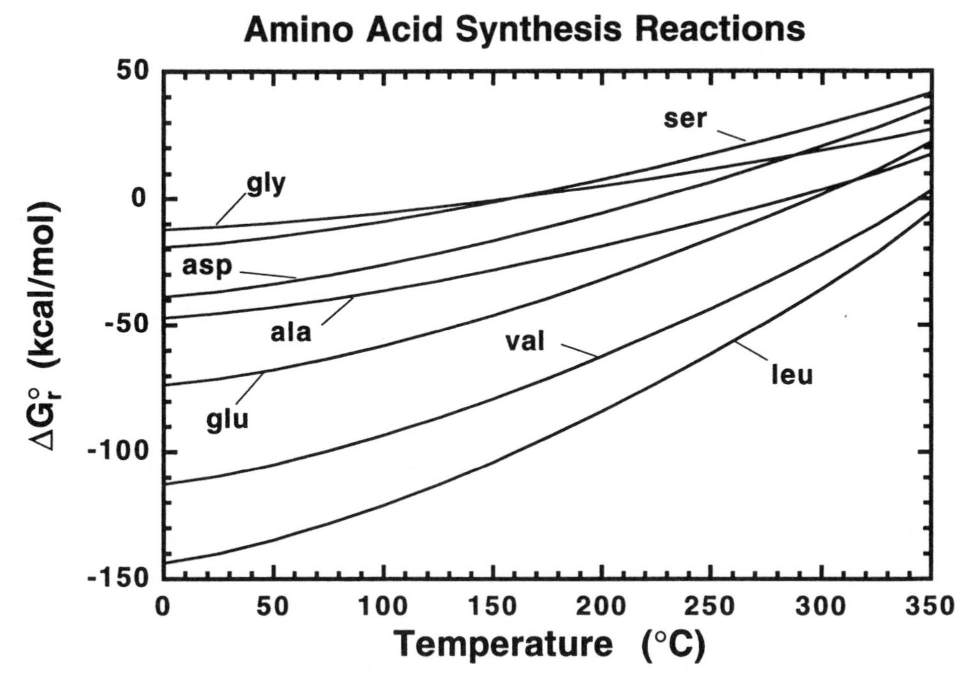

Figure 5.5 Standard-state Gibbs free energies of amino acid synthesis reactions (5.6–5.12) in aqueous solution as functions of temperature at P_{sat}.

Standard-state Gibbs free energies of these reactions at high temperatures and pressures can be calculated with data and parameters from Shock and Helgeson (1990), Shock *et al.* (1989, 1997), and Amend and Helgeson (1997). Values at pressures and temperatures consistent with the boiling curve for H_2O (P_{sat}) are shown in Figure 5.5, where it can be seen that in many cases ΔG_r° is negative over wide ranges of temperature and that values become less negative (and even become positive for all but valine and leucine) as temperature increases. To calculate values of the overall Gibbs free energy of reaction, values of ΔG_r° need to be combined with activity products calculated for hydrothermal solutions.

Again, we will adopt the values of $a(CO_2(aq))$ and $a(H_2(aq))$ from McCollom and Shock (1997), together with their pH for a submarine hydrothermal solution at 100 °C (5.72) obtained by mixing cold sea water with vent fluid. In addition, we need estimates of the activities of NH_4^+ and the various amino acids in this solution. Concentrations of these compounds in sea water have been documented, but are unknown in vent fluids. It is safe to assume that these compounds are probably not at higher concentrations in the vent fluid than they are in sea water, as their main source is likely to be sea water. In fact, they may be at considerably lower concentrations. With this in mind, setting the concentrations of NH_4^+ and amino acids to zero in the hot vent fluid will provide a conservative estimate of the concentrations in the 100 °C solution. A likely concentration for NH_4^+ in sea water is 10 μmol/l (Millero, 1996), and the 1:2.55 ratio of vent fluid to sea water to reach 100 °C leads to log $(a(NH_4^+)) = -5.14$. Taking the total concentration of free amino acids to be 40 nmol/l in sea water (Lee and Bada, 1975, 1977; Liebezeit *et al.*, 1980; Thurman, 1985), and the mole percentages of free amino acids from Dawson and Gocke (1978) (ala 10%; asp 6%; glu 7%; gly 20%; leu 2%; ser 20%; val 2%) leads to the log activities listed in Table 5.1. Also listed in this table are values of ΔG_r at 100 °C for amino acid synthesis according to the reactions listed above.

Table 5.1 Hydrothermal synthesis of amino acids

Amino acid	Sea water concentration (nmol/l)	Calculated log activity in 100 °C solution	Overall Gibbs free energy (kcal/mol)
Alanine	4	−8.40	−5.087
Aspartic acid	2.5	−8.74	8.943
Glutamic acid	2.8	−8.69	−1.639
Glycine	8	−8.24	4.756
Leucine	0.8	−9.24	−27.213
Serine	8	−8.24	16.998
Valine	0.8	−9.24	−20.877

Synthesis reactions for alanine, glutamic acid, valine and leucine are all exergonic at 100 °C in submarine hydrothermal systems. This means that these amino acids are more stable than the mixture of CO_2, H_2 and NH_4^+ that results from mixing sea water and vent fluid to reach 100 °C in these systems. These amino acids could join acetic acid in the middle dip of Figure 5.4. In contrast, reactions to form aspartic acid, glycine and serine all require energy. Note that the reactions to form the largest molecules tend to release the greatest amounts of energy. This can be understood after considering the relative hydrogen to carbon ratios of these compounds. If this ratio is high, then the carbon in the amino acid is more reduced relative to the carbon in an amino acid with a lower H:C ratio. Because the oxidation state of the fluid at 100 °C strongly favours the production of reduced compounds from CO_2 (see below), *the larger amino acids are actually more stable thermodynamically than the smaller amino acids.* These relations are consistent with relative stabilities of amino acids, which have been used to interpret amino acid alteration experiments (Shock, 1990a; Helgeson and Amend, 1994).

One implication of these results is that the overall synthesis of amino acids may be close to a metastable equilibrium process for hyperthermophiles. As an example, synthesis of equal activities of these seven amino acids at 100 °C yields a combined ΔG_r of −101.332 kJ/mol (−24.219 kcal/mol) (sum of values in Table 5.1). So, despite the fact that some of the synthesis reactions require energy at 100 °C, amino acid synthesis in general may not cost much metabolic energy. In addition, peptide bond formation requires less energy at high temperatures than at low (Shock, 1992a). These observations may help to explain the extremely short doubling times of hyperthermophile populations at their optimum growth temperatures in the laboratory. Some species of hyperthermophiles completely reproduce their genetic information and much of their membranes, enzymes and other biomolecules in an hour or so (Stetter, 1996; Blöchl *et al.*, 1997). Monomer synthesis is thermodynamically favoured because the energy for synthesis is provided by the concentrations of CO_2(aq), H_2(aq) and other constituents in the hydrothermal solution. It follows that H_2(aq) can be both the energy source and the electron donor in these reactions. Nevertheless, these favourable reactions do not occur rapidly on their own. Kinetic barriers must exist or organisms would not be able to benefit from the favourable reactions.

Without kinetic barriers to redox reactions, life would be impossible. However, microbes are not the only catalysts in natural systems. Synthesis of organic compounds may be facilitated by mineral or other geochemical catalysts present in hydrothermal systems. The following sections of this chapter illustrate many types of organic transformation and synthesis reactions that are thermodynamically favoured in hydrothermal systems.

5.5 The potential for hydrothermal organic synthesis

The favourable thermodynamic conditions that support chemolithoautotrophs are simple consequences of ordinary geological processes, and the energy is present even if the microbes are not. Hot rocks crack as they cool; water circulates through the cracked-rock system, acts as a refrigerant, and transports heat from the rocks towards space. As a consequence, water–rock reactions change the composition of both, and the hot fluid mixes dynamically with the cooler overlying sea water. These circulation, reaction and mixing processes generate the potential for organic synthesis that microbes tap to drive their metabolism. This potential can yield high concentrations of a wide variety of aqueous organic compounds, as illustrated by several theoretical studies (Shock, 1990b, 1992b, 1996, 1997; Shock and Schulte, 1997). A brief summary is presented here to illustrate two points. First, many chemolithoautotrophic reactions that generate organic compounds are exergonic in hydrothermal systems, and second, the mixing of fluids in hydrothermal systems may have provided the disequilibrium necessary to drive the first metabolism on earth.

Calculating the distribution of species in a metastable state, like that represented by the middle dip in Figure 5.4, is analogous to a stable equilibrium calculation. Take, for example, the acetogenesis reaction discussed above. If we want to calculate the maximum amount of acetic acid that can be produced, we need to determine the concentrations (m = molalities) of $CO_2(aq)$, $H_2(aq)$, H_2O and $CH_3COOH(aq)$ in metastable equilibrium for a known bulk composition at a given temperature and pressure. The four concentrations provide four unknowns. Four equations obtained from the three mass balance constraints and the one mass action expression provide the means to solve for these four unknowns:

$$C_{total} = mCO_2(aq) + 2\,mCH_3COOH(aq) \tag{5.13}$$

$$H_{total} = 2\,mH_2(aq) + 4\,mCH_3COOH(aq) + 2\,mH_2O \tag{5.14}$$

$$O_{total} = mH_2O + 2\,\mathrm{m}CH_3COOH(aq) + 2\,mCO_2(aq) \tag{5.15}$$

$$\log K = \log\,[a(CH_3COOH(aq))] + 2\log\,[a(H_2O)] - 2\log\,[(aCO_2(aq))] - 4\log\,[a(H_2(aq))] \tag{5.16}$$

where K stands for the equilibrium constant given by

$$\Delta G^{\circ}_{r} = -2.303\,RT\log K \tag{5.17}$$

All of the terms on the left of these four expressions are known, and activities can be related to molalities through activity coefficients. This approach can be generalised to illustrate the potential for organic synthesis by recognising that the addition of another organic compound requires the addition of another mass action expression involving the equilibrium constant for the analogue to reaction (5.5), and modifying the mass balance expressions to include the new compound. Calculations of this type that include carboxylic acids (mono- and di-acids), alcohols, aldehydes, ketones and alkenes were conducted by Shock and Schulte (1997) for hydrothermal systems using various mass balance constraints derived from geological and geochemical arguments for the early Earth and Mars.

The consequences of mixing fluids with different temperatures, compositions and oxidation states are readily calculated with geochemical models that simultaneously account for mass balance and mass action constraints. In the calculations described here and elsewhere we have used the EQ3/6 package (Wolery, 1992; Wolery and Daveler, 1992) with customised databases consistent with Helgeson *et al.* (1978), Shock and Helgeson (1988, 1990), Shock *et al.* (1989, 1992, 1997), Schulte and Shock (1993), Shock and Koretsky

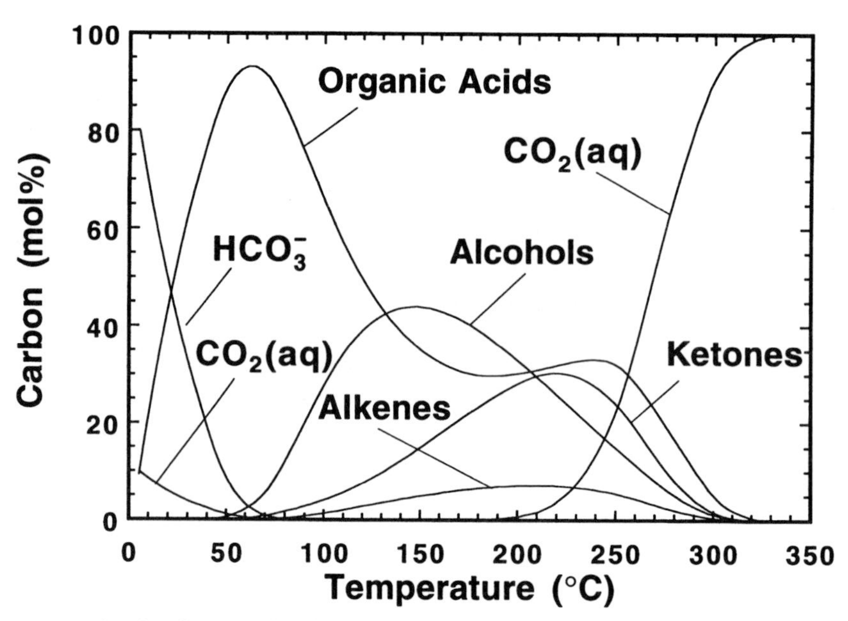

Figure 5.6 The distribution of carbon in metastable equilibrium states obtained by mixing hydrothermal fluids with sea water. This particular example refers to a calculation in which the sea water has had its dissolved O_2 removed, and the initial oxidation state of the hydrothermal fluid is one half log unit in oxygen fugacity below the value set by the mineral assemblage fayalite-magnetite-quartz at 350 °C (Shock and Schulte, 1997). Both of these constraints are meant to simulate conditions that may have prevailed in submarine hydrothermal systems on the early earth. Note that at the highest and lowest temperatures carbon is stable as inorganic forms (CO_2(aq) at high temperature, and HCO_3^- and CO_2(aq) at low temperature). At intermediate temperatures carbon is more stable as a variety of organic compounds, as indicated by the labelled curves. Curves correspond to the summation of the concentrations of individual compounds depicted in Figure 5.7.

(1993, 1995) and Sverjensky *et al.* (1997). The EQ3 code evaluates the chemical speciation of an aqueous solution of known bulk composition, temperature and pressure, and the EQ6 code predicts the consequences as that solution reacts with a rock or is mixed with another solution. Mixing calculations were conducted with and without mineral precipitation to simulate variations in the kinetics of disequilibrium states. Calculations that include methane in the speciation of aqueous carbon were used to assess the energetics of methanogenesis and methanotrophy from 0 °C to 350 °C (McCollom and Shock, 1997), and those that exclude methane characterise the metastable states in which organic compounds are preserved and in which organic synthesis can occur (Shock, 1990b, 1992b; Shock and Schulte, 1997). With reference to Figure 5.4, the latter are analogous to allowing reactions to proceed to the local minimum (acetic acid + H_2O) rather than the global minimum (methane + H_2O).

Results of calculations of this type are summarised in Figure 5.6 for mixing hydrothermal fluids with cold sea water that lacks dissolved oxygen. Although other models of Archaean sea water are possible (Karhu and Holland, 1996; Ohmoto, 1996; Russell and Hall, 1997; Watanabe *et al.*, 1997), modern sea water without O_2(aq) provides a well-constrained initial condition for these calculations. The metastable equilibrium distribution of carbon is shown in this plot as a function of temperature. At the far right side of the diagram the carbon in the pure hydrothermal fluid is predominantly present as CO_2(aq) and at stable

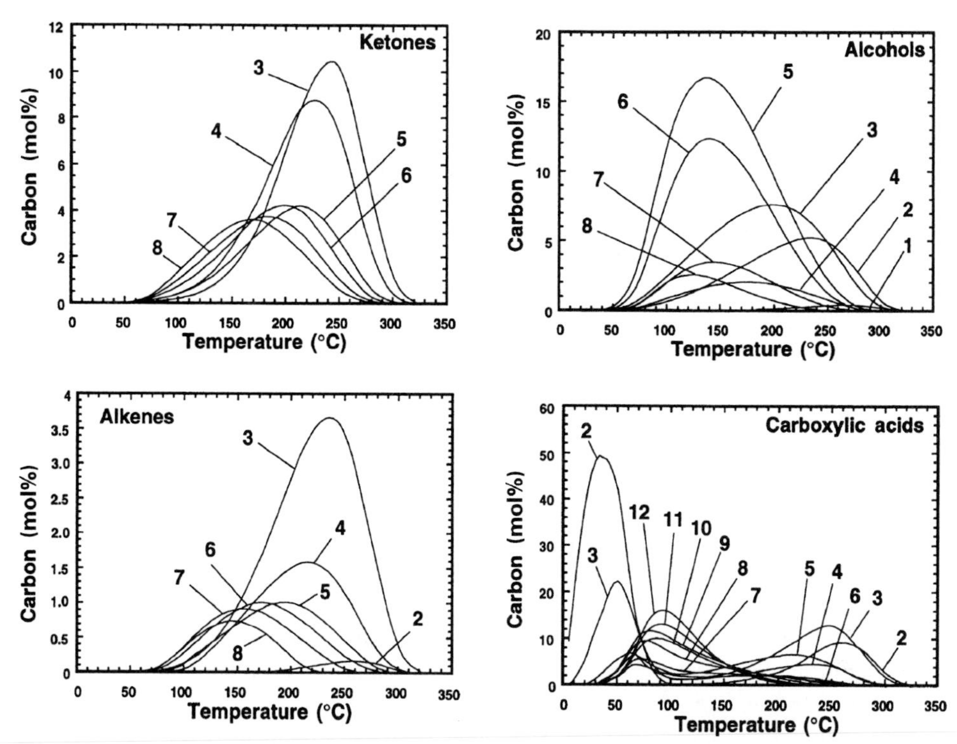

Figure 5.7 The distribution of ketones, alkanes, alcohols and carboxylic acids corresponding to the curves shown in Figure 5.6. Curves show the percentage of carbon in the system that would be present as each compound if they all formed in a metastable equilibrium state in response to fluid mixing in hydrothermal systems (Shock and Schulte, 1997). The labels refer to the number of carbon atoms in each compound. In the case of ketones, curves for acetone through octanone are shown. Among the alkenes, which contribute a few per cent of the total dissolved carbon, propene dominates at high temperatures where ethene is a minor constituent, and a mix of the larger molecules prevails at lower temperatures. Results for methanol to octanol are shown for the alcohols, and curves for acetic to dodecanoic acid are shown for the carboxylic acids (formic was included in the calculations but contributes insignificantly to the total carbon in the metastable equilibrium state). In each case there is a shift from short compounds at higher temperatures to longer compounds at lower temperatures. This shift is consistent with changes in the relative oxidation state of the mixed fluid (see text).

equilibrium with an oxidation state set by the hot volcanic rocks that host the hydrothermal system. As the calculation proceeds, cold sea water is, in effect, 'titrated' into the hydrothermal fluid. The temperature drops in the resulting mixture as more and more sea water is added. With this drop in temperature, the distribution of carbon shifts towards more reduced forms: ketones, alkenes, alcohols and organic acids (aldehydes are highly unstable under these conditions and do not account for a meaningful percentage of the dissolved carbon in the metastable states). *All of these organic compounds are* more *stable than the mixture of* CO_2 *and* H_2 *that results when hydrothermal fluids and sea water mix.* Therefore, if the free energy of the system can be lowered, organic compounds are likely products.

Each of the curves in Figure 5.6 that bears an organic compound label shows the sum of several compounds of that type to the total percentage of aqueous carbon. These total percentages for ketones, alkenes, alcohols and carboxylic acids are broken down by individual compounds in Figure 5.7. Note that in each case there is a drive to form longer compounds as the temperature decreases. This can be understood by considering the H:C ratio of these organic compounds and the oxidation state of the mixed fluid. As the alkyl chain increases in length, the overall H:C ratio increases, which is consistent with the notion that the carbon is relatively more reduced with increasing alkyl chain length. The point at which CO_2 and H_2 in the mixed fluid are farthest from equilibrium is the point of greatest relative reduction and falls between 150 °C and 100 °C. Approaching this range from higher temperatures, the most stable ketones and alkenes are the longer molecules. Similarly, pentanol and hexanol are more stable than ethanol or propanol in this temperature range. Also, among the carboxylic acids, the longer-chain compounds (octanoic, nonanoic, decanoic, undecanoic, dodecanoic, etc.) account for most of the aqueous carbon in this temperature range. As temperature is decreased further (by adding more sea water), the overall oxidation state of the mixture becomes more oxidised, leading to the predominance of the short-chain carboxylic acids, and finally to inorganic carbon in the form of bicarbonate (see Figure 5.6).

5.5.1 Implications for hyperthermophiles

Results of the type shown in Figures 5.6 and 5.7 reveal that there is an enormous thermodynamic drive for organic synthesis as fluids mix in hydrothermal systems, complementary to the results for acetogenesis and amino acid synthesis described above. It follows that organisms can take advantage of this situation by synthesising many organic compounds at no overall energetic cost. This does not mean that there is not an activation energy to be overcome. Catalysts such as enzymes are still essential to lower those kinetic barriers. However, as the kinetic barriers are lowered, synthesis can be exergonic for many reactions that yield organic compounds. It is completely possible that there is sufficient free energy released in these reactions to offset the energy required to overcome the activation barriers. If so, then synthesis of monomers as well as biopolymers may be energy-yielding processes for hyperthermophiles. This would help to explain the extremely rapid doubling times for hyperthermophilic organisms in laboratory cultures.

The energetic payback from organic synthesis in hydrothermal ecosystems is also profoundly different from photosynthesis of organic compounds at the surface of the earth. There, photosynthetic organisms are synthesising organic compounds in a 20% O_2 atmosphere. This requires work (in the thermodynamic sense) against the chemical system. Light energy is extremely powerful and provides the kick to drive organic synthesis uphill. No thermodynamic work of this magnitude is required in hydrothermal ecosystems. There, organisms can take advantage of favourable thermodynamics to build the constituents of their cells without an additional source of energy required. As a consequence, life may be considerably easier at high temperatures than at temperatures we consider hospitable.

5.5.2 Implications for an emergent metabolism

Contrary to conventional wisdom, the results summarised here show that organic compounds are more stable at high temperatures in hydrothermal systems than they are at lower temperatures. Their enhanced stability is a direct consequence of the instability of the mixture of CO_2 and H_2 engendered by geological processes. If lower energy states were catalysed in early hydrothermal systems, organic compounds would be the products.

The nature of minerals and/or inorganic solutes capable of catalysing organic reactions at realistic hydrothermal conditions is largely unknown. There are indications that clay minerals like montmorillonite (Bell *et al.*, 1994), the iron oxide magnetite (Palmer and Drummond, 1986; McCollom *et al.*, 1998), and certain iron sulfides (Huber and Wächtershäuser, 1997) may be capable of catalysing reactions involving aqueous organic compounds. If these and other catalysts also work in hydrothermal systems, then reactions leading to simple organic compounds may occur widely in these systems. If such reactions were catalysed through coupled inorganic and organic redox reactions in hydrothermal systems on the early earth, then something like a simple metabolism may be a geochemical process. In any event, it is probably profitable to study how metabolism could have its roots in hydrothermal geochemistry.

Recent results of modelling the energetics of the tricarboxylic acid cycle at high temperatures and pressures show that some of the reactions can be exergonic when considered in the context of the surrounding hydrothermal environment (Amend and Shock, 1998). For example, the synthesis of fumarate from oxaloacetate and the formation of succinate from fumarate are both exergonic in present-day hydrothermal systems (assuming equal activities of the two acids). In fact, there is enough energy released by these two reactions at some temperatures to drive the synthesis of 2-oxoglutarate and citrate. However, it appears that the reaction of citrate with CO_2(aq) and H_2(aq) to yield pyruvate and oxaloacetate may require energy at all temperatures. On the other hand, if the concentrations of either CO_2(aq) or H_2(aq) were higher in hydrothermal systems on the early earth (both likely possibilities), then these steps in the cycle might also yield energy. This might not be a requirement if the energy-consuming reactions could be linked to reactions likely to yield energy. Excellent candidates for exergonic hydrothermal reactions include sulfur and sulfate reduction, methanogenesis, and sulfate reduction coupled to Fe^{2+} oxidation to magnetite or methane oxidation to CO_2 (McCollom and Shock, 1997). Coupling exergonic reduction reactions to endergonic reduction reactions may require intermediate reactions that produce something like ATP that can serve as the energy currency in these coupled processes.

5.6 Hydrothermal organic synthesis versus organic destruction in hot water

If organic compounds are stable relative to CO_2 and H_2 in hydrothermal solutions, why do they decompose in hot water experiments? The short answer is that the overwhelming majority of experiments on the survivability of organic compounds in water at high temperatures are not conducted at conditions that are reached in hydrothermal systems. Numerous intensive variables other than temperature and pressure must be similarly controlled to simulate a realistic hydrothermal fluid. The single most important variable is the oxidation state of the solution, although the activities of CO_2, NH_4^+ and many other simple inorganic and organic constituents can greatly affect the results of such experiments (Shock, 1990a, 1992c; Helgeson and Amend, 1994).

The reactions discussed above are overall reactions and do not imply any mechanism. The thermodynamic properties of reactions are completely independent of the reaction mechanisms, and apply even if there is no pathway for the reaction to proceed on its own. The lack of such pathways allows organisms to conduct chemosynthetic lifestyles. This does not mean that kinetics and thermodynamics of reactions are independent. Rates of reactions are notoriously functions of how far the reactions are from equilibrium. Nevertheless, this does not seem to enter into the design of experiments in which organic compounds are subjected to high temperatures in aqueous solution.

The rate at which reactions proceed is typically a function of the concentration of the products. Often this is discussed in terms of a schematic reaction like

$$A \rightarrow B + C \quad (5.18)$$

The rate at which the concentration of A decreases will depend on the concentrations of B and C in the system. An equilibrium concentration of A is possible even if it cannot be detected. If the concentrations of B and C are at the equilibrium values, then the reaction will not appear to proceed at all. Conversely, if they are absent, the concentration of A may drop rapidly in an effort to produce B and C and reach the equilibrium state. The connection between rates and the position of equilibrium for simple reactions of this type is the foundation of transition-state theory and other efforts to quantify rates in terms of energetics. Poorly designed experiments in which reactants are studied in the absence of products cannot be placed into rigorously quantitative frameworks provided by theory. Nevertheless, many investigators choose to do a few poorly constrained experiments rather than to carefully examine the system they purport to study. Why is this?

Most hot water experiments on organic compounds are not conducted in the presence of the reaction products. As a result, it is not possible to determine the rate of reaction in reference to how far the system is from the equilibrium state. In fact, most experiments of this type are conducted with only the reactant present, with no evaluation (or appreciation, apparently) of what the equilibrium concentration with respect to the products would be. The thermodynamic drive to generate reaction products is enormous in such experiments, and so the rates obtained are extremely high. Experiments are generally terminated long before steady-state or equilibrium conditions prevail. In numerous cases the reaction products are not even identified, only the disappearance of the reactant! 'Rates' obtained from experiments in which there is not a complete accounting for mass balance are useless, especially if the goal is to understand natural systems. Is that the goal of these experiments?

The design of the vast majority of experiments on the reactivity of organic compounds in hot water is guaranteed to lead to the rapid destruction of these compounds. This is a perplexing state of affairs that will continue to baffle those trying to understand life at high temperatures until useful and well-constrained experiments become the norm. There is a general belief that organic compounds are easily and 'quantitatively' destroyed (a concept devoid of thermodynamic meaning) at high temperatures in water, despite numerous experiments that show production of aqueous organic compounds at high temperature (Siskin and Katritzky, 1991; Katritzky *et al.*, 1997; McCollom *et al.*, 1998) or their transformation at equilibrium concentrations (Seewald, 1994). This belief persists even though it has been shown to be based on unjustifiable assumptions through calculations of the type described above. Luckily, human ignorance is not shared by hyperthermophiles, which continue to conduct rapid and efficient biosynthesis in hydrothermal fluids as they have for more than 4 billion years.

5.7 Conclusions

Geochemical modelling of the energetics of reactions that hyperthermophiles use as energy sources in submarine hydrothermal system indicates that reductive strategies like methanogenesis, sulfate reduction and sulfur reduction can provide energy at elevated temperatures. At 100 °C, where hyperthermophiles are known to live, methane is considerably more stable than the combination of H_2 and CO_2 (or bicarbonate) provided by fluid mixing. Therefore, methanogenesis lowers the free energy of the system and supplies energy to metabolic processes. If temperatures are controlled by fluid mixing, then oxidative reactions like methanotrophy, sulfur oxidation and sulfide oxidation can only release energy at temperatures ≤ 50 °C. On the early earth this temperature would have been lower given the likelihood that O_2 was at much lower concentrations in the atmosphere and oceans.

There is also an enormous thermodynamic drive to generate organic compounds in submarine hydrothermal systems as hydrothermal fluids mix with sea water. Hyperthermophiles take advantage of this thermodynamic drive to synthesise many of the small organic compounds that they require at no energetic cost. In fact, they are likely to *gain* energy in the process. The result is a geochemically provided benthic boondoggle in which microbes are given a free lunch that they are paid to eat. These conditions exist at present in submarine hydrothermal systems, but are likely to have been more extreme on the early earth where the atmosphere and sea water were probably less oxidised than at present, where host-rocks of hydrothermal systems may have been less silica rich and therefore able to generate more hydrogen from reduction of water, and where heat flow and the intensity and distribution of hydrothermal systems were greater. Hydrothermal systems are therefore the most likely location near the surface of the earth for the emergence of metabolism from coupled inorganic and organic redox reactions.

5.8 Summary

Submarine hydrothermal fluids are far from thermal or chemical equilibrium with sea water, but mix dynamically with sea water in sea floor hydrothermal systems. The resulting disequilibrium states provide sources of geochemical energy that chemolithoautotrophic hyperthermophiles tap to drive biosynthesis of organic molecules and other metabolic processes. The amounts of energy can be evaluated with thermodynamic calculations that account for the difference between equilibrium and the states that occur in nature. As an example, methanogenesis from CO_2 and H_2 provides up to 105 kJ/mol (25 kcal/mol) in a mixed solution at 100 °C. Organic synthesis reactions are similarly exergonic. Synthesis of acetic acid from CO_2 and H_2 can yield nearly 67 kJ/mol (16 kcal/mol) at 100 °C, and reactions that produce several amino acids also release energy. *It follows that hyperthermophiles can generate many of the simple organic compounds that they require for cell growth and gain energy in the process*. These conditions differ greatly from organic synthesis at present surface conditions, where the composition of the atmosphere requires thermodynamic work on the system. Nevertheless, the extremely conducive conditions of hydrothermal ecosystems are provided by the common geological processes of magma generation, sea floor spreading and hydrothermal circulation, all of which are common throughout earth's history. Additional model calculations of the potential for organic synthesis in submarine hydrothermal systems on the early earth indicate that the formation of organic compounds would lower the free energy of these systems, enhancing their likelihood as environments for the emergence of metabolism.

Acknowledgements

Thanks to Jan Amend, Melanie Summit, Anna-Louise Reysenbach, Laura Wetzel, John Baross, and Mike Russell for many helpful discussions. Technical support from Barb

Winston, Rachel Lindvall, and Doug LaRowe is greatly appreciated. This chapter (GEOPIG contribution no. 149) reflects research funded by NSF grants OCE-9220337, and EAR-9418500, and NASA grant NAG5-4002.

References

AMEND, J. P. and HELGESON, H. C. (1997) Calculation of the standard molal thermodynamic properties of aqueous biomolecules at elevated temperatures and pressures I. L-α-Amino acids. *J. Chem. Soc. Faraday Trans.*, **93** (in press).

AMEND, J. P. and SHOCK, E. L. (1998) Geochemical constraints on the energetics of the tricarboxylic acid cycle in hydrothermal ecosystems (in preparation).

ANDERSON, G. M. (1996) *Thermodynamics of Natural Systems* (New York: Wiley).

BELL, J. L. S., PALMER, D. A., BARNES, H. L. and DRUMMOND, S. E. (1994) Thermal decomposition of acetate. III. Catalysis by mineral surfaces. *Geochim. Cosmochim. Acta*, **58**, 4155–4177.

BLÖCHL, E., RACHEL, R., BURGGRAF, S., HAFENBRADL, D., JANNASCH, H. W. and STETTER, K. O. (1997) *Pyrolobus fumarii*, gen. and sp. nov., represents a novel group of archaea, extending the upper temperature limit for life to 113 °C. *Extremophiles*, **1**, 14–21.

DAWSON, R. and GOCKE, K. (1978) Heterotrophic activity in comparison to the free amino acid concentrations in Baltic sea water samples. *Oceanol. Acta*, **1**, 45–54.

ELDERFIELD, H. and SCHULTZ, A. (1996) Mid-ocean ridge hydrothermal fluxes and the chemical composition of the oceans. *Annu. Rev. Earth Planet. Sci.*, **24**, 191–224.

HELGESON, H. C. and AMEND, J. P. (1994) Relative stabilities of biomolecules at high temperatures and pressures. *Thermochim. Acta*, **245**, 89–119.

HELGESON, H. C., DELANEY, J. M., NESBITT, H. W. and BIRD, D. K. (1978) Summary and critique of the thermodynamic properties of rock-forming minerals. *Am. J. Sci.*, **278-A**, 1–229.

HUBER, C. and WÄCHTERSHÄUSER, G. (1997) Activated acetic acid by carbon fixation on (Fe,Ni)S under primordial conditions. *Science*, **276**, 245–247.

JOHNSON, J. W., OELKERS, E. H. and HELGESON, H. C. (1992) SUPCRT92: A software package for calculating the standard molal thermodynamic properties of minerals, gases, aqueous species, and reactions from 1 to 5000 bar and 0 to 1000 °C. *Computers Geosci.*, **18**, 899–947.

KARHU, J. A. and HOLLAND, H. D. (1996) Carbon isotopes and the rise of atmospheric oxygen. *Geology*, **24**, 867–870.

KATRITZKY, A. R., IGNATCHENKO, E. S., ALLIN, S. M., SISKIN, M., FERRUGHELLI, D. L. and RÁBAI, J. (1997) Aqueous high-temperature chemistry of carbo- and heterocycles. 30. Aquathermolysis of phenyl-substituted hydroxyquinolines. *Energy Fuels*, **11**, 174–182.

LEE, C. and BADA, J. L. (1975) Amino acids in equatorial Pacific ocean water. *Earth Planet. Sci. Lett.*, **26**, 61–68.

LEE, C. and BADA, J. L. (1977) Dissolved amino acids in the equatorial Pacific, the Sargasso Sea, and Biscayne Bay. *Limnol. Oceanogr.*, **22**, 502–510.

LIEBEZEIT, G., BOLTER, M., BROWN, I. F. and DAWSON, R. (1980) Dissolved free amino acids and carbohydrates at pycnocline boundaries in the Sargasso Sea and related microbial activity. *Oceanol. Acta*, **3**, 357–362.

MCCOLLOM, T. M. and SHOCK, E. L. (1997) Geochemical constraints on chemolithoautotrophic metabolism by microorganisms in sea floor hydrothermal systems. *Geochim. Cosmochim. Acta*, **61**, 4375–4391.

MCCOLLOM, T. M., RITTER, G. and SIMONEIT, B. R. T. (1998) Lipid synthesis under hydrothermal conditions by Fischer–Tropsch-type reactions. *Origins of Life and Evolution of the Biosphere* (submitted).

MILLERO, F. J. (1996) *Chemical Oceanography*, 2nd edn (Boca Raton, FL: CRC Press).

OHMOTO, H. (1996) Evidence in pre-2.2 Ga palaeosols for the early evolution of atmospheric oxygen and terrestrial biota. *Geology*, **24**, 1135–1138.

PALMER, D. A. and DRUMMOND, S. E. (1986) Thermal decarboxylation of acetate. Part I. The kinetics and mechanism of reaction in aqueous solution. *Geochim. Cosmochim. Acta*, **50**, 813–823.

RUSSELL, M. J. and HALL, A. J. (1997) The emergence of life from iron monosulfide bubbles at a hydrothermal redox front. *J. Geol. Soc.*, **154**, 377–402.

SCHULTE, M. D. and SHOCK, E. L. (1993) Aldehydes in hydrothermal solutions: standard partial molal thermodynamic properties and relative stabilities at high temperatures and pressures. *Geochim. Cosmochim. Acta*, **57**, 3835–3846.

SEEWALD, J. S. (1994) Evidence for metastable equilibrium between hydrocarbons under hydrothermal conditions. *Nature*, **370**, 285–287.

SHOCK, E. L. (1990a) Do amino acids equilibrate in hydrothermal fluids? *Geochim. Cosmochim. Acta*, **54**, 1185–1189.

SHOCK, E. L. (1990b) Geochemical constraints on the origin of organic compounds in hydrothermal systems. *Origins of Life and Evolution of the Biosphere*, **20**, 331–367.

SHOCK, E. L. (1992a) Stability of peptides in high temperature aqueous solutions. *Geochim. Cosmochim. Acta*, **56**, 3481–3491.

SHOCK, E. L. (1992b) Chemical environments in submarine hydrothermal systems. In *Marine Hydrothermal Systems and the Origin of Life*, ed. N. Holm, a special issue of *Origins of Life and Evolution of the Biosphere*, **22**, 67–107.

SHOCK, E. L. (1992c) Hydrothermal organic synthesis experiments. In *Marine Hydrothermal Systems and the Origin of Life*, ed. N. Holm, a special issue of *Origins of Life and Evolution of the Biosphere*, **22**, 135–146.

SHOCK, E. L. (1995) Organic acids in hydrothermal solutions: standard molal thermodynamic properties of carboxylic acids, and estimates of dissociation constants at high temperatures and pressures. *Am. J. Sci.*, **295**, 496–580.

SHOCK, E. L. (1996) Hydrothermal systems as environments for the emergence of life. In *Evolution of Hydrothermal Ecosystems on Earth (and Mars?)*, Ciba Foundation Symposium 202, pp. 40–60 (Chichester: Wiley).

SHOCK, E. L. (1997) High temperature life without photosynthesis as a model for Mars. *J. Geophys. Res.*, **102**, 23687–23694.

SHOCK, E. L. and HELGESON, H. C. (1988) Calculation of the thermodynamic and transport properties of aqueous species at high pressures and temperatures: correlation algorithms for ionic species and equation of state predictions to 5 kb and 1000 °C. *Geochim. Cosmochim. Acta*, **52**, 2009–2036.

SHOCK, E. L. (1990) Calculation of the thermodynamic and transport properties of aqueous species at high pressures and temperatures: standard partial molal properties of organic species. *Geochim. Cosmochim. Acta*, **54**, 915–945.

SHOCK, E. L. and KORETSKY, C. M. (1993) Metal–organic complexes in geochemical processes: calculation of standard partial molal thermodynamic properties of aqueous acetate complexes at high pressures and temperatures. *Geochim. Cosmochim. Acta*, **57**, 4899–4922.

SHOCK, E. L. and KORETSKY, C. M. (1995) Metal–organic complexes in geochemical processes: Estimation of standard partial molal thermodynamic properties of aqueous complexes between metal cations and monovalent organic acid ligands at high pressures and temperatures. *Geochim. Cosmochim. Acta*, **59**, 1497–1532.

SHOCK, E. L. and SCHULTE, M. D. (1997) Organic synthesis during fluid mixing in hydrothermal systems. *Origins of Life and Evolution of the Biosphere* (in press).

SHOCK, E. L., HELGESON, H. C. and SVERJENSKY, D. A. (1989) Calculation of the thermodynamic and transport properties of aqueous species at high pressures and temperatures: standard partial molal properties of inorganic neutral species. *Geochim. Cosmochim. Acta*, **53**, 2157–2183.

SHOCK, E. L., MCCOLLOM, T. and SCHULTE, M. D. (1995) Geochemical constraints on chemolithoautotrophic reactions in hydrothermal systems. *Origins of Life and Evolution of the Biosphere*, **25**, 141–159.

SHOCK, E. L., OELKERS, E. H., JOHNSON, J. W., SVERJENSKY, D. A. and HELGESON, H. C. (1992) Calculation of the thermodynamic properties of aqueous species at high pressures and temperatures: effective electrostatic radii, dissociation constants, and standard partial molal properties to 1000 °C and 5 kb. *J. Chem. Soc., Faraday Trans.*, **88**, 803–826.

SHOCK, E. L., SASSANI, D. C., WILLIS, M. and SVERJENSKY, D. A. (1997) Inorganic species in geologic fluids: correlations among standard molal thermodynamic properties of aqueous ions and hydroxide complexes. *Geochim. Cosmochim. Acta*, **61**, 907–950.

SISKIN, M. and KATRITZKY, A. R. (1991) Reactivity of organic compounds in hot water: geochemical and technological implications. *Science*, **254**, 231–237.

STETTER, K. O. (1996) Hyperthermophilic prokaryotes. *FEMS Microbiol. Rev.*, **18**, 149–158.

SVERJENSKY, D. A., SHOCK, E. L. and HELGESON, H. C. (1997) Prediction of the thermodynamic properties of aqueous metal complexes to 1000 °C and 5 kb. *Geochim. Cosmochim. Acta*, **61**, 1359–1412.

THURMAN, E. M. (1985) *Organic Geochemistry of Natural Waters* (Dordrecht: Martinus Nijhoff/ Dr W. Junk).

WATANABE, Y. NARAOKA, H., WRONKIEWICZ, D. J., CONDIE, K. C. and OHMOTO, H. (1997) Carbon, nitrogen, and sulfur geochemistry of Archaean and Proterozoic shales from the Kaapvaal Craton, South Africa. *Geochim. Cosmochim. Acta*, **61**, 3441–3459.

WOESE, C. R., KANDLER, O. and WHEELIS, M. L. (1990) Towards a natural system of organisms: proposal for the domains Archaea, Bacteria, and Eukarya. *Proc. Natl Acad. Sci. USA*, **87**, 4576–4579.

WOLERY, T. J. (1992) *EQ3NR, A computer program for geochemical aqueous speciation-solubility calculations: Theoretical manual, user's guide, and related documentation (version 7.0)*, UCRL-MA-110662-PT-III (Livermore, CA: Lawrence Livermore National Laboratory).

WOLERY, T. J. and DAVELER, S. A. (1992) *EQ6, A computer program for reaction path modeling of aqueous geochemical systems: Theoretical manual, user's guide, and related documentation (version 7.0)*, UCRL-MA-110662-PT-IV. (Livermore, CA: Lawrence Livermore National Laboratory).

6

The Emergence of Life from FeS Bubbles at Alkaline Hot Springs in an Acid Ocean

MICHAEL J. RUSSELL, DAN E. DAIA AND ALLAN J. HALL

Department of Geology and Applied Geology, University of Glasgow, Scotland, UK

6.1 Introduction

Corliss *et al.* (1981) and Barross and Hoffman (1985) have argued cogently that hot spring sites on the deep sea floor provide the only stable environment in which life could have emerged and become established. Woese *et al.* (1990) and Lake (1988) have independently proposed the universal ancestor to have been a hyperthermophile. Schwartzman *et al.* (1993) have also emphasised that life could not have had anything but hot beginnings because of the temperature of the early earth. And Shock (1996) has established a theoretical window for the viability of a geochemical metabolism on the mixing of relatively oxidised cool sea water with reduced hydrothermal solutions, between about 80 °C and 200 °C. In particular, Shock demonstrates the hypothetical potential of hydrothermal solutions to generate organic polymers from dissolved carbon dioxide on mixing with sea water. The organic metastabilities depend crucially on the initial redox state of the hydrothermal solution. If the hydrothermal solution were in equilibrium with the quartz–fayalite–magnetite (SiO_2+Fe_2SiO_4+Fe_3O_4, or QFM) buffer, as in the crust, then organic molecules would be generated. But if the solution equilibrated with the more oxidised pyrite–pyrrhotite–magnetite (FeS_2+FeS+Fe_3O_4, or PPM) buffer, which happens to approximate the redox state on the outside of the membrane (Russell *et al.*, 1994), then carbonate would be stable. Thus the conditions envisaged by Shock (1996), of a geochemical metabolism in waiting, are realisable at certain putative hot seepage or spring sites.

Our aim is to go some way towards bridging the gap between what is known or surmised about the chemical and physical states of an energetic early earth, and what look to be the minimum requirements for the first hyper- or even super-thermophilic metabolist. In attempting this we draw on the aforesaid evidence and that gleaned from the recent investigations of the ferredoxins carried out by Adams (1992), Qiu *et al.* (1994), Johnson (1994), Volbeda *et al.* (1995), Daniel and Danson (1995); Cammack (1996) and Kletzin and Adams (1996): that is, we look at the processes involved in the emergence of life. But the gap is dauntingly wide. One of the most taxing questions to answer is how the large activation energies required for the reduction of carbon oxides in aqueous conditions were bypassed. Apart from the important recent experiments of Huber and Wächtershäuser (1997) and Heinen and Lauwers (1996), demonstrations of carbon fixation in such conditions have not been encouraging. Thus in attempting to bridge this gap we also re-examine the chemical literature, relying substantially on Behr (1988), with a view to guiding further experiments on the abiotic production of organic molecules in the conditions analogous to this workshop's central assumption.

6.2 Scale and approach

Wächtershäuser's (1988, 1990a, 1992) 'retrodictive' approach to seeking out how life might have originated is reminiscent of the 'onion heuristic' of Hartman (1975), peeling back layer after layer of life's processes until one supposedly finds the oldest, and central, process. As Grannick (1957) would have it, 'biosynthesis recapitulates biopoiesis'. The approach is also comparable to Hunter's (1792, published in 1859) 'uniformitarian' concept generally applied to understanding earth history, formulated in the principle – 'the present is the key to the past'. These similar methods of detection have been extraordinarily successful in establishing geological history and mapping evolution. However, they extrapolate to fundamentally conservative and rather unremarkable scenarios of a relatively 'unpunctuated' past. Of course, these methods are safe but they do not allow us to appreciate evolutionary jumps to higher levels of complexity (cf. Schwartzman *et al.*, 1994). And they fall short of exposing absolute beginnings. The tendency is for a reductionist and restrictive assumption to hold sway. For example, the scale of the immediately preliving entities is presumed to be nanoscopic. This leads to the entertainment of limited and contradictory notions such as the RNA world and the iron–sulfur world (see Orgel and Crick, 1993; Wächtershäuser, 1992). Reasons why such ideas are attractive to chemists, biochemists and microbiologists are that they employ scales and approaches with which they are familiar. But life presumably emerged in response to physical and chemical energy potentials across scales megascopic through to microscopic, and involved the complex interactions of large numbers of chemical species in self-organising autocatalytic cycles.

We prefer a 'bottom up' approach: to start from a consideration of conditions when the earth was young, when thermal and chemical energy potentials were high. Working our way from the Hadean through to the Archaean Eon we attempt, by interpolation from likely geochemical conditions through to what is known of the metabolic processes in the thermophiles, to consider the nature of the catalysts likely to have been involved in protometabolic steps, pathways and cycles. Of course there lies the danger of mythologising creation, so we add that other technique of the detective, the reconstruction. Let us investigate what it must have been like on the early earth; what were the potential tensions and through which agencies could the extraordinary thermal and chemical energies available be spent?

During and following the early formation of the earth's core, huge quantities of thermal energy were continually generated frictionally by gravitation. Apart from gravitational differentiation, heat was also generated by bolide impacts and lunar tides as well as radioactive decay. Very early in earth history there was probably a magma ocean at ~1300 °C (Ballhaus and Ellis, 1996). Even by 4.2 Ga, heat flow was at least five times that obtaining today (Patchett, 1996). The earth was more reduced than now and far from equilibrium with its super-hot secondary atmosphere or 'volatisphere' (comprised essentially of water vapour, carbon oxides and nitrogen). The carbon oxides escaped from a hot mantle where they were high-temperature equilibrium phases, but because of a cross-over of the iron and carbon redox boundaries at ~250 °C, then atmospheric carbon oxides found themselves out of equilibrium (though metastable) with respect to the mantle and crust in latest Hadean times (Shock, 1992; and see Hall, 1986). (Even so it was not as far from chemical equilibrium as today's atmosphere with its load of biogenic dioxygen.)

Because of the low conductivity of silicates, liquids and gases, most of the earth's heat must have been initially dissipated by mass transfer; by convection in the mantle and in the 'volatisphere', and eventually by convection of the newly condensed and subsequently cooled ocean water in the fracture-permeable crust. And because of the focusing of chemical energy by convection, then a protometabolism (and subsequently life) would have been coupled with this physical process, quickening the pace of chemical equilibration. At this simple level, chemolithotrophy can be viewed as an 'attractor' (cf. Schwartzman *et al.*, 1994), simply in the terms

$$3\ Fe_2SiO_4 + H_2O + CO_2 \longrightarrow 2\ Fe_3O_4 + 3\ SiO_2 + CH_2O$$

where Fe_2SiO_4 is fayalite (olivine), a significant constituent of oceanic crust, and Fe_3O_4 is magnetite, a mineral in which two-thirds of the iron is in the ferric state, found in hydrated and oxidised portions of the oceanic crust.

The emergence of life depends not only on chemical interactions but also, as we imply, on the physics of the large-scale kinetic system described above, i.e. the convection cell, which could focus chemical energy at points on the ocean floor. But hydrothermal convection had to await radiative cooling of the Hadean ocean. Under the high pressure of the volatisphere, the first proper ocean would have condensed at ~374 °C, the critical point of water. Nevertheless, the development of a steep temperature gradient would have been required before hydrothermal convection could occur within the crust, i.e. a dense downdraft fed from a relatively cool ocean would have been necessary to help drive the convective process. Magmatic intrusions would have driven the first high-enthalpy hydrothermal convection cells, and black-smoker-like exhalations therefrom would have carried Fe, Ni, Co, Zn and other metal ions (cf. Von Damm, 1990; Russell and Hall, 1997) to this newly condensed ocean, made acid by high atmospheric pressures of carbon oxides and hydrogen chloride (Lafon and Mackenzie, 1974; Walker, 1990).

Notwithstanding the affinity that hyperthermophiles have for black smokers today, arguing from differing premises both Shock (1992) and ourselves (Russell *et al.*, 1988) have concluded that life was more likely to have emerged from lower-temperature (medium-enthalpy) ocean floor springs and seepages away from magmatic intrusions. The development of this medium-enthalpy type of open convective system must have awaited still further cooling of the ocean, probably to about 100 °C, to allow a thermal contrast of 100–150 °C (see Section 6.4.1).

6.3 Model

The medium-enthalpy hydrothermal convection systems operating in the upper crust coincidentally transferred chemical, along with thermal, energy to the Hadean ocean. The earliest protometabolism is conceived as comprising three processes (Figure 6.1). The first took place during the modification of sea water as it convected at 200–250 °C through the Hadean oceanic crust. We consider this to have been concomitant with a protoanabolic pathway. During convection a proportion of CO, CO_2, bicarbonate, formate and a still smaller proportion of the water, were reduced to organic acids, alcohols and aldehydes of low molecular mass (Shock, 1992, 1996). Other 'electron rich' entities such as H_2, OH^-, SH^-, NH_4^+ and possibly CN^-, along with bisulfide or carbonyl complexes of trace elements such as W, Zn and Co (Barnes and Czamanske, 1967; Coveney *et al.*, 1987; Ferris,

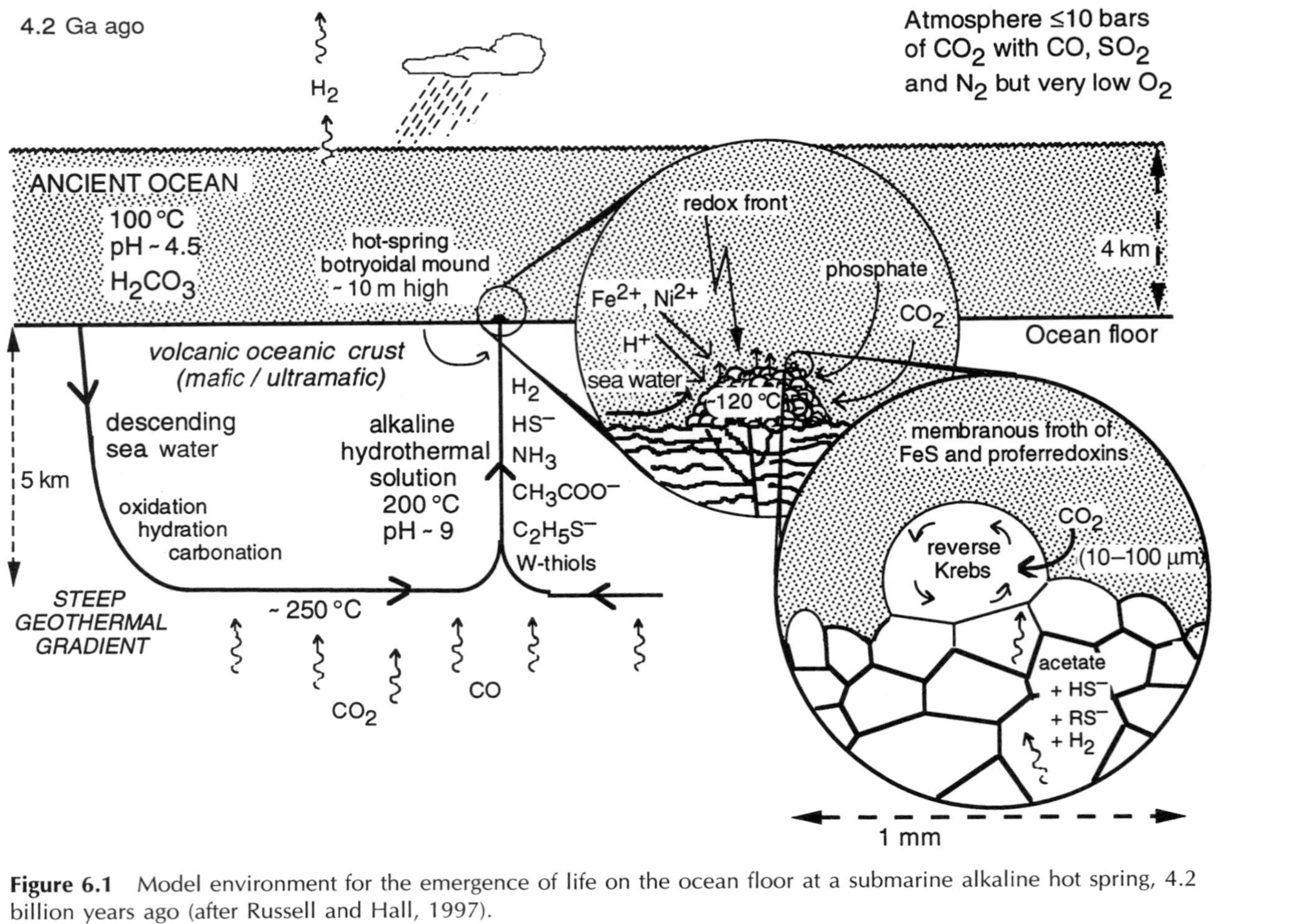

Figure 6.1 Model environment for the emergence of life on the ocean floor at a submarine alkaline hot spring, 4.2 billion years ago (after Russell and Hall, 1997).

1992; Shock, 1992; Macleod *et al.*, 1994; Seward and Barnes, 1997) would also have been present in the fluid and, in the company of the abiotic organic molecules, would have been borne towards the exhalative site. Note that iron is insoluble in these conditions (Crerar *et al.*, 1978).

The second process was also protoanabolic, one which happened on the mixing of these highly reduced hydrothermal solutions with the relatively cool, oxidised carbonic ocean. Such mixing is envisaged as having generated a variety of longer-chained organic molecules (the theoretical calculations of Shock and Schulte, 1995; Shock, 1996), thiols, amines (catalysed by $MgCl_2$ – Kimoto and Fujinaga, 1990; Cole *et al.*, 1994; or an FeS slurry – Heinen and Lauwers, 1996), acetic acid and activated methyl thioester (catalysed by (Fe,Ni)S; Huber and Wächtershäuser, 1997), and possibly amino acids (catalysed by FeS_2+FeS+Fe_3O_4 or illite; Hennet *et al.*, 1992). Of the experiments, those of Huber and Wächtershäuser (1997) are particularly relevant. These authors demonstrate that, assuming a natural source of CO and methane thiol (which is provided by our first process; and see Heinen and Lauwers, 1996), a relatively high yield of acetic acid is generated in a narrow pH range around 6.5, catalysed by a fresh FeS/NiS precipitate (made in a not dissimilar way to our membranes (Russell *et al.*, 1994; Russell and Hall, 1997)). Such a pH would be met within our putative (Fe,Ni,Co)S membrane, which separated the acid ocean from the alkaline hydrothermal solution. We assume then that all the above molecules were synthesised either within the hydrothermal cell at ~200 °C, or between 100° and 150 °C within, or on the surface of, the membranes comprising the iron monosulfide bubbles developed at the exhalative centres (Russell *et al.*, 1989). The bubbles acted as the catalytic culture chambers within which the first 'organic soup' was generated and constrained.

The third process was protocatabolic, and began as the freshly generated organic molecules were oxidised at low redox potentials by tungsten(VI) in the membrane.

We picture the FeS bubbles inflating at the interface between the HS^--bearing hydrothermal solution and hot acid Hadean sea water containing Fe, Ni and Co. These metals were contributed to the ocean by the very high-temperature black-smoker-like springs at oceanic spreading centres and/or over the top of mantle plumes. The meteoritic flux would have maintained high tenors of dissolved iron and other transition metals in the ocean. Some of the ferrous iron was photooxidised to aqueous ferric iron complexes at the ocean surface (Cairns-Smith, 1978; Braterman *et al.*, 1984):

$$Fe^{II} + H^+ + h\nu \longrightarrow Fe^{III} + \tfrac{1}{2} H_2$$

$$Fe^{III} + 2\,H_2O \longrightarrow FeOOH + 3\,H^+$$

But most of the ferrous iron would have reacted with the bisulfide delivered to the sea floor by the medium-temperature mineralising solutions. The gelling and nanocrystallisation of the colloidal FeS caused the pressurised membrane to age and split. This allowed the mineralising solution to escape to the sea where a new, flexible membrane grew immediately from the 'parent' at the solution–sea water interface to generate an iron monosulfide daughter (Figure 6.1). Such iron monosulfide chambers can be produced in the laboratory, for example by injecting a 0.25 mol/l solution of HS^- into a 0.025 mol/l concentration of ferrous iron at STP (Figure 6.2). Similar membranes are generated at a variety of molarities of each reactant down to 20 mmol/l and at temperatures up to 90 °C (Russell and Hall, 1997). Our X-ray diffraction analysis of the membrane reveals small, broad peaks consistent with nanocrystalline mackinawite ($Fe_{(1+x)}S$). As we will see in Section 6.8.3, mackinawite can contain up to 20% nickel (or cobalt). Judging from our own

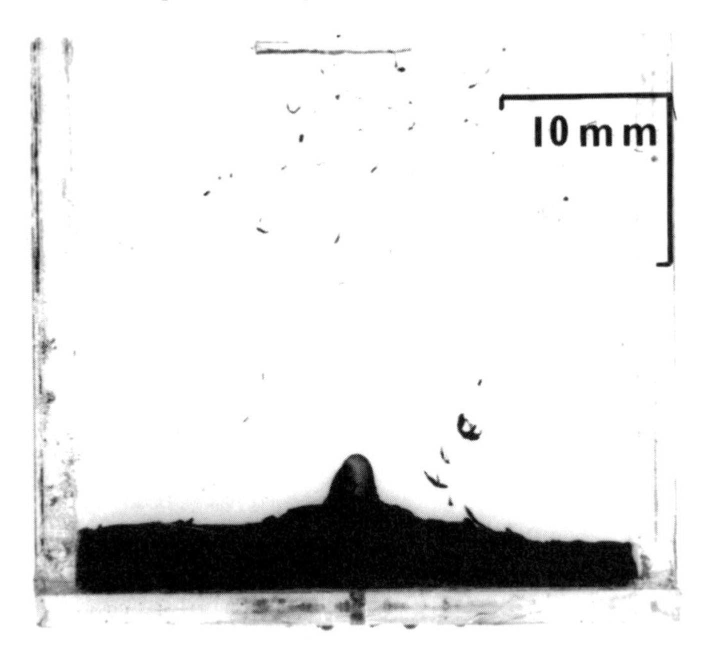

Figure 6.2 Photograph of a FeS membrane generated in the laboratory at STP by injecting 250 mmol/l HS^- (representing the alkaline hydrothermal solution) through the base of a visijar into 25 mmol/l Fe^{2+} (representing the Hadean ocean).

experiments, it may be that Huber and Wächtershäuser's (1997) catalytic FeS precipitate not only consists of mackinawite but, where nickel and iron were added in equal parts to sodium sulfide, also violarite, $FeNi_2S_4$ (Section 6.8.4, cf. greigite Fe_3S_4) (Vaughan and Craig, 1985).

Although it was the pressure of the hydrothermal solution that led to the continued inflation/rupturing and new development of iron monosulfide bubbles to produce a sulfide mound, we suggest (Russell *et al.*, 1994) that, in time, osmotic pressure took over from the hydrodynamic pressure at the submarine seepage site. The osmotic pressure was induced by the catabolic process which increased the effective solute concentrations. Eventually, protometalloenzymes and organic polymers would have displaced the mackinawite comprising the membrane (so preventing the FeS from growing to crystalline proportions and/or being sulfidised) as explained in Section 6.9. Then, the three processes were concerted in this complex of, by now, essentially organic bubbles, so engendering the first metabolising cooperative.

6.4 Clues from hydrothermal orebodies

To gain some insight into convective processes developing away from magmatic intrusion in the Hadean, we examine orebodies known to have been generated in broadly comparable, though more recent, circumstances. Giant metalliferous orebodies of this type are usually coeval in any one mineral province, and can contain as much as 10^{13} g of base metal as sulfide (Russell and Skauli, 1991). They are known as the sedimentary-exhalative (or SEDEX) type of mineral deposit, though it is acknowledged that much of the ore is developed just beneath the sea floor (Russell, 1996; Anderson *et al.*, 1998). At least 10^{17} g of sea water must course through the crust in a million years or so to account for the accumulation of such masses, assuming solubilities of between 10 and 100 ppm for the base metals. We presume somewhat comparable bodies formed on the stressed ocean floor in the Hadean. However, there were important differences. In the Hadean the mineralising solutions interacted with a basaltic or komatiitic, rather than a felsic, crust. Thus the metalliferous bodies were fed from alkaline rather than acid solutions and would have lacked lead and iron but would have carried up to 20 mmol of sulfur, mainly as HS^-, to the ocean bottom (Macleod *et al.*, 1994).

The giant SEDEX deposits are discrete, standing at least 20 km from their nearest neighbours (Russell, 1978, 1988). They are generated in strained crust and may be broadly associated with some basaltic volcanism, though the volcanics provide neither the metal nor much of the heat to the hydrothermal system. Given that generation of SEDEX deposits requires such huge masses of hydrothermal fluid at 200–250 °C (even now there are only ~1.4×10^{24} g of water on the entire planet) and that their lead and sulfur isotope ratios are strongly suggestive of basement involvement, their genesis must involve free thermal convection of sea water circulating 5 km or more into the crust in the form of cells about 20 km in diameter (Samson and Russell, 1987; Mills *et al.*, 1987; Boyce *et al.*, 1993; Banks and Russell, 1993; Russell and Skauli, 1991).

It is the pressure imposed by the supernatent sea water that causes an opening of stressed, uneven fractured surfaces, creating space for further fluid invasion and thereby triggering a runaway feedback (Pine and Batchelor, 1984). We imagine that as water migrates down and away from stress fractures in the upper crust it reaches rock hot enough to respond to stress in a plastic manner so that the space between surfaces can no longer be maintained. This ‘cracking front’ controls the ‘floor’ of the hydrodynamic system (Lister, 1975; Russell *et al.*, 1995). Exothermic hydration of the crust also produces further heat (Fyfe, 1974; Haack and Zimmermann, 1996). The floor to the convection cell deepens with time as heat is ‘mined’ from the crust. The system can operate for at least a million years. Strong support for this idea comes from studies of the 9.1 km deep drill hole in southeastern Germany. Here water under normal pressures is found to occupy fractures throughout its length, showing that it is connected through to the surface (Zoback, 1995). Drilling ceased at 9.1 km, where the temperature was 270 °C and the rock was too plastic to stay open, a finding predicted by this hypothesis (Russell *et al.*, 1995).

6.4.1 Temperature of first metabolising systems

Schwartzman *et al.* (1993) have argued that particular major evolutionary developments could only take place as surface temperatures dropped sufficiently to allow the emergence of new kingdoms or phyla, presumably with more complex biochemistries (and see Hoyle, 1972). This argument can be applied to the emergence of life itself. Empirical evidence suggests that certain organisms now can operate at up to about 170 °C (Cragg and Parks, 1994) and perhaps beyond (Deming and Baross, 1992). Everett Shock shows that organic molecules are stable in water at low redox conditions close to the quartz–fayalite–magnetite buffer with respect to carbonate below about 200 °C (Shock, 1996). Under still lower redox conditions, i.e. where the mineral awaruite (Ni_3Fe), common in ultramafic rocks, is stable (Krishnarao, 1964), the upper temperature for reduction could be around 250 °C (Shock, 1992). Therefore the ‘onset’ of a chemical metabolism had to await the development of hydrothermal convection cells of the type discussed above that operated between 200 °C and 250 °C. For steady-state convection to operate at such a temperature requires that the Hadean ocean water replenishing the inflowing limbs be substantially denser and therefore cooler. Just how much cooler might be estimated from the Raleigh equation as applied to fluid flow in porous media:

$$R = \frac{K \alpha g H \Delta T}{K_m \nu}$$

where R is the Raleigh number (which must exceed 30 if convection is to set in); K is permeability; α is the coefficient of cubic expansion; g is the acceleration due to gravity; H is the thickness of the permeable layer; ΔT is the temperature contrast between the overlying freestanding ocean and the hydrothermal fluid, K_m is the thermal diffusivity of the water-saturated medium; and ν is the kinematic viscosity of the fluid. Of these terms it is (fracture) permeability that is the significant variable in these circumstances (Lister, 1975), fluctuating over orders of magnitude. The action of the deviatoric stress constantly reopens fractures which have been narrowed by the hydration and pressure solution of minerals constituting the crust (Strens *et al.*, 1987). Accordingly, we have to make many assumptions regarding the values inserted into the equation, the chief being fracture width and spacing as well as the heat flow in crust at a distance from oceanic spreading centres, a flow augmented by exothermic hydration and oxidation reactions (Fyfe, 1974; Haack and Zimmermann, 1996). Given a strong horizontal differential stress in the upper crust, brought about by lithospheric motions driven by mantle convection, and supposing the permeability to be sufficient (i.e. $\sim 10^{-16}$ m^2) then we may assume a ΔT of about 100 °C (Davis *et al.*, 1980). As medium-enthalpy convection operates in the crust at 200–250 °C (e.g. Samson and Russell, 1987; Banks and Russell, 1992), we speculate that the ocean had to have cooled to around, or just above, 100 °C for steady hydrothermal convection systems of this kind to ensue. Currents of ocean water at this temperature would have cooled the sulfide mound precipitated over the seepage site to between about 120 °C and 150 °C, encompassing the first point for Schwartzman's curve (Schwartzman *et al.*, 1993). Thus we surmise that metabolism would have 'kicked in' somewhere within this temperature range. We now turn to the chemical tensions likely to have sparked the geochemical stage of the metabolic process.

6.5 Clues from metabolism

6.5.1 *Shock's geochemical metabolism*

The suggestion that life emerged at hot springs requires the reduction of the carbon oxides (Corliss *et al.*, 1981; Hall, 1986; Ferris, 1992; Wächtershäuser, 1992; Shock, 1992, 1996). In calculating the metastabilities of a variety of organic compounds, Shock (1992, 1996) has demonstrated the potential for synthesis of carboxylates, ketones and alcohols as inorganic CO_2 and carbonate in a sea water mix with hydrothermal solution. As we have seen, the temperature window for organic synthesis during mixing lies between 50 °C and 250 °C (Shock, 1996). In Shock's calculations the fugacity of the hydrothermal solution is buffered by the quartz–magnetite–fayalite (QFM) mineral suite. The fugacity of carbon dioxide is taken as 10 bars, the presumed atmospheric pressure in the Hadean (Walker, 1985; Kasting *et al.*, 1993), a conservative value since mass balance calculations allow a pressure of more than 50 bars (Walker, 1985). Shock (1996) also demonstrates theoretically that within the upper half of this temperature range ketones, alcohols and the shorter-chained carboxylic acids would be synthesised, whereas in lower half of the temperature range long-chained carboxylic acids are favoured.

Shock (1996) shows that these various organic molecules could be generated when a strongly reduced hydrothermal fluid mixes with carbonic sea water by entrainment immediately beneath the medium-enthalpy hot spring, as well as in the mixing zone at the spring site itself (Figure 6.1). The fact that above about 150 °C the synthesis of small organic molecules is favoured, whereas below this temperature the longer-chained organic molecules feature, is consistent with the idea that hydrothermal fluids can provide organic 'primers' to a system which uses these to fix further carbon and hydrogen as well as other available radicals. We surmise that ethane thiolate ($CH_3CH_2S^-$), likely to have been a constituent of the sulfide-rich fluids (de Duve, 1991; Kaschke *et al.*, 1994; Cole *et al.*, 1994; Heinen and Lauwers, 1996), acted as one of the primers. As a temperature of 120–150 °C has been argued for the first metabolist on physical and empirical grounds, there is a seeming convergence with geochemical theory (Section 6.4.1). Once metabolists had an information system guaranteeing some of their offspring independence from a simple geochemical metabolism, it was open to certain living entities to evolve and exploit energies available at somewhat higher (e.g. Cragg and Parkes, 1994), as well as lower temperatures (cf. Schwartzman *et al.*, 1993).

6.6 Clues from hydrothermal geochemistry

As there are no remaining ore deposits related to hydrothermal systems operating in the oceanic crust away from ocean floor spreading centres, and as ocean water chemistry has changed significantly from Hadean times, we have to surmise the likely composition of such ore-bearing fluids. Numerical modelling and laboratory experiments with sulfide-bearing ultramafic rocks indicate that a reduced alkaline solution will dissolve up to ~750 ppm or ~23 mmol/l of sulfide as HS^- at 250 °C (Macleod *et al.*, 1994). Organic thiolates (RS^-) may also be generated (Kaschke *et al.*, 1994; Heinen and Lauwers, 1996).

Barnes and his coworkers (e.g. Barnes and Czamanske, 1967; Seward and Barnes, 1997) have made extensive studies of the solubilities of the sulfhydryl metal complexes soluble in alkaline hydrothermal solutions. The alkaline hydrothermal fluid we envisage as the fount of protometabolism would have contained not only sulfur and a proportion of the metallic elements (Na, K, Mg, Zn, W) required by life, but also most of what we tend to think of as the poisonous, rarer metals of high atomic number. However, most of these 'transition' elements are only poisonous to aerobic life: they are generally benign in anaerobes where the mid-point redox potentials would have exceeded the *E*h range obtaining in the Hadean ocean, or they are insoluble near the hydrogen potential at neutral pH. Mercury, soluble as the bisulfide (Barnes and Czamanske, 1967), is a good example. The genes that code for complexes of this element are of ancient lineage, and protect aerobes by synthesising enzymes that reduce Hg^{II} to the metal (Osborn *et al.*, 1995). In this state it cannot invade the cell and replace the 'vulnerable' metals in the active sites of enzymes.

One transition element that has a particularly low mid-point redox potential, when complexed with sulfide, is tungsten. Because it can act as a redox switch from W^{IV} to W^{VI} and back near the hydrogen potential, it appears to be required by all hyperthermophiles (Kletzin and Adams, 1996). Tungsten is frequently enriched close to gold ore in ultramafic terrains of Archaean age (Mueller, 1991; Phillips *et al.*, 1984). From theory and experiment, such gold is assumed to be transported as a sulfide complex (e.g. $Au(HS)_2^-$) in alkaline solutions (Seward, 1973; Boyle *et al.*, 1975; Phillips and Groves, 1983; Seward and Barnes, 1997). The ultramafic rocks hosting such gold deposits have a chemistry similar to the likely composition of the Hadean ocean floor (Macleod *et al.*, 1994). Thus it is probable that the off-ridge alkaline hydrothermal solutions had the capacity to carry tungsten as well as gold.

Further support for this idea comes from the discovery of tungsten in the rare form of the sulfide tungstenite (WS_2) in ultramafic rocks in Tamvatnei in northeasternmost Siberia (Voevodin *et al.*, 1979). In this recent ore deposit it comprises the main tungsten ore mineral and is associated with cryptocrystalline quartz, carbonate, graphite and the metals mercury (as cinnabar), arsenic and antimony as well as iron sulfides. Nekrasov and Konyushok (1982) have argued that it was transported as sulfide complexes, probably as $W^{VI}S_4^{2-}$ or $W^{IV}S_3^{2-}$. Multianion complexes such as $WO_{4-x}S_x^{2-}$ are also possible (Seward and Barnes, 1997).

6.7 Overcoming the kinetic barrier to carbon dioxide reduction

Although Everett Shock (1996) and ourselves (Russell and Hall, 1997) converge on a model for the emergence of life consistent with the geological and geochemical conditions likely to have obtained on the early earth, experiments undertaken in the laboratory to fix carbon in comparable conditions have generally resulted in rather low yields (Heinen and Lauwers, 1996). The one notable exception is the experimental generation of acetic acid by Huber and Wächtershäuser (1997), although these authors used CO rather than the more refractory CO_2. We have discussed in Section 6.3 how their experiment could be viewed within the context of an FeS membrane. In this section we consider the difficulties involved in the abiotic fixation of carbon (particularly in the reduction of CO_2), that we might design better, or at least appropriate, experiments to further test the hypothesis. (Of course, as Wächtershäuser (1990b) implies, we would expect reduction of CO_2 to be difficult, otherwise there would be no call for life.)

As with the other nonmetals from the second period, carbon and oxygen have a tendency to form covalent bonds using their outer shell *s* and *p* orbitals to form hybridised *sp*, sp^2 or sp^3 orbitals. In contrast with similar elements from the higher periods, they easily form $\pi(pp)$ double bonds. These features are apparent in the two oxides of carbon, CO and CO_2. Both contain an *sp* hybridised carbon atom that in carbon dioxide determines the linear shape of the molecule, leading to a relatively simple IR spectrum (ν_3 (asymmetric stretch), at 2349 cm^{-1} and ν_2 (C—O deformation) at 667 cm^{-1}) and an overall nonpolar character ($\mu = 0$ D) (Creutz, 1993; Lide and Frederikse, 1993–1994; Nenitescu, 1984). The molecule displays a significant degree of resonance stabilisation (–214.5 kJ/mol relative to two C═O bonds as in formaldehyde), illustrated by the following canonical structures:

$$^{-}{:}\ddot{\underset{..}{O}}{-}C{\equiv}O{:}^{+} \longleftrightarrow \ddot{O}{=}C{=}\ddot{O} \longleftrightarrow {}^{+}{:}O{\equiv}C{-}\ddot{\underset{..}{O}}{:}^{-}$$

Carbon dioxide is the highest oxidation product of carbon and a very stable molecule at ordinary temperatures. Its C═O bond strength (D°_{298} = 532.2 kJ/mol) is second only to the CO bond strength in carbon monoxide (D°_{298} = 1076.5 kJ/mol) (Lide and Frederikse, 1993–1994). From the perspective of the molecular orbital theory, the energy of the LUMO (lowest unoccupied molecular orbital) has been estimated at ~3.8 eV, indicating a high electron affinity associated with the carbon atom (Keene, 1993). Berdnikov's calculations, which were made on the assumption that the entropy and heat of hydration of $CO_2^-\cdot$ are similar to those of NO_2^-, have produced a value of the electron affinity,

Table 6.1 Standard enthalpies ($\Delta H°$), free energies ($\Delta G°$), and redox potentials ($\Delta E°$) for CO_2 reductions at 25 °C and 1 bar (Lide and Frederikse, 1993–1994). Aqueous media values corrected for pH 7 (O'Connell *et al.*, 1987)

Reaction	$\Delta H°$ (kJ/mol)	$\Delta G°$ (kJ/mol)	$E°$(V) (vs NHE)
$CO_2(g) + H_2(g) \xrightarrow{2e^-} HCOOH(l)$	−31.2	+33.0	
$CO_2(g) + H^+ \xrightarrow{2e^-} HCOO^-(aq)$			−0.41
$CO_2(g) + 2\,H_2(g) \xrightarrow{4e^-} HCOH(l) + H_2O(l)$	−0.9	+54.8	
$CO_2(g) + 4\,H^+ \xrightarrow{4e^-} HCOH(aq) + H_2O(l)$			−0.48
$CO_2(g) + 3\,H_2 \xrightarrow{6e^-} CH_3OH(l) + H_2O(l)$	−131.4	−9.3	
$CO_2(g) + 6\,H^+ \xrightarrow{6e^-} CH_3OH(aq) + H_2O(l)$			−0.38
$CO_2(g) + 4\,H_2(g) \longrightarrow CH_4(g) + 2\,H_2O(l)$	−252.5	−130.1	
$2\,CO_2(g) + 2e^- \longrightarrow C_2O_4^{2-}(aq)$			−0.90
$2\,CO_2(g) + 4\,H_2 \longrightarrow CH_3COOH(l) + 2\,H_2O(l)$	−269.1	−75.3	
$3\,CO_2(g) + 7\,H_2 \longrightarrow C_2H_5COOH(l) + 4\,H_2O(l)$	−473.4	−183.21	
$4\,CO_2(g) + 10\,H_2 \longrightarrow C_3H_7COOH(l) + 6\,H_2O(l)$	−674.6	−269.12	

$I = -0.5$ eV (Berdnikov, 1975). The first ionisation potential is high (13.7 eV/molecule, comparable to that of the hydrogen atom) indicating the pronounced electrophilicity of the central carbon atom (Keene, 1993). This is expressed in the capacity of nucleophiles such as H_2O, NH_3, HO^-, R^- to attack CO_2.

6.7.1 *Thermodynamic aspects of CO_2 reactivity*

Several standard enthalpies ($\Delta H°$), free energies ($\Delta G°$) and redox potentials ($\Delta E°$) for CO_2 reductions, leading to various products, are listed in Table 6.1 (values at 25 °C and 1 bar calculated from Lide and Frederikse (1993–1994), aqueous media values corrected for pH 7 from O'Connell *et al.* (1987)). It is immediately apparent that CO_2 reduction is not spontaneous for C_1 and C_2 products less reduced than methanol, methane or acetic acid. The formation of one or more equivalents of water tips the balance towards reduction and it favours the spontaneous production of compounds with a longer carbon chain.

Standard values are of limited utility in the study of complex geochemical systems such as those relevant to the emergence of life. In their critique of the use of standard Gibbs free energies in the chemical and biochemical literature, Shock *et al.* (1995) emphasised the special features of hydrothermal systems, in which the activities of various reactants are governed by geochemical constraints. These must be taken into account when calculating the overall Gibbs free energy available to the system and thus ascertaining the thermodynamic tendency towards CO_2 reduction and the identity of the most stable organic products (Shock *et al.*, 1995).

6.7.2 *Kinetic aspects of CO_2 reactivity*

Carbon dioxide is far from being an inert compound. It reacts with a variety of chemicals, and this reactivity is employed in industry as well as in the laboratory. The bulk of CO_2 consumed in industry as a chemical reagent is transformed into urea (reaction 2, Figure 6.3).

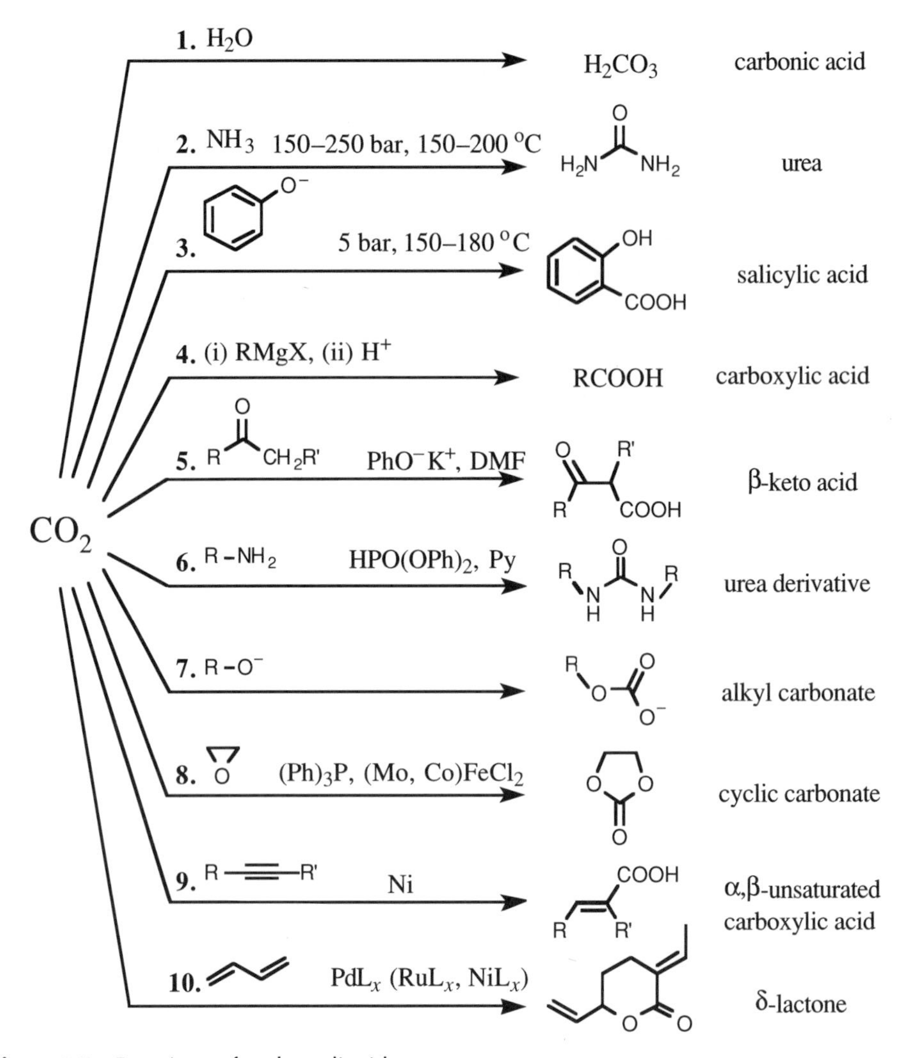

Figure 6.3 Reactions of carbon dioxide.

The familiar reaction of CO_2 with water (reaction 1, Figure 6.3) to produce carbonic acid (or dihydrogen carbonate) is an integral part of the transformations that govern the major chemical cycles on our planet. It reveals the acid anhydride character of CO_2. Carbon dioxide is also an important reagent in organic synthesis, in such reactions as the carboxylation of phenoxides (the Kolbe–Schmidt reaction; reaction 3, Figure 6.3), the synthesis of carboxylic acids from organometallic compounds (reaction 4, Figure 6.3), in reaction with active hydrogen compounds (reaction 5, Figure 6.3), the carboxylation of phosphorus ylides, the synthesis of urea derivatives from amines (reaction 6, Figure 6.3), and the preparation of carbonates from alkoxide ions (reaction 7, Figure 6.3) or from epoxides (reaction 8, Figure 6.3). It can react with alkynes (reaction 9, Figure 6.3) and several other unsaturated compounds (reaction 10, Figure 6.3), often in the presence of transition metal catalysts (Behr, 1988; March, 1992; Xiaoding and Moulijn, 1996). Some of these reactions are of great importance, salicylic acid (used for pharmaceuticals), and ethyl carbonate (used as a solvent and reaction intermediate) being high tonnage industrial chemicals (Behr, 1988). From a mechanistic point of view it is apparent that in most organic reactions CO_2 requires nucleophilic attack by a carbanion or a double bond, sometimes activated by a transition metal catalyst.

Despite the relatively high electron affinity of CO_2, electrochemical studies have shown the cathodic reduction to be characterised by large overpotentials, indicating a significant kinetic barrier to the reaction:

$$CO_2 + e^- \longrightarrow CO_2^{-\cdot}$$

Thus the standard potential for the $CO_2/CO_2^{-}\cdot$ couple in dry dimethyl formamide (DMF) on a dropping mercury electrode was determined as −2.21 V vs SCE, and the most positive potentials reported, on ruthenium electrodes, were in the range −0.55 to −0.65 V vs SCE (O'Connell *et al.*, 1987). On accepting an electron, CO_2 changes shape from linear to bent in the $CO_2^{-}\cdot$ radical anion (Keene, 1993). Similarly, in complexes of CO_2 with transition metals, the molecule always takes a bent shape and no examples of complexes containing linear CO_2 ligands have yet been found (Creutz, 1993). As mentioned above, the linear structure is very stable and its transformation into an angular structure involves a change in hybridisation at carbon, from *sp* to sp^2, requiring energy of activation in the process. Even water, which attacks CO_2 to form carbonate, does so relatively slowly ($k = 6.7 \times 10^{-4}$ mol^{-1} l s^{-1} at 25 °C), a much stronger nucleophile, like OH^-, being necessary to ensure high rates of reaction ($k = 8.5 \times 10^3$ mol^{-1} l s^{-1} at 25 °C). Since most living organisms function at an internal pH close to 7, OH^- cannot play a significant role. Life has developed the zinc-containing enzyme carbonic anhydrase, one of the most efficient biological catalysts known, achieving an increase in the CO_2 hydrolysis second-order rate constant up to the value of $k = 7.5 \times 10^7$ mol^{-1} l s^{-1} at 25 °C (Allen, 1988).

The kinetic barrier is even greater for the direct reactions of CO_2 with hydrogen, the reduction requiring high temperatures (800–1000 °C for the uncatalysed water-gas shift reaction (Behr, 1988; O'Connell *et al.*, 1987)), a concentrated form of energy, or a suitable catalyst. The activation energy can be delivered electrochemically, as in the numerous experiments of cathodic CO_2 reduction. It can also be attained photochemically, or even with the help of high-energy ionising radiation. Electrochemical CO_2 reduction experiments suffer from several disadvantages at the very low reduction potentials frequently used in these experiments, such as the competition between the CO_2 reduction and proton reduction with H_2 evolution. Most often the products are formic acid or formate (in aqueous solutions), oxalate and carbon monoxide (at low water activity), and methanol, and methane (in low yields on ruthenium, molybdenum or copper electrodes) (O'Connell *et al.*, 1987; Sullivan *et al.*, 1988). Few electrochemical reductions of CO_2 resulting in higher molecular mass products have been reported, and often vital data on the methods of identification and quantification have been omitted. The difficulties related to the competition with proton reduction, high overpotentials, and low current efficiencies have been partly overcome by careful choice of electrode material (Frese and Summers, 1988) or special treatment and construction of the electrode (Ayers and Farley, 1988; Schwartz *et al.*, 1994), by selecting other solvents than water, and by the use of electrocatalysis with transition metal complexes (Behr, 1988; O'Connell *et al.*, 1987; Sullivan *et al.*, 1993). Photochemical reduction of CO_2 has also been achieved using various catalytic materials: semiconductors, suspensions of metal oxides, and transition metal complexes (Lewis and Shreve, 1993). There is at least one reported experiment where CO_2 reduction was carried out in water under the influence of γ-irradiation (Kolomnikov *et al.*, 1982). Although

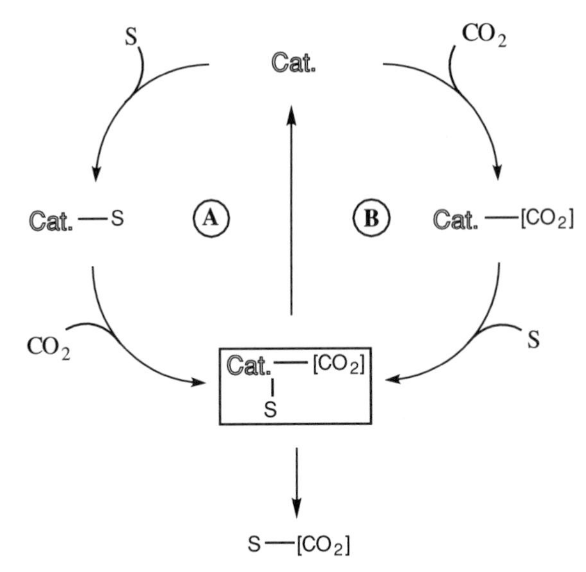

Figure 6.4 Mechanistic pathways of CO_2 reduction (Behr, 1988).

little is known about the mechanism of these reductions, they must rely on the production of reactive intermediates by the delivery of an electron or a photon to a reducible substrate, often directly to CO_2. In the case of submarine hydrothermal systems, however, the most likely mechanism of CO_2 activation is by heterogeneous or homogenous catalysis. The likely absence of light precludes a photochemical mechanism and, although electrochemical processes do occur in nature in which chemical discontinuities give rise to electrochemical cells capable of generating electrical currents, it seems unlikely that such cells supplying the potential required for CO_2 activation could form in submarine hydrothermal systems. On the other hand, these systems, especially 4.2 Ga years ago when the oxidation state of the ocean was much lower than today, are characterised by pressures and temperatures higher than those used in most aqueous CO_2 reduction experiments, and these rather extreme conditions should aid the catalytic activation of carbon dioxide.

6.7.3 *A rational approach to carbon dioxide activation*

A catalyst can enhance the rate of CO_2 reaction with a substrate S by two possible mechanistic pathways, illustrated in Figure 6.4 (Behr, 1988). Route A illustrates the activation of the substrate, with formation of a substrate–catalyst reactive intermediate. This subsequently reacts with CO_2 yielding an activated complex which eliminates the product, regenerating the catalyst. In route B, CO_2 is the first to combine with the catalyst, and the reactive intermediate is a CO_2–catalyst complex.

The existence of stoichiometric reactions leading to relatively stable complexes along routes A or B gives a good indication of the ability of a putative catalyst to activate CO_2. A logical approach to the design of homogeneous catalysts for a reaction would consist of combining these stoichiometric steps into a catalytic cycle (Behr, 1988; DuBois and Miedaner, 1988).

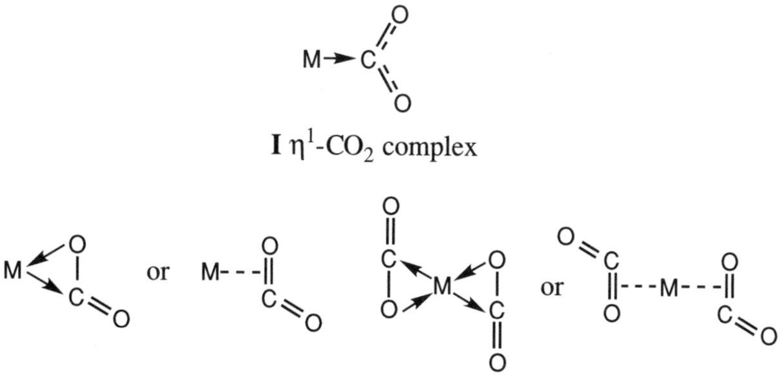

I η^1-CO_2 complex

II mono (η^2-CO_2 complex) **III** bis (η^2-CO_2 complex)

Figure 6.5 Binding modes of CO_2 to transition metal M.

6.7.4 *Transition metals, possible catalysts by the route B of CO_2 activation*

Of particular relevance to route B of activation in Figure 6.4 is the ability of CO_2 to bind to certain transition metals, affording CO_2–metal complexes. In a few cases their structure has been determined by X-ray analysis. Behr (1988) has systematised the possible binding modes of CO_2 to transition metals but only 3 of the 24 possible structures have been shown to exist in real compounds (Figure 6.5; M = transition metal).

Creutz explains the paucity of CO_2 complexes by the inherent instability of the strongly reducing metal centres that participate in their formation. This makes them difficult to handle in presence of air, moisture and proton sources. However, this lability can, up to a point, be a bonus when the complexes are included in catalytic cycles. A theoretical treatment of the two binding modes of CO_2 using complexes of Rh and Co as models, showed that in the η^1-complexes of these d^8 metal centres the metal–carbon bond is formed by the overlap of the filled dz^2 (σd) orbital of the metal with the empty π^* orbital of CO_2, whereupon the electron density shifts from the metal to CO_2, an oxidative addition (Creutz, 1993). A survey of the transition metals from the point of view of their capacity to bind CO_2 could help uncover the most promising candidates for CO_2 activation. We have restricted our survey to those elements likely to be present either in Hadean sea water or medium-enthalpy alkaline hydrothermal convection cells and seepages therefrom. Table 6.2 contains examples of transition metal–CO_2 complexes described in the literature.

The preparation of these complexes can be carried out by substitution of labile ligands as in the following example (where N_2 is the substituted ligand) (Alvarez *et al.*, 1984):

$$Mo(N_2)_2(PMe_3)_4 + 2\ CO_2 \longrightarrow Mo(CO_2)_2(PMe_3)_4 + 2\ N_2$$

and nickel–CO_2 compounds can be prepared from high electron density phosphine complexes at normal pressures and temperatures:

$$Ni(PCy_3)_3 + CO_2 \xrightarrow{\text{normal } T,\, p} Ni(CO_2)(PCy_3)_2 + PCy_3 \quad (Cy = \text{cyclohexyl})$$

The cobalt–CO_2 complex, $Co(CO_2)Na(np_3)THF$ (where THF = tetrahydrofuran, np = neopentyl) probably contains a CO_2 molecule coordinated to cobalt through the carbon

Table 6.2 Examples of CO_2 complexes described in the literature

Group	Period	CO_2 complexes	Structure	Reference
6	5	$Mo(CO_2)_2(PMe_2Ph)_4$	III	Chatt *et al.* (1974)
		$Mo(CO_2)_2dppe_2$	III	Chatt *et al.* (1983)
		$Mo(CO_2)_2(PMe_3)_4$	III	Carmona *et al.* (1983)
		$Mo(CO_2)_2(PMe_3)_3(CNPr^i)$	III	Alvarez *et al.* (1984)
		$Mo(CO_2)Cp_2$	I/II	Gambarotta *et al.* (1985)
	6	$W(CO_2)(CO)_5^{2-}$	—	Maher *et al.* (1982)
8,9,10	4	$Fe(CO_2)(PMe_3)_4$	—	Karsch (1977)
		$Fe(CO_2)(dmpp)_2$	—	Karsch (1984)
		$Fe(CO_2)Cp(CO)_2^-$	—	Bodnar *et al.* (1982)
		$[Co(CO_2)(PPh_3)_2]_n$	—	Speier *et al.* (1977)
		$[Co(CO_2)(pr\text{-}salen)K{\cdot}THF]_n$	—	Floriani and Fachinetti (1974)
		$Co(CO_2)HL_n$	—	Bianco *et al.* (1978)
		$Co(CO_2)Na(np_3)THF$	—	Bianchini and Meli (1984)
		$Ni(CO_2)(PR_3)_4$	—	Jolly *et al.* (1971)
		$Ni(CO_2)(PCy_3)_2$	I/II	Aresta *et al.* (1975)
		$Ni(CO_2)(PEt_3)_2$	—	Aresta and Nobile (1977)
		$Ni(CO_2)(PBu_3^n)_2$	—	Aresta and Nobile (1977)
11	4	$Cu(CO_2)(CH_3COO)(PPh_3)_2$	—	Miyashita and Yamamoto (1973)
		$Cu_2(CO_2)(C_6H_4PPh_2)_2(PPh_3)_2$	—	Miyashita and Yamamoto (1976)

atom and to a sodium cation through the two oxygen atoms. CO_2 activation is therefore being achieved by a combination of acidic and basic metal centres (Bianco *et al.*, 1978). A similar mode of CO_2 activation was achieved in the complex below, which by virtue of its weakly bound CO_2 could function as a carrier (Floriani and Fachinetti, 1974):

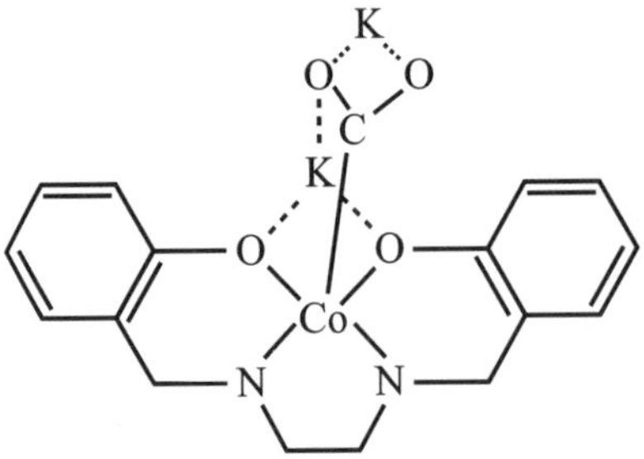

Complexes like [Co(pr-salen)K(CO_2)THF] (where pr-salen = *N,N'*-bis[(2 hydroxyphenyl)-methylene]-1,2-propanediamine), and the one above are prepared by insertion of CO_2 between the two metals, for example in the binuclear complex [Co(pr-salen)K] (Floriani and Fachinetti, 1974).

6.7.5 *Transition metals, possible catalysts by the route A of CO_2 activation*

The starting materials in the preparation of CO_2 complexes are highly reactive transition metal complexes which often reduce CO_2 after ligation. One of the most common mechanisms of reduction is insertion into a metal–ligand bond, leading to the transfer of a ligand to the CO_2 molecule (Behr, 1988). Theoretically the ligand can become attached either to the carbon or to the oxygen of the CO_2. The last alternative is rarely encountered and in general the reaction can be depicted as

$$M-L+CO_2 \longrightarrow M\langle O,O \rangle C-L \quad \left(\text{or } M-O-C(=O)-L \right)$$

These reactions are good candidates for steps in a catalytic cycle by route A of activation. The ligand L can be hydrogen, an organic group, oxygen or nitrogen but the last two are less relevant for a discussion of carbon dioxide reduction.

A different mechanism of reaction for CO_2 complexes, particularly relevant to carbon–carbon bond formation, is oxidative coupling between a transition metal complex, an unsaturated compound (which can be an alkene, alkyne, allene), and CO_2, resulting in the formation of a metallaheterocycle (Behr, 1988):

$$L_nM + X{=}Y + O{=}C{=}O \longrightarrow L_nM(-X-Y-C(=O)-O-)$$

Finally, CO_2 can be assisted in binding to the transition metal by a ligand containing a nucleophilic group (C or N) affording a five-membered ring product (Creutz, 1993):

$$M(-P-C(N)-C(=O)-O-)$$

After decomposition, these transition metal complexes will give products incorporating reduced CO_2.

Group 5

Vanadium Organovanadium compounds have been shown to react with CO_2. Insertion into trialkyl or triaryl complexes of vanadium yields, V^{II} and V^{IV}, disproportionation products (Figure 6.6) (Razuvaev *et al.*, 1981).

Allyl vanadocene is η^1-bound and reacts to form a η^1-bound monocarboxylato complex (see below) (Nieman and Teuben, 1985); η^3-allyl vanadium compounds like $CpV(\eta^3\text{-}C_3H_5)_2$ (where Cp = 1,3-cyclopentadiene) react only at 60 bar, forming bridging carboxylate species (Nieman *et al.*, 1984):

$$Cp_2V-CH_2CH{=}CH_2 + CO_2 \xrightarrow{\text{room temp.}} Cp_2V-O-C(=O)-CH_2CH{=}CH_2$$

Group 6

Chromium Darensbourg and his coworkers (1988) have studied the reaction with CO_2 of the chromium complexes *cis*-$HCr(CO)_4L^-$, where L = CO, PMe_3, $P(OMe)_3$. These complexes are very suitable for CO_2 insertion at normal temperature and pressure. They are

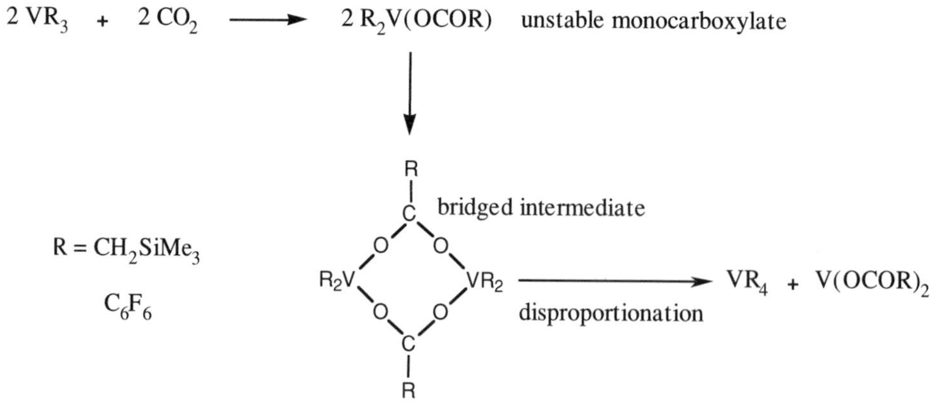

$$2\,VR_3 + 2\,CO_2 \longrightarrow 2\,R_2V(OCOR)$$ unstable monocarboxylate

R = CH_2SiMe_3

C_6F_6

bridged intermediate ($R_2V(\mu\text{-}O_2CR)_2VR_2$)

$$\longrightarrow VR_4 + V(OCOR)_2$$ disproportionation

Figure 6.6 Reaction of CO_2 with organovanadium compounds (Razuvaev *et al.*, 1981).

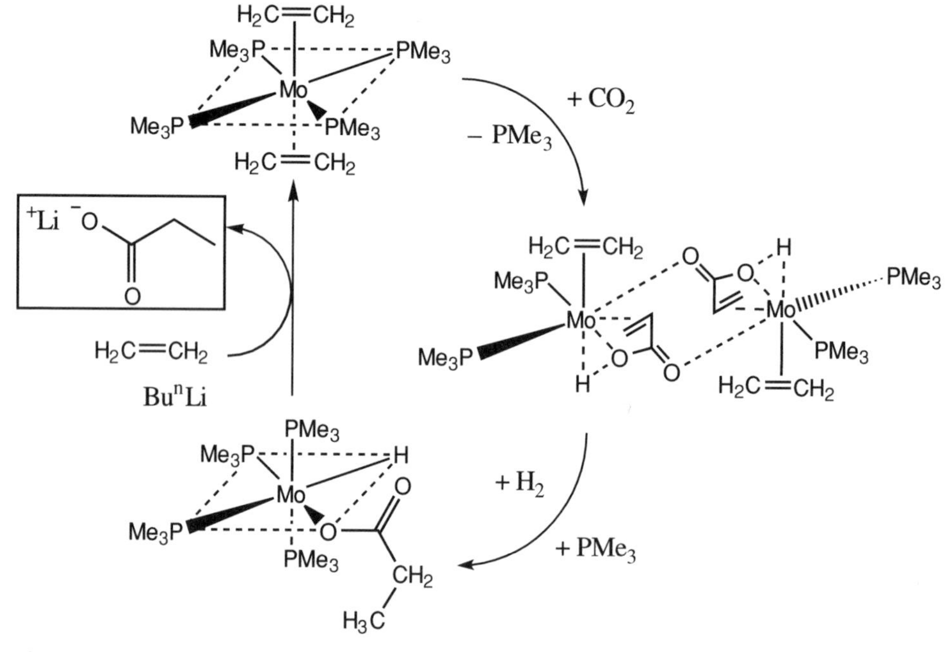

Figure 6.7 Example of a possible catalytic cycle for the synthesis of propionic (or acrylic) acid from CO_2 and ethene (Behr, 1988).

hydridic hydrides and easily interact with electrophilic CO_2. Similar alkyl complexes undergo insertion into the Cr—C bond:

$$[L(CO)_4Cr]\text{—}R^- + CO_2 \longrightarrow [L(CO)_4Cr]\text{—}O_2CR^-$$

where R = CH_3, C_2H_5, C_6H_5, $CH_2C_6H_5$; L = CO, PMe_3, $P(OMe)_3$

The electron density at the metal centre is the essential factor in the rate of insertion (the higher the electron density, the higher the rate). Thus, good electron-donating ligands (like PMe_3, $P(OMe)_3$) increase the insertion rate (Darensbourg *et al.*, 1988).

Molybdenum The bis(ethene)molybdenum complex in Figure 6.7, which has an electron-rich metal centre surrounded by labile trimethylphosphine ligands, reacts with CO_2, ethene,

H_2, and butyllithium affording lithium propionate. The starting complex is recovered unchanged. The reaction sequence outlines a possible future catalytic synthesis of propionic acid or (in absence of H_2) acrylic acid from ethene and CO_2 (Behr, 1988; Sullivan *et al.*, 1993).

Tungsten An unusual reaction of a tungsten(II) complex with CO_2 is the following oxidative addition of CO_2 involving the insertion of the metal in the C—O bond:

$$WCl_2(PMePh_2)_4 + CO_2 \xrightarrow{\text{normal } T \text{ and } p} W(O)Cl_2(CO)(PMePh_2)_2$$

Carbon dioxide is effectively reduced to CO in this reaction and the yield is > 95% (Bryan *et al.*, 1987). By insertion into a tungsten-phosphacyclopentane, CO_2 becomes part of a seven-membered metallaheterocycle (Darensbourg *et al.*, 1985):

$$\left[(CO)_4W\overline{\quad PPh_2 \quad}\right]^- {}^+PPh_4 + CO_2 \longrightarrow \left[(CO)_4W(Ph_2P\text{—}\cdots\text{—}C(=O)\text{—}O)\right]^- {}^+PPh_4$$

In a major contribution, Darensbourg, Bauch and Ovalles (1988) demonstrate that insertion into an anionic alkyl(aryl)-tungsten complex *cis*-$RW(CO)_4L^-$ proceeds smoothly:

$$[L(CO)_4W]\text{—}R^- + CO_2 \longrightarrow [L(CO)_4W]\text{—}O_2CR^-$$

where R = CH_3, C_2H_5, C_6H_5, $CH_2C_6H_5$; L = CO, PMe_3, $P(OMe)_3$

As mentioned in the case of the similar chromium complexes, the electron density at the metal centre is the essential factor in the rate of insertion. Also nonanionic (neutral) complexes which are isoelectronic with *cis*-$RW(CO)_4L^-$, such as $CH_3Re(CO)_5$, do not undergo insertion. The mechanism is an associative interchange, I_a with the transition state:

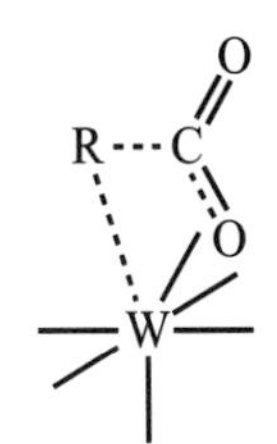

The reaction is first order in CO_2 and in metal substrate and, in the presence of CO, it runs parallel to, and independently from, an insertion of CO which proceeds by a different mechanism.

CO_2 insertion into the W—H bond of *cis*-$HW(CO)_4L^-$ hydrides, (L=CO, PMe_3, $P(OMe)_3$) is very easy. These are hydridic hydrides (electron density located at the H) which easily attack the electrophilic C of CO_2 (Darensbourg *et al.*, 1988).

Alkyl formate production is catalysed by tungsten complexes like μ-$H[W_2(CO)_{10}]^-$. The overall reaction is

$$CO_2 + H_2 + ROH \xrightarrow{[\text{cat.}]} HCO_2R + H_2O$$

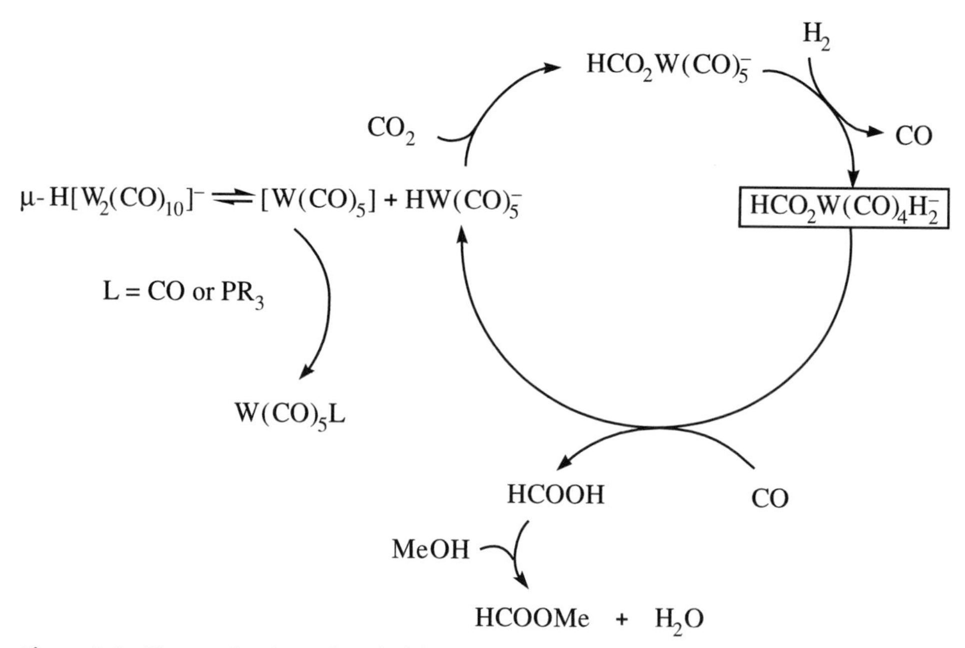

Figure 6.8 The mechanism of methyl formate synthesis catalysed by the tungsten complex $H[W_2(CO)_{10}]^-$ (Darensbourg *et al.*, 1988).

The reaction pathway is shown in Figure 6.8.

The metalloformate intermediate is extremely CO-labile. The addition of CO inhibits the hydrogenation of CO_2 and it was therefore suggested that the intermediate has the structure in Figure 6.8 (outlined formula). Tungsten in the intermediate undergoes no change of oxidation state. A ligand-assisted heterolytic splitting of dihydrogen takes place (this is accomplished owing to the basicity of the distal oxygen of the monodentate formate ligand, otherwise demonstrated by its ability to interact with kryptofix encapsulated sodium):

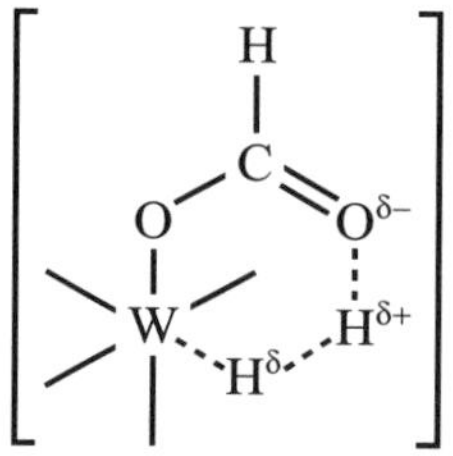

In methanol the turnover rate is 15 in 24 h at 35 bars (CO_2/H_2) and 125 °C. The turnover decreases with higher molecular mass alcohols owing to a solvent inhibition of the addition of molecular hydrogen to the formate complex (Darensbourg *et al.*, 1988).

Group 7

Manganese Few reactions of CO_2 with manganese complexes have been reported. Thus carboxylates have been synthesised from organomanganese halides (RMnX), dialkylmanganese (R_2Mn) and lithium or magnesium organomanganates (R_3MnLi, R_3MnMgX), where R = alkyl, alkenyl or aryl (Normant and Cahiez, 1983):

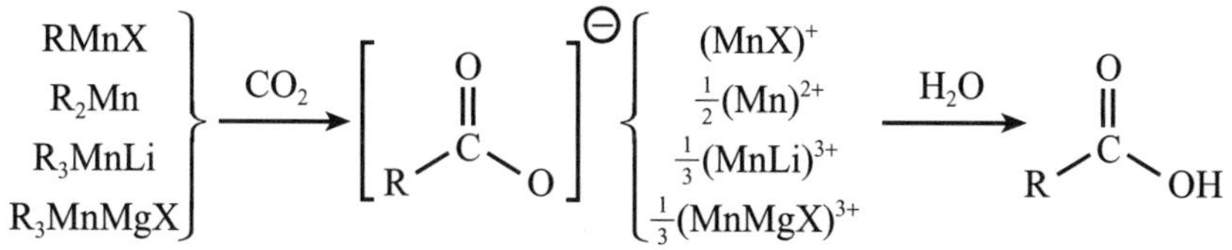

$$\left.\begin{matrix} RMnX \\ R_2Mn \\ R_3MnLi \\ R_3MnMgX \end{matrix}\right\} \xrightarrow{CO_2} \left[R{-}C(=O){-}O \right]^{\ominus} \left\{ \begin{matrix} (MnX)^{+} \\ \frac{1}{2}(Mn)^{2+} \\ \frac{1}{3}(MnLi)^{3+} \\ \frac{1}{3}(MnMgX)^{3+} \end{matrix} \right. \xrightarrow{H_2O} R{-}C(=O){-}OH$$

Behr *et al.* (1984) have reported a facile insertion of CO_2 into the manganese phosphacyclobutane

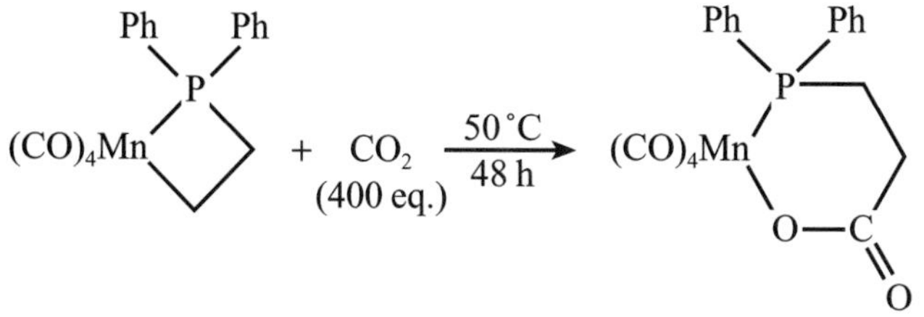

$$(CO)_4Mn(\text{Ph}_2\text{P-phosphacyclobutane}) + \underset{(400\ \text{eq.})}{CO_2} \xrightarrow[48\ \text{h}]{50\,^{\circ}\text{C}} (CO)_4Mn(\text{Ph}_2P{-}CH_2CH_2{-}C(=O){-}O)$$

Group 8

Iron Hydrides of iron have been reported to undergo insertion leading to formate complexes (Bianco *et al.*, 1972). An insertion into the coordinatively unsaturated iron complex shown below results in an acetato complex (Ikariya and Yamamoto, 1976):

$$Fe(CH_3)_2(dppe)_2 + CO_2 \xrightarrow{\text{pyridine}} Fe(OOCCH_3)_2(py)_n(CO_2)_m$$

where dppe = 1,2-bis(diphenylphosphino)ethane, py = pyridine

At room temperature, a THF solution of bis(ethene)bis(triethylphosphine)iron(0), a phosphine ligand (PMe_3, dmpe or dcpe), and equimolar amounts of ethene and CO_2 afforded iron carboxylate complexes (ν_{max} 1580 cm^{-1}) which, on treatment with HCl and MeOH, gave dimethyl succinate or dimethyl methylmalonate (depending on the ligand used) (Figure 6.9). No monocarboxylic acids were formed (Hoberg *et al.*, 1987).

Fischer–Tropsch syntheses with CO_2/H_2 mixtures, resulting in C_2 to C_5 alkanes and alkenes, are catalysed by iron catalysts doped with copper or by Fe/Al_2O_3 catalysts. The reactions probably proceed via CO (Barrault *et al.*, 1981; Pijolat and Perichon, 1982).

Group 9

Cobalt Kolomnikov and Volpin (Kolomnikov *et al.*, 1972) have described an unusual insertion reaction of CO_2 into cobalt complexes of the type $L_2Co(CO)Et$, ($L = PPh_3$) whereby, besides the carboxylate product $LCo(OCOEt)_2$, the carbon-coordinated $LCo(CO_3)_nCOOEt$ has also been obtained. Schrauzer and Sibert (1970) achieved the insertion of CO_2 into the Co—CH_3 bond of methylcobaloxime/1,4-butanedithiol, obtaining small amounts of acetic acid in an attempt to simulate the microbial CO_2 fixation of methylcorrinoids (Figure 6.10). The mechanism involves reduction by dithiol of the Co^{III} centre in cobaloxime [depicted as (Co)] to Co^{I}.

Group 10

Nickel There are numerous examples of insertion reactions into the Ni—C bond of organonickel complexes. Thus (bipy)$NiEt_2$ (where bipy = 2,2′-bipyridyl) can suffer

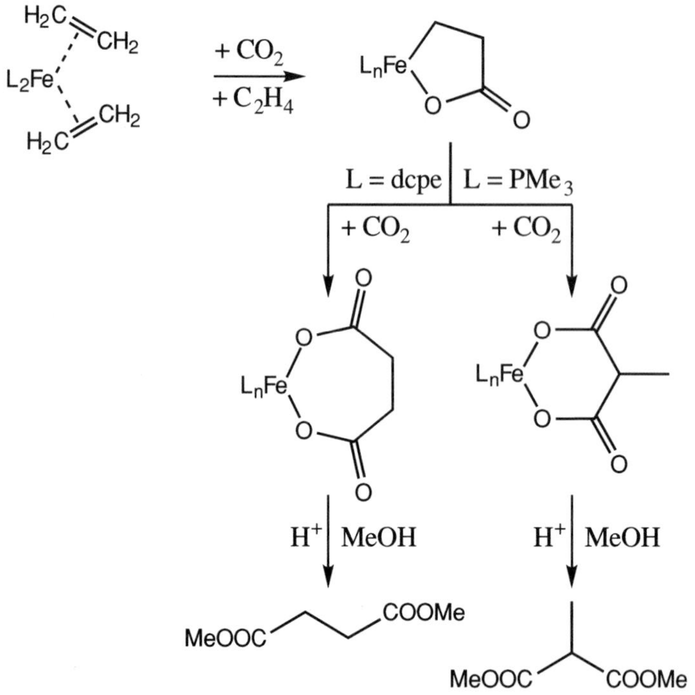

H_2C CH_2 L_2Fe CH_2 H_2C $+ CO_2$ $+ C_2H_4$ L_nFe O O

L = dcpe | L = PMe_3

$+ CO_2$ $+ CO_2$

H^+ | MeOH H^+ | MeOH

MeOOC COOMe MeOOC COOMe

Figure 6.9 The reaction of bis(ethene)bis(triethylphosphine)iron(0) with CO_2 (Hoberg *et al.*, 1987).

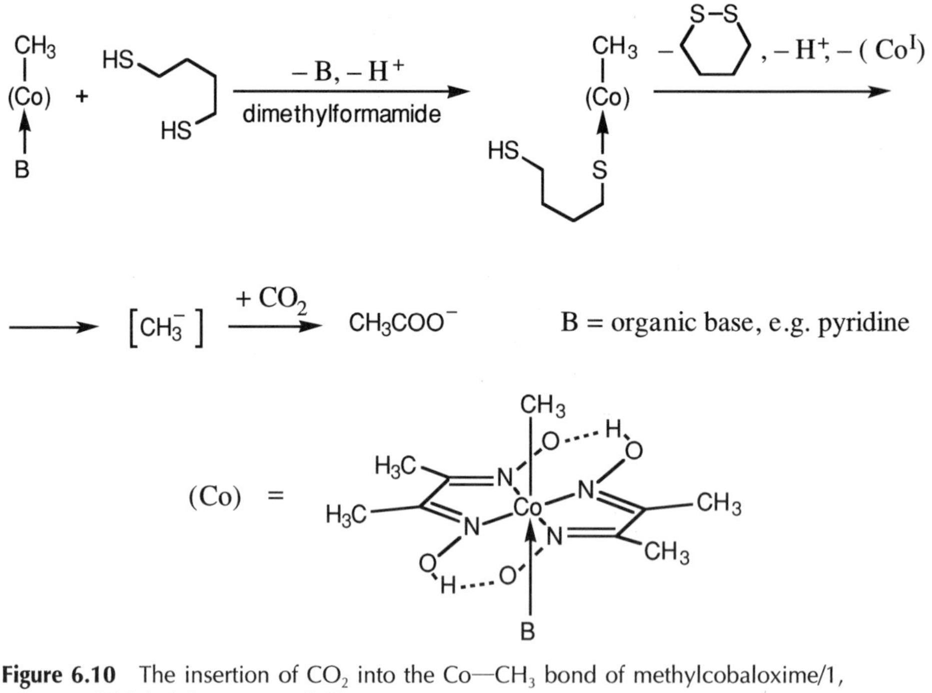

CH_3 (Co) B + HS HS $-B, -H^+$ dimethylformamide CH_3 (Co) HS S S–S $, -H^+, -(Co^I)$

$[CH_3^-]$ $\xrightarrow{+CO_2}$ CH_3COO^- B = organic base, e.g. pyridine

(Co) = CH_3 H_3C H_3C N Co N O H O CH_3 CH_3 B

Figure 6.10 The insertion of CO_2 into the Co—CH_3 bond of methylcobaloxime/1, 4-butanedithiol (Schrauzer and Sibert, 1970).

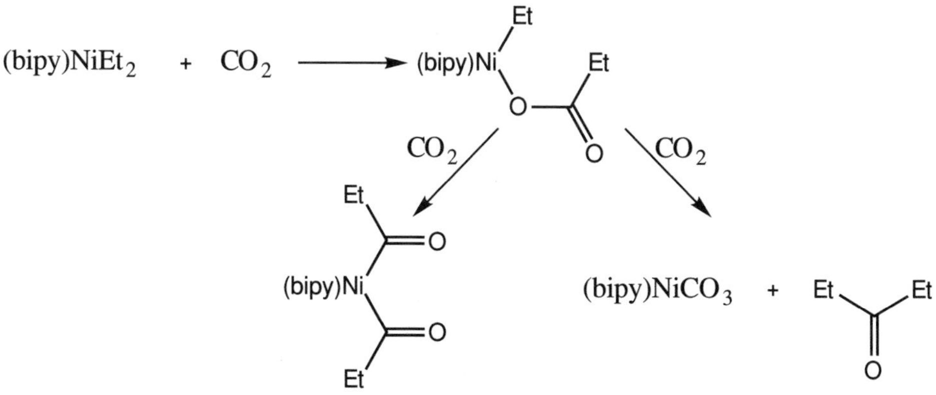

$(bipy)NiEt_2 + CO_2 \rightarrow (bipy)Ni(Et)(OCOEt)$; with CO_2 → $(bipy)Ni(COEt)_2$; with CO_2 → $(bipy)NiCO_3$ + Et_2CO

Figure 6.11 The insertion of CO_2 into the Ni—Et bond of $(bipy)NiEt_2$ (Yamamoto and Yamamoto, 1978).

monoinsertion of CO_2, giving rise to the monocarboxylate. It can also insert a second CO_2 molecule or react with CO_2, affording the carbonate and diethylketone (Figure 6.11) (Yamamoto and Yamamoto, 1978).

Complexes formed by nickel with unsaturated compounds also react with CO_2. An example is bis(η^3-allyl)nickel which reacts at room temperature with CO_2 yielding η^3-allyl(vinylacetato)nickel. This releases butyrolactone on heating to 140 °C (Tsuda *et al.*, 1979).

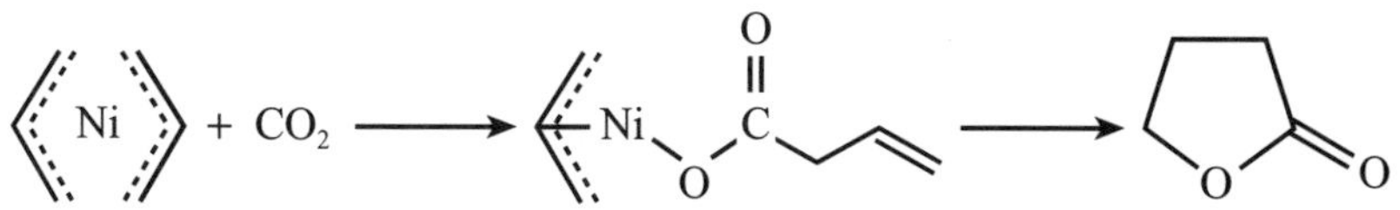

Ni + CO_2 → Ni–O–C(=O)–$CH_2CH=CH_2$ → butyrolactone

Nickel complexes containing easily dissociating phosphine or arsine ligands, a strongly bound chelate ligand, and a nickel–aryl bond (known to catalyse the oligomerisation of ethene) react with CO_2 at 20–30 bar giving low yields of benzoates. When ethene is added to the system and the reaction is performed around 0 °C, a mixture containing nickel 3-phenylpropionate is obtained. It was concluded that insertion of CO_2 into nickel–alkyl bonds proceeds more easily than into nickel–aryl bonds. The ligand field has a great influence on reactivity, basic ligands favouring the insertion of CO_2 (Behr, 1988).

Carbon dioxide was also reacted with nickel metallacycles like the nickelacyclopentane

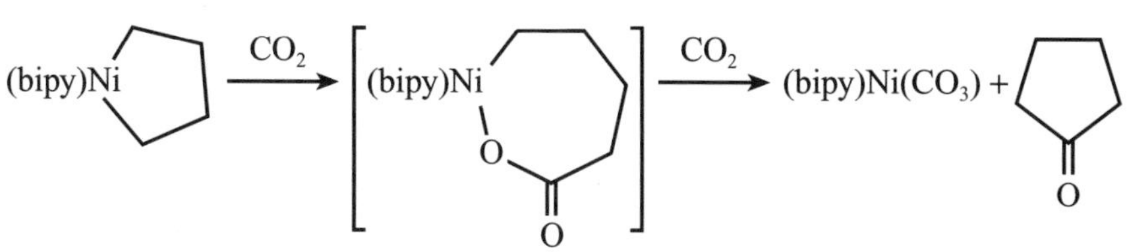

(bipy)Ni(cyclopentane) $\xrightarrow{CO_2}$ [(bipy)Ni–O–C(=O) seven-membered ring] $\xrightarrow{CO_2}$ $(bipy)Ni(CO_3)$ + cyclopentanone

The seven-membered ring intermediate was not isolated, as it reacted with a further CO_2 molecule yielding the carbonate $(bipy)Ni(CO_3)$ and cyclopentanone (Behr, 1988). The insertion was only possible due to the strongly basic chelate ligand bipyridyl. Likewise four-membered ring nickel metallacycles reacted with CO_2 prompted by the tendency to relieve the ring strain. By oxidative coupling with alkynes or alkenes and CO_2, nickel complexes containing highly basic ligands can form oxanickelacyclopentanes that give the corresponding carboxylic acids by acidic decomposition (see a similar example at

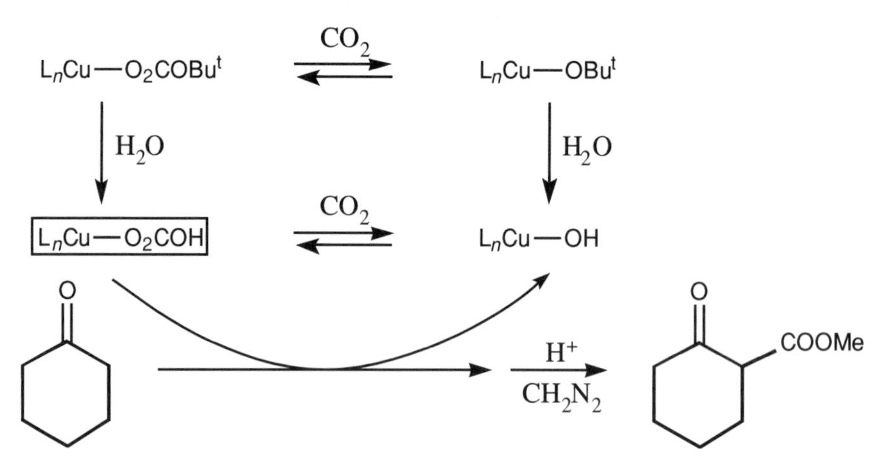

Figure 6.12 Carboxylation of cyclohexanone by a copper bicarbonate complex obtained by CO_2 insertion into a complex copper alkoxide (Tsuda and Saegusa, 1972).

Iron, Section 6.7.5). In different conditions, two ethene molecules and an equivalent of CO_2 can be made to react with one molecule of nickel complex, affording seven-membered rings and the corresponding unsaturated carboxylic acids.

Behr (1988) also shows that dienes react with nickel complexes through an η^2-intermediate, yielding dianionic fragments bound to nickel through a carboxylate bond and an η^3-allyl moiety. The propensity of nickel complexes to react with alkynes and CO_2 has been put to work in the reaction of 1-hexyne with CO_2, in the presence of nickel complexes containing bidentate phosphine ligands. The product is 4,6-dibutyl-2-pyrone and the mechanism proceeds by the formation of an oxanickelacyclopentene (Inoue *et al.*, 1977). Nickel complexes derived from $Ni(cod)_2$ (where cod = 1,5-cyclooctadiene) and phosphine ligands catalyse the reaction of butadiene with CO_2 (see Figure 6.3) (Behr, 1988).

Nickel is one of the most effective catalysts in the Sabatier process. Thus high yields of methane have been obtained by using Ni—Os clusters on γ-Al_2O_3 (Bardet *et al.*, 1978; Gray, 1983):

$$CO_2 + 4\,H_2 \longrightarrow CH_4 + 2\,H_2O$$

Group 11

Copper Copper alkoxides can incorporate a CO_2 molecule, giving alkyl carbonates. These can be hydrolysed to copper–bicarbonate complexes that have the ability to deliver CO_2 to organic reagents, thus functioning as carriers with a possible role in catalysis (Figure 6.12) (Tsuda and Saegusa, 1972).

The insertion of CO_2 into the Cu—C bond has been investigated extensively. Ligand-free arylcopper(I) complexes only insert CO_2 if phosphines are present in the reaction mixture. Alkyl complexes of the type CH_3—$Cu(PCy_3)$ can insert CO_2, yielding the carboxylate, but they can also attach a further molecule of CO_2, giving copper carboxylate–CO_2 adducts (De Pasquale and Tamborski, 1969; Ikariya and Yamamoto, 1974; Marisch *et al.*, 1982). The reaction with vinylcopper compounds occurs without additives and produces the corresponding α,β-unsaturated carboxylic acid (Cahiez *et al.*, 1975; Normant *et al.*,

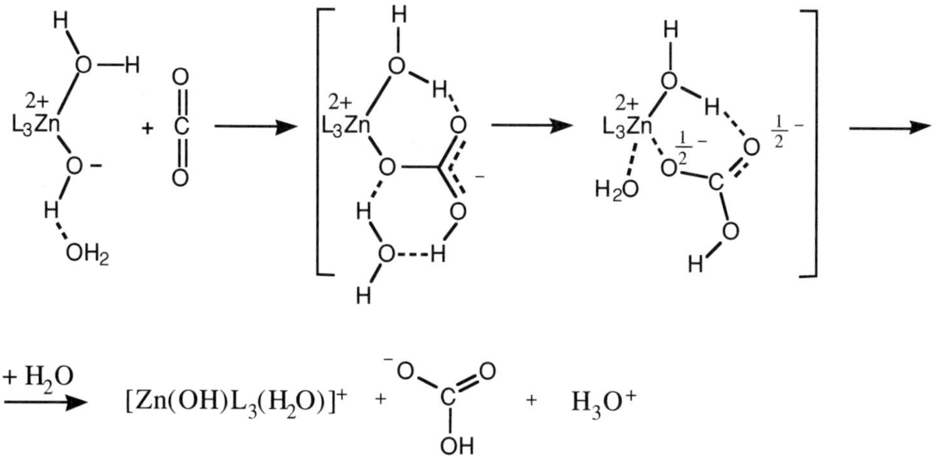

$$+H_2O \longrightarrow [Zn(OH)L_3(H_2O)]^+ + HCO_3^- + H_3O^+$$

L = histidine ligands in the protein

Figure 6.13 The reaction mechanism at the zinc centre in carbonic anhydrase (Allen, 1988).

1973, 1974). Vinyl lithiocuprates have been reacted with acetylene and CO_2, producing similar unsaturated carboxylic acids (Alexakis and Normant, 1982). The CO_2 insertion reaction was also performed successfully with copper acetylides; the final product, after treatment with MeI, being the methyl alkynoate (Tsuda *et al.*, 1974).

Certain ligands, such as cyanomethyl (in $NCCH_2—Cu(PBu^n_3)_x$) promote the formation of a CO_2 adduct at room temperature. This subsequently transfers the carbon dioxide molecule to organic substrates at 70 °C, functioning as a CO_2 carrier (see above) (Tsuda *et al.*, 1975, 1976, 1978). A copper/zinc catalyst is commonly used in the synthesis of methanol from $CO/CO_2/H_2$ mixtures. It has been suggested that CO and CO_2 are reduced in parallel in the process (Denise *et al.*, 1982).

Group 12

Zinc Zinc features in CO_2 chemistry mainly through its role in the enzyme carbonic anhydrase (Figure 6.13) (see Section 6.7.2). There it plays the role of a strong Lewis acid, stabilising the attacking HO^- group and the subsequent cyclic transition state, thereby facilitating CO_2 hydration by an 'around the corner' S_N2 mechanism (Allen, 1988).

In conjunction with copper it catalyses the conversion of $CO/CO_2/H_2$ mixtures to methanol. As diethylzinc, in equimolar mixtures with water or aromatic dicarboxylic acids, it is a highly effective catalyst for the copolymerisation of CO_2 with epoxides, producing polycarbonates (Behr, 1988).

$$n\,ZnEt_2 + n\,H_2O \longrightarrow Et(Zn—O)_n—H + (2n-1)\,C_2H_6$$

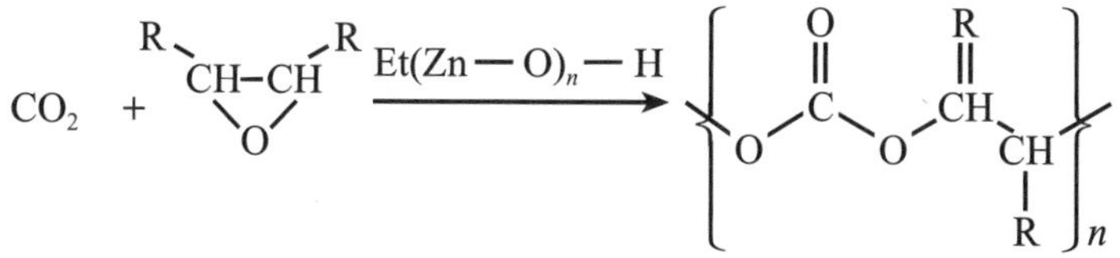

$$CO_2 + \text{RCH–CHR (epoxide)} \xrightarrow{Et(Zn—O)_n—H} \left[O–C(=O)–O–CH(R)–CH(R) \right]_n$$

6.7.6 Clues from CO_2 reduction experiments

In order for a transition metal to play a part in a catalytic cycle A (Figure 6.4), the product of insertion must not be too strongly bound. Practically all insertion reactions result in oxygen-bound complexes, with the notable exception of the carbon-coordinated $(PPh_3)Co(CO_3)_nCOOEt$ obtained from $(PPh_3)_2Co(CO)Et$ and CO_2. It is desirable therefore that the metal–oxygen bond is weak enough to allow the departure of the product and the regeneration of the catalyst. Consequently metals from the earlier groups 3, 4 and perhaps 5 are less likely to make suitable catalysts for CO_2 reduction, and on this basis DuBois and Miedaner (1988) chose the later transition metals Fe, Co, Ni, and the platinic metals as subjects of their efforts to design catalysts for CO_2 reduction. The stoichiometric insertion of CO_2 into the M—H bond as well as the direct reduction of CO_2 with H_2, catalysed by transition metal complexes, lead mainly to formate. The difficulty of further reducing formate is well known (Miles and Fletcher, 1988).

In examining route B of activation, information can be derived from the synthesis and reactivity of transition metal–CO_2 complexes. In his review of the CO_2 binding to transition metal centres Creutz (1993) underlines the necessity for the metal to be in a reduced state within the starting complex. Thus η^1-CO_2 complexes of Fe, Co, Rh, Ir and W are prepared from metals in the d^8 configuration, which implies exceedingly low oxidation states (Fe^0, Co^{1+}, Rh^{1+}, Ir^{1+} and W^{2-}). The η^2-complexes of Nb, Mo and Ni, possessing d^2, d^4, d^6, and d^{10} configurations require these metals in the oxidation states 0 or 2+. Bound CO_2 is often stabilised by a second metal centre or by hydrogen bonding to an appropriate ligand, and the metal precursor can be a coordinatively unsaturated 16-electron complex or an 18-electron complex possessing a good leaving group (Creutz, 1993). Sullivan and his coworkers (1988) remark that the dominant reaction of CO_2 complexes is with acidic acceptors such as water. By transferring oxygen to the acidic acceptor the CO_2 complexes yield carbonyl complexes:

$$M(\eta\text{-}CO_2) + H_2O \longrightarrow [M(CO)]^{2+} + 2\,OH^-$$

$$OH^- + CO_2 \longrightarrow CO_3^{2-} + H^+$$

This effectively results in the disproportionation of two CO_2 molecules into one of carbon monoxide and one of carbonate, a reaction that has also been observed during electrolytic reduction of CO_2 (Sullivan *et al.*, 1988). Thus complexes of group 4, and also Mo, Fe, Co, Rh and Cu can bring about the deoxygenation of CO_2 leading to the formation of carbonyl complexes (Behr, 1988).

Other modes of reaction, like the deoxygenation of CO_2 to CO with oxygen abstraction by a phosphine ligand, or the intramolecular reaction of bound CO_2 with an organic ligand within a transition metal–CO_2 complex (see *trans*-$Mo(\eta^2\text{-}C_2H_4)_2(PMe_3)_4$), are uncommon (Sullivan *et al.*, 1988). Clearly the research in transition metal–CO_2 complexes and their chemistry points to CO as the main product of the reactions by route B of CO_2 activation.

As sites for the emergence of life, hydrothermal systems generate a variety of organic compounds (Shock, 1996), but their synthesis requires C—C bond formation. As shown in earlier paragraphs, with the assistance of transition metals, carbon dioxide is amenable to C—C bond formation by the two mechanisms of insertion and oxidative addition. These call for the presence of an organic ligand, or of an unsaturated organic compound, and a coordinatively unsaturated transition metal. The hydrothermal system has potential to supply a small number of organic species which could act as a basis (a primer) for the synthesis of larger compounds through the catalysed addition of CO_2.

6.7.7 Likely natural catalysts

From this overview of the CO_2–transition metal complex chemistry, a substantial number of metals emerge as potential candidates for the catalysis of CO_2 reduction understood in the broader sense as resulting in C—C as well as C—H formation. While the metals of groups 3, 4 and 5 can be generally considered too alkaline to give effective catalysts, especially in the presence of water, vanadium cannot be completely excluded as a candidate for catalysis considering its excellent redox properties. The metals of groups 6 to 10 all have a high catalytic potential. However, the involvement of exceedingly rare metals like Re and of the platinum group elements (Ru, Rh, Pd, Os, Ir, Pt) in CO_2 catalysis in the context of the emergence of life is probably unlikely. Nevertheless, they are the best catalysts overall for reduction and demonstrate a rich CO_2 chemistry. In group 11, copper is the only metal with a significant CO_2 chemistry, featuring also as a catalyst in methanol synthesis, together with zinc, from group 12. The list of transition metals with a possible role as catalysts for CO_2 reduction in the first stages of the emergence of life must therefore contain Cr, Mo, W, Mn, Fe, Co, Ni, Cu and Zn.

6.7.8 Surface catalysis: clues from homogeneous catalysis

Surface catalysts are of a greater complexity than their molecular counterparts because of the sheer variety of compositions and structures that catalytic surfaces can present to the reacting species. Only a few mechanisms of reactions taking place on the surface of solid catalysts are known in detail, although the effort of research in this field is substantial. Despite the difficulties, important advances have been made and a large amount of experimental evidence has been accumulated which shows that in essence the nature of catalysis at the surface is similar to that happening on metal complexes. The theory of active centres of H. S. Taylor (1925) has found ample experimental confirmation and it has since been shown that these centres are most often steps and kinks on different faces of the catalyst crystals. Such centres contain atoms of transition metal which, although coordinatively unsaturated, are nevertheless stabilised by the bulk of the solid of which they form a part. This stabilisation is not unlike that conferred by ligands upon a transition metal complex. Even more similarities can be found between metal-cluster complexes and some of these active centres. Although no straightforward predictions can be made as to the catalytic activity of transition metal surface catalysts from the chemistry of their complexes, experience shows that there are strong analogies between them (Gates, 1992).

Transition metals are likely to be largely precipitated as sulfides at the sites where life might have emerged, and it is important that the possibility of CO_2 and CO reduction by these potentially catalytically active precipitates is examined. In this context, the reported catalytic activity of the iron–sulfur clusters (Tezuka *et al.*, 1982) in electrochemical CO_2 reduction gains a special significance.

6.7.9 CO_2 and CO reduction compared

The structural and chemical differences between the two oxides of carbon are significant. With transition metals, carbon monoxide forms an important class of compounds, the metal carbonyls, and the transition metal–CO bond has been well studied owing to the numerous chemical processes in which it plays a part. This bond involves carbon coordination to the metal and back-donation from the occupied *d* orbitals of the metal to an unoccupied nonbonding π^* orbital of CO, being quite different from the bond found

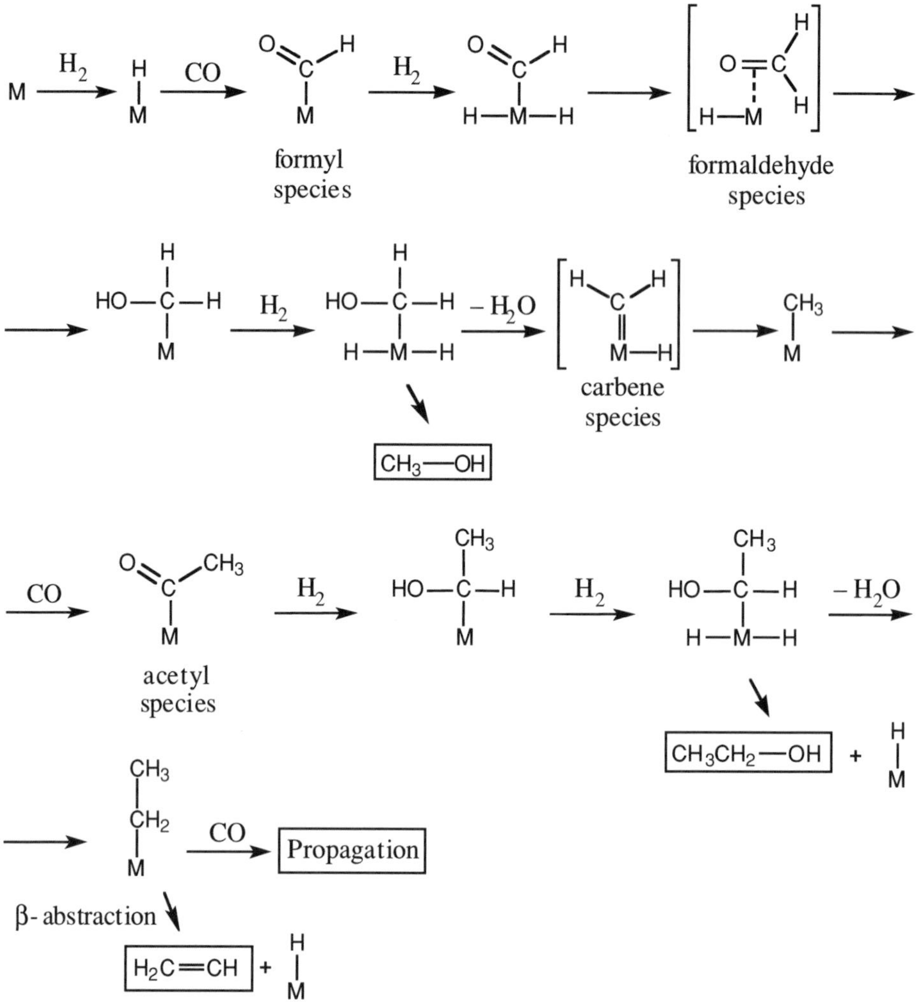

Figure 6.14 The mechanism of the Fischer–Tropsch synthesis of hydrocarbons (Henrici-Olivé and Olivé, 1984).

in η^1-bonded CO_2 complexes. The catalysed reduction of carbon monoxide in the form of synthesis gas (CO/H_2) lies at the basis of several important industrial processes such as the hydroformylation of alkenes (Oxo synthesis), the Fischer–Tropsch synthesis of hydrocarbons and the syntheses of methanol, methane and acetic acid. The research effort in the field has been stimulated by the economic importance of CO reduction and has resulted in a good understanding of the chemistry involved. The literature on the subject is vast and has been summarised in numerous books and review papers (Falbe, 1980; Henrici-Olivé and Olivé, 1984; Sheldon, 1983). Nevertheless, in the context of the emergence of life in hydrothermal systems, the following observations can be made.

Most of the basic chemical building blocks of life have been synthesised abiotically in origin-of-life experiments with the exception of the fatty acids essential to biological membranes. The Fischer–Tropsch synthesis is essentially a polymerisation reaction capable of supplying long-chain aliphatic hydrocarbons and oxgenated compounds, including fatty acids (Ferris, 1992). The mechanism, according to Henrici-Olivé and Olivé (1984) is presented in Figure 6.14.

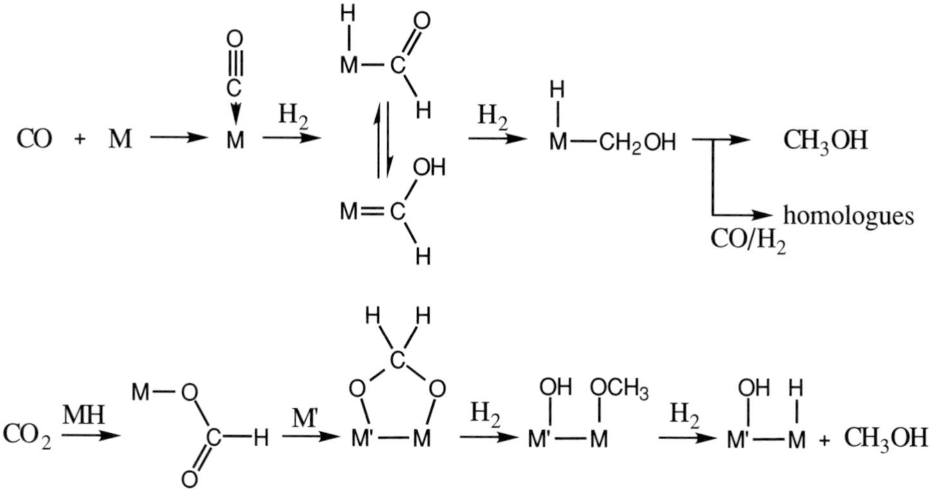

Figure 6.15 Possible mechanisms of CO and CO_2 reduction to methanol on Cu/Zn catalysts (Denise *et al.*, 1982).

In hydrothermal systems the presence of water could have an inhibiting effect over the dehydration step of the reaction, halting the chain building sequence and diverting the process towards methanol and methane formation (but see Shock, 1996). The industrial acetic acid synthesis from methanol and CO (Crabtree, 1994) informs the recently published work of Huber and Wächtershäuser (1997), who demonstrate the synthesis of acetic acid from methane thiol and CO on a NiS/FeS catalyst (see Sections 6.3 and 6.8.4). This is the first account of a high-yield hydrothermal C_2 compound synthesis from simple C_1 precursors in conditions relevant to the emergence of life.

Differences of reactivity between CO and CO_2 have been mentioned earlier in Section 6.7.5 (Group 6) (Darensbourg *et al.*, 1988). Here a further example will be presented (Denise *et al.*, 1982). The industrial synthesis of methanol from synthesis gas, on Cu/Zn catalysts, requires, for optimal yields, the addition of CO_2 to the stream of reactants. The reaction is thought to involve parallel reduction of CO and CO_2 by the different mechanisms shown in Figure 6.15.

Denise *et al.* (1982) propose that CO_2 reduction proceeds by an oxygen-bound intermediate while CO binds only through carbon. It is known that in this process CO always forms small amounts of the higher alcohols and alkanes, and adding methanol to the synthesis gas causes an increase in their concentration. However, in CO_2/H_2 mixtures, addition of methanol does not result in the formation of such homologous products. This is explained by the fact that the oxygen-bound CO_2 intermediate is incapable of C—C bond formation, reacting only to afford the desired C_1 product (Denise *et al.*, 1982).

6.8 Indications from mineralogy

Given the likely paucity of abiotic organic molecules in the Hadean, many researchers have suggested inorganic beginnings for life and the first membranes (e.g. Bernal, 1960; Cairns-Smith, 1982). If this were so, then we might expect minerals involved in prebiotic metabolism to contain those 'labile . . . elements . . . which undergo the greatest transformations in the mineral world' (Bernal, 1960). Bernal had in mind iron, phosphorus and sulfur, although he was drawn to the clays as the likely destination for most of these elements and as a basis for a catalytic surface and information system (Bernal, 1951, 1960). But the mineral series that can offer the greatest number of transformations and is, at the same time, the most accommodating of different transition metals, is the mackinawite–smythite–greigite group. In fact, like the clays, mackinawite itself is a layered or sheet mineral along the (001) structural plane. This plane remains intact throughout the oxidative transition through smythite to greigite, though in the latter mineral it is (111) (Krupp, 1994; Posfai *et al.*, 1998). We concentrate below on the minerals from this series because many of the metals concluded to be significant in Section 6.7.5 can be accommodated in one or more of these structures. Therefore, as well as comprising the first membrane, it may be that the mackinawite/greigite reaction sequence also acted as the substratum to what Edwards (1996) has termed the 'universal ancestral metabolic complex.'

We also consider djerfisherite and the like as possible sources of sulfide, awaruite as a catalyst for reduction of carbon oxides in the crust, and pyrite as it features in the Wächtershäuser (1990a) hypothesis.

6.8.1 Djerfisherite ($K_3[Na,Cu][Fe,Ni]_{12}S_{14}$) as sulfur source

Minerals such as djerfisherite (Genkin *et al.*, 1970), found in mineralised mafic and ultramafic complexes and enstatite chondrites, are possible sources of bisulfide or sulfhydryl complexes which may have been generated during medium-enthalpy convection of ocean water in the early mafic crust. Desulfidisation of djerfisherite could have left awaruite (see below) as a residuum. Comparable minerals found in similar circumstances are rasvumite (KFe_2S_3) (Sokolova *et al.*, 1970; Czamanske *et al.*, 1979) and erdite ($NaFeS_2{\cdot}2H_2O$) (Konnert and Evans, 1980; Czamanske *et al.*, 1980; and see Godlevskiy, 1959).

6.8.2 Awaruite (Ni_3Fe) as crustal 'hydrogenase'

Awaruite is found commonly in ultramafic rocks which have suffered relatively low-temperature (<375 °C) hydration or serpentinisation (Ulrich, 1890; Nickel, 1959; Krishnarao, 1964). Awaruite is isometric with a face-centred cubic lattice and a cell dimension of 3.5522 Å (Nickel, 1959; Botto and Morrison, 1976). Iron atoms occupy the four corners of a primitive cube and the three nickel atoms are situated in the face-centred sites (Leech and Sykes, 1939; Botto and Morrison, 1976). We imagine this filamentous mineral catalysing the hydrogenation of carbonic acid to aldehyde and alcohols (and in this milieu, the sulfide equivalents) at ~250 °C as broadly envisaged by Shock (1992, 1996). Hydrogen, bridging a Ni—Fe site, could react with CO or CO_2 transiently bonded to either metal.

6.8.3 Mackinawite's ($Fe(Ni,Co)_{1+x}S$) 'vital' role

Our X-ray diffraction analysis of the initially amorphous, colloidal FeS membrane that is generated as the sulfhydryl and the ferrous solutions react in the laboratory invariably reveals a mackinawite trace. The crystallites comprising the membrane are a hundred

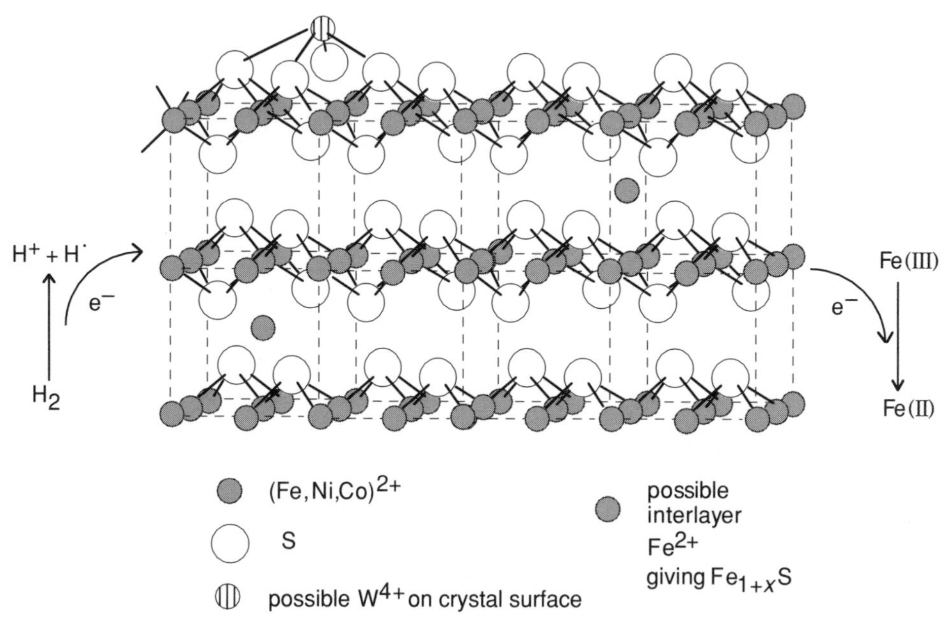

Figure 6.16 Mackinawite, FeS, which can contain some Ni and Co replacing Fe. Note the layered structure with electron transport possible along the metal-rich layers permitting mackinawite to act as an electron transfer agent catalysing the hydrogenation of carboxylic acids with aqueous Fe(III) acting as an electron acceptor. Interlayer iron may be accommodated by electrons contributing to metallic bonding in the iron-rich layer. Tungsten (cf. Kletzin and Adams, 1996; Das *et al.*, 1996), as well as nickel (cf. Volbeda *et al.*, 1995), bonded to sulfur on the crystal surface, may act as catalytic sites.

or so nanometres across (Russell and Hall, 1997). Lennie and Vaughan (1996) have demonstrated the formation of a mackinawite-type structured phase within one second of reaction. Mackinawite is a nonstoichiometric tetragonal mineral often containing some small excess of iron, or perhaps a deficiency of sulfur (Vaughan and Craig, 1978; Lennie and Vaughan, 1996). It is platy, like a clay, because two interfacing sulfur layers are arranged along (001), parallel to the tetragonal base (Figure 6.16). Like a clay, it may be somewhat hydrated as it grows in aqueous conditions, a factor to be considered in modelling likely catalytic activities. The iron atoms are in the low-spin state and are arranged in tetrahedral sites in fully occupied layers alternating with empty layers (Vaughan and Tossell, 1981; Krupp, 1994). They are metallically bonded (Vaughan and Ridout, 1971). Natural mackinawite can accommodate up to 20% nickel, cobalt and copper in its metal layer, and perhaps a small excess of these elements and iron is also bonded tetrahedrally between the sulfur–sulfur layers (Kouvo *et al.*, 1963; Vaughan 1969, 1970) (Figure 6.16). Nickel (or cobalt) and iron atoms anchored to four sulfurs on the surface of a growing mackinawite (Morse and Arakaki, 1993) could control a hydrogen atom just above the (001) plane (see below).

Mackinawite is potentially a major temporary sink for many trace metals during early diagenesis in anoxic conditions, even for calcium (Morse and Arakaki, 1993). Morse and Arakaki (1993) demonstrate that the surface affinity of mackinawite during adsorption of Mn^{2+}, Cr^{3+}, Co^{2+}, Ni^{2+}, Zn^{2+}, Cd^{2+} and Cu^{2+} is in the order of their decreasing solubility as sulfides. The adsorption of W^{4+} was not considered in these experiments. We speculate that WS_4^{2-} ions would present themselves to the metal layer. A tungsten atom at the apex of this pyramidal structure could hold a carbon oxide entity in the same way as effected by tungsten enzymes and their model compounds (Chan *et al.*, 1995; Kletzin and Adams, 1996; Sarkar and Das, 1992; Das *et al.*, 1996) (Figure 6.16).

A tetrahedrally coordinated zinc complex such as $Zn(OH)(SH)_3^{2-}$ in the alkaline hydrothermal solution, sorbed on the mackinawite surface, might have acted as a prebiotic carbonic anhydrase (cf. Hayashi *et al.*, 1990; Allen, 1988) (and see Section 6.7.5 and Figure 6.12).

Metal–metal distances in the (111) plane could be as short as 2.6 Å (Taylor and Finger, 1970; Vaughan and Craig, 1978). This is comparable to the 2.5 Å spacing between Ni and Fe in the crystal structure of the model complex $[Fe_3Ni(CO)_{12}]^{2-}$ synthesised by Ceriotti *et al.* (1984). These authors show how bimetallic Fe–Ni carbonyl clusters can ligate hydrogen atoms just above two iron atoms and one nickel atom in this complex. So we suggest mackinawite to be the first hydrogenase and recall that the Ni–Fe distance in the hydrogenase from *Desulfovibrio gigas* is 2.55 Å or 2.7 Å, depending on whether it is activated or not (Volbeda *et al.*, 1995). Moreover the Ni–Fe in the active site is bonded through two cysteine sulfur atoms. Nickel–iron 'pairs' on the surface or edge of mackinawite crystallites could cleave H_2 as a stage towards hydrogenation of carbon oxides, rather as the hydrogenase operates. An electron could be lost to the Ni^{II} centre, to be conducted along the iron layers situated along (001), ultimately finding its way to an electron acceptor such as Fe^{III} on the outside of the membrane.

In marked contrast, metal–metal distances normal to (001) are 5.03 Å and, because of the two intervening sulfur layers, the mineral offers strong electrical resistance in this 'direction' (Vaughan and Craig, 1978; Krupp, 1994). The sulfur layers themselves are held together by weak Van der Waals bonds (Vaughan and Ridout, 1971).

Although we concentrate on nickel, tungsten and zinc here, the presence of other metals adsorbed on to the mackinawite surface confers, at least in theory, a heterogeneous catalytic capability to this mineral. This is particularly the case for elements having propensities for attracting various anionic ligands. One possibility is that calcium adsorbed on the surface would ligate phosphate, as it does in the mineral apatite $Ca_5(PO_4)_3(F,OH,Cl)$. Whether diphosphates could act as ligands is unknown though worthy of investigation. As Keefe and Miller (1995) point out, the only known diphosphate mineral is canaphite $CaNa_2PO_7{\cdot}4H_2O$ (Peacor *et al.*, 1985; Rouse *et al.*, 1988), which occurs in vugs in basalts associated with zeolites. It could indicate a disposition for pyrophosphate generation during phosphorylation of adjacent calcium atoms on the mackinawite surface in the relatively anhydrous conditions within the membrane in the presence of protons. Or it may be that iron itself acted as a site for the diphosphate on the mineral surface.

6.8.4 Greigite (Fe_3S_4) and violarite, ($FeNi_2S_4$)

In this subsection we consider two thiospinels with comparable formulae though with rather different properties (Figure 6.17a). Greigite is considered here to indicate what might take place upon minor oxidation of the iron monosulfide membrane. Nanocrystalline violarite, on the other hand, may have been intrinsic to the sulfide membrane and played a more important catalytic role.

Greigite is generated in our experiments only when we subject a membrane concentrate to more than 7 bars of H_2S at 90 °C. When this is done, the reaction continues to form some pyrite, a mineral with a very different structure (see Figure 6.18) (and see Drobner *et al.*, 1990). The growth of both greigite and pyrite is blocked by CO_2, presumably because of the growth of a carbonate patina. Perhaps the mackinawite–greigite

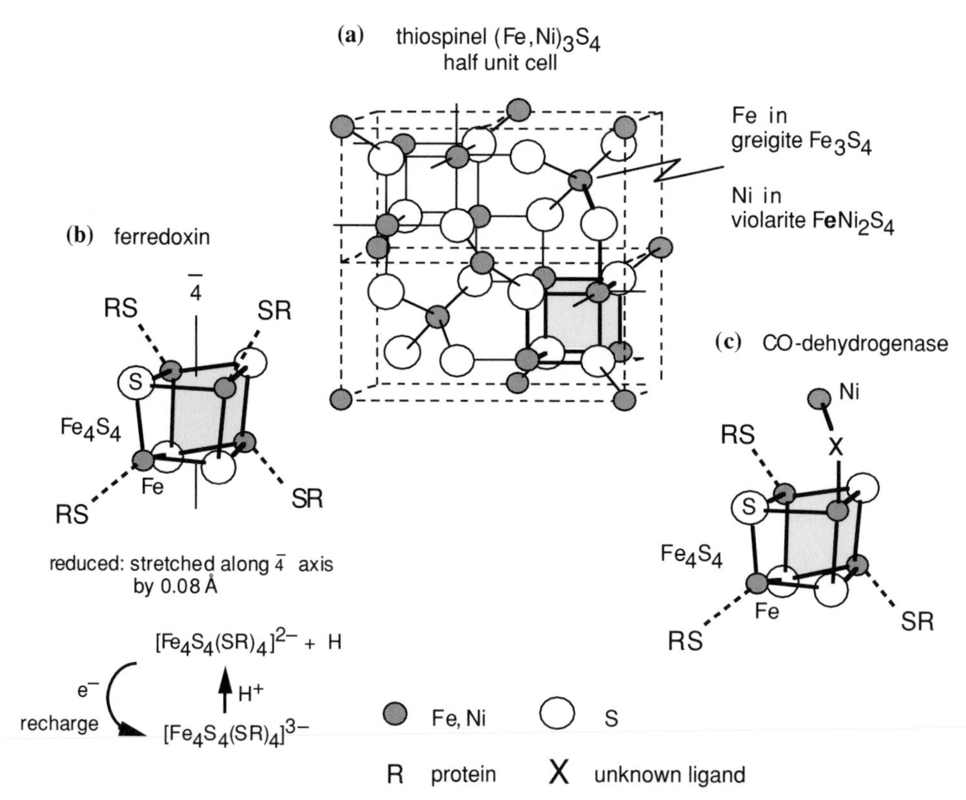

Figure 6.17 Similarity in structural relationship of the $Fe_4^{2.5+}S_4$ 'cubane' unit of the thiospinel greigite Fe_3S_4 (a) with the Fe_4S_4 'thiocubane' unit in protoferredoxins and ferredoxins (b), and the iron–sulfide (Fe_4S_4) core and Ni atom (c) as found in CO-dehydrogenase (Qiu *et al.*, 1994). The latter is also comparable to the $Fe_2Ni_2S_4$ 'cubane' group and tetrahedral Ni site in the thiospinel violarite $FeNi_2S_4$ (a). Fe_4S_4 is variably distorted in ferredoxins depending on the oxidation state (Berg and Holm, 1982). The protoferredoxin would be intermediate in oxidation state between greigite (more oxidised) and mackinawite (more reduced) (modified from Russell and Hall, 1997).

transition was significant in the fixation of carbon oxides at the cell dimension level. Greigite, an inverse spinel, is more ionic than normal spinel (Figure 6.17a) (Vaughan and Craig, 1978). Theoretically, mackinawite should easily oxidise to greigite (through smythite (Fe_9S_{11})) by loss of iron. In this process the remaining one-third of the iron atoms would shift from tetrahedral to octahedral sites (Krupp, 1994). This might be a simple and even reversible process as the cubic close-packed sulfur sublattices would remain undisturbed (Krupp, 1994).

Another aspect of the mineral is that it contains moieties structurally similar to both the cubane $[Fe_4S_4]^{2+}$ structure (where the formal iron oxidation state is $Fe^{2.5+}$) in ferredoxins (Figure 6.17b) and the Fe (formally as Fe^{3+}) centres tetrahedrally coordinated to four sulfurs typically found in the rubredoxins, which have higher redox midpoint potentials. Given these high oxidation states of iron, we surmise that a reduced protoferredoxin $[Fe_4S_4(SR)_4]^{3-}$, self-assembled from hydrothermal thiolates (cf. Bonomi *et al.*, 1985), would be generated at a lower redox potential and would act as a buffer preventing oxidation of mackinawite to greigite.

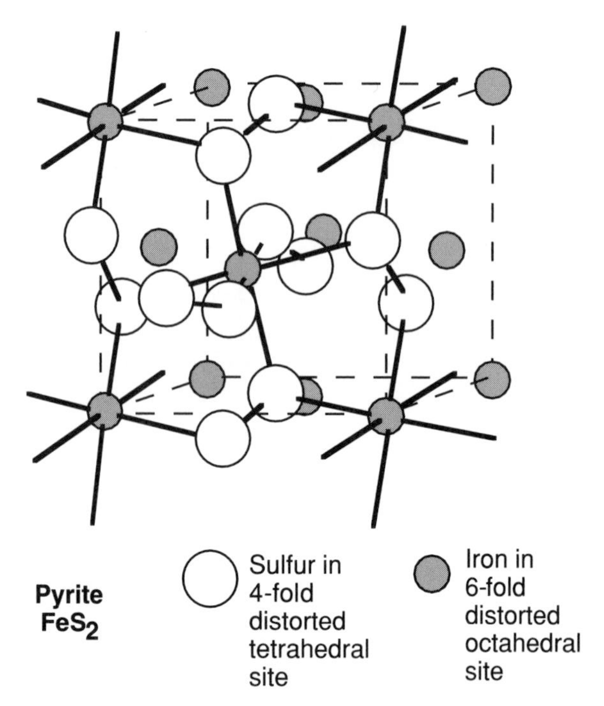

Figure 6.18 Atomic structure of pyrite drawn to show the ferrous iron ligated to six sulfur pairs, $Fe^{2+}(S_2^{2-})_6$. The complex structure makes the mineral difficult to nucleate and very difficult to reduce (Finklea *et al.*, 1976).

In another variation of our membrane experiments, and stimulated by Huber and Wächtershäuser's (1997) work, we added 50% Ni^{2+} to the Fe^{2+} solution presented to bisulfide. In this case violarite ($FeNi_2S_4$) was generated (Figure 6.17a) (along with mackinawite). The monosulfide millerite (NiS), the expected product in these experiments, was not found. Unlike greigite, violarite is a normal thiospinel with extensively delocalised valence electrons (Vaughan and Craig, 1978). In violarite nickel occurs in both the tetrahedral and octahedral sites, whereas the iron is probably restricted the octahedral site in low-spin state (i.e. in the 'cubane' $Fe_2Ni_2S_4$) (Figure 6.17a). With this catalyst we have successfully repeated Huber's and Wächtershäuser's generation of acetic acid. Thus violarite nanocrystals within a predominantly mackinawitic membrane may have the potential to act as a carbon monoxide dehydrogenase (in place of the putative pentlandite of Huber and Wächtershäuser, 1997). After all, in the Ni—X—Fe_4S_4 centre in CO dehydrogenase (Figure 6.17c) the electrons are also extensively delocalised (Qiu *et al.*, 1994). And in this sulfide mineral the sulfurs on the surface are likely to be hydrogenated, leaving the nickel and iron sites open (cf. Qiu *et al.*, 1994).

6.8.5 ***Pyrite (FeS_2)***

Pyrite is the most oxidised of the iron sulfides (Figure 6.18). It has a complex structure. Ferrous iron is ligated to six sulfur pairs, $Fe(S_2^{2-})_6$, making the mineral difficult to nucleate and very difficult to reduce (Finklea *et al.*, 1976; Schoonen and Barnes, 1991). The ferrous iron itself cannot release an electron in these circumstances. For these reasons we do not consider the mineral to have been deeply implicated in the first biochemical cycles (*pace* Wächtershäuser, 1992). The energy available from fluid mixing is theoretically entirely adequate for driving an early metabolism (Shock, 1996; Russell and Hall, 1997). Moreover pyrite's catalytic activity is low when compared to that of mackinawite and the thiospinels.

6.9 Clues from metalloenzymes

Although there are many types of metalloenzyme, here we consider only those which have a metal sulfide centre, i.e. those likely to have had a mineral precursor. Thus we exclude the macrocyclic corrin ligand and the kind (Eschenmoser, 1988), because though ancient (Pratt, 1993), they probably were not synthesised before protometabolism was well under way.

6.9.1 *Simple ferredoxins*

Ferredoxins are the only electron transfer agents which are common to the membranes of all prokaryotes. In 'primitive' forms they play the role of NADP(H) (Daniel and Danson, 1995). They are versatile redox catalysts and some act as hydrogenases. This propensity often requires the incorporation of nickel (Albracht, 1994). Simpler versions, ligated with short polymers of prebiotic organic compounds, would also have been extremely effective (Müller and Schladerbeck, 1985). Moreover, because of their thermodynamic stability, they can spontaneously self-assemble in water (Hagen *et al.*, 1981). Indeed Bonomi *et al.* (1985) have shown that it is possible for $[Fe_4S_4(SC_6H_5)_4]^{2-}$ to assemble in water from $FeCl_2$, sulfur and thiophenol (C_6H_5SH) at STP. Mercaptoethanol ($HOCH_2CH_2SH$) has a similar ligating ability, producing $[Fe_4S_4(SCH_2CH_2OH)_4]^{2-}$, though in this case the sulfurtransferase rhodanese was required to effect self-assembly. Thus such 'protoferredoxins' are likely to have contributed to the early membrane, and played a part in reversing the citric acid cycle (Hartman, 1975; Wächtershäuser, 1990a, Russell and Hall, 1997).

We have synthesised organic molecules which might fill the role of ligands, in the absence of peptides, for such 'protoferredoxins'. For example, dissolving formaldehyde (0.1 mol/l) and magnesium chloride (0.1 mol/l) in a 5 mol/l aqueous solution of ammonium formate and bubbling hydrogen sulfide at a rate of 50 ml/min, we have generated methylaminomethanethiol, carbon/sulfur ring compounds and insoluble polymers which, combined, had the elemental analyses corresponding to the formula $C_8H_{16}S_5N_2$ (Cole *et al.*, 1994; after Kimoto and Fujinaga, 1990). In modern iron sulfide proteins the various operative redox potentials, which range from −0.4 to +0.2 V at pH 7, 25 °C, are obtained largely through the tuning effect the particular protein has on the iron sulfide centre. But as the molecular masses of these early simple ligands were low, the redox potential in the probotryoidal membrane would have been more restricted and approximated that of the mineral pair (FeS/Fe_3S_4 ca. −0.5 V at pH 7, see Figure 6.19).

6.9.2 *Polyferredoxins*

Polyferredoxins are relatively evolved ferredoxins with up to twelve $[Fe_4S_4]^{2+/+}$ centres, a configuration important to methanogenesis (Reeve *et al.*, 1989; Steigerwald *et al.*, 1990;

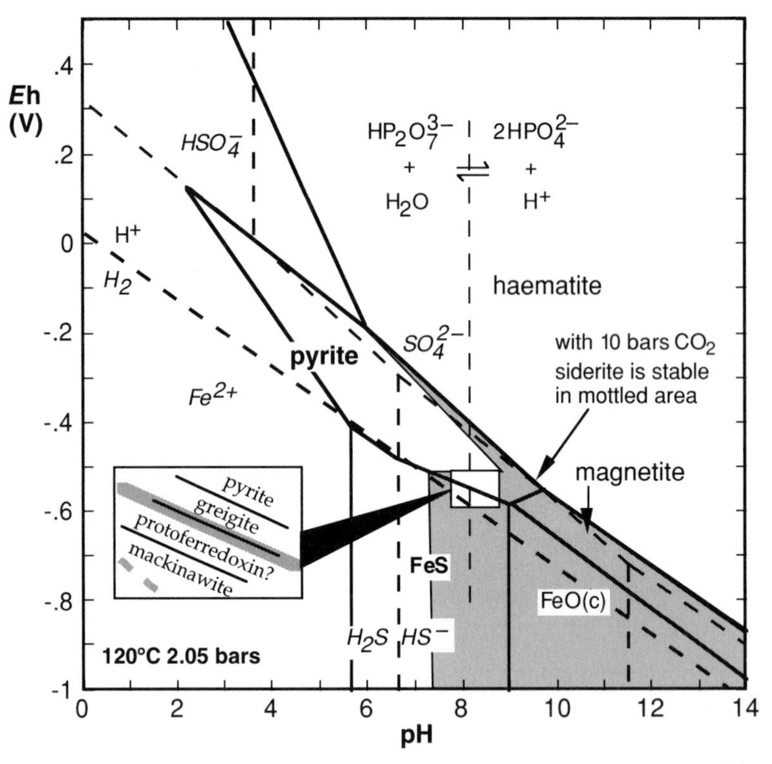

Figure 6.19 *E*h–pH diagram (modified from Russell and Hall, 1997) produced for activities of $H_2S(aq) = 10^{-3}$, and $Fe^{2+} = 10^{-6}$, using Geochemist's Workbench (Bethke, 1996) to illustrate the very low oxidation state required for the stability of iron monosulfide, FeS and the general redox relationships of FeS and pyrite, FeS_2. The inset shows notional phase relations emphasising the intermediate oxidation state of the FeS component of membrane protoferredoxins. The inset is positioned to indicate the *E*h–pH conditions pertaining to alkaline hydrothermal fluid as it enters Hadean sea water. Also illustrated is the influence of high CO_2 atmospheric pressure which leads to a decrease, but not an elimination, of the stability field of FeS in favour of siderite. However, FeS precipitates forming at Hadean submarine hot-spring sites would be prone to carbonation externally to siderite, as well as oxidation to pyrite, unless protected by, for example, adsorbed and bonded abiogenic organic molecules. The pH boundary of monophosphate/polyphosphate is shown. Polyphosphate entering a protocell from acidic sea water could drive condensations, and then be recharged from the monophosphate by the ambient protonmotive force.

Hedderich *et al.*, 1992; Cammack, 1996). They allow electron transfer from one Fe_4S_4 centre to the next, presumably passing electrons through the membrane (Reeve *et al.*, 1989). We suggest that the polyferredoxins are mimicking an earlier 'abiotic' configuration where such centres were perhaps sequestered by the sulfur- and nitrogen-bearing polymers constituting a portion of the early membrane (Kimoto and Fujinaga, 1990; Cole *et al.*, 1994). In this earlier case, electrons could have been translocated through the membrane to the outside where protons and Fe^{III} in the ocean could have accepted them. This mechanism would have displaced the less controlled conduction along (001) in mackinawite mentioned above.

6.9.3 *Complex ferredoxins*

Nickel–iron hydrogenase

Lindahl *et al.* (1990) have demonstrated that carbon monoxide dehydrogenase in anaerobic autotrophic bacteria can reduce CO_2 to CO at a negligible overpotential. Beginning with carbon monoxide and a methyl group, this same enzyme can catalyse the production of an acetyl group. After iron, nickel is one of the most important transition metals for bacterial anabolism. This element often plays a part in hydrogenations (Cammack, 1988). Volbeda *et al.* (1995) have determined the structure of the nickel–iron hydrogenase from *Desulfovibrio gigas*, a bacterium that reduces sulfate in the presence of molecular hydrogen. In this enzyme a nickel atom is ligated to a protein through four cysteines ($HOOCNH_2CHCH_2S^-$). An iron atom between 2.55 Å and 2.7 Å away from the nickel atom is coordinated to two cysteines as well as one carbon monoxide and two cyanide molecules (Happe *et al.*, 1997). Coupled to these active sites are two $[Fe_4S_4]$ and one $[Fe_3S_4]$ sites, arranged in series which allow transfer of electrons to an electron acceptor. The distance between the redox centres averages ~6 Å, and the nickel site is separated from the proximal $[Fe_4S_4]$ centre by about 7 Å (Volbeda *et al.*, 1995). The active catalytic site comprising the nickel–iron centre cleaves molecular hydrogen to a proton (H^+) and a hydride (H^-). The hydrogen may be held initially as an Fe—H—Ni bond (see Ceriotti *et al.*, 1984). The proton can be neutralised by reaction with hydroxide. Then reactive atomic hydrogen $H^{\cdot}$ remains as one of the electrons is conducted away from the hydride to an electron acceptor via the FeS centres (Volbeda *et al.*, 1995). This atomic hydrogen can then contribute to the generation of formaldehyde by reducing the CO_2 or CO bound to the adjacent iron centre. In Section 6.8.3 we speculated that mackinawite may have played this role prior to the development of the enzyme.

Tungstoenzymes

A further biological catalyst or enzyme involved in redox catalysis in highly reduced conditions (i.e. at the hydrogen potential, see Figure 6.19) contains tungsten (Kletzin and Adams, 1996). In fact, tungsten is a required element in the metabolism of all hyperthermophilic bacteria that live at the base of the evolutionary tree (Stetter, 1996). In this enzyme a single tungsten atom is linked through four sulfurs to nitrogen-bearing organic rings (two pterin molecules) which are themselves bridged by a magnesium ion (Kletzin and Adams, 1996, figure 7). This enzyme can oxidise aldehyde (Chan *et al.*, 1995). We have suggested (Russell and Hall, 1997; and Section 6.6) that traces of complexes such as $[W^{VI}S_4]^{2-}$ were carried in the alkaline hydrothermal solution to the hot spring mounds where they reacted with freshly generated organic ligand to form clusters of the type to become significant in some hyperthermophilic archaebacteria (e.g. Mukund and Adams, 1992). Sarkar and Das (1992) have generated a functional model of a tungsten enzyme in the laboratory $[W^{IV}O(S_2C_2(CN)_2)_2]^{2-}$ using simple organic ligands. Given this work, we imagine tungsten(VI) to have been ligated in the early membrane through carbon/sulfur/nitrogen ring complexes of the form generated by Cole *et al.* (1994), to produce $[W^{VI}O(S_2R_2)_2]^0$ (cf. Das *et al.*, 1996). Prior to the synthesis of such organic molecules, tungsten may have been sorbed directly on the (001) surface of mackinawite (Section 6.8.3 and Figure 6.16).

6.10 Tensions across the membrane

De Duve (1991) points out that redox tension amounting to 300 mV is sufficient to drive cellular metabolism. We have shown theoretically that the pH and the *E*h contrasts (see Section 6.4) between a medium-enthalpy hydrothermal solution with the Hadean ocean both amount to about 300 mV and total about 600 mV (Russell and Hall, 1997). This finding holds whatever the nature of the first membrane (Figure 6.19). A combined potential of at least half a volt may be sufficient for chemical reactions to self-organise into a continually metabolising system.

Intriguingly, the chemical potential difference (ΔpH) (Figure 6.19) could be thought of as a natural protonmotive force (pmf). Such a force drives the energy transduction system in all cells. In living systems it is generated from electrons and/or protons acting on the cell membrane. In the model considered here, protons translocating across the membrane could re-dimerise the monophosphate to the diphosphate. Iron, or even calcium, on surface of mackinawite, might act as the docking point for phosphate in the membrane (Section 6.8.3). In Figure 6.19 we show how, in theory, oxidative phosphorylation with inorganic phosphate could be driven by this natural protonmotive force (and see Wood, 1977; Baltscheffsky and Baltscheffsky, 1992). The pH-dependent boundary between the mono- and diphosphate fields intersects the (primarily) *E*h-dependent iron sulfide fields (Figure 6.19). This 'convergence' is an important feature of the hypothesis (Hall and Russell, 1998).

6.11 Discussion

Kinetic or dissipative structures are usually coupled one to another (Prigogine and Stengers, 1984), as in the stacking of convection cells. A particular example is the way a convective magmatic updraft focuses heat which in turn drives a black-smoker-type hydrothermal convection cell within the ocean crust. And a physical/biochemical coupling obtains at black smokers themselves because of the redox potential provided at these hot spring sites (McCollom and Shock, 1997). In the Hadean, a geochemical protometabolism would have been even more strongly coupled and reliant on hydrothermal convection, albeit of the 'manageable', medium-enthalpy variety operating away from oceanic spreading centres.

Yet any attempt to represent the myriad of interactions involved in even a protometabolist is impossible in a linear text. Instead we must imagine that, just as water can find every fracture, channel and interlinked pore space in an attempt to achieve its own level, so physicochemical energy will eventually find all the available pathways open to it and, where some entities and agencies are common to different pathways, will find accommodation by a self-organised pulsation and autocatalytic cycling of matter. And it is the unattainability of equilibrium that keeps the process going.

Understanding life's emergence might be a question of experimenting with recipes and conditions and attempting to track interactions during fluid mixing in pressure vessels. Here we have attempted to list the likely available ingredients and energy sources in order to carry out further 'cook and look' experiments.

6.12 Conclusions

All the energy and chemicals required for protometabolism to be triggered appear to *converge* at a Hadean submarine alkaline hydrothermal seepage (Russell *et al.*, 1994;

Shock, 1996; and see Huber and Wächtershäuser, 1997). A protometabolism, involving transition metals as catalysts, would effect the more rapid spending of the earth's chemical and electrochemical energy 4 billion or so years ago. Thermal convection sees to the dissipation of most of the earth's thermal energy. And in our view the thermal and chemical processes were coupled as life emerged.

We have attempted a semiquantitative assessment of some of the tensions at the surface of the early earth as a prerequisite to understanding how the energies could be spent in a self-organising way, i.e. by a protometabolising system. There is no doubt that the potential for protometabolism exists (McCollom and Shock, 1997). Yet the problem remains, how can the energies actually effect the hydrogenation and amination of the carbon oxides?

In this chapter we have sought to investigate possible pathways towards the fixation of carbon, and to identify the likely available transition metals and their mineral hosts that might be used to synthesise abiotic organic molecules. While nanocrystals of mackinawite and violarite present themselves as likely protometalloenzymes, organic and inorganic models of particular recurring structures in living systems may also help in the elucidation of the mechanisms used by the earliest metabolists (cf. Schrauzer and Sibert, 1970; Karlin, 1993; Das *et al.*, 1996). Given the convergence of views of microbiologists, biochemists and geochemists that a submarine hydrothermal self-organising heat and chemical exchanger could have evolved into life, we anticipate that transdisciplinary research teams intent on elucidating life's origins will focus further upon inorganic–organic interfaces at about 300 K and 300 bars. For, to borrow a phrase from David Garner, 'it is the inorganic elements that bring organic chemistry to life'.

6.13 Summary

The emergence of life was contingent upon the onset of medium-enthalpy hydrothermal convection in the basalts and komatiites comprising the Hadean ocean floor. These convection cells operated at an upper temperature of about 250 °C, controlled by the cracking front in the crust. The initial sink for this (hydro)thermal energy was a recently condensed but by then relatively cool (~100 °C) ocean. There was chemical disequilibrium between the hot seepage waters and the ocean. The sinks for hydrothermal sulfide, HS^-, were oceanic Fe, Co and Ni in a reaction that produced a catalytic membrane. This colloidal FeS membrane, precipitated spontaneously at the seepage site, maintained chemical disequilibrium and an energy potential. Also, the membrane controlled redox reactions, hydrogenations, aminations and condensations as well as catabolic reactions. The ultimate sinks for the hydrothermal H_2 produced by the exothermic reaction between water and ferrous iron silicates of the crust were the CO_2 and CO in the Hadean ocean. The sinks for electrons lost during oxidation were Fe^{3+} and SO_4^{2-}; the sink for oceanic protons was monophosphate and RS^-.

Thus in this hypothetical system the overall thermal contrast (ΔT) amounted to ~150 °C and the ΔpH (4 units) amounted to ~240 mV. The redox potentials (ΔEh) depended on a variety of pairs of complexes in disequilibrium: potentials ranging up to ~700 mV, but assumed to average ~300 mV. Reactions taking place within the catalytic FeS membrane, driven by these potentials, organised themselves via feedback mechanisms into reproducing eobiological cycles that controlled the dissipation of much of the energy available at the hot seepages.

We argue that anabolic processes, such as pre-enzyme hydrogenations and aminations, may have been catalysed by awaruite (Ni_3Fe) in the crust, and Ni-, Co- and W-bearing mackinawite (Fe(Ni,Co,W?)$_{1+x}$S) nanocrystals comprising the membrane at the hot alkaline spring or seepage. Electrons were transferred to Fe^{III} complexes sorbed on the membrane's exterior, along the close-packed Fe^{II} layer in the mackinawite. Localised reductions were encouraged by the mackinawite-greigite [$Fe_{1+x}S \rightarrow Fe_3S_4$] transition in a quarter cell. Nanocrystalline violarite ($FeNi_2S_4$) in the membrane might have acted as a prebiotic CO dehydrogenase. Condensations were mediated by phosphorylations on the mackinawite surfaces. Catabolic processes were catalysed by tungsten adsorbed on the mackinawite. Thiols generated in the system would also have ligated $[Fe_4S_4]^{2+/+}$ cuboids (as $[Fe_4S_4(SHC_2H_5)_4]^{3-/2-}$) which otherwise, on oxidation, would have comprised a portion of greigite. These protoferredoxins would have imparted more control to the protometabolist, a first step in evolution. However, as experimental support for the hypothesis is somewhat meagre, the catalytic propensities of the transition metals are reviewed to inform fresh approaches to the reduction of carbon oxides in the further investigation of this model.

Acknowledgements

We thank Everett Shock, David Vaughan, Mike Adams and David Garner for discussions. Financial support from the NERC grant GR3/09926 is gratefully acknowledged.

References

ADAMS, M. W. W. (1992) Novel iron–sulfur centers in metalloenzymes and redox proteins from extremely thermophilic bacteria. *Adv. Org. Chem.*, **38**, 341–396.

ALBRACHT, S. P. J. (1994) Nickel hydrogenases: in search of the active site. *Biochem. Biophys. Acta*, **1188**, 167–204.

ALEXAKIS, A. and NORMANT, J. F. (1982) Vinyl copper reagents 6. Synthesis of conjugated dienes via the addition of vinyl cuprates to acetylene. *Tetrahedron Lett.*, **23**(49), 5151–5154.

ALLEN, L. C. (1988) Enzymatic activation of carbon dioxide. In *Catalytic Activation of Carbon Dioxide*, ed. W. M. Ayers, vol. 363, pp. 93–101 (Washington, DC: American Chemical Society).

ALVAREZ, R., CARMONA, E., GUTTIEREZ-PUEBLA, E., MARIN, M. J., MONGE, A. and POVEDA, M. L. (1984) Synthesis and X-ray crystal structure of $[Mo(CO_2)_2(PMe_3)_3(CNPr^i)]$. The first structurally characterised bis(carbon dioxide) adduct of a transition metal. *J. Chem. Soc., Chem. Commun.*, 1326–27.

ANDERSON, I. K., ASHTON, J. H., BOYCE, A. J., FALLICK, A. E. and RUSSELL, M. J. (1998) Ore depositional processes in the Navan Zn+Pb deposit, Ireland. *Econ. Geol.*, **93** (in press).

ARESTA, M. and NOBILE, C. F. (1977) (Carbon dioxide)bis(trialkylphosphine)nickel complexes. *J. Chem. Soc., Dalton Trans.*, (7), 708–711.

ARESTA, M., NOBILE, C. F., ALBANO, V. G., FORNI, E. and MANASSERO, M. (1975) New nickel–carbon dioxide complex: synthesis, properties, and crystallographic characterisation of (carbon dioxide)-bis(tricyclohexylphosphine)nickel. *J. Chem. Soc., Chem. Commun.*, 636–637.

AYERS, W. M. and FARLEY, M. (1988) Carbon dioxide reduction with an electric field assisted hydrogen insertion reaction. In *Catalytic Activation of Carbon Dioxide*, ed. W. M. Ayers, vol. 363, pp. 147–154 (Washington, DC: American Chemical Society).

BALLHAUS, C. and ELLIS, D. J. (1996) Mobility of core melts during earth's accretion, *Earth Planet. Sci. Lett.*, **143**, 137–145.

BALTSCHEFFSKY, M. and BALTSCHEFFSKY, H. (1992) Inorganic pyrophosphate and inorganic pyrophosphatases. In *Molecular Mechanisms in Bioenergetics*, ed. L. Ernster, pp. 331–348.

BANKS, D. A. and RUSSELL, M. J. (1992) Fluid mixing during ore deposition at the Tynagh base-metal deposit, Ireland. *Eur. J. Mineral.*, **4**, 921–931.

BARDET, R., PERRIN, M., PRIMET, M. and TRAMBOUZE, Y. (1978) Effect of support in the methanation of carbon dioxide on nickel catalysts. *J. Chimie Phys. Physico-Chim. Biol.*, **75**(11–12), 1079–1083.

BARNES, H. L. and CZAMANSKE, G. K. (1967) Solubilities and transport of ore minerals. In *Geochemistry of Hydrothermal Ore Deposits*, ed. H. L. Barnes, pp. 236–333.

BAROSS, J. A. and HOFFMAN, S. E. (1985) Submarine hydrothermal vents and associated gradient environments as sites for the origin and evolution of life. *Origins of Life and Evolution of the Biosphere*, **15**, 327–345.

BARRAULT, J., FORQUY, C., MENEZO, J. C. and MAUREL, R. (1981) Selective hydrocondensation of carbon monoxide to light olefins with alumina-supported iron catalysts. *React. Kinet. Catal. Lett.*, **15**(2), 153–158.

BEHR, A. (1988) *Carbon Dioxide Activation by Metal Complexes* (Weinheim: VCH).

BEHR, A., KANNE, U. and THELEN, G. (1984) Insertion of carbon dioxide into manganaphosphacycloalkanes. *J. Organomet. Chem.*, **269**, C1–3.

BERDNIKOV, V. M. (1975) The energy relations in the reduction of carbon dioxide in aqueous solution and electron affinity of carbon dioxide. *Russian J. Phys. Chem.*, **49**(11), 1771–1772.

BERG, J. M. and HOLM, R. H. (1982) Structures and reactions of iron–sulfur protein clusters and their synthetic analogues. In *Iron–sulfur Proteins*, ed. T. G. Spiro, pp. 1–66 (New York: Wiley-Interscience).

BERNAL, J. D. (1951) *The Physical Basis of Life* (London: Routledge and Kegan Paul).

BERNAL, J. D. (1960) The problem of stages in biopoesis. In *Aspects of the Origin of Life*, ed. M. Florkin, pp. 30–45 (New York: Pergamon Press).

BETHKE, C. (1996) *Geochemical Reaction Modeling* (Oxford: Oxford University Press).

BIANCHINI, C. and MELI, A. (1984) Bifunctional activation of CO_2: a case where the basic and acidic sites are not held in the same structure. *J. Am. Chem. Soc.*, **106**, 2698–2699.

BIANCO, V. D., DORONZO, S. and ROSSI, M. (1972) The reaction of carbon dioxide with hydrido- and dinitrogen complexes of iron. *J. Organomet. Chem.*, **35**, 337–339.

BIANCO, V. D., DORONZO, S. and GALLO, N. (1978) Reaction between carbon dioxide and hydride complexes of cobalt (I). *J. Inorg. Nucl. Chem.*, **40**(10), 1820–1821.

BODNAR, T., COMAN, E., MENARD, K. and CUTLER, A. (1982) Homogenous reduction of ligated carbon dioxide and carbon monoxide to alkoxymethyl ligands. *Inorg. Chem.*, **21**, 1275–1277.

BONOMI, F., WERTH, M. T. and KURTZ, D. M. (1985) Assembly of $Fe_nS_n(SR)^{2-}$(n=2,4) in aqueous media from iron salts, thiols and sulfur, sulfide, thiosulfide plus rhodonase. *Inorg. Chem.*, **24**, 4331–4335.

BOTTO, R. I. and MORRISON, G. H. (1976) Josephinite: a unique nickel–iron. *Am. J. Sci.*, **276**, 241–274.

BOYCE, A. J., COLEMAN, M. L. and RUSSELL, M. J. (1983) Formation of fossil hydrothermal chimneys and mounds from Silvermines, Ireland. *Nature*, **306**, 545–550.

BOYLE, R. W., ALEXANDER, W. M. and ASLIN, G. E. M. (1975) Some observations on the solubility of gold. *Geol. Surv. Can.*, Paper 75–24.

BRATERMAN, P. S., CAIRNS-SMITH, A. G., SLOPER, R. A., TRUSCOTT, T. G. and CRAW, M. (1984) Photo-oxidation of iron(II) in water between pH 7.5 and 4.0. *J. Chem. Soc., Dalton Trans.*, 1441–1445.

BROCK, T. D. (1985) Life at high temperatures. *Science*, **230**, 132–138.

BRYAN, J. C., GEIB, S. J., RHEINGOLD, A. L. and MAYER, J. M. (1987) Oxidative addition of carbon dioxide, epoxides and related molecules to $WCl_2(PMePh_2)_4$ yielding tungsten (IV) oxo, imido, and sulfido complexes. Crystal and molecular structure of $W(O)Cl_2(CO)(PMePh_2)_2$. *J. Am. Chem. Soc.*, **109**, 2826–2828.

CAHIEZ, G., NORMANT, J. F. and BERNARD, D. (1975) Carbonation des divers organocuivreaux. *J. Organomet. Chem.*, **94**, 463–468.

CAIRNS-SMITH, A. G. (1982) *Genetic Takeover and the Mineral Origins of Life* (Cambridge: Cambridge University Press).

CAMMACK, R. (1988) Nickel in metalloproteins. *Adv. Org. Chem.*, **32**, 297–333.

CAMMACK, R. (1995) Splitting molecular hydrogen. *Nature*, **373**, 556–557.

CAMMACK, R. (1996) Iron and sulfur in the origin and evolution of biological energy conversion systems. In *Origin and Evolution of Biological Energy Conversion*, ed. H. Baltscheffsky, pp. 43–69 (Deerfield Beach, FL: VCH).

CARMONA, E., GONZALEZ, F., POVEDA, M. L., MARIN, J. M., ATWOOD, J. L. and ROGERS, R. D. (1983) Reaction of *cis*-$[Mo(N_2)_2(PMe_3)_4]$ with CO_2. Synthesis and characterisation of products of disproportionation and the X-ray structure of a tetrametallic mixed-valence Mo^{II}–Mo^{V} carbonate with a novel mode of carbonate binding. *J. Am. Chem. Soc.*, **105**, 3365–3366.

CERIOTTI, A., CHINI, P., FUMAGALLI, A., KOETZLE, T. F., LONGONI, G. and TAKUSAGAWA, F. (1984) Synthesis of bimetallic Fe–Ni carbonyl structures: crystal structure of $[N(CH_3)3CH_2Ph]$-$[Fe_3Ni(CO)8(\mu\text{-}CO)4(\mu3\text{-}H)]$. *Inorg. Chem.*, **23**, 1363–1368.

CHAN, M. K., MUKUND, S., KLETZIN, A., ADAMS, M. W. W. and REES, D. C. (1995) Structure of a hyperthermophilic tungsopterin enzyme, aldehyde ferredoxin oxidoreductase. *Science*, **267**, 1463–1469.

CHATT, J., HUSSAIN, W. and LEIGH, G. J. (1983) A possible bis(carbon dioxide) adduct of molybdenum (0). *Transition Metal Chem. (Weinheim.)*, **8**(6), 383–384.

CHATT, J., KUBOTA, M., LEIGH, G. J., MARCH, F. C., MASON, R. and YARROW, D. J. (1974) A possible carbon dioxide complex of molybdenum and its rearrangement product di-m-carbonato-bis{carbonyltris(dimethylphenylphosphine)molybdenum}: X-ray crystal structure. *J. Chem. Soc., Chem. Commun.*, 1033–1034.

COLE, W. J., KASCHKE, M., SHERRINGHAM, J. A., CURRY, G. B., TURNER, D. and RUSSELL, M. J. (1994) Can amino acids be synthesised by H_2S in anoxic lakes? *Marine Chem.*, **45**, 243–256.

CORLISS, J. B., BAROSS, J. A. and HOFFMAN, S. E. (1981) An hypothesis concerning the relationship between submarine hot springs and the origin of life on earth. *Oceanol. Acta*, Special issue, 59–69.

COVENEY, R. M., GOEBEL, E. D., ZELLER, E. J., DRESCHHOFF, G. A. M. and ANGINO, E. E. (1987) Serpentinisation and the origin of hydrogen gas in Kansas. *Bull. Am. Assoc. Petroleum Geol.*, **71**, 39–48.

CRABTREE, R. H. (1994) *The Organometallic Chemistry of the Transition Metals*, 2nd edn (New York: Wiley).

CRAGG, B. A. and PARKES, R. J. (1994) Bacterial profiles in hydrothermally active deep sediment layers from Middle Valley (NE Pacific), sites 857 and 858. In *Proceedings of the Ocean Drilling Program, Scientific Results*, ed. M. J. Mottl, E. E. Davis, A. T. Fisher and J. F. Slack, **139**, pp. 509–516.

CRERAR, D. A., SUSAK, N. J., BOROSIK, M. and SCHWARTZ, S. (1978) Solubility of the buffer assemblage pyrite+pyrrhotite+magnetite in NaCl solutions from 200–350 °C. *Geochim. Cosmochim. Acta*, **42**, 1427–1437.

CREUTZ, C. (1993) Carbon dioxide binding to transition metal centres. In *Electrochemical and Electrocatalytic Reactions of Carbon Dioxide*, ed. B. P. Sullivan, K. Krist and H. E. Guard, pp. 19–67 (Amsterdam: Elsevier).

CZAMANSKE, G. K., ERD, R. C., SOKOLOVA, M. N., DOBROVOL'SKAYA, M. G. and DMITRIEVA, M. T. (1979) New data on rasvumite and djerfisherite. *Am. Mineral.*, **64**, 776–778.

CZAMANSKE, G. K., LEONARD, B. F. and CLARK, J. R. (1980) Erdite, a new hydrated sodium iron sulfide mineral. *Am. Mineral.*, **65**, 509–515.

DANIEL, R. M. and DANSON, M. J. (1995) Did primitive microorganisms use nonhaem iron proteins in place of NAD/P? *J. Mol. Evol.*, **40**, 559–563.

DARENSBOURG, D. J., KUDAROSKI, R. and DELORD, T. (1985) Synthesis and X-ray structure of anionic chelating phosphine-acyl derivative of tungsten [cyclic]$[PPh_4][W(CO)_4C(O)$-$CH_2CH_2CH_2PPh_2]$ and the reactivity of its decarbonylated analogue with carbon dioxide. *Organometallics*, **4**, 1094–1097.

DARENSBOURG, D. J., BAUCH, C. G. and OVALLES, C. (1988) Metal-induced transformations of carbon dioxide. In *Catalytic Activation of Carbon Dioxide*, ed. W. M. Ayers, vol. 363, pp. 26–41 (Washington, DC: American Chemical Society).

DAS, S. K., BISWAS, D., MAITI, R. and SARKAR, S. (1996) Modeling the tungsten sites of inactive and active forms of hyperthermophilic *Pyrococcus furiosus* aldehyde ferredoxin oxidoreductase. *J. Am. Chem. Soc.*, **118**, 1387–1397.

DAVIS, E. E., LISTER, C. R. B., WADE, U. S. and HYNDMAN, R. D. (1980) Detailed heat flow measurements over the Juan de Fuca ridge system. *J. Geophys. Res.*, **85**, 299–310.

DE DUVE, C. (1991) *Blueprint for a Cell: The Nature and Origin of Life* (Burlington, NC: Neil Patterson Publishers).

DE PASQUALE, R. J. and TAMBORSKI, C. (1969) Reactions of pentafluorophenylcopper reagent. *J. Org. Chem.*, **34**(6), 1736–1740.

DEMING, J. W. and BAROSS, J. A. (1993) Deep-sea smokers: windows to a subsurface biosphere? *Geochim. Cosmochim. Acta*, **57**, 3219–3230.

DENISE, B., SNEEDEN, R. P. A. and HAMON, C. (1982) Hydrocondensation of carbon dioxide. IV. *J. Mol. Catal.*, **17**(2–3), 359–366.

DROBNER, E., HUBER, H., WÄCHTERSHÄUER, G., ROSE, D. and STETTER, K. O. (1990) Pyrite formation linked with hydrogen evolution under anaerobic conditions. *Nature*, **346**, 742–744.

DUBOIS, D. L. and MIEDANER, A. (1988) Use of stoichiometric reactions in the design of redox catalyst for carbon dioxide reduction. In *Catalytic Activation of Carbon Dioxide*, ed. W. M. Ayers, vol. 363, pp. 42–51 (Washington, DC: American Chemical Society).

EDWARDS, M. R. (1996) Metabolite channeling in the origin of life. *J. theor. Biol.*, **179**, 313–322.

ESCHENMOSER, A. (1988) Vitamin B12: Experiments concerning the origin of its molecular structure. *Angew. Chem. Int. Ed. Engl.*, **27**, 5–39.

FALBE, J. (ed.) (1980) *New Syntheses with Carbon Monoxide* (New York: Springer-Verlag).

FERRIS, J. P. (1992) Chemical markers of prebiotic chemistry in hydrothermal systems. *Origins of Life and Evolution of the Biosphere*, **22**, 109–134.

FINKLEA, S., CATHEY, S. and AMMA, E. L. (1976) Investigation of the bonding mechanism in pyrite using the Mössbauer effect. *Acta Crystallogr.*, **A32**, 529–537.

FLORIANI, C. and FACHINETTI, G. (1974) Sodium [*N,N′*-ethylenebis(salicylideneiminato)-cobaltate(I)], a reversible carbon dioxide carrier. *J. Chem. Soc., Chem. Commun.*, 615–616.

FRESE, K. W. and SUMMERS, D. P. (1988) Electrochemical reduction of aqueous carbon dioxide at electroplated Ru electrodes. In *Catalytic Activation of Carbon Dioxide*, ed. W. M. Ayers, vol. 363, pp. 155–170 (Washington, DC: American Chemical Society).

FYFE, W. S. (1974) Heat of chemical reactions and submarine heat production. *Geophys. J. Roy. Astron. Soc.*, **37**, 213–215.

GAMBAROTTA, S., FLORIANI, C., CHIESI-VILLA, A. and GUASTINI, C. (1985) Carbon dioxide and formaldehyde coordination on molybdenocene to metal and hydrogen bonds of the C_1 molecule in the solid state. *J. Am. Chem. Soc.*, **107**, 2985–2986.

GATES, B. C. (1992) *Catalytic Chemistry* (New York: Wiley).

GENKIN, A. D., TRONEVA, N. V. and ZHURAVLEV, N. N. (1970) The first occurrence in ores of the sulfide of potassium, iron, and copper, djerfisherite. *Geochem. Int.*, **7**, 693–701.

GODLEVSKIY, M. N. (1959) The problem of the genesis of Cu–Ni sulfide deposits in the Siberian platform. *Geol. Rudn. Mestorozd*, No. 2.

GRANICK, S. (1957) Speculations on the origins and evolution of photosynthesis. *Ann. NY Acad. Sci.*, **69**, 292–308.

GRAY, T. J. (1983) Methanation process and Raney catalyst for it. *Eur. Pat. Appl.*, EP 87.771 (7.9.1983), C.A. 99:215603d.

HAACK, U. K. and ZIMMERMANN, H. D. (1996) Retrograde mineral reactions: a heat source in continental crust? *Geol. Runds.*, **85**, 130–137.

HAGEN, K. S., REYNOLDS, J. G. and HOLM, R. H. (1981) Definition of reaction sequences resulting in self-assembly of $[Fe_4S_4(SR)_4]^{2-}$ clusters from simple reactants. *J. Am. Chem. Soc.*, **103**, 4054–4063.

HALL, A. J. (1986) Pyrite–pyrrhotine reactions in nature. *Mineral. Mag.*, **50**, 223–229.

HALL, A. J. and RUSSELL, M. J. (1998) Chemical and redox constraints on the emergence of life. *Nature* (submitted).

HARTMAN, H. (1975) Speculations on the origin and evolution of metabolism. *J. Mol. Evol.*, **4**, 359–370.

HAPPE, R. P., ROSEBOOM, W., PLERIK, A. J., ALBRACHT, S. P. J. and BAGLEY, K. A. (1997) Biological activation of hydrogen. *Nature*, **385**, 126.

HAYASHI, K., SUGAKI, A. and KITAKAZE, A. (1990) Solubility sphalerite in aqueous sulfide solutions at temperatures between 25–240 °C. *Geochim. Cosmochim. Acta*, **54**, 715–725.

HEDDERICH, R., ALBRACHT, S. P. J., LINDER, D., KOCH, J. and THAUER, R. K. (1992) Isolation and characterisation of polyferredoxin from *Methanobacterium thermoautotrophicum*: the *mvhB* gene product of the methylviologen-reducing hydrogenase operon. *FEBS Lett.*, **298**, 65–68.

HEINEN, W. and LAUWERS, A. M. (1996) Organic sulfur compounds resulting from the interaction of iron sulfide, hydrogen sulfide and carbon dioxide in an anaerobic aqueous environment. *Origins of Life and Evolution of the Biosphere*, **26**, 131–150.

HENNET, R. J.-C., HOLM, N. G. and ENGEL, M. H. (1992) Abiotic synthesis of amino acids under hydrothermal conditions and the origin of life: a perpetual phenomenon? *Naturwissenschaften*, **79**, 361–365.

HENRICI-OLIVÉ, G. and OLIVÉ, S. (1984) *The Chemistry of the Catalysed Hydrogenation of Carbon Monoxide* (New York: Springer-Verlag).

HOBERG, H., JENNI, K., ANGERMUND, K. and KRUGER, C. (1987) CC-linkages of ethene with CO_2 on an iron(0) complex – synthesis and crystal structure analysis of $[(PEt_3)_2Fe(C_2H_4)_2]$. *Angew. Chem., Int. Ed. Engl.*, **26**(2), 153–155.

HOYLE, F. (1972) The history of the earth. *Q. J. R. Astron. Soc.*, **13**, 328–345.

HUBER, C. and WÄCHTERSHÄUSER, G. (1997) Activated acetic acid by carbon fixation on (Fe,Ni)S under primordial conditions. *Science*, **276**, 245–247.

HUNTER, J. (1859) *Observations and Reflections on Geology* (London: Taylor and Francis).

ICHIKAWA, M. (1979) Catalyst carrying cluster compounds. *Jpn Kokai Tokkyo Koho, Jpn Pat.*, 79.41.291 (2.4.1979), C.A. 91:10046n.

IKARIYA, T. and YAMAMOTO, A. (1974) Preparation and properties of ligand-free methylcopper and of copper alkyls coordinated with 2,2′-bipyridyl and tricyclohexyl phosphine. *J. Organomet. Chem.*, **72**, 145–151.

IKARIYA, T. and YAMAMOTO, A. (1976) Preparation and properties of methyliron complexes with tertiary phosphine ligands and their decomposition pathways through the formation of carbenoid intermediates. *J. Organomet. Chem.*, **118**, 65–78.

INOUE, Y., ITOH, Y. and HASHIMOTO, H. (1977) Incorporation of carbon dioxide in alkyne oligomerisation catalysed by nickel (0) complexes. Formation of substituted 2-pyrones. *Chem. Lett.*, (8), 855–856.

JOHNSON, M. K. (1996) Iron–sulfur proteins. In *Encyclopedia of Inorganic Chemistry*, ed. R. B. King, vol 4, pp. 1896–1915 (Chichester: Wiley).

JOLLY, P. W., JONAS, K., KRÜGER, C. and TSAY, Y.-H. (1971) The preparation, reactions and structure of bis[bis-(tricyclohexylphosphine)nickel] dinitrogen. *J. Organometal. Chemi.*, **33**, 109–122.

KARLIN, K. D. (1993) Metalloenzymes, structural motifs, and inorganic models. *Science*, **261**, 701–708.

KARSCH, H. H. (1977) Functional derivatives of trimethylphosphine. III. Ambivalent behavior of tetrakis(trimethylphosphine)iron: reaction with carbon dioxide. *Chemi. Beri.*, **110**(6), 2213–2221.

KARSCH, H. H. (1984) Complexes with phosphinomethanes and -methanides as ligands. VIII. Iron (0) complexes with $Me_2PCH_2PMe_2$ and $Me_2P[CH_2]_3PMe_2$ ligands: C–H activation versus pentacoordination. *Chem. Ber.*, **117**, 3123–3133.

KASCHKE, M., RUSSELL, M. J. and COLE, J. W. (1994) $[FeS/FeS_2]$. A redox system for the origin of life. *Origins of Life and Evolution of the Biosphere*, **24**, 43–56.

KASTING, J. F., EGGLER, D. H. and RAEBURN, S. P. (1993) Mantle redox evolution and the oxidation state of the Archaean atmosphere. *J. Geol.*, **101**, 245–257.

KEEFE, A. D. and MILLER, S. L. (1995) Are polyphosphates or phosphate esters prebiotic reagents? *J. Mol. Evol.*, **41**, 693–702.

KEENE, F. R. (1993) Thermodynamic, kinetic, and product considerations in carbon dioxide reactivity. In *Electrochemical and Electrocatalytic Reactions of Carbon Dioxide*, ed. B. P. Sullivan, K. Krist and H. E. Guard, pp. 118–144 (Amsterdam: Elsevier).

KIMOTO, T. and FUJINAGA, T. (1990) Non-biotic synthesis of organic polymers on H_2S-rich seafloor. A possible reaction in the origin of life. *Marine Chem.*, **30**, 179–192.

KLETZIN, A. and ADAMS, M. W. W. (1996) Tungsten in biological systems. *FEMS Microbiol. Rev.*, **18**, 5–63.

KOLOMNIKOV, I. S., STEPOVSKA, G., TYRLIK, S. and VOL'PIN, M. E. (1972) Introduction of carbon dioxide at the cobalt–carbon and cobalt–hydrogen bonds. *Zh. Obs. Khim.*, **42**(7), 1652.

KOLOMNIKOV, I. S., LISYAK, T. V., KONASH, E. A., RUDNEV, A. V., KALYAZIN, E. P. and KHARITONOV, U. J. (1982) Reduction of CO_2 in aqueous solutions in presence of titanium dioxide and γ-radiation. *Zh. Org. Khim.*, 528–529.

KONNERT, J. A. and EVANS, H. T. (1980) Crystal structure of erdite, $NaFeS_2{\cdot}2H_2O$. *Am. Mineral.*, **65**, 516–521.

KOUVO, O., VUORELAINEN, Y. and LONG, J. V. P. (1963) A tetragonal iron sulfide. *Am. Mineral.*, **48**, 511–524.

KRISHNARAO, J. S. R. (1964) Native nickel–iron alloy, its mode of occurrence, distribution and origin. *Econ. Geol.*, **59**, 443–448.

KRUPP, R. E. (1994) Phase relations and phase transformations between the low-temperature iron sulfides mackinawite, greigite, and smythite. *Eur. J. Mineral.*, **6**, 265–278.

KRUPP, R. E., OBERTHUR, TH. and HIRDES, W. (1994) Composition of the early precambrian atmosphere and hydrosphere, thermodynamic constraints from mineral deposits. *Mineral. Mag.*, **58A**, 499–500.

LAFON, G. M. and MACKENZIE, F. T. (1974) Early evolution of the oceans: a weathering model. In *Studies in Paleo-oceanography*, ed. W. W. Hay, pp. 205–218 (Tulsa: Society of Economic Paleontolgists and Mineralogists, Special Publication 20).

LAKE, J. A. (1988) Origin of eukaryotic nucleus determined by rate-invariant analysis of rRNA sequences. *Nature*, **331**, 184–186.

LEECH, P. and SYKES, C. (1939) The evidence for a superlattice in the nickel–iron alloy Ni_3Fe. *Phil. Mag.*, **27**, 948–960.

LENNIE, A. R. and VAUGHAN, D. J. (1996) *Spectroscopic studies of iron sulfide formation and phase relations at low temperatures*. Geochemical Society Special Publication No. 5117–5131.

LEWIS, N. S. and SHREVE, G. A. (1993) Photochemical and photoelectrochemical reduction of carbon dioxide. In *Electrochemical and Electrocatalytic Reactions of Carbon Dioxide*, ed. B. P. Sullivan, K. Krist and H. E. Guard, pp. 263–289 (Amsterdam: Elsevier).

LIDE, D. R. and FREDERIKSE, H. P. R. (1993–1994) *CRC Handbook of Chemistry and Physics*, 74th edn (London: CRC Press).

LINDAHL, P. A., MÜNCK, E. and RAGSDALE, S. W. (1990) CO dehydrogenase from *Clostridium thermoacetium*. *J. Biol. Chem.*, **265**, 3873–3879.

LISTER, C. R. B. (1975) On the penetration of water into hot rock. *Geophys. J. Roy. Astron. Soc.*, **39**, 465–509.

MACLEOD, G., MCKEOWN, C., HALL, A. J. and RUSSELL, M. J. (1994) Hydrothermal and oceanic pH conditions of possible relevance to the origin of life. *Origins of Life and Evolution of the Biosphere*, **23**, 19–41.

MAHER, J. M., LEE, Y. R. and COOPER, N. J. (1982) Evidence for oxide transfer from coordinated CO_2 to coordinated CO in anionic CO_2 complex. *J. Am. Chem. Soc.*, **104**, 6797–6799.

MARCH, J. (1992) *Advanced Organic Chemistry, Reactions Mechanism and Structure*, 4th edn (New York: Wiley).

Marisch, N., Camus, A. and Nardin, G. (1982) Reaction of carbon dioxide with arylcopper (I) complexes containing tertiary phosphines. *J. Organometal. Chem.*, **239**, 429–437.

McCollom, T. M. and Shock, E. L. (1997) Geochemical constraints on chemolithoautotrophic metabolism by microorganisms in sea floor hydrothermal systems. *Geochim. Cosmochim. Acta*, **61**, 4375–4391.

Miles, M. H. and Fletcher, A. N. (1988) Electrochemical studies of carbon dioxide and sodium formate in aqueous solutions. In *Catalytic Activation of Carbon Dioxide*, ed. W. M. Ayers, vol. 363, pp. 171–178 (Washington, DC: American Chemical Society).

Mills, H., Halliday, A. N., Ashton, J. H., Anderson, I. K. and Russell, M. J. (1987) Origin of a giant orebody at Navan, Ireland. *Nature*, **327**, 223–225.

Miyashita, A. and Yamamoto, A. (1973) Insertion of carbon dioxide into the Cu–CH_3 bond of methylbis(triphenylphosphine)copper etherate. *J. Organometal. Chem.*, **49**, C57–58.

Miyashita, A. and Yamamoto, A. (1976) Insertion reactions of carbon dioxide and carbon disulfide into alkyl-copper bonds of alkylcopper (I) complexes having tertiary phosphine ligands. *J. Organometal. Chem.*, **113**, 187–200.

Morse, J. W. and Arakaki, T. (1993) Adsorption and coprecipitation of divalent metals with mackinawite (FeS). *Geochim. Cosmochim. Acta*, **57**, 3635–3640.

Mueller, A. G. (1991) The Savage Lode magnesian skarn in the Marvel Loch gold–silver mine, Southern Cross greenstone belt, Western Australia. Part 1: Structural setting, petrography, and geochemistry. *Can. J. Earth Sci.*, **28**, 659–685.

Mukund, S. and Adams, M. W. W. (1991) The novel tungsten–iron–sulfur protein of the hyperthermophilic Archaebacterium, *Pyrococcus furiosus*, is an aldehyde ferredoxin oxireductose. *J. Biol. Chem.*, **266**, 14208–14216.

Müller, A. and Schladerbeck, N. (1985) Systematik der Bildung von Elektronentransferclusterzentrum $\{Fe_nS_n\}^{m+}$ mit Relevanz zur Evolution der Ferredoxine. *Chimia*, **39**, 23–24.

Nekrasov, I. Ya. and Konyushok, A. A. (1982) The physicochemical conditions of tungstenite formation. *Mineral. Zh.*, **4**, 33–40 (in Russian).

Nenitescu, C. D. (1984) *Chimie Generala*, 5th edn (Bucharest: Editura Didactica si Pedagogica).

Nickel, E. H. (1959) Nickel–iron in the serpentine rock of the eastern townships of Quebec Province. *Can. Mineral.*, **6**, 307–319.

Nieman, J. and Teuben, J. (1985) Reactions of vanadocene-carbyls with carbon monoxide, xylylisocyanide and carbon dioxide. *J. Organometal. Chem.*, **287**, 207–204.

Nieman, J., Pattiasina, J. W. and Teuben, J. (1984) Synthesis and characterisation of bisbenzyl and bis-allyl complexes of titanium (III) and vanadium (III); catalytic isomerisation of alkenes with $CpV(h^3\text{-}C_3H_5)_2$. *J. Organometal. Chem.*, **262**, 157–169.

Normant, J. F. and Cahiez, G. (1983) Organomanganous reagents: their use in organic synthesis, Part 2. In *Modern Synthetic Methods*, ed. R. Scheffold, vol. 3, pp. 173–216 (Frankfurt am Main: Verlag Salle/Sauerländer).

Normant, J. F., Cahiez, G., Chuit, C. and Villieras, J. (1973) Stereospecific synthesis of di- and tri-substituted acids by carbonation of vinyl-copper reagents. *J. Organometal. Chem.*, **54**, C53–56.

Normant, J. F. and Cahiez, G. (1974) Organocuivreaux vinyliques III. Carbonatation: preparation stereospecifique d'acides a-ethyleniques et de quelques derives. *J. Organometal. Chem.*, **77**, 281–287.

O'Connell, C., Hommeltoft, S. I. and Eisenberg, R. (1987) Electrochemical approaches to the reduction of carbon dioxide. In *Carbon Dioxide as a Source of Carbon*, ed. M. Aresta and G. Forti, pp. 33–54 (Dordrecht: Reidel).

Orgel, L. E. and Crick, F. H. C. (1993) Anticipating an RNA world, some past speculations on the origin of life: Where are they today? *Fed. Proce. Fed. Am. Soc. Exp. Biol.*, **7**, 238–239.

Osborn, A. M., Bruce, K. D., Strike, P. and Ritchie, D. A. (1995) Sequence conservation between regulatory mercury resistance genes in bacteria from mercury polluted and pristine environments. *System. Appl. Microbiol.*, **18**, 1–6.

Patchett, P. J. (1996) Scum of the earth after all. *Nature*, **382**, 758–759.

PEACOR, D. R., DUNN, P. J., SIMMONS, W. B. and WICKS, F. J. (1985) Canaphite, a new calcium phosphate hydrate from the Paterson area, New Jersey. *Mineral. Rec.*, **16**, 467–468.

PHILLIPS, G. N. and GROVES, D. I. (1983) The nature of Archaean gold fluids as deduced from gold deposits in Western Australia. *Geol. Soc. Austr. J.*, **30**, 25–29.

PHILLIPS, G. N., GROVES, D. I. and MARTYN, J. E. (1984) An epigenetic origin for Archaean banded iron-formation-hosted gold deposits. *Econ. Geol.*, **79**, 162–171.

PIJOLAT, M. and PERICHON, V. (1982) Demonstration of three steps in the hydrogenation of carbon dioxide on iron/aluminum oxide catalyst. *C. R. Seances Acad. Sci., Serie 2*, **295**(3), 343–346.

PINE, R. J. and BATCHELOR, A. S. (1984) Downward migration of shearing in a jointed rock during hydraulic injections. *Int. J. Rock Mech., Miner. Sci. Geomech. Abstr.*, **21**, 249–263.

PÓSFAI, M., BUSECK, P. R., BAZYLINSKI, D. A. and FRANKEL, R. B. (1988) Reaction sequence of iron sulfide minerals in bacteria and their use as biomarkers. *Science*, **280**, 880–883.

PRATT, J. M. (1993) Nature's design and use of catalysts based on Co and the macrocycle corrin ligand: 4×10^9 years of coordination chemistry. *Pure Appl. Chem.*, **65**, 1513–1520.

PRIGOGINE, I. and STENGERS, I. (1984) *Order Out of Chaos* (London: Heinemann).

QIU, D., KUMAR, M., RAGSDALE, S. W. and SPIRO, T. G. (1994) Nature's carbonylation catalyst: Raman spectroscopic evidence that carbon monoxide binds to iron, not nickel, in CO dehydrogenase. *Science*, **264**, 817–819.

RAZUVAEV, G. A., LATAYAEVA, V. N., VYSHINSKAYA, L. I. and DROBOTENKO, V. V. (1981) Synthesis and properties of covalent tri- and tetravalent vanadium. *J. Organometal. Chem.*, **208**, 169–182.

REEVE, J. N., BECKLER, G. S., CRAM, D. S. *et al.* (1989) A hydrogenase-linked gene in *Methanobacterium thermoautotrophicum* strain DH encodes a polyferredoxin. *Proc. Nat. Acad. Sci. USA*, **86**, 3031–3035.

ROUSE, R. C., PEACOR, D. R. and FREED, R. L. (1988) Pyrophosphate groups in the structure of canaphite, $CaNa_2P_2O_7{\cdot}4H_2O$: the first occurrence of a condensed phosphate as a mineral. *Am. Mineral.*, **73**, 168–171.

RUSSELL, M. J. (1978) Downward-excavating hydrothermal cells and Irish-type ore deposits: importance of an underlying thick Caledonian Prism. *Trans. Inst. Mining Metall., (Appl. Earth Sci. sect. B)*, **89**, B168–171.

RUSSELL, M. J. (1988) Chimneys, chemical gardens and feldspar horizons±pyrrhotine in some SEDEX deposits: aspects of alkaline environments of deposition. In *Proceedings of the Seventh IAGOD Symposium*, ed. E. Zachrisson, pp. 183–190 (Schweizerbartsche Verlagsbuchhandlung).

RUSSELL, M. J. (1996) The generation at hot springs of ores, microbialites and life. *Ore Geol. Rev.*, **10**, 199–214.

RUSSELL, M. J. and HALL, A. J. (1997) The emergence of life from iron monosulfide bubbles at a submarine hydrothermal redox and pH front. *J. Geol. Soc. London*, **154**, 377–402.

RUSSELL, M. J. and SKAULI, H. (1991) A history of theoretical developments in carbonate-hosted base metal deposits and a new tri-level enthalpy classification. *Econ. Geol.: Monogr.*, **8**, 96–116.

RUSSELL, M. J., HALL, A. J., CAIRNS-SMITH, A. G. and BRATERMAN, P. S. (1988) Submarine hot springs and the origin of life [Correspondence]. *Nature*, **336**, 117.

RUSSELL, M. J., HALL, A. J. and TURNER, D. (1989) In vitro growth of iron sulfide chimneys: possible culture chambers for origin-of-life experiments. *Terra Nova*, **1**, 238–241.

RUSSELL, M. J., DANIEL, R. M., HALL, A. J. and SHERRINGHAM, J. (1994) A hydrothermally precipitated catalytic iron sulfide membrane as a first step toward life. *J. Mol. Evol.*, **39**, 231–243.

RUSSELL, M. J., COUPLES, G. D. and LEWIS, H. (1995) SEDEX genesis and super-deep boreholes: do hydrostatic pore pressures exist down to the brittle-ductile boundary? In *Mineral Deposits: From Their Origin to Their Environmental Impact*, ed. J. Pasava, B. Kribek and K. Zak, pp. 315–318 (Rotterdam: A. A. Balkema).

SAMSON, I. M. and RUSSELL, M. J. (1987) Genesis of the Silvermines zinc–lead–barite deposit, Ireland: fluid inclusion and stable isotope evidence. *Econ. Geol.*, **82**, 371–394.

SARKAR, S. and DAS, S. K. (1992) CO_2 fixation by $[W^{IV}O(S_2C_2(CN)_2)_2]^{2-}$: functional model for the tungsten-formate dehydrogenase of *Clostridium thermoaceticum, Proce. Ind. Acad. Sci. (Chem. Sci.)*, **104**, 533–534.

SCHOONEN, M. A. A. and BARNES, H. L. (1991) Mechanisms of pyrite and marcasite formation from solution: III. Hydrothermal processes. *Geochim. Cosmochim. Acta*, **55**, 3491–3504.

SCHRAUZER, G. N. and SIBERT, J. W. (1970) Acetate synthesis from carbon dioxide and methylcorrinoids. Simulation of the microbial carbon dioxide fixation reaction in a model system. *J. Am. Chem. Soc.*, **92**, 3509–3510.

SCHWARTZ, M., VERCAUTEREN, M. E. and SAMMELS, A. F. (1994) Fischer–Tropsch electrochemical CO_2 reduction to fuels and chemicals. *J. Electrochem. Soc.*, **141**(11), 3119–3127.

SCHWARTZMAN, D., VOLK, T. and MCMENAMIN, M. (1993) Did surface temperatures constrain microbial evolution? *Bioscience*, **43**, 390–393.

SCHWARTZMAN, D., SHORE, S. N., MCMENAMIN, M. and VOLK, T. (1994) Self-organisation of the earth's biosphere – geochemical or geophysiological? *Origins of Life and Evolution of the Biosphere*, **24**, 435–450.

SEWARD, T. M. (1973) Thio complexes of gold and the transport of gold in hydrothermal ore solution. *Geochim. Cosmochim. Acta*, **37**, 379–399.

SEWARD, T. M. and BARNES, H. L. (1997) Metal transport in hydrothermal ore fluids. In *Geochemistry of Hydrothermal Ore Deposits*, 3rd edn, ed. H. L. Barnes, pp. 435–486 (New York: Wiley).

SHELDON, R. A. (1983) *Chemicals from Synthesis Gas: Catalytic Reactions of CO and H_2* (Boston: D. Reidel).

SHOCK, E. L. (1992) Chemical environments of submarine hydrothermal systems. *Origins of Life and Evolution of the Biosphere*, **22**, 67–107.

SHOCK, E. L. (1996) Hydrothermal systems as environments for the emergence of life. In *Evolution of Hydrothermal Ecosystems on Earth (and Mars?)*, ed. G. R. Bock and J. A. Goode, pp. 40–60 (CIBA Foundation Symposium No. 202).

SHOCK, E. L. and SCHULTE, M. D. (1995) Hydrothermal systems as locations of organic synthesis on the early Earth and Mars. *EOS, Trans. Am. Geophys. Union*, **76**/46 Supplement (P21A-6), F335.

SHOCK, E. L., MCCOLLOM, T. and SCHULTE, M. D. (1995) Geochemical constraints on chemolithoautotrophic reactions in hydrothermal systems. *Origins of Life and Evolution of the Biosphere*, **25**, 141–159.

SOKOLOVA, M. N., DOBROVOL'SKAYA, M. G., ORGANOVA, N. I., KAZAKOVA, M. E. and DMITRIK, A. L. (1970) A sulfide of iron and potassium – the new mineral rasvumite, *Vses. Mineralog Obschch. Zap*, **99**, 712–720.

SPEIER, G., SIMON, A. and MARKO, L. (1977) The reaction of carbon dioxide and dioxigen with dinitrogentris(triphenylphosphine)cobalt (0). *Acta Chim. Acad. Sci. Hung.*, **92**, 169.

STEIGERWALD, V. J., BECKLER, G. S. and REEVE, J. N. (1990) Conservation of hydrogenase and polyferredoxin structures in the hyperthermophilic Archaebacterium *Methanothermus fervidus. J. Bacteriol.*, **172**, 4715–4718.

STETTER, K. O. (1996) Hyperthermophilic prokaryotes. *FEMS Microbi. Rev.*, **18**, 149–158.

STRENS, M. R., CANN, D. L. and CANN, J. R. (1987) A thermal balance model of the formation of sedimentary-exhalative lead–zinc deposits. *Econ. Geol.*, **82**, 1192–1203.

SULLIVAN, B. P., BRUCE, M. R. M., O'TOOLE, R. *et al.* (1988) Electrocatalytic carbon dioxide reduction. In *Catalytic Activation of Carbon Dioxide*, ed. W. M. Ayers, vol. 363, pp. 52–90 (Washington DC: American Chemical Society).

SULLIVAN, B. P., KRIST, K. and GUARD, H. E. (1993) *Electrochemical and Electrocatalytic Reactions of Carbon Dioxide* (Amsterdam: Elsevier Sciencae).

TAYLOR, L. A. and FINGER, L. W. (1970) Structural refinement and composition of mackinawite. *Carnegie Inst. Wash. Geophys. Lab. Annu. Rep.*, **69**, 318–322.

TAYLOR, H. S. (1925) *Treatise on Physical Chemistry* (New York: Van Nostrand).

TEZUKA, M., YAJIMA, T., TSUCHIYA, A., MATSUMOTO, Y., UCHIDA, Y. and HIDAI, M. (1982) Electroreduction of carbon dioxide catalysed by iron–sulfur clusters $[Fe_4S_4(SR)_4]^{2-}$. *J. Am. Chem. Soc.*, **104**, 6834–6836.

TSUDA, T. and SAEGUSA, T. (1972) Reaction of cupric methoxide and carbon dioxide. *Inorg. Chem.*, **11**(10), 2561–2563.

TSUDA, T., UEDA, K. and SAEGUSA, T. (1974) Carbon dioxide insertion into organocopper and organosilver compounds. *J. Chem. Soc., Chem. Commun.*, 380–381.

TSUDA, T., CHUJO, Y. and SAEGUSA, T. (1975) Reversible carbon dioxide fixation by organocopper complexes. *J. Chem. Soc., Chem. Commun.*, 963–964.

TSUDA, T., CHUJO, Y. and SAEGUSA, T. (1976) Copper (I) cyanoacetate as a carrier of activated carbon dioxide. *J. Chem. Soc., Chem. Commun.*, 415–416.

TSUDA, T., CHUJO, Y. and SAEGUSA, T. (1978) Copper complex acting as a reversible carbon dioxide carrier. *J. Am. Chem. Soc.*, **100**, 630–632.

TSUDA, T., CHUJO, Y. and SAEGUSA, T. (1979) Formation of γ-lactones by the reaction of p-allylnickel complexes with carbon dioxide. *Synth. Commun.*, **9**, 427–429.

ULRICH, G. H. F. (1890) On the discovery, mode of occurrence and distribution of nickel–iron alloy awaruite on the west coast of South Island, New Zealand. *Q. J. Geol. Soc. London*, **46**, 619–633.

VAUGHAN, D. J. (1969) Nickelian mackinawite from Vlakfontein, Transvaal. *Am. Mineral.*, **54**, 1190–1193.

VAUGHAN, D. J. (1970) Nickelian mackinawite from Vlakfontein, Transvaal: a reply. *Am. Mineral.*, **55**, 1807–1808.

VAUGHAN, D. J. and CRAIG, J. R. (1978) *Mineral Chemistry of Natural Sulfides* (Cambridge: Cambridge University Press).

VAUGHAN, D. J. (1985) The crystal chemistry of iron–nickel thiospinels. *Am. Mineral.*, **70**, 1036–1043.

VAUGHAN, D. J. and RIDOUT, M. S. (1971) Mössbauer studies of some sulfide minerals. *J. Inorg. Nucl. Chem.*, **33**, 741–746.

VAUGHAN, D. J. and TOSSELL, J. R. (1981) Electronic structure of thiospinel minerals: results from MO calculations. *Am. Mineral.*, **66**, 1250–1253.

VOEVODIN, V. N., GARAN, V. I., ZHITKOV, N. G., PERMYAKOV, A. P. and TSOPANOV, O. KH. (1979) Tungsten mineralisation in listvenites of the Tamvatnei ore cluster. *Geology Rudnykh Mestorozhdeniy*, **3**, 43–55 (in Russian).

VOLBEDA, A., CHARON, M. H., PIRAS, C., HATCHIKIAN, E. C., FREY, M. and FONTECILLA-CAMPS, J. C. (1995) Crystal structure of the nickel–iron hydrogenase from *Desulfovibrio gigas*. *Nature*, **373**, 580–587.

VON DAMM, K. L. (1990) Sea floor hydrothermal activity: black smoker chemistry and chimneys. *Ann. Rev. Earth Planet. Sci.*, **18**, 173–204.

WÄCHTERSHÄUSER, G. (1988) Pyrite formation, the first energy source for life: a hypothesis. *System. Appl. Microbiol.*, **10**, 207–210.

WÄCHTERSHÄUSER, G. (1990a) Evolution of the first metabolic cycle. *Proc. Nat. Acad. Sci. USA*, **87**, 200–204.

WÄCHTERSHÄUSER, G. (1990b) The case for the chemoautotrophic origin of life in an iron–sulfur world. *Origins of Life and Evolution of the Biosphere*, **20**, 173–176.

WÄCHTERSHÄUSER, G. (1992) Groundworks for an evolutionary biochemistry: the iron–sulfur world. *Prog. Biophys. Mol. Biol.*, **58**, 85–201.

WALKER, J. C. G. (1985) Carbon dioxide on the early earth. *Origins of Life and Evolution of the Biosphere*, **16**, 117–127.

WALKER, J. C. G. (1990) Precambrian evolution of the climate system. *Paleogeog. Paleoclimatol. Paleoecolol.*, **82**, 261–289.

WOESE, C. R., KANDLER, O. and WHEELIS, M. L. (1990) Towards a natural system of organisms: proposals for the domains Archaea, Bacteria, and Eukarya. *Proc. Nat. Acad. Sci. USA*, **87**, 4576–4579.

WOOD, H. G. (1977) Some reactions in which inorganic pyrophosphate replaces ATP and serves as a source of energy. *Fed. Proc. Fede. Am. Soc. Exp. Biol.*, **36**, 2197–2205.

XIAODING, X. and MOULIJN, J. A. (1996) Mitigation of CO_2 by chemical conversion: plausible chemical reactions and promising products. *Energy and Fuels*, **10**, 305–325.

YAMAMOTO, T. and YAMAMOTO, A. (1978) Insertion of carbon dioxide into the Ni–R bond of NiR_2(bpy) (R=CH_3, C_2H_5) and elimination of diethyl ketone from the intermediate product $Ni(C_2H_5)(OCOC_2H_5)$(bpy). *Chem. Lett.*, (6), 615–616.

ZOBACK, M. D. (1995) Frictional analysis of induced earthquakes near 9 km depth at the KTB site: evidence for the brittle/ductile transition? *Geol. Soc. Am. Abstr.*, 1995 Fall Meeting, F533.

7

Facing Up to Chemical Realities: Life Did Not Begin at the Growth Temperatures of Hyperthermophiles

STANLEY L. MILLER[1] AND ANTONIO LAZCANO[2]

[1] *Department of Chemistry and Biochemistry University of California, San Diego, La Jolla, California, 92093–0506, USA*

[2] *Facultad de Ciencias, UNAM, Cd. Universitaria, México, 04510, DF, Mexico*

7.1 Introduction

Although considerable efforts have been made to understand the emergence of the first living systems, we still do not know when and how life originated. Since it is sometimes possible to correlate major evolutionary changes with environmental conditions, several attempts have been made to infer the conditions in which life arose by studying the oldest known organisms. An example of the fruitfulness of this approach may sometimes have is found in the ideas of Oparin (1938), whose suggestion that anaerobic heterotrophy was the primordial metabolism led to the hypothesis of chemical evolution and, eventually, to the development of prebiotic chemistry and other related origins-of-life research (Miller *et al.*, 1997).

Today we know that such extrapolations into the distant past merit considerable caution. We are very far from understanding the origin and characteristics of the first living beings, which may have lacked even the most familiar features found in extant cells, such phosphodiester-backbone genetic macromolecules (Lazcano and Miller, 1996). Thus, proposals that the basal position of anaerobic sulfur-reducing chemosynthetic hyperthermophiles in molecular trees indicate that the first living systems originated in a high-temperature environment, such as those found today in deep-sea hydrothermal vents (Holm, 1992) should not be taken for granted. As discussed throughout this chapter, this idea is a simple extrapolation of the growth temperature of extant hyperthermophiles to the origin of life (Figure 7.1, broken line tree). *There is no more justification for this extrapolation than for a mesophilic origin* (Figure 7.1, solid line tree), *or an even higher temperature origin (not shown).*

7.2 The antiquity of hyperthermophiles

A thermophilic origin of life is not a new idea. It was suggested by Harvey (1924) that the first forms of life were heterotrophic thermophiles that had originated in hot springs. Examination of the prokaryotic branches of unrooted rRNA trees also suggested that the

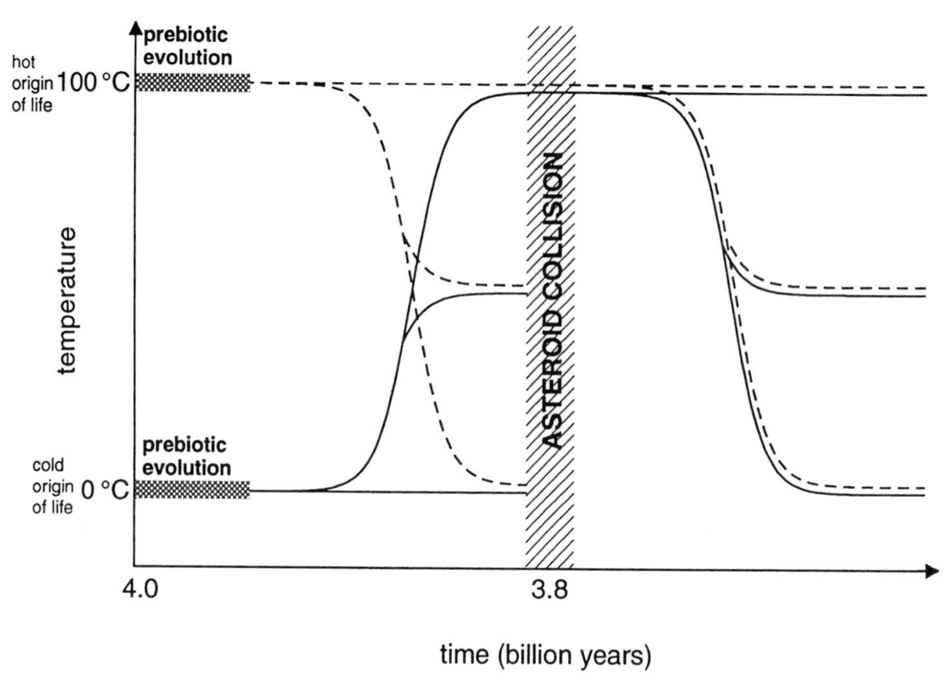

Figure 7.1 Some alternative temperature regimes for the origin and early evolution of life. The broken line shows a hot origin of life followed by adaptation to several lower temperatures. Only the hyperthermophiles would survive the last asteroid impact which boiled the ocean. The solid line shows a low-temperature origin of life followed by adaptation to higher temperatures, with survival of only the secondarily adapted hyperthermophiles after the asteroid collision. Not shown are earlier collisions that may have frustrated the origin of life or its survival.

ancestors of both eubacteria and archaebacteria were extreme thermophiles (Achenbach-Richter *et al.*, 1987). Today the antiquity of hyperthermophiles is widely accepted not only for archaebacteria but also for the less well-known eubacterial extremophiles (Stetter, 1994). It is sometimes overlooked that the eubacterial rooting of the universal tree (Gogarten *et al.*, 1989; Iwabe *et al.*, 1989; Brown and Doolittle, 1994) implies that the hyperthermophilic bacteria such as *Thermotoga* and *Aquifex* are closer to the last common ancestor than the oldest hyperthermophilic archaea, including the recently described korarchaeota, which branch below the euryarchaeota/crenarchaeota split (Barns *et al.*, 1996). However, it should be kept in mind that there are alternative opinions. For example, some hyperthermophile sequences are displaced from their basal position if molecular markers other than elongation factors or ATPase subunits are employed; that is, other molecular trees also open up the possibility that the last common ancestor of all living beings was not a thermophilic prokaryote (cf. Forterre, 1996).

7.3 Hyperthermophiles may be ancient, but they are hardly primitive

It is important to distinguish between ancient organisms and primitive organisms. Hyperthermophiles may be cladistically ancient, but they are hardly primitive relative to the first living organisms. In fact, the archaeal transcriptional and translational apparatus are very similar to those of eukaryotes (Keeling and Doolittle, 1995), and neither do the bacterial or the archaeal hyperthermophiles seem to be more primitive in their metabolic abilities than their mesophilic prokaryotic and eukaryotic counterparts (Adams, 1993). Primitive living systems, according to some current opinion, would initially refer to the pre-RNA world, in which life was first based on polymers using backbones other than ribose-phosphate and possibly bases different from AUGC. This was followed by a stage in which life was based on RNA as both the genetic material and catalysts. The RNA world was followed by a DNA/protein world with rather limited biosynthetic capabilities and, in the subsequent stages of biological evolution, the basic characteristics of metabolic pathways were established. In view of this enormous metabolic development, a constant-temperature extrapolation is hardly justified. These considerations apply if the alternative hypothesis is correct that mesophiles are the most ancient, as suggested by some phylogenetic trees. In this case, constant-temperature extrapolation would point to a low-temperature origin of life, but a high-temperature regime or an even colder one would be equally justified.

The antiquity of hyperthermophiles fits in with the plausible hypothesis of impact frustration of the origin of life (Maher and Stevenson, 1988; Sleep *et al.*, 1989), for which, however, there is no direct geological evidence. There is indirect evidence from lunar craters but no evidence that these impacts on the earth frustrated the origin of life. If the last large asteroid to hit the earth was 400 km in diameter or larger, it would have converted the entire ocean to steam. This would have killed off most organisms, but both the archaeal and bacterial hyperthermophiles would have been selected for, thereby explaining their basal position in phylogenetic trees. Such extreme thermophiles are sometimes said to be submarine-vent organisms (Gogarten-Boekels *et al.*, 1995), but any hyperthermophiles that had already colonised extreme habitats would have survived. Evidence of non-homologous biochemical adaptations to high temperatures between the archaeal and the bacterial hyperthermophiles would provide support to this catastrophic scheme and to the hypothesis of a mesophilic last common ancestor.

7.4 What was the physical environment of the origin of life?

Many strong statements have been made about the primitive earth, but there is no direct geological evidence for any of these hypotheses, since there are no sedimentary rocks older than 3.8×10^9 years. However attractive it may be to some, the idea that the acidic, high-temperature submarine hydrothermal vents mimic the prebiotic environment is without direct support. The temperature of the primitive earth during the period of the origin of life is unknown. On the basis of planetary formation models, it is generally thought, without direct evidence, that the entire earth remained molten for several hundred million years after its formation 4.6×10^9 years ago (Wetherill, 1990). The very old sedimentary rocks in the Greenland Isua formation have been heated to 500 °C, so the evidence on the conditions at that time has largely been destroyed. There is some evidence for life at 3.85×10^9 years in the Akilia formation near Isua, which is based largely on light isotopic carbon (Mojzsis *et al.*, 1996). The Apex sediments in the Australian Warrawoona formation 3.5×10^9 years old contain very convincing cyanobacteria-like microfossils (Schopf, 1993). Thus, life is thought to have originated some time between 4.0 and 3.5 billion years ago, but there is no direct evidence for the temperature conditions.

7.5 High temperatures give higher reaction rates, but the lifetimes of organic molecules are drastically reduced

The one advantage of high temperatures is that the chemical reactions could go faster and the primitive enzymes could have been less efficient (Harvey, 1924), but the price paid is loss of organic compounds by decomposition and diminished stability of the genetic material. The problem with monomers is bad enough, but it is worse with polymers, e.g. RNA and DNA (Lindahl, 1993), whose stability in the absence of efficient repair enzymes would be too low to maintain genetic integrity in hyperthermophiles. RNA and DNA are clearly too unstable to exist in a hot prebiotic environment. The existence of an RNA world with ribose appears to be incompatible with the idea of a hot origin of life. The stability of ribose and other sugars is the worst problem, but pyrimidines and purines and some amino acids are nearly as bad. The half-life of ribose at 100 °C and pH 7 is only 73 minutes, and other sugars have comparable half-lives (Larralde *et al.*, 1995). The half-life for decomposition of cytosine at 100 °C is 19 days, and for adenine is 370 days at 100 °C (Garrett and Tsau, 1972; Shapiro, 1995; Levy and Miller, unpublished data).

Amino acids are stable (e.g. alanine with a half-life for decarboxylation of approximately 19 000 years at 100 °C), but serine decarboxylates to ethanolamine with a half-life of 320 days (Vallentyne, 1964), with dealdolization and dehydration as additional decomposition routes (Bada *et al.*, 1995). Similar considerations show that the growth of organisms at 250 °C or 350 °C and the origin of life at such temperatures (Corliss *et al.*, 1981) is very unlikely (White, 1984; Miller and Bada, 1988). It is clear that if the origin of life took place at 100 °C or higher temperatures, then the organic compounds involved must have been used immediately after their prebiotic synthesis. *Thus, a high-temperature origin of life may be possible, but it cannot involve adenine, uracil, guanine, cytosine, a ribose-phosphate backbone, nor even most of the 20 amino acids.*

7.6 Hyperthermophily may be a derived character

A possibility that has not been given enough attention is that hyperthermophiles are now at the base of some trees simply because they outcompeted older mesophiles when they adapted to lower temperatures, rather than being the sole survivors of an impact event. Some of the molecular features that are adaptations to hot environments could have enhanced the survival chances of hyperthermophiles and their immediate descendants under less extreme temperature conditions. It is possible, for instance, that the development of operons allowed the rapid channelling of thermolabile biochemical intermediates between biosynthetic enzymes, carrying with it the selective advantage of coordinated regulation (N. Glansdorf, personal communication). Another possibility is that the heat-shock response, which is universally distributed, can be interpreted as a remnant of the hyperthermophilic ancestors of extant life (Miller and Lazcano, 1995). Heat-shock proteins are not only involved in thermotolerance but also in protection against other stress-inducing agents and environmental insults including starvation conditions, UV-irradiation, DNA-damaging agents, alcohol, amino acid analogues, etc. (Watson, 1990). Accordingly, it is possible to envision that heat-shock genes evolved in ancient hyperthermophiles, 'pre-adapting' them to other stress-inducing conditions at low temperatures, allowing them successfully to outcompete mesophiles (Miller and Lazcano, 1995).

7.7 Conclusions

Although the origin-of-life problem is riddled with many unanswered questions, perhaps one of the most important is that of the nature of the first genetic molecule. Given the difficulties associated with the prebiotic synthesis and accumulation of RNA, we believe that important effort should be devoted to the developing of models of a pre-RNA world, i.e. of a stage in which phenotype and genotype also resided in the same polymer, so that no protein or related catalysts were required to be synthesised.

We have addressed the possibility that neither the first living systems nor the last common ancestor were hyperthermophiles. Evidence of analogous thermoadaptation mechanisms in archaeal and bacterial hyperthermophiles would provide support for the hypothesis of a mesophilic last common ancestor to both prokaryotic branches, but once more this would tell us nothing about the temperature at which the origin of life took place. While the schemes discussed in this chapter may not necessarily be correct, they suggest that additional explanations can be advanced to explain the phylogenetic distribution of hyperthermophiles. Even if they are the oldest, a straight-line temperature extrapolation back in time to the origin of life is not warranted. Prebiotic chemistry points towards a low-temperature regime for the emergence of living systems. If this conclusion is valid, it merits a search for mesophiles older than hyperthermophiles.

7.8 Summary

A high temperature (~100 °C) origin of life has been proposed, largely for the reason that the hyperthermophiles are claimed to be the last common ancestor of modern organisms. Even if they are the oldest extant organisms, they can say nothing about the temperatures of the origin of life or about the RNA world. There is no geological evidence for the physical setting of the origin of life because there are no unmetamorphosed rocks from that period. Prebiotic chemistry points to a low-temperature origin because most biochemicals decompose rather rapidly at temperatures of 100 °C (e.g. half-lives are 73 minutes for ribose, 19 days for cytosine, and 370 days for adenine). Hyperthermophiles may appear at the base of some phylogenetic trees because they out-competed the mesophiles when they adapted to lower temperatures, perhaps owing to enhanced production of heat shock proteins, or more regulated biosynthetic pathways.

Acknowledgements

We thank Jason P. Dworkin and Matthew Levy for helpful comments. Support was provided by the NSCORT (NASA Specialized Center for Research and Training) in Exobiology at the University of California, San Diego. This paper is largely based on a previously published article (Miller and Lazcano, 1995).

References

ACHENBACH-RICHTER, L., GUPTA, R., STETTER, K. O. and WOESE, C. R. (1987) Were the original eubacteria thermophiles? *System. Appl. Microbiol.*, **9**, 34–39.

ADAMS, M. W. W. (1993) Enzymes and proteins from organisms that grow near and above 100 °C. *Annu. Rev. Microbiol.*, **47**, 627–658.

BADA, J. L., MILLER, S. L. and ZHAO, M. (1995) The stability of amino acids at submarine hydrothermal vent temperatures. *Origins of Life and Evolution of the Biosphere*, **25**, 111–118.

BARNS, S. M., DELWICHE, C. F., PALMER, J. D. and PACE, N. R. (1996) Perspectives on archaeal diversity, thermophily and monophily from environmental rRNA sequences. *Proc. Natl Acad. Sci. USA*, **93**, 9188–9193.

BROWN, J. R. and DOOLITTLE, W. F. (1994) Root of the universal tree based on ancient aminoacyl-tRNA synthetase gene duplications. *Proc. Natl Acad. Sci. USA*, **92**, 2441–2445.

CORLISS, J. B., BAROSS, J. A. and HOFFMAN, S. E. (1981) An hypothesis concerning the relationship between submarine hot springs and the origin of life on earth. *Oceanol. Acta*, **4** (suppl), 59–69.

FORTERRE, P. (1996) A hot topic: the origin of hyperthermophiles. *Cell*, **85**, 789–792.

GARRETT, E. R. and TSAU, J. (1972) Solvolyses of cytosine and cytidine. *J. Pharm. Sci.*, **61**, 1052–1061.

GOGARTEN-BOEKELS, M., HILARIO, E. and GOGARTEN, J. P. (1995) The effects of heavy meteoritic bombardments of the early evolution – the emergence of the three domains of life. *Origins of Life and Evolution of the Biosphere*, **25**, 251–264.

GOGARTEN, J. P., KIBAK, H., DITTRICH, P. *et al.* (1989) Evolution of the vacuolar H^+-ATPase: implications for the origin of eukayotes. *Proc. Natl Acad. Sci. USA*, **86**, 6661–6665.

HARVEY, R. B. (1924) Enzymes of thermal algae. *Science*, **60**, 481–482.

HOLM, N. G. (ed.) (1992) *Marine Hydrothermal Systems and the Origin of Life* (Dordrecht: Kluwer Academic). Also a special issue of *Origins of Life and Evolution of the Biosphere*, **22**, 1–241.

IWABE, N., KUMA, K., HASEGAWA, M., OSAWA, S. and MIYATA, T. (1989) Evolutionary relationship of archaebacteria, eubacteria, and eukaryotes inferred from phylogenetic trees of duplicated genes. *Proc. Natl Acad. Sci. USA*, **86**, 9355–9359.

KEELING, P. J. and DOOLITTLE, W. F. (1995) Archaea: narrowing the gap between prokaryotes and eukaryotes. *Proc. Natl Acad. Sci. USA*, **92**, 5761–5764.

LARRALDE, R., ROBERTSON, M. P. and MILLER, S. L. (1995) Rates of decomposition of ribose and other sugars: implications for chemical evolution. *Proc. Natl Acad. Sci. USA*, **92**, 8158–8160.

LAZCANO, A. and MILLER, S. L. (1996) The origin and early evolution of life: prebiotic chemistry, the pre-RNA world, and time. *Cell*, **85**, 793–798.

LINDAHL, T. (1993) Instability and decay of primary structure of DNA. *Nature*, **362**, 709–715.

MAHER, K. A. and STEVENSON, D. J. (1988) Impact frustation of the origins of life. *Nature*, **331**, 612–614.

MILLER, S. L. and BADA, J. L. (1988) Submarine hot springs and the origin of life. *Nature*, **334**, 609–611.

MILLER, S. L. and LAZCANO, A. (1995) The origin of life – did it occur at high temperatures? *J. Mol. Evol.*, **41**, 689–692.

MILLER, S. L., SCHOPF, J. W. and LAZCANO, A. (1997) Oparin's 'Origin of Life': sixty years later. *J. Mol. Evol.*, **44**, 351–353.

MOJZSIS, S. J., ARRHENIUS, G., MCKEEGAN, K. D., HARRISON, T. M., NUTMAN, A. P. and FRIEND, C. R. L. (1996) Evidence for life on earth before 3,800 million years ago. *Nature*, **384**, 55–59.

OPARIN, A. I. (1938) *The Origin of Life* (New York: Macmillan).

SCHOPF, J. W. (1993) Microfossils of the early Apex chert: new evidence of the antiquity of life. *Science*, **260**, 640–646.

SHAPIRO, R. (1995) The prebiotic role of adenine: a critical analysis. *Origins of Life and Evolution of the Biosphere*, **25**, 83–98.

SHAPIRO, R. and KLEIN, R. S. (1966) The deamination of cytidine and cytosine by acidic buffer solutions. Mutagenic implications. *Biochemistry*, **5**, 2358–2362.

SLEEP, N. H., ZAHNLE, K. J., KASTING, J. F. and MOROWITZ, H. J. (1989) Annihilation of ecosystems by large asteroid impacts on the early earth. *Nature*, **342**, 139–142.

STETTER, K. O. (1994) The lesson of archaebacteria. In *Early Life on Earth*, Nobel Symposium No. 84, ed. S. Bengtson, pp. 114–122 (New York: Columbia University Press).

VALLENTYNE, J. R. (1964) Biogeochemistry of organic matter II: thermal reactions and transformation products of amino compounds. *Geochim. Cosmochim. Acta.*, **28**, 157–188.

WATSON, K. (1990) Microbial stress proteins. *Adv. Microbiol. Physiol.*, **31**, 183–223.

WETHERILL, G. W. (1990) Formation of the earth. *Annu. Rev. Earth Planet. Sci.*, **18**, 205–256.

WHITE, R. H. (1984) Hydrolytic stability of biomolecules at high temperatures and its implications for life at 250 °C. *Nature*, **310**, 430–432.

PART THREE

Nucleic Acid-based Phylogenies

8

Were our Ancestors Actually Hyperthermophiles? Viewpoint of a Devil's Advocate

PATRICK FORTERRE

Institut de Génétique et Microbiologie, Université Paris-Sud, CNRS, URA 1354, Orsay, France

8.1 Introduction

The hypotheses that life originated in a very hot environment and that present-day hyperthermophiles are direct descendants of a 'hot loving' last universal common ancestor are very appealing, especially for the community of people dealing with these fascinating microbes. If these ideas are correct, working on hyperthermophiles means studying the origin of life – an inspiring field open to public scrutiny, journalistic enquiry and profound philosophical meaning. Of course, the study of hyperthermophiles is rewarding in itself for many reasons, especially in understanding how life managed to evolve in extremely diverse physical environments, but the hot origin of life story is the 'cherry on the cake'. As a consequence, we should be aware of this prejudice and try to falsify the above hypotheses as well as supporting them. Thus, I will focus here on several points that are, in my opinion, the Achiles heels of the hot origin of life scenario, and that certainly warrant further investigation (see also Forterre, 1996).

8.2 The conflict between primitiveness and hyperthermophily?

The hypothesis of a direct link between a hot origin of life and a hyperthermophilic last universal common ancestor (LUCA) supposes the previous existence at one time of very primitive hyperthermophiles. In particular, since DNA most likely originated after RNA (i.e. after the invention of ribonucleotide reductases), one should imagine primitive hyperthermophilic cells with RNA genomes. If peptide bonds are indeed produced by rRNA, as suggested by the work of Noller (1992), one can even think of RNA-based hyperthermophiles without 'modern' proteins. This is a most difficult point because RNA is a very fragile molecule (discussed in Forterre, 1995). Indeed, the reactivity of the hydroxyl in 2′ carbon of ribose can promote hydrolysis of phosphodiester bonds, and this reaction becomes very rapid at high temperature (Ginoza *et al.*, 1964). Furthermore, RNA thermodegradation is greatly accelerated by physiological concentrations of magnesium (Lindhal, 1966). Magnesium was probably already present in living systems of the RNA world since this salt is essential for ribozyme activities.

In modern hyperthermophiles, stable RNAs might be protected against thermodegradation by their elaborated tertiary structures, interactions with proteins, and possibly methylation of the critical 2′ OH at the surface of the molecule. As a matter of fact, extensive ribose methylation of tRNA has been reported in hyperthermophiles (Edmonds *et al.*, 1991). Similarly, the problem of mRNA thermodegradation in present-day prokaryotes could be bypassed by the high turnover of these short-lived molecules, which is made possible by the coupling of transcription and translation (Forterre, 1995). This could even explain why eukaryotes, which require stable mRNA, cannot adapt to hyperthermophilic conditions (Forterre, 1995). However, advocates of the hot origin of life hypothesis have yet to propose reasonable mechanisms that might have prevented RNA thermodegradation in primitive RNA-based cells without paralysing ribozyme activities, the latter being mainly linked to the critical reactivity of the 2′ ribose hydroxyl. Otherwise, it would remain more appropriately cautious to conclude like Joyce (1988) that 'without the benefit of evolutionary improvement, such an entity [a primitive organism which has not yet embarked on a particular evolutionary pathway] must have been biochemically inept in the extreme'.

Finally, the hypothesis of a direct connection between a hot origin of life and today's hyperthermophiles is also challenged by the presence, in the latter, of elaborated molecular devices that seem to be specific for thermoadaptation. One of them, reverse gyrase, is indeed a 'modern' enzyme, which most likely originated from the fusion of a helicase and a topoisomerase module, each of them being the product of a long evolutionary pathway in their own respective protein superfamily (Forterre *et al.*, 1995). If reverse gyrase is indeed essential for a hyperthermophilic lifestyle (which remains to be demonstrated), one can argue that organisms living before this fusion were probably not hyperthermophiles.

If one supports the hypothesis that hyperthermophilic microbes originated from mesophilic, or at least from much less thermophilic, microorganisms, it still remains to determined at which stage in the evolution of life adaptation to hyperthermophily took place. In particular, did it occur before the separation of Archaea, Bacteria and Eukarya (in other words, was LUCA itself a hyperthermophile?) or after the separation of the three domains? In the latter case, did it occur in a common lineage to Archaea or Bacteria, such as in the thermoreduction hypothesis for the origin of prokaryotes (Forterre, 1995), or independently, either once or several times in each domain, and when?

8.3 What does the universal tree of life say?

The main argument put forward to support the idea of a hyperthermophilic LUCA is based on rRNA sequence comparison. This shows that all branches leading to hyperthermophiles in the universal tree of life are short and located at the base of the tree (Woese, 1987), with the root of this tree between hyperthermophilic Archaea and Bacteria (Woese *et al.*, 1990; Stetter, 1992). However, these data and their interpretations are disputable.

The short branches in the rRNA tree are usually taken to indicate that hyperthermophiles are less evolved than their mesophilic counterparts, i.e., they more greatly resemble LUCA than do mesophiles. To be valid, this conclusion would require that rRNA evolved at the same rate in all lineages and that all main cellular and molecular features of a given organism evolve at the same rate as its rRNA. The first assumption (a perfect clock) is inconsistent with the geometry of the tree (the relative lengths of the branches), and the second one can only be an article of faith. The short length of the branches within the hyperthermophilic lineages only means that their rRNA sequences evolve less rapidly

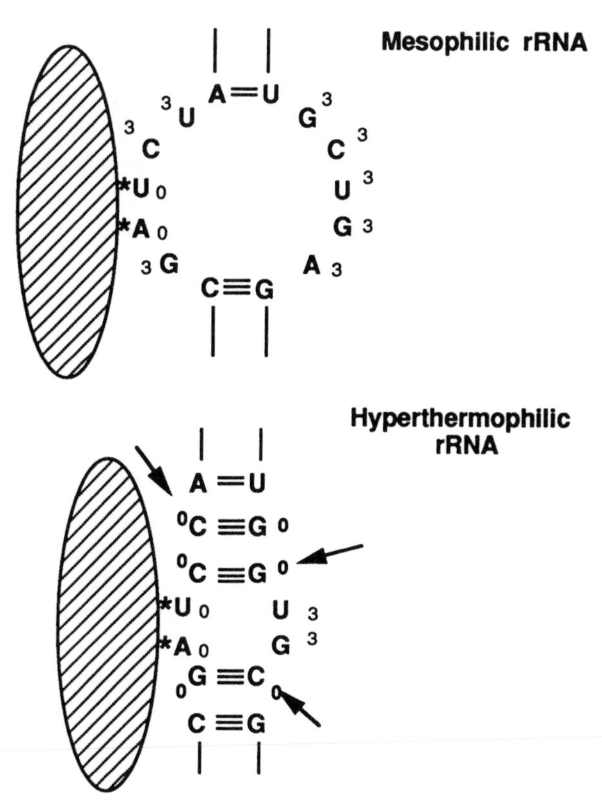

Figure 8.1 Reduction of the number of variable sites in rRNA from hyperthermophiles by formation of additional base pairing. The drawing illustrates a theoretical situation. The stars indicate nucleotides which interact with a ribosomal protein and cannot change during evolution. The arrows pinpoint the three differences between the mesophilic (50% GC in the loop) and the hyperthermophilic molecules (70% GC in the corresponding region). Figures indicate the possible number of variations for a given nucleotide: 24 changes are possible in the mesophilic molecule, while only 6 can be tolerated in the hyperthermophilic one.

than those of mesophiles and does not say anything about the phenotype of LUCA. Either the rate of rRNA evolution has been reduced during the transition from mesophiles to hyperthermophiles, or it has been speeded up during the transition from hyperthermophiles to mesophiles.

A plausible explanation of the shortness of branches of hyperthermophiles, which has not yet been fully addressed by current methods of tree evaluation, is that the higher GC contents of their rRNA (by about 10%) reduces the number of possible nucleotide substitutions, slowing down their evolutionary rate (see Figure 8.1 and legend). This increase in GC content is a general phenomenon in hyperthermophilic stable RNA (Kowalak *et al.*, 1994; Dalgaard and Garrett, 1993) and is likely required for stabilisation of existing secondary structures and possible formation of new ones (see two examples in Burggraf *et al.*, 1992). Major problems still remain concerning the topology of the universal rRNA tree, making the position of hyperthermophiles a very open question. In particular, Zillig and coworkers reported that the hyperthermophilic bacterium *Aquifex pyrophilus*, which is located at the base of the 16S rRNA bacterial tree, is grouped with Proteobacteria in an RNA polymerase tree (Klenk *et al.*, 1984). Interestingly, Burggraf *et al.* (1992) previously noticed that the 16S rRNA helix of *A. pyrophilus*, lying between (*E. coli*) positions 198 and 219, has an unusual form encountered previously only in γ-Proteobacteria. It is thus possible that *Aquifex* is truly related to Proteobacteria. Burggraf *et al.* (1992) detected four signatures in the *A. pyrophilus* rRNA shared with Archaea that, according to them, could have been primitive characters retained in the deepest branch of the archaeal tree. However, three of them correspond to replacement of AU pairs by GC pairs, and can be better interpreted as thermoadaptation in the *A. pyrophilus* lineage.

Another challenge to the validity of rRNA trees came recently from studies of eukaryotic tubulin phylogeny (Edlind *et al.*, 1996; Keeling and Doolittle, 1996), as well as the presence of chitin (Canning, 1990) and of similar meiosis (Flegel and Pasharawipas, 1995) that strongly suggest that microsporidia, one of the first lineages branching off the eukaryotic 18S rRNA tree (Vossbrinck *et al.*, 1986), could actually be fungi. In fact, the grouping of microsporidia and fungi is also supported by a cladistic analysis of the 18S rRNA (H. Phillipe, personal communication) and a common insertion of 12 amino acids in the elongation factor EF1α (Kamaishi *et al.*, 1996). In conclusion, the problem of GC content and secondary structures in rRNA has to be adressed before anything can be said about the position of hyperthermophiles in the tree of life. This should require the elucidation of which GC positions in hyperthermophilic rRNA are related to thermoadaptation in order to remove them from the analysis and to evaluate the remaining phylogenetic signal (if any) after such correction.

Rooting the tree of life in the prokaryotic branch is also repeatedly used as an argument to support the hypothesis of a hyperthermophilic LUCA. Of course, the controversial nature of the position of hyperthermophiles at the base of the universal tree means that the rooting of the tree itself is not as significant in the question of a hyperthermophilic origin. Furthermore, this rooting is still a matter of debate (Forterre *et al.*, 1993; Forterre, 1997). The trees of duplicated genes used to support the bacterial rooting are plagued by several problems. One of them is that bacterial branches are much longer than the eukaryal and archaealories in both the ATPase and the elongation factor trees. This can produce the artefactual grouping of bacteria at the base of the combined universal tree according to the phenomenon of long branch attraction (Felsenstein, 1978). In the case of the ATPase tree, the finding of eukaryotic-like ATPases in several species of bacteria has raised the possibility that this tree is plagued by a problem of paralogy (Forterre *et al.*, 1993). These eukaryotic-like ATPases might have been introduced to Bacteria by lateral gene transfer, but alternatively, they could indeed testify for an ancestral duplication of the ATPase genes. In the latter case, the bacterial rooting obtained by mixing paralogous proteins in the same tree would be confusing. The same problem also occurs in the ileu-tRNA synthetase tree which has also been used to root the tree of life (H. Phillipe, personal communication), since new bacterial sequences are grouped between archaeal and eukaryal ones. Finally, cladistic analyses of the elongation factor and ileu-tRNA synthetase amino acid sequence alignments indicate that the number of slowly evolving positions that can be polarized to support a particular rooting is extremely low (Forterre *et al.*, 1993, Forterre, 1997), such that, in my opinion, the results obtained with classical phylogenetic analyses cannot be trusted.

The main problem in sequence-based molecular phylogeny (especially when dealing with very ancient divergences) is that only a few sites in a protein sequence have evolved slowly enough to avoid the problem of complete saturation by multiple substitutions, but not so slowly as to retain the trace of events which have occurred during the common history of the two domains which once shared a specific common ancestor (synapomorphies in the cladistic nomenclature; Hennig, 1966). All programmes of tree construction presently used to recover some hidden phylogenetic treasures from the variable sites present in the molecule presume incomplete saturation. However, a new method used to measure saturation (Philippe *et al.*, 1994) indicates that protein sequences are indeed very often saturated for such ancient divergences (this is the case for the elongation factors, see Philippe and Adoutte, 1995). This saturation effect well explains confusions observed in many phylogenetic trees without resorting to *ad hoc* hypotheses, such as networks of gene transfers, fusion scenarios or extensive gene duplication (multiple paralogues).

8.4 A perspective from comparative biochemistry

If we cannot trust sequence-based molecular phylogeny in deciding the position of hyperthermophiles in the universal tree of life, what can we do? In my opinion, the possibility remains of carefully examining the distribution and nature of 'hyperthermophilic' features in the two prokaryotic domains in order to identify some trends and put forward some hypotheses.

A case can be made for a hyperthermophilic ancestor of all archaea since their putative hyperthermophilic features such as reverse gyrase, ether-linked lipids, or similar modified nucleosides in rRNA are homologues and (more importantly) some of them are also present in mesophilic archaea, suggesting that they are relics of an ancient adaptation to high temperature. This is the case for ether-linked lipids which are the only type of glycerolipids found in archaea (the ether bond is more stable than the ester one); for the presence in all archaeal tRNA of the modified nucleoside archaeosine (Gregson *et al.*, 1993), which could play a role in tRNA stabilisation; and for the absence of the heat-labile dihydrouridine in most tRNA from mesophilic archaea (Edmonds *et al.*, 1991). However, we should be aware that such observations might still be circumstantial and need further investigation (for example, it has not yet been demonstrated that archaeosine actually help to stabilise tRNA at high temperature, and archaeal lipids are apparently also well adapted to low temperatures).

At the moment, the only argument supporting the hypothesis of a common hyperthermophilic ancestor of Archaea and Bacteria is reverse gyrase, which is present (and homologuous) in the two prokaryotic domains (Bouthier de la Tour *et al.*, 1998). However, it remains possible that this enzyme was laterally transferred from Archaea to Bacteria. Furthermore, the role of reverse gyrase in thermoadaptation is presently unclear. Indeed, while hyperthermophilic archaea have reverse gyrase and relaxed or positively supercoiled DNA (Lopez-Garcia and Forterre, 1997), the hyperthermophilic bacterium *Thermotoga* has both a reverse gyrase and a normal gyrase, as well as negatively supercoiled DNA (Guipaud *et al.*, 1997).

On the other hand, the independent emergence of hyperthermophily in the two prokaryotic domains is suggested by the fact that lipids from hyperthermophilic bacteria mimic those of archaea but are structurally very different, i.e. they have an opposite stereochemistry and are made from fatty acids or alcohol instead of isoprenol (Langworthy and Pond, 1986). Lipids from hyperthermophilic bacteria and archaea thus exhibit some analogies but are not homologues. If the common ancestor of bacteria and archaea was a hyperthermophile, it is difficult to understand why such hyperthermophilic features evolved independently in the two domains. One explanation would be that this ancestor was devoid of glycerolipids! However, this is unlikely, since this ancestor was probably a highly elaborated organism, considering all metabolic pathways, enzymes and mechanisms for gene expression shared by archaea and bacteria.

What about the common ancestor of Bacteria? In that case too, the lipid data suggest that adaptation to hyperthermophily occured independently in different lineages, since *Aquifex pyrophilus, Thermotoga maritima* (maximal growth temperatures of 95 °C and 90 °C, respectively) and extreme thermophiles such as *Thermomicrobium roseum* (maximal growth temperatures of 80 °C) all exhibit very different patterns of lipid composition (Langworthy and Pond, 1986). In all cases, we have bacterial lipids trying to mimic archaeal ones but using different strategies: ether link with fatty alcohol in one case, tetraester formation in another, formation of long chain diols in a third. Further comparative analyses of hyperthermophilic features in *Aquifex* and *Thermotoga* and of their relationships with their less thermophilic relatives (*Geotogales*, *Hydrogenobacter*) should be very informative in this regard.

A case study has recently emerged from the work on DNA polymerases which suggests that secondary adaptation to thermophily has occurred twice independently in the bacterial lineage. The *Taq* DNA polymerases from *Thermus aquaticus*, as well as a DNA polymerase recently isolated from a newly identified thermophilic strain of *Bacillus stearothermophilus* both belong to the family A of DNA polymerase (prototype *E. coli* pol I) but lack the 3′–5′ exonuclease editing activity present in their homologous counterparts (for a recent review on thermophilic DNA polymerases, see Perler *et al.*, 1996). This editing activity is present in most DNA polymerases belonging to the three major DNA polymerase families (A, B and C) and is widely distributed in the three domains of life (Forterre *et al.*, 1994). Interestingly, in all cases, this activity is due to homologous proteins, either present as a DNA polymerase subunit or at the N-terminus region of the polypeptide bearing the DNA polymerase activity. All these data indicate that this exonuclease-editing activity is very ancient and has been lost in thermophilic DNA polymerases mentioned above. Indeed, crystallographic analyses have revealed that an exonuclease module is also present (structurally conserved) in *Taq* and *B. stearothermophilus* DNA polymerases, but has been modified in such a way as to eliminate its exonuclease activity (Korolev *et al.*, 1995; Kiefer *et al.*, 1997). One could even speculate that elimination of the editing activity of pol I might have been necessary to increase the mutation rate during the process of thermoadaptation in these two species (a testable hypothesis). Interestingly, some of the modifications that have eliminated the 3′–5′ exonuclease activity of *Taq* and *B. stearothermophilus* DNA polymerases can be tentatively related to thermoadaptation and are different in the two enzymes. In the *Taq* polymerase the crevice which normally accommodates the single-stranded DNA to be edited is filled by hydrophobic residues, which should enhance the enzyme's stability, while in the *B. stearothermophilus* enzyme entry to this crevice is prevented by the folding back of a particular polypeptide loop (Kiefer *et al.*, 1997). These observations argue in favour of the independent adaptation to thermophily of pol I in *Thermus* and *Bacillus* from mesophilic ancestors.

In my opinion, this polymerase story indicates that comparative structural and phylogenetic analyses of psychrophilic, mesophilic, thermophilic and hyperthermophilic proteins might be a very useful tool in determining the sense of thermoadaptation, either from cold to hot or from hot to cold. In the first case, one would expect that 'mesophilic features' (defined according to the point of view of an organism growing at 100 °C) are primitive

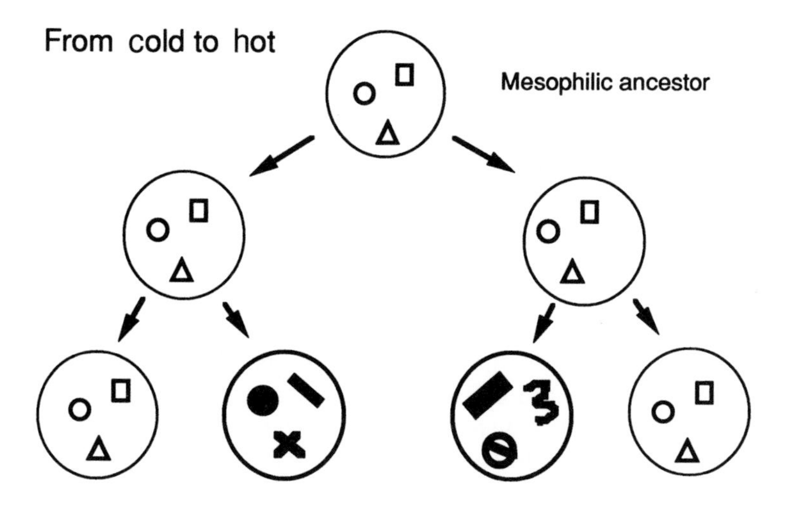

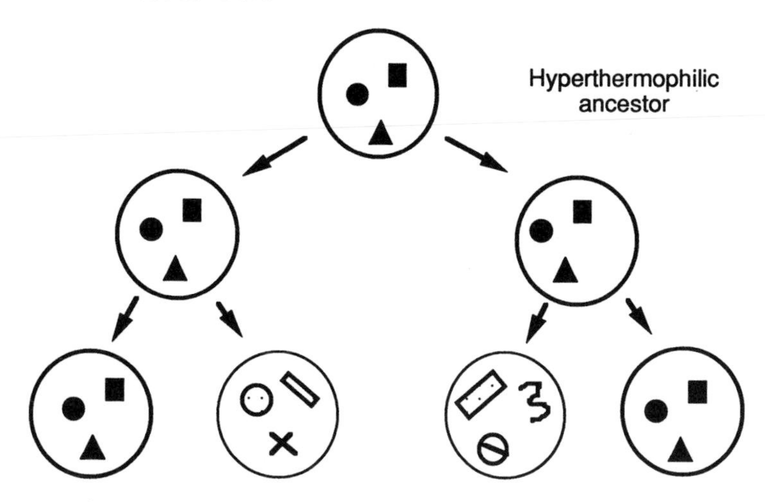

Figure 8.2 Evolutionary relationships between thermoadaptative features in two different scenarios. *From cold to hot*: homologous characters related to adaptation at high temperature in hyperthermophiles (solid symbols) are different in the two lineages since they originated independently from characters originally adapted to a cold environment; their mesophilic versions (open) are similar in these two lineages, since they descended from characters originally adapted to mesophily. *From hot to cold*: homologous characters related to adaptation at low temperature in mesophiles (open features) are different in the two lineages since they originated independently from characters originally adapted to a hot environment; their hyperthermophilic versions (solid) are similar in these two lineages, descending from characters originally adapted to hyperthermophily.

traits which should be more generally distributed and similar in all domains and lineages, while hyperthermophilic features (defined according to our point of view) are derived traits which have evolved independently in different domains and/or lineages. In the second case (from hot to cold) the opposite trend is expected (Figure. 8.2).

8.5 Conclusions

It will be certainly difficult to prove or disprove the hot origin of life hypothesis *per se*. Several geophysical arguments support the idea of a primitive hot environment, but this also is a hot and controversial topic at usual origin-of-life meetings. For example, taking into account the lower intensity of the sun 4 billion years ago, Bada and coworkers even imagine a primitive earth with frozen oceans regularly thawed by the impacts of giant meteorites and warmed by hydrothermal activity (Bada *et al.*, 1994), a quite complex biotope with all temperature possibilities. However, even if life appeared at high temperature, this does not imply that all subsequent evolution up to LUCA occurred in a hot environment, considering that LUCA was the product of a long evolutionary period (Forterre *et al.*, 1995; Miller and Lazcano, 1995).

The task of unravelling the sense of thermoadaptation in the history of life, from hot to cold or from cold to hot, might be less difficult. I think that it will be possible to identify the actual evolutionary pathways leading to present-day hyperthermophiles by combining sequence-based molecular phylogenies (at the protein level) and comparative biochemistry and molecular biology, at least if we examine these analyses with open minds and not try to force them into previously designed scenarios based on rRNA phylogeny. The rapid accumulation of data on hyperthermophiles, especially during the last two years, and the entry to the field of many new teams, attracted by the scientific challenges still to be adressed in this area, make me reasonably optimistic. One can seriously hope that some of the questions discussed here will be solved or close to being solved in the near future.

8.6 Summary

I review and discuss here some arguments against the hypothesis that our ancestors were hyperthermophiles. The idea of a direct link between a putative hot origin of life and present-day hyperthermophiles is challenged by the intrinsically low stability of RNA at high temperature (in the context of the RNA-world scenario), and by the presence of sophisticated molecular features of thermoadaptation in hyperthermophiles. On the other hand, the evolutionary arguments in favour of the ancestry of hyperthermophiles are shaken by new data which put rRNA-based phylogenies into question. Considering the sense of thermoadaptation, either from hot to cold or from cold to hot, I argue that only the combination of comparative biochemistry and molecular phylogeny will be informative. For example, the analysis of DNA polymerase structures, activities and phylogeny clearly indicates that adaptation to moderate thermophily occurred at least twice independently in the bacterial domain.

References

Bada, J. L., Bigham, C. and Miller, S. L. (1994) Impact melting of frozen oceans on the early earth – implications for the origin of life. *Proc. Natl Acad. Sci. USA*, **91**, 1248–1250.

Bouthier de la Tour, C., Portemer, C., Kaltoum, H. and Duguet, M. (1998) Reverse gyrase from the hyperthermophilic bacterium, *Thermotoga maritima*: properties and gene structure. *J. Bacteriol.*, **180**, 274–281.

Burggraf, S., Olsen, G. J., Stetter, K. O. and Woese, C. R. (1992) A phylogenetic analysis of *Aquifex pyrophilus*. *System. Appl. Microbiol.*, **15**, 352–356.

CANNING, E. U. (1990) Phylum Microsporidiae. In *Handbook of Protista*, ed. L. Margulis, J. O. Corliss, M. Melkonian and D. J. Chapman, pp. 53–72 (Boston: Jones and Bartlett).

DALGAARD, J. Z. and GARRETT, R. A. (1993) Archaeal hyperthermophile genes. In *The Biochemistry of Archaea*, ed. M. Kates, Dj Kushner and A. T. Matheson, pp. 535–564 (Amsterdam: Elsevier).

EDLIND, T. D., LI, J., VISVESVARA, G. S., VODKIN, M. H., MCLAUGHLIN, G. L. and KATIYAR, S. K. (1996) Phylogenetic analysis of β-tubulin sequences from amitochondrial protozoa. *Mol. Phylogeet. Evol.*, **5**, 359–367.

EDMONDS, C. G., CRAIN, P. F., GUPTA, R. *et al.* (1991) Post-transcriptional modification of transfer RNA in thermophilic Archaea (Archaebacteria). *J. Bacteriol.*, **173**, 3138–3148.

FELSENSTEIN, J. (1978) Cases in which parsimony or compatibility methods will be positively misleading. *System Zool.*, **27**, 401–410.

FLEGEL, T. W. and PASHARAWIPAS, T. (1995) A proposal for typical eukaryotic meiosis in microsporidians. *Can. J. Microbiol.*, **41**, 1–11.

FORTERRE, P. (1995) Thermoreduction, a hypothesis for the origin of prokaryotes. *C. R. Acad. Sci. Paris*, **318**, 415–422.

FORTERRE, P. (1996) A hot topic: the origin of hyperthermophiles. *Cell*, **85**, 789–792.

FORTERRE, P. (1997) Protein versus rRNA: rooting the universal tree of life? *ASM News*, **63**, 89–95.

FORTERRE, P., BENACHENOU, N., CONFALONIERI, F., DUGUET, M., ELIE, C. and LABEDAN, B. (1993) The nature of the last universal ancestor and the root of the tree of life, still open questions. *Biosystem*, **28**, 15–32.

FORTERRE, P., BERGERAT, A., GADELLE, D. *et al.* (1994) Evolution of DNA topoisomerases and DNA polymerases: a perspective from archaea. *System. Appl. Microbiol.*, **16**, 746–758.

FORTERRE, P., CONFALONIERI, F., CHARBONNIER, F. and DUGUET, M. (1995) Speculations on the origin of life and thermophily: review of available information on reverse gyrase suggest that hyperthermophilic prokaryotes are not so primitive. *Origin of Life and Evolution of the Biosphere*, **25**, 235–249.

GINOZA, W., CAROL, J., VESSEY, K. and CAMARK, C. (1964) Mechanisms of inactivation of single-stranded virus nucleic acids by heat. *Nature*, **203**, 606–609.

GREGSON, J. M., CRAIN, P. F., EDMONDS, C. G. (1993) Structure of the Archaeal transfer RNA nucleoside G-asterisk-15 (2-amino-4,7-dihydro-4-oxo-7-beta-D-ribofuranosyl-1H-pyrrolo [2,3-d] pyrimidine-5-carboximidamide (Archaeosine). *J. Biol Chem.*, **268**, 10076–10086.

GUIPAUD, O., MARGUET, E., NOLL, K., BOUTHIER DE LA TOUR, C. and FORTERRE, P. (1997) Both DNA gyrase and reverse gyrase are present in the hyperthermophilic bacterium *Thermotoga maritima*. *Proc. Natl Acad. Sci.*, **94**, 10606–10611.

HENNIG, W. (1966) *Phylogenetic Systematics* (Urbana, IL: University of Illinois Press).

JOYCE, G. (1988) Hydrothermal vents too hot? *Nature*, **334**, 564.

KAMAISHI, T., HASHIMOTO, T., NAKAMURA, Y. *et al.* (1996) Protein phylogeny of translation elongation factor EF-1α suggests microsporidians are extremely ancient eukaryotes. *J. Mol. Evol.*, **42**, 257–263.

KEELING, P. J. and DOOLITTLE, W. F. (1996) Alpha-tubulin from early-diverging eukaryotic lineages and the evolution of the tubulin family. *Mol. Biol. Evol.*, **13**, 1297–1305.

KIEFER, J. R., MAO, C., HANSEN, C. J. *et al.* (1997) Crystal structure of a thermostable *Bacillus* DNA polymerase I large fragment at 2.1 Å resolution. *Structure*, **5**, 95–108.

KLENK, H. P., PALM, P. and ZILLIG, W. (1994) DNA-dependent RNA polymerases as phylogenetic marker molecules. *System. Appl. Microbiol.*, **16**, 638–647.

KOROLEV, S., NAYAL, M., BARNES, W. M., DICERA, E. and WAKSMAN, G. (1995) Crystal structure of the large fragment of *Thermus aquaticus* DNA polymerase I at 2.5-angstrom resolution: structural basis for thermostability. *Proc. Natl Acad. Sci. USA*, **92**, 9264–9268.

KOWALAK, J. A., DALLUGE, J. J., MCCLOSKEY, J. A. and STETTER, K. O. (1994) The role of posttranscriptional modification in stabilisation of transfer RNA from hyperthermophiles. *Biochemistry*, **33**, 7869–7876.

LANGWORTHY, T. A. and POND, J. L. (1986) Archaebacterial ether lipids and chemotaxonomy. In *Archaebacteria '85* ed. O. Kandler and W. Zillig, pp. 278–285 (New York: Gustav Fischer).

LINDHAL, T. (1966) Irreversible heat inactivation of transfer ribonucleic acids. *J. Biol. Chem.*, **242**, 1970–1973.

LOPEZ-GARCIA, P. and FORTERRE, P. (1997) DNA topology in hyperthermophilic archaea: reference states and their variation with growth phase, growth temperature, and temperature stresses. *Mol. Microbiol.*, **23**, 1267–1279.

MILLER, S. L. and LAZCANO, A. (1995) The origin of life at high temperature? *J. Mol. Evol.*, **41**, 689–692.

NOLLER, H. F., HOFFARTH, V. and ZIMNIAK, L. (1992) Unusual resistance of peptidyl transferase to protein extraction procedures. *Science*, **256**, 1416–1419.

PERLER, F. B., KUMAR, S. and KONG, H. (1996) Thermostable DNA polymerases. *Adv. Protein Chem.*, **48**, 377–435.

PHILIPPE, H. and ADOUTTE, A. (1995) How reliable is our curent view of eukaryotic phylogeny? In *Protistological Actualities*, ed. G. Brugerolle and J.-P. Mignot, pp. 17–33.

PHILIPPE, H., SÔRHANNUS, U., BAROIN, A., PERASSO, R., GASSE, F. and ADOUTTE, A. (1994) Comparison of molecular and palaeontological data in diatoms suggests a major gap in the fossil record. *J. Evol. Biol.*, **7**, 247–265.

STETTER, K. (1992) Life at the upper temperature border. In *Frontiers of Life*, ed. J. Trân Thanh Van, K. Trân Thanh Van, J. C. Mounolou, J. Schneider and C. McKay, pp. 195–219 (Gif-sur-Yvetette: Editions Frontières).

VOSSBRINCK, C. R., MADDOX, J. V., FRIEDMAN, S., DEBRUNNER-VOSSBRINCK, B. A. and WOESE, C. R. (1986) Ribosomal RNA sequence suggests microsporidia are extremely ancient eukaryotes. *Nature*, **326**, 411–414.

WOESE, C. (1987) Bacterial evolution. *Microbiol. Rev.*, **51**, 221–271.

WOESE, C., KANDLER, O. and WHEELIS, M. (1990) Towards a natural system of organisms: proposal for the domains *Archae, Bacteria, and Eukarya. Proc. Natl Sci. USA*, **87**, 4576–4579.

9

Hyperthermophilic and Mesophilic Origins of the Eukaryotic Genome

JAMES A. LAKE, RAVI JAIN, JONATHAN E. MOORE AND MARIA C. RIVERA

Molecular Biology Institute and MCD Biology, University of California, Los Angeles, Los Angeles, California, USA

9.1 Introduction

The origin of the eukaryotic cell was a milestone in the evolution of life. Eukaryotes differ from prokaryotes in a number of ways. A principal difference is that they contain an extensive system of internal membranes that enclose the nucleus and define organelles, and this compartmentalisation has required a number of unique eukaryotic innovations. The nucleus and its origin are intimately connected with the origin of eukaryotes since a nucleus is present in all eukaryotes, and only in eukaryotes. It is, in fact, the defining character for which eukaryotes are named (eu = good or true, karyote = kernel, as in nucleus).

Many questions concerning the prokaryotic origins of nuclear genes are presently incompletely understood, and current ideas concerning their origins are changing rapidly with the advent of completely sequenced genomes. The most widely known view holds that eukaryotes are derived from an evolutionary ancestor common to all Archaea. The eocyte view, favoured by our laboratory, differs from this view in two ways. First, it posits that the Archaea are not a true (monophyletic) evolutionary group because a major hyperthermophilic group, the eocytes (crenarchaeotes), is phylogenetically closer to the eukaryotes. Second, it is becoming clear that ribosomal RNA molecule does not represent the evolution of life as claimed by the Archaeal theory, but is representative of only a limited part of the genome. It is becoming clearer and clearer that eukaryotic genomes have significant contributions from eubacteria. Even the mechanism by which the nucleus was formed may have involved an endosymbiosis between two bacterial types.

9.2 Prokaryotic diversity

Before discussing eukaryotic origins and attempting to identify the prokaryotic group(s) that is (are) most closely related to the eukaryotic nucleus (sister taxon), it is useful to survey briefly their phenotypic and phylogenetic diversity. In principle, there are several groups with prokaryotic organisation any one of which might be related to the eukaryotic nucleus. These include (1) the eubacteria, (2) the halobacteria, (3) the methanogens and relatives of the methanogens, and (4) sulfur-metabolising, high-temperature organisms known as the eocytes.

(1) The eubacteria are a diverse group that includes all the photosynthetic bacteria (except for the halobacteria) as well as many nonphotosynthetic groups. Most eubacteria are mesophiles; however, the eubacteria also include extreme thermophiles, such as *Thermotoga maritima* and *Aquifex pyrophilus*, which can grow up to 90 °C and 95 °C, respectively (Huber and Stetter, 1992). The lipids of eubacteria are primarily of the ester type, although *Thermotoga*, *Aquifex*, and their relatives also contain branched ether lipids.

(2) The halobacteria are extreme halophiles. They are carbon heterotrophs that can use an unusual photosynthesis system, namely a light-driven proton pump based on bacteriorhodopsin. Like eubacteria, they contain the biochemical pathways for the synthesis of C_{40} and C_{50} carotenoids (Goodwin, 1980).

(3) The methanogens are a phylogenetically diverse group, despite the fact that they share a common phenotype, namely that they are strict anaerobes with the ability to chemically reduce carbon compounds to methane to provide energy. Associated with the methanogens is a phenotypically diverse group of organisms represented by such organism as *Thermococcus celer*, and *Archaeoglobus fulgidus*, and, possibly, *Methanopyrus kandleri*. This last organism grows at temperatures up to 112 °C. Like methanogens, it reduces carbon compounds to methane; however, it is not closely related to other methanogens but instead is intermediate between them and the next group of organisms (Burggraf *et al.*, 1991; Rivera and Lake, 1996).

(4) The fourth prokaryotic group, the eocytes, consists of thermophilic, mostly sulfur-metabolising organisms, many of which can grow at temperatures in excess of 100 °C. The eocytes include *Sulfolobus*, *Desulforococcus*, *Thermoproteus*, *Pyrodictium*, *Pyrobaculum*, etc. *Sulfolobus sulfataricus*, oxidises sulfur to H_2S. Others, such as *Acidianus infernus*, can oxidise or reduce sulfur to H_2SO_4 or to H_2S. The organisms with the highest maximum growth temperatures are *Pyrodictium occultum* (112 °C), *Pyrodictium abyssum* (112 °C) and *Pyrolobus fumarii* (113 °C). The group is metabolically diverse, uniformly thermophilic, and phylogenetically monophyletic.

9.3 Origin of the eukaryotes

9.3.1 *Theories for the origin of eukaryotes*

The ability of eukaryotes and prokaryotes to transfer genes laterally is well known (Syvanen and Kado, 1997). Numerous genes originally contained in the mitochondrial and chloroplast genomes have been transported and are now encoded in the nucleus (Gray, 1993; Baldauf and Palmer, 1990; for a thoughtful review see Smith *et al.*, 1992). Thus not all nuclear genes are indicative of the ancestry of the bulk of the nuclear genes, and even the mechanism of formation of the nucleus is unknown.

Several theories exist for the origin of the nucleus. Most scientists still subscribe to the *karyogenic hypothesis*, shown in Figure 9.1. In this theory the nucleus and its enclosing membranes were gradually acquired through some (unspecified) segregating process. The competing theory, the *endokaryotic hypothesis*, is less well known. This theory is based on the idea that the nucleus, like the other eukaryotic organelles enclosed in double membranes (the chloroplast and mitochondrion), has been derived through capture by an engulfing bacterium (Lake, 1982). The latter proposal is simple and would parsimoniously explain the origin of all double-membrane organelles through a single mechanism

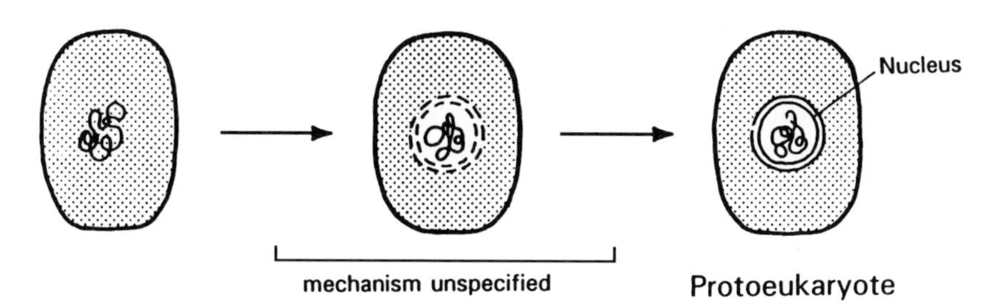

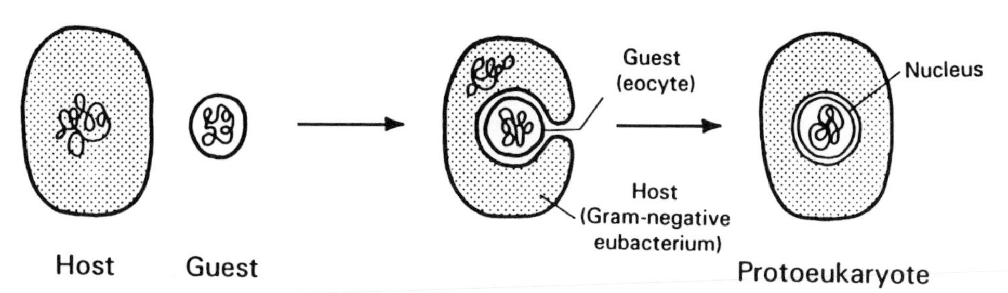

Figure 9.1 Competing hypotheses for the evolution of the nucleus (Lake and Rivera, 1994).

rather than requiring two different mechanisms (one for the mitochondrion and chloroplast and another for the nucleus). Until recently there were few data to support either the autogenous or the fusion theories.

Now, however, fusion and chimeric theories for the origin of eukaryotes are increasingly gaining support from molecular data. Fusion theories have been proposed by Martin and coworkers (Henze *et al.*, 1995), by Sogin (1991), by Zillig *et al.* (1992), by Gupta and coworkers (Gupta *et al.*, 1994; Golding and Gupta, 1994), and by us (Lake, 1982; Lake and Rivera, 1994). All of these theories are prompted by the incongruence of phylogenetic trees reconstructed from different types of data. The earliest chimeric proposal (Lake, 1982), was motivated by the incongruence between the distributions of membrane lipids (ester in eubacteria and eukaryotes, versus ether in halobacteria, methanogens, and eocytes), and ribosomal structures (present in eocytes and eukaryotes but absent in eubacteria, halobacteria, and methanogens; discussed in Rivera and Lake, 1992). Influenced by these data, and also by the previously mentioned observation that the nucleus was enclosed by double membranes, it was proposed that a fusion eukaryote could have been obtained by the engulfment of one prokaryote by another.

Sogin (1991), concerned with the inconsistency of branch lengths in trees reconstructed from 18S rDNA and ATPase genes, thoughtfully proposed a chimeric model in which the rRNA was contributed by an ancient lineage with an RNA-based metabolism and the proteins were contributed by a sulfur-dependent prokaryote (eocyte). Since the model posits different origins for rRNA and proteins it can explain the rate differences observed in trees reconstructed from these two genes types.

Golding and Gupta (1994) presented a strong case using the HSP70 heat shock protein found in eukaryotes to advocate a fusion of prokaryotes to form a protoeukaryote. They found a close relationship between the eukaryotic HSP70 (not the HSP contributed by a mitochondrion) and the Gram-negative eubacterium *E. coli*. Furthermore, the distribution of an insert (present in eukaryotes and Gram-negative eubacteria, and absent in Gram-positive eubacteria, halobacteria, and methanogens) argued for a tree in which the eubacteria are not monophyletic. The list of proteins considered by Golding and Gupta includes aspartate aminotransferase, HSP70, ferredoxin, glutamate dehydrogenase, glutamine synthetase, pyroline-5-carboxylate reductase, and deoxyribodipyrimidine photo-lyase. Zillig (Zillig *et al.*, 1992) have also made a case for a fusion of two prokaryotes through consideration of glyceraldehyde-phosphate dehydrogenase, L-malate dehydrogenase, aspartate aminotransferase, and phosphoglycerate kinase. Others, notably Martin and coworkers (Martin, 1997) regard 'the nuclear genes of several glycolytic enzymes of the eukaryotic cytosol as being acquired from eubacteria early in eukaryotic evolution, probably through endosymbiosis'.

Clearly there is evidence that the origins of the eukaryotes are complex and probably involve at least two prokaryotic antecedents, an *E. coli*-like eubacterium and a hyperthermophilic sulfur-metabolising eocyte, and the process could be even more complex. It is apparent that *individual genes are no longer suitable to describe this process* and one needs to follow the inheritance of genomes, particularly looking for classes of genes (Riley, 1993) that may have been inherited as a group. Thus, the study of entire genomes is necessary in order to understand this process in any comprehensive way.

9.3.2 Genomic evidence for a chimeric origin of eukaryotes

Although complete genomes are only now becoming available from diverse prokaryotes and eukaryotes, their analysis may soon provide comprehensive answers. Preliminary results from our laboratory, for example, are already suggesting that different groups of genes may have different origins. In Figure 9.2 (work in progress), the analysis of the genomes of *Saccharomyces cerevisiae* (a eukaryote), *Methanococcus jannaschii* (a methanogen), and *Synechocystis* PCC 6803 (a cyanobacterium) is illustrated. Each square represents the analysis of a different homologous gene. The solid squares represent individual genes functioning in translation and the open squares are genes functioning in the biosynthesis of cofactors, prosthetic groups, and carriers. When a measure of the distance from a specific methanogen gene to the homologous gene in the cyanobacterium is compared with the distance from the methanogen gene to the yeast homologue, one finds a dramatic separation of these genes into two regions of the graph. The translation genes occupy positions above the diagonal and the biosynthetic genes occupy positions below the diagonal. Hence, *the evidence for several pathways of inheritance of eukaryotic genes is becoming very clear* and some of these genes are from eubacterial mesophiles (Rivera, Jain, Moore, and Lake, 1998).

9.3.3 Genomic evidence for two functionally distinct lineages within prokarytoes

Prokaryotic and eukaryotic evolution has long been viewed primarily through the perspective of a single molecule, ribosomal RNA (rRNA), but the availability of complete genomes now offers the possibility of reconstructing a more balanced picture of cellular

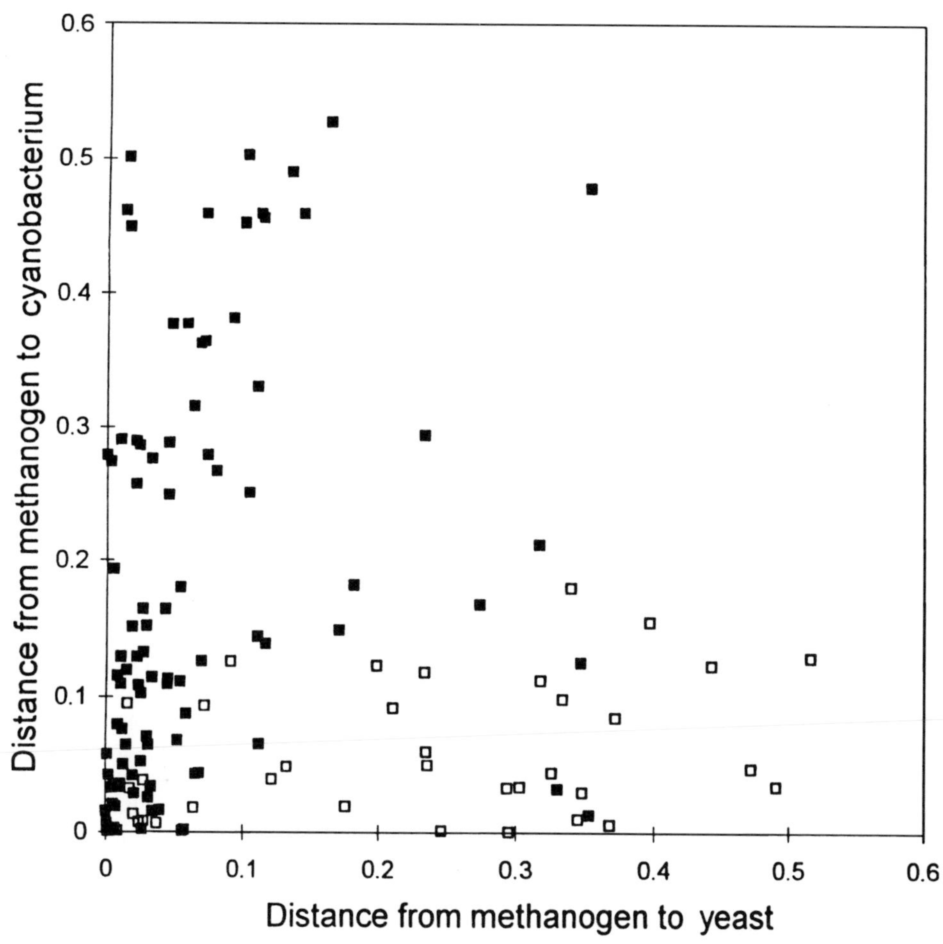

Figure 9.2 A distance plot illustrating the alternative inheritances for genes from different functional groups. Each dot represents the analysis of a different gene. The solid squares represent individual genes functioning in translation and the open squares are genes functioning in the biosynthesis of cofactors, prosthetic groups, and carriers (Riley, 1993).

evolution based on entire functional classes of molecules. Owing to the past emphasis on analyses from a single molecule, rRNA, it is generally assumed that prokaryotes have evolved as pure lineages relatively uncorrupted by horizontal gene transfer (but see Syvanen and Kado, 1997). New evidence from the analysis of complete genomes now suggests that the evolution of prokaryotes appears to be as complex as, if not more so than, that of eukaryotes.

In recent work from our laboratory (Rivera, Jain, Moore and Lake, 1998), analyses of complete genomes indicates that a major prokaryotic reorganisation preceded the formation of the eukaryotic cell. In comparisons of the entire set of *Methanococcus jannaschii* genes with their homologues from *Escherichia coli*, *Synechocystis* 6803, and the yeast *Saccharomyces cerevisiae*, it was shown that prokaryotic genomes are derived from two different ancestral lineages which correspond to broadly defined functional classes of genes. The deepest diverging lineage, the *informational lineage*, codes for genes which function in translation, transcription, and replication, and also includes GTPases, tRNA synthetases,

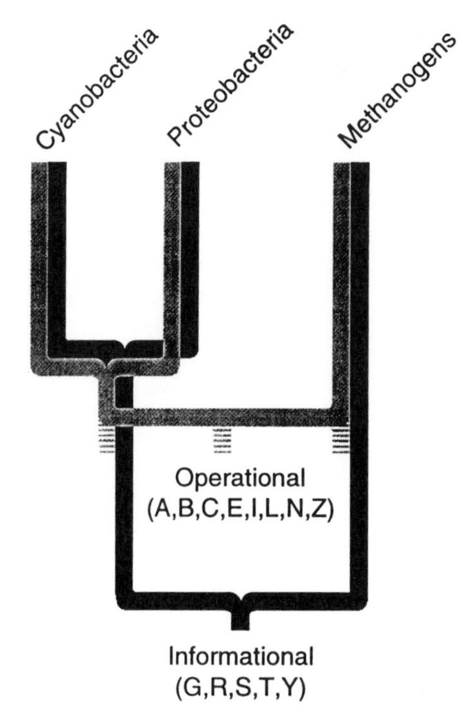

Figure 9.3 Evolution of the operational and informational lineages. The prokaryotic evolution of the operational and informational lineages is shown in light grey and dark grey, respectively. The functional categories are: amino acid synthesis (A), biosynthesis of cofactors (B), cell envelope proteins (C), energy metabolism (E), GTPases and homologues of vacuolar ATPases (G), intermediary metabolism (I), fatty acid and phospholipid biosynthesis (L), nucleotide biosynthesis (N), replication (R), transcription (S), translation (T), transport (X), tRNA synthetases (Y), and regulatory genes (Z). Three possible origins of the operational lineage are shown as dashed light grey lines.

and vacuolar ATPase homologues. The more recently diverging, *operational lineage*, codes for amino acid synthesis, the biosynthesis of cofactors, the cell envelope, energy metabolism, intermediary metabolism, fatty acid and phospholipid biosynthesis, nucleotide biosynthesis, and regulatory functions. The broad outlines of this exchange are summarised in Figure 9.3. Prior to these studies, the horizontal transfer of operational genes, consisting of more than half a genome, between distantly related prokaryotes was totally unforeseen!

At present, we are not sure where the operational lineage originated. It may have come from (i) the genome of an early eubacterium, (ii) the genome of a methanogen precursor, or (iii) some, as yet undiscovered or now extinct, alternative type of prokaryotic life. Furthermore, the transfer may have occurred as a single massive exchange or as a series of multiple, smaller exchanges. We do not currently know. We do know, however, that the presence of such massive exchange greatly complicates the analysis of eukaryotic origins. Accordingly, in the remainder of this chapter we will restrict our discussion to genes involved in translation (i.e. informational genes) since these seem to have been inherited through a single pathway, and will ask, specifically, which prokaryotic group contributed their translation genes to the eukaryotic nucleus.

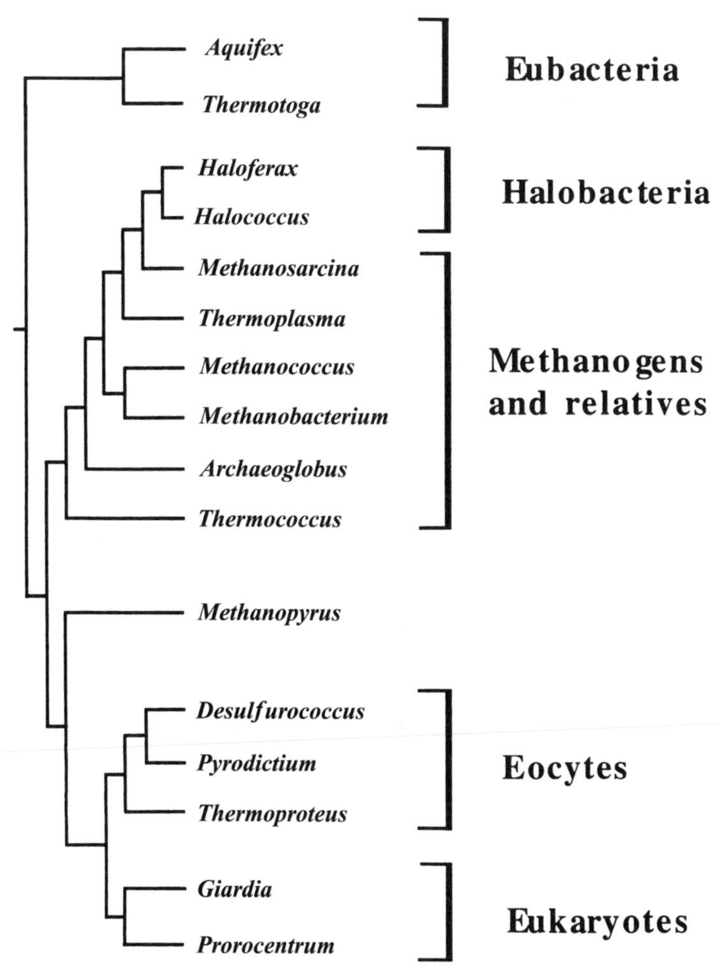

Figure 9.4 The rooted phylogenetic tree relating prokaryotic and eukaryotic organisms reconstructed from 16S and 18S ribosomal RNA sequences using paralinear distances (Lockhart *et al.*, 1994; Lake, 1994) and the bootstrapper's gambit multi-taxon tree reconstruction algorithm (Lake, 1995). Two hundred bootstrap replicates supported the halobacteria + methanogens and relatives clade at the 99% level and supported the eocytes + eukaryotes clade at the 100% level.

9.3.4 *Which group of prokaryotes is the closest relative of the eukaryotic nucleus?*

The two most intensively studied prokaryotic genes are the 16/18S ribosomal RNA (rRNA) genes and the genes of protein synthesis factor EF-Tu (EF-1α in eukaryotes). The results of reconstructing the phylogenetic tree of life from 16/18S rRNA genes is shown in Figure 9.4. For the taxa in this analysis, we find that the eukaryotes, at the bottom of the figure, are most closely related to the eocyte prokaryotes, and the halobacteria + methanogens form a separate clade. Since the results of such analyses produce unrooted trees, we have arbitrarily rooted the tree in the branch leading to the eubacteria in accord with the results of others (Gogarten *et al.*, 1989; Iwabe *et al.*, 1989). When one uses

Table 9.1 Recent sequence analyses of EF-1α sequences

Supports Archaea	Supports Eocyta	References
	+	Baldauf *et al.* (1996)
	+	Hashimoto and Hasegawa (1996)
	+	Runnegar (1993)
	+	Lake (1994)
	+	Hasegawa *et al.* (1993)
	+	Cousineau *et al.* (1992)
+		Creti *et al.* (1991)

EF-1α genes to reconstruct the these relationships, we and others similarly find virtually identical results. In contrast, Woese and Olsen (1986) using identical 16/18S sequences but different alignments and tree reconstruction programmes, find a different placement for eukaryotes and have based their claim for the existence of Archaea on their finding. In support of the eocyte theory, virtually every recent analysis of EF-1α sequence has supported the eocyte theory and rejected the archaeal theory (see Table 9.1).

One must ask, 'Why are such differences found?'. The answer is almost certainly due to artefacts in alignment and tree reconstruction known as *long branch attraction*. Such attraction occurs when taxa are evolving at different rates. The effect of this artefact is to cause rapidly evolving taxa to join with other rapidly evolving taxa in an evolutionary tree, and slowly evolving taxa to join with other slow ones in the tree, whether or not the taxa are phylogenetically related. Some authors, ourselves included, have concluded that it is the attraction of the long branches of the eubacteria and eukaryotes that causes them to be placed together in the archaeal tree, and that this tree is therefore simply an artefact (see Volters and Erdmann, 1988).

9.4 Using sequence inserts and operon organisation to map eukaryotic origins

9.4.1 Elongation factor genes

Because of the long branch attraction artefacts, we searched for molecular sequences which contained structural features, such as inserted segments, that would evolve much more slowly than individual nucleotides, be more easily interpreted, and therefore be much less sensitive to long branch artefacts.

The molecule we chose to study was protein synthesis elongation factor EF-Tu (EF-1α in eukaryotes), (Rivera and Lake, 1992). EF-Tu is a ubiquitous protein that transports aminoacyl-tRNAs to the ribosome and participates in their selection by the ribosome. Within the GDP-binding domain of EF-Tu, the amino acid sequence, $KNMITG_{94}$, which is strictly conserved in EF-1α and EF-Tu sequences, terminates an α-helix and is followed by a β-strand that is terminated by $GPMP_{113}$ at the GDP binding site. The sequence $QTREH_{118}$ then starts a 3_{10} helix. The amino acid motifs of the eukaryotic EF-1α are similar, except that the 4-amino-acid sequence $GPMP_{113}$ in prokaryotes is replaced by the 11-amino-acid sequence GEFEAGISKDG and its variants in eukaryotes (see Table 9.2).

Since the eukaryotic 11-amino-acid insert is so well conserved among eukaryotic sequences, we thought that eocyte sequences might also contain the 11-amino-acid insert.

Table 9.2 Comparison of the *Methanopyrus kandleri* EF-1 sequence to the sequences from methanogens, halobacteria, eubacteria, eocytes, and eukaryotes. Four-amino-acid and 11-amino-acid segments are underlined. Small letters represent sequences from the PCR primers. The sequences compared are the following. The methanogens and their relatives are Mp. kand., *Methanopyrus kandleri*, T. celer, *Thermococcus celer*; P. woesei, *Pyrococcus woesei*; A. fulg., *Archaeoglobus fulgidus*; Mc. van., *Methanococcus vannielii*; and T. acido., *Thermoplasma acidophilum*. The halobacteria: H. maris., *Halobacterium marismortui*. The eubacterial sequences are E. coli, *Escherichia coli*, the thermophilic eubacteria Th. mar., *Thermotoga maritima* and the halophilic cyanobacteria D. sal, *Dactylococcopsis salina*. The eocytes are D. muco, *Desulfurococcus mucosus*; Td. mar., *Thermodiscus maritimus*; P. occu., *Pyrodictium occultum*; A. infe., *Acidianus infernus*; Su. acid., *Sulfolobus acidocaldarius*. The eukaryotes are: Giardia, *Giardia lamblia*; Tetrahy., *Tetrahymena pyriformis*; yeast, *Saccharomyces cerevisiae*; tomato, *Lycopersicon esculentum*; Droso., *Drosophila melanogaster*; rat, *Rattus norvegicus*; and human, *Homo sapiens*. Original sources for the sequences are listed in Rivera and Lake (1996)

Taxon	Organism		11-amino-acid segment	4-amino-acid segment	
Methanogens and relatives					
	Mp. kand.	KNMITGASQADAAILVVAADD		---GVMP	QTREH
	T. celer	KNMITGASQADAAVLVVAVTD		---GVMP	QTKEH
	P. woes.	KNMITGASQADAAVLVVAATD		---GVMP	QTKEH
	A. fulg.	KNMITGASQADAAVLVMDVVE		---KVQP	QTREH
	Mc. vann.	KNMITGASQADAAVLVVNVDD		AKSGIQP	QTREH
	T. acido.	KNMITGTSQADAAILVISARD		-GEGVME	QTREH
Halobacteria					
	H. maris.	KNMITGASQADNAVLVVAADD		---GVQP	QTQEH
Eubacteria					
	Th. mar.	KNMITGAAQMDGAILVV AATD		---GPMP	QTREH
	D. sal.	KNMITGAAQMDGAIIVCSA A D		---GPMP	QTREH
	E. coli	KNMITGAAQMDGAILVV AATD		---GPMP	QTREH
Eocytes					
	Su. acid.	KNMITGASQADAAILVVSAKK	GEYEAGMSAEG		QTREH
	Td. mari.	KNMITGASQADAALLVVSARK	GEFEAGMSAEG		QTREH
	P. occu.	KNMITGASQADAAILVVSARK	GEFEAGMSAEG		QTREH
	D. muco.	KNMITGASQADAAILVVSARK	GEFEAGMSAEG		QTREH
	A. infe.	KNMITGASQADAAIIAVSAKK	GEFEAGMSEEG		QTREH
Eukaryotes					
	Giardia	KNMITGTSQADVAILVVAAGQ	GEFEAGISKDG		QTREH
	Tetrahy.	KNMITGTSQADVAILMIASPQ	GEFEAGISKDG		QTREH
	Yeast	KNMITGTSQADCAILIIAGGV	GEFEAGISKDG		QTREH
	Tomato	KNMITGTSQADCAVLIIDSTT	GGFEAGISKDG		QTREH
	Droso.	KNMITGTSQADCAVQIDAAGT	GEFEAGISKND		QTREH
	Rat	KNMITGTSQADCAVLIVAAGV	GEFEAGISKNG		QTREH
	Human	KNMITGTSQADCAVLIVAAGV	GEFEAGISKNG		QTREH

Using the polymerase chain reaction and DNA primers designed for use with the KNMITG and QTREH sites, we amplified, cloned, and sequenced the insert region with the results shown in Table 9.2. The eocyte amino acid sequences, translated from DNA, shared the eukaryotic motif (11-amino-acids) rather than that found in methanogens, halobacteria, and eubacteria (4 amino acids). The longer 11-amino-acid segment present in eocytes and eukaryotes shares little similarity with the shorter, 4-amino-acid segment found in other prokaryotes.

Based on these results, we could directly test the eocyte and archaeal theories for the origin of the nucleus. The trees corresponding to both theories, reconstructed from 16/18S rRNA sequences, are shown in Figure 9.5. The fundamental difference between these two theories is that in the eocyte tree (at the top of the figure) the eukaryotic nucleus shares a most recent common ancestor solely with the eocytes, whereas in the archaebacterial theory the eukaryotes are most closely related to an ancestral organism that gave rise to the halobacteria, the methanogens and their relatives, and the crenarchaea (= eocytes). Since these two theories are based on the *topology* of the phylogenetic trees, the two are *mutually exclusive* and hence eminently testable. Although both theories could be incorrect, both cannot be correct. Hence if one theory is correct the other must be incorrect.

Our conclusions are shown in Figure 9.5. We have mapped the changes onto the trees representative of both theories. Starting from the 4-amino-insert at the root of the tree, each solid box indicates a change from the 4-amino-acid segment to the 11-amino-acid form. The eocyte tree is favoured because it requires only a single change, whereas the archaebacterial tree requires two independent but identical changes. (The archaeal tree could also be explained by one appearance of the 11-amino-acid form and one reappearance of the 4-amino-acid form, but even so, two changes would be required.)

Several lines of reasoning buttress the interpretation that eocytes are the closest relatives of the eukaryotes. First, the 11-amino-acid segments present in eocytes and eukaryotes are very likely homologous. Eight of eleven amino acids (seven in *Sulfolobus* and *Acidianus*) are identical to the consensus eukaryotic sequence. Amino acid shuffling of the segments produced random alignments that score 6–7 standard deviations lower than those found for the eukaryotic–eocyte alignment, thereby implying homology (Waterman and Eggert, 1987). Second, the alignments are well defined. No gaps are needed to align the eukaryotic and eocytic EF-1α sequences, and no gaps are needed to align the eubacteria, methanogen, and halobacterial sequences. Third, the sequences encoding EF-1α are not likely to have been laterally transferred between organisms, since EF-1α is present in all cells and, during protein synthesis, interacts with cellular components encoded by genes dispersed throughout the bacterial genome, including aminoacyl-tRNAs, ribosomal proteins, elongation factor EF-Ts, and 16S and 18S ribosomal RNAs (Hill *et al.*, 1990). Thus these results lend strong support to the proposal that the eukaryotes and eocytes are sister taxa within the 'tree of life'.

9.4.2 Organisation of ribosomal operons

A number of fundamental molecular properties have been thought to have an idiosyncratic distribution on the tree of life, principally because they did not fit the archaeal tree. Yet these same molecular properties fit the eocyte theory perfectly. This is particularly true for the organisation of ribosomal rRNA operons.

Because small subunit ribosomal RNA sequences are the standard for defining the phylogenetic positions of organisms, a large database of ribosomal RNAs exists and one knows far more about the organisation of ribosomal operons than about any other operons. Eubacteria, halobacteria, methanogens, and eocytes contain three rRNAs – 16S, 23S, and 5S – which are homologous to the eukaryotic 18S, 5.8S+28S, and 5S. (For simplicity we will refer to both the eukaryotic and prokaryotic homologues using the prokaryotic

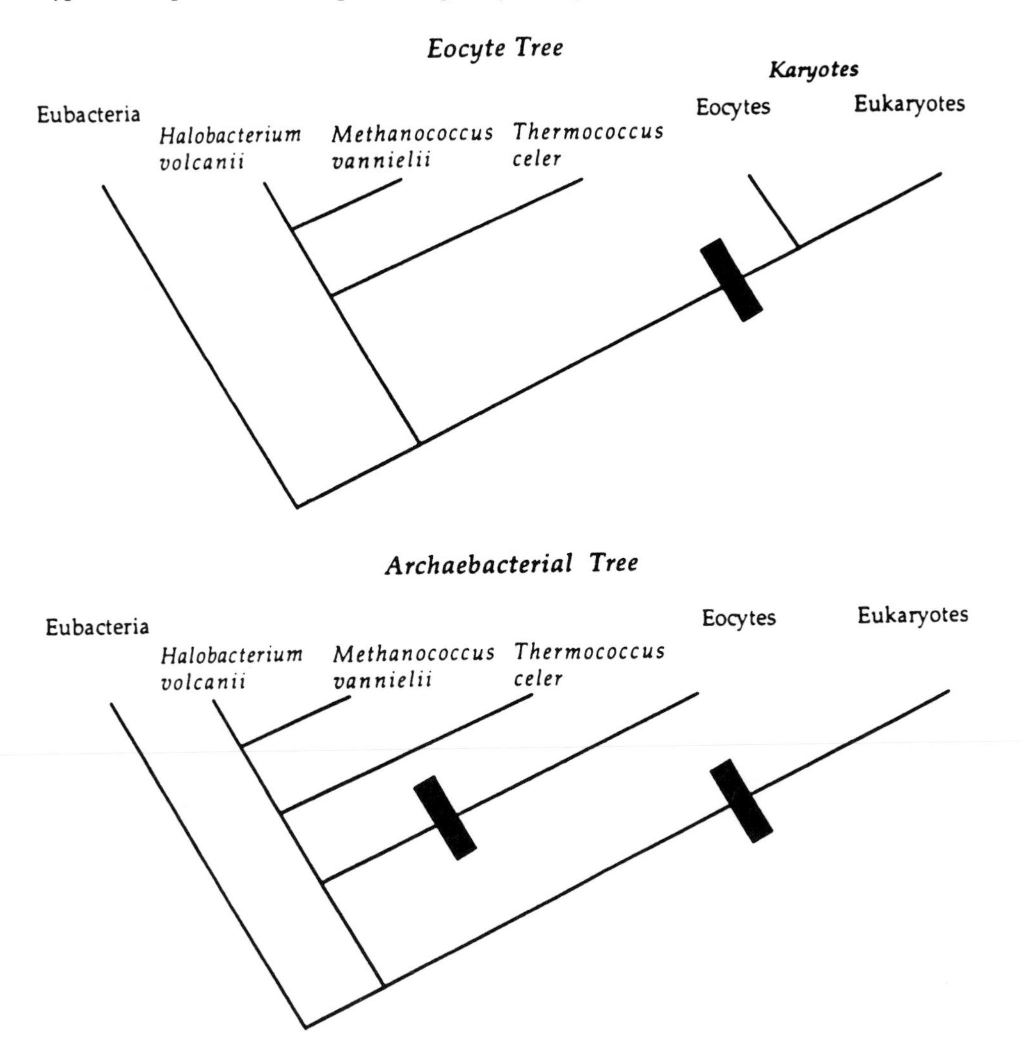

Figure 9.5 Rooted trees illustrating the two theories proposed to explain the origin of the eukaryotic nucleus. The trees corresponding to both theories are reconstructed from 16S/18S sequences and rooted in the branch leading to the eubacteria. The solid boxes indicate changes from the 4-amino-acid segment to the 11-amino-acid form. The eocyte tree is favoured by parsimony since it requires only a single change to the 11-amino-acid segment whereas the archaebacterial tree is opposed since it requires two independent changes (the same distribution could also be explained by one appearance of the 11-amino-acid form and reappearance of the 4-amino-acid form, but still two changes would be required).

labels.) The number of ribosomal rRNA transcriptional units varies between one and four in the halobacteria and the methanogens. Ribosomal operons are arranged in the same general pattern in eubacteria, halobacteria, and methanogens, namely 16S–tRNA–23S–5S. Occasionally an additional tRNA gene will be found between the 16S and 23S genes or following the 5S gene (reviewed in Brown *et al.*, 1989). *Thermoplasma*, which routinely clusters in phylogenetic trees with the methanogens (see Figure 9.5), is an exception to this general rule and, unlike any other prokaryote, contains unlinked 16S, 23S, and 5S

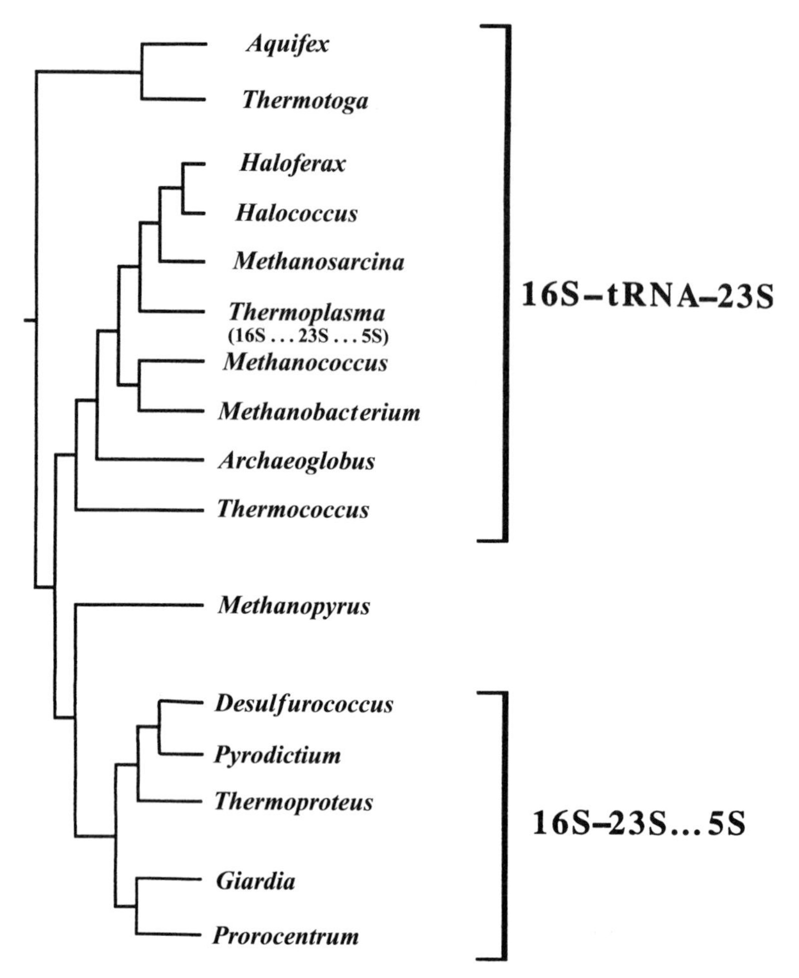

Figure 9.6 The phylogenetic tree from Figure 9.4 illustrating the distribution of ribosomal operon types. In the 16S–tRNA–23S type the operon contains 16S rRNA + tRNA (possibly two) + 23S rRNA + 5S rRNA. In the 16S–23 . . . 5S type the operon contains the 16S rRNA + 23S rRNA with the 5S rRNA frequently transcribed separately.

genes (Tu and Zillig, 1982). The pattern in eocytes and eukaryotes is different from that in the eubacteria, halobacteria, and methanogens. In the eocytes, the 16S–23S genes are linked without a tRNA spacer and there is a variable linkage of 5S rRNA encoding genes to the 16S–23S unit. The non-operon-associated 5S rRNA gene of *Desulfurococcus mobilis* forms its own transcriptional unit (Kjems and Garrett, 1988), but those of many other eocytes contain a 16S–23S–5S unit. The eukaryotic pattern is similar with a 16S–23S (equivalent) transcription unit lacking tRNA spacers and with the 5S either separately transcribed or linked (Gerbi, 1985). An exception to this rule is found among the Cryptomonads, where the rRNA genes are unlinked (Gray, 1992).

Although it cannot easily be explained by the archaebacterial theory, this pattern of rRNA operon organisation easily fits the eocyte tree well. In Figure 9.6, the tree is labelled with the types of operons. *To accommodate this distribution on the eocyte tree only a single change of operon type is required*, namely, the 16S–tRNA–23S–5S pattern found in eubacteria, halobacteria, and methanogens is substituted by the derived 16S–23S.

Depending upon the operon organisation in *Methanopyrus* (presently unknown), the site will be either before or after *Methanopyrus* branches. In either case, still only a *single* change will be required.

9.5 Conclusions

Of all the genes functioning in translation that have been sequenced to date, the EF-Tu molecule seems to offer the most reliable indication of early divergences. Genome analyses show that it is the slowest-evolving sequence of its class and should, therefore, be the most reliable for phylogenetic purposes. It is also unlikely to be laterally transferred between organisms because it is present in all cells and, during protein synthesis, interacts with cellular components that are dispersed throughout the bacterial genome. Furthermore, direct phylogenetic analyses of this molecule by almost all authors support the eocyte tree. Significant support for the eocyte tree also comes from the observations that eukaryotic ribosomal operons are organised like those of *Sulfolobus, Desulfurococcus*, and *Thermoproteus* and not organised like the tRNA-containing rRNA operons of halobacteria, methanogens, and eubacteria.

We are just starting to fathom the diversity of life on earth through the complete analyses of genomes. Although the relationship of the nuclear genes of eukaryotes to those of prokaryotes are almost certain to be more complicated than that found for translation genes, *it now appears that the translation apparatus of eukaryotes has been derived from that of hyperthermophilic eocyte prokaryotes*. Thus the eukaryotic cell is appearing to be more of a chimera between mesophilic eubacteria and hyperthermophilic prokaryotes. Indeed, even the organism at the very root of the tree of life (see Figure 9.6) is most parsimoniously interpreted as being a hyperthermophilic prokaryote (Lake, 1988). The tremendous progress that is being made towards determining the relationships among bacteria and eukaryotes using gene sequences and complete genomes gives us optimism that many of the major events that occurred during the evolution of life, including the role of hyperthermophily, will soon be understood.

9.6 Summary

The origin of the eukaryotic cell has been enigmatic. Within the last decade, with the availability of DNA sequences and entire genomes from diverse organisms, it has become possible to investigate the remote origins of eukaryotes. This review describes the current understanding of relationships among the major groups of prokaryotes and eukaryotes and emphasises an ongoing search for determining which prokaryotes contributed genes to eukaryotes. It now appears that nuclear genes have originated from both *mesophilic* eubacteria and (particularly genes involved in translation) from *hyperthermophilic* eocyte prokaryotes (crenarchaea).

References

Baldauf, S. L. and Palmer, J. D. (1990) Evolutionary transfer of the chloroplast tufA gene to the nucleus. *Nature*, **344**, 262–263.

Baldauf, S., Palmer, J. and Doolittle, W. F. (1996) The root of the universal tree and the origin of eukaryotes based on elongation factor phylogeny. *Proc. Natl Acad. Sci. USA*, **93**, 7749–7754.

BROWN, J. W., DANIELS, C. J. and REEVE, J. N. (1989) Gene structure, organisation, and expression in archaebacteria. *CRC Crit. Rev. Microbiol.*, **16**, 287–338.

BURGGRAF, S., STETTER, K. O., ROUVIERE, P. and WOESE, C. R. (1991) *Methanopyrus kandleri*: An archaeal methanogen unrelated to all other known methanogens. *System. Appl. Microbiol.*, **14**, 346–351.

COUSINEAU, B., CERPA, C., LEFEBVRE, J. and CEDERGREN, R. (1992) The sequence of the gene encoding elongation factor Tu from *Chlamydia trachomatis* compared with those of other organisms. *Gene*, **120**, 33–41.

CRETI, R., CITARELL, F., TIBONI, O., SANANGELANTONI, A., PALM, P. and CAMMARANO, P. (1991) Nucleotide sequence of a DNA region comprising the gene for elongation factor 1α from the ultrathermophilic Archaeote *Pyrococcus woesei*: phylogenetic implications. *J. Mol. Evol.*, **33**, 332–342.

GERBI, S. A. (1985) Evolution of ribosomal DNA. In *Molecular Evolutionary Genetics*, ed. R. J. MacIntyre, chap. 7 (New York: Plenum Publishing).

GOGARTEN, J. P., KIBAK, H. and DITTRICH, P., *et al.* (1989) Evolution of the vacuolar H^+-ATPase: implications for the origin of Eukaryotes. *Proc. Natl Acad. Sci. USA.*, **86**, 6661–6665.

GOLDING, G. B. and GUPTA, R. S. (1994) Protein-based phylogenies support a chimeric origin for the eukaryotic genome. *Mol. Biol. Evol.*, **12**, 1–6.

GOODWIN, T. W. (1980) *The Biochemistry of the Carotenoids*, volume I (London: Chapman and Hall).

GRAY, M. W. (1992) The endosymbiont hypothesis revisited. *Int. Rev. Cytol.*, **141**, 233–357.

GRAY, M. W. (1993) Origin and evolution of organelle genomes. *Curr. Opin. Genet. Dev.*, **3**, 884–890.

GUPTA, R. S., AITKEN, K., FALAH, M. and SINGH, B. (1994) Cloning of *Giardia lamblia* heat shock protein HSP70 homologues: implications regarding origin of eukaryotic cells and of endoplasmic reticulum. *Proc. Natl Acad. Sci. USA*, **91**, 2895–2899.

HASEGAWA, M., HASHIMOTO, T. and ADACHI, J. (1993) Origin and evolution of eukaryotes as inferred from protein sequence data. In *The Origin and Evolution of Prokaryotic and Eukaryotic Cells*, ed. Hartman and Matsuno (Singapore: World Science).

HASHIMOTO, T. and HASEGAWA, M. (1996) Origin and early evolution of eukaryotes inferred from the amino acid sequences of translation elongation factors 1α/Tu and 2/G. *Adv. Biophys.*, **32**, 73–120.

HENZE, K., BADR., A., WETTERN, M., CERFF, R. and MARTIN, W. (1995) A nuclear gene of eubacterial origin in *Euglena gracilis* reflects cryptic endosymbioses during protist evolution. *Proc. Natl Acad. Sci. USA*, **92**, 9122–9126.

HILL, W. E., DAHLBERG, A., GARRETT, R. A., MOORE, P. B., SCHLESSINGER, D. and WARNER, J. R. (eds) (1990) *The Ribosome, Structure, Function, and Evolution* (Washington, DC: American Society of Microbiology Press).

HUBER, R. and STETTER, K. O. (1992) The *Thermotogales*: hyperthermophilic and extremely thermophilic bacteria. In *Thermophilic Bacteria*, ed. J. K. Kristansson, pp. 185–194 (Boca Raton, FL: CRC Press).

IWABE, N., KUMA, K., HASEGAWA, M., OSAWA, S. and MIYATA, T. (1989) Evolutionary Relationship of Archaebacteria, Eubacteria, and Eukaryotes inferred from phylogenetic trees of duplicated genes. *Proc. Natl Acad. Sci. USA*, **86**, 9355–9359.

KJEMS, J. and GARRETT, R. A. (1988) Novel expression of the ribosomal RNA genes in the extreme thermophile and archaebacterium, *Desulfurococcus mobilis. EMBO J.*, **6**,3521–3527.

LAKE, J. A. (1982) Mapping evolution with ribosome structure: intralineage constancy and interlineage variation. *Proc. Natl. Acad. Sci. USA*, **79**, 5948–5952.

LAKE, J. A. (1988) Origin of the eukaryotic nucleus determined by rate-invariant analysis of rRNA sequences. *Nature*, **331**, 184–186.

LAKE, J. A. (1994) Reconstructing evolutionary trees from DNA and protein sequences: paralinear distances. *Proc. Natl Acad. Sci. USA*, **91**, 1455–1459.

LAKE, J. A. (1995) Calculating the probability of multitaxon evolutionary trees: bootstrappers Gambit. *Proc. Natl Acad. Sci. USA*, **92**, 9662–9666.

LAKE, J. A. and RIVERA, M. C. (1994) Was the nucleus the first endosymbiont? *Proc. Natl Acad. Sci. USA*, **91**, 2880–2881.

LOCKHART, P. J., STEEL, M. A., HENDY, M. D. and PENNY, D. (1994) Recovering evolutionary trees under a realistic model of sequence evolution. *Mol. Biol. Evol.*, **11**, 605–612.

MARTIN, W. (1997) Endosymbiosis and the origins of chloroplast–cytosol isoenzymes; revising the gene transfer corollary. In *Horizontal Gene Transfer*, ed. M. Syvanen and C. Kado (London: Chapman and Hall).

RILEY, M. (1993) Functions of the gene products of *Escherichia Coli. Microbiol. Rev.,* **57**, 862–893.

RIVERA, M. C. and LAKE, J. A. (1992) Evidence that eukaryotes and eocytes prokaryotes are immediate relatives. *Science*, **257**, 74–76.

RIVERA, M. C. and LAKE, J. A. (1996) The phylogeny of *Methanopyrus kandleri. Int. J. System. Bacteriol.*, **46**, 348–351.

RIVERA, M. C., JAIN, R., MOORE, J. and LAKE, J. A. (1998) Genomic evidence for two funtionally distinct gene classes. *Proc. Natl Acad. Sci. USA*, **95**, 6239–6244.

RUNNEGAR, B. (1993) Proterozoic eukaryotes: evidence from biology and geology. In *Early Life on Earth*, ed. S. Bengtson (Cambridge: Cambridge University Press).

SMITH, M. W., FENG, D. F. and DOOLITTLE, R. F. (1992) Evolution by acquisition: the case for horizontal gene transfers. *Trends Biochem. Sci.*, **17**, 489–493.

SOGIN, M. L. (1991) Early evolution and the origin of eukaryotes. *Curr. Opin. Genet. Dev.*, **1**, 457–463.

STETTER, K. (1997) In 'Random Samples'. *Science*, **275**, 933.

SYVANEN, M. and KADO, C. (eds) (1997) *Horizontal Gene Transfer* (London: Chapman and Hall).

TU, J. and ZILLIG, W. (1982) Organisation of rRNA structural genes in the archaebacterium *Thermoplasma acidophilum. Nucleic Acids Res.*, **10**, 7231–7227.

VOLTERS, J. and ERDMANN, V. A. (1989) The structure and evolution of Archaebacterial ribosomal RNAs. *Can. J. Microbiol.*, **35**, 43–51.

WATERMAN, M. S. and EGGERT, M. (1987) A new algorithm for best subsequence alignments with application to tRNA–rRNA comparisons. *J. Mol. Biol.*, **197**, 723–728.

WOESE, C. R. and OLSEN, G. J. (1986) Archaebacterial phylogeny: perspectives on the urkingdoms. *System. Appl Microbiol.* **7**, 161–177.

ZILLIG, W., PALM, P. and KLENK, H.-P. (1992) A model of the early evolution of organisms: the arisal of the three domains of life from the common ancestor. In *The Origin and Evolution of the Cell*, ed. H. Hartman, and K. Matsuno, pp. 47–78 (Singapore: World Scientific).

PART FOUR

Gene Exchange and Evolution

10

Deciphering the Molecular Record for the Early Evolution of Life: Gene Duplication and Horizontal Gene Transfer

LORRAINE OLENDZENSKI AND J. PETER GOGARTEN

Department of Molecular and Cell Biology, University of Connecticut, Connecticut, USA

10.1 Introduction

The evolutionary history of microorganisms can be reconstructed using evidence from the fossil and geological records (Schopf, 1993; Schopf and Cornelius, 1992). However, the microfossil record is far from complete and unique morphological characters do not exist for most microbial groups. In lieu of these data, two types of molecular data found in extant organisms have proved useful in the study of early evolution: comparative analysis of biochemical pathways (Wächtershäuser, 1992; Morowitz, 1992) and analysis of macromolecular sequence data. The best-studied molecule for sequence analysis is undoubtedly the small subunit ribosomal RNA (rRNA), but many other macromolecules, especially proteins, are elucidating the relationships among prokaryotic kingdoms and domains.

At best, however, phylogenetic analysis based on extant macromolecules can only yield information on the phylogeny of the molecules under study. Comparison of phylogenies based on different molecular markers sometimes gives conflicting tree topologies. These confilicting topologies can arise for a number of reasons, but some cases can be attributed to horizontal gene transfer (Gogarten, 1995; Gogarten *et al.*, 1996). Thus, even if molecular phylogenies could be resolved without ambiguity, the step from molecular phylogeny to species phylogeny would remain complicated because genetic information has been transferred horizontally between independent evolutionary lineages. The best-corroborated example of this is transfer of genetic material from the eubacteria that evolved into chloroplasts and mitochondria to the nucleus of the host cells in which they now reside (Margulis, 1993). It is therefore not surprising that comparative analysis of different molecular phylogenies reveals a net-like structure of the species phylogeny (Hilario and Gogarten, 1993).

To date, more than 10 000 different 16S rRNA sequences are available through the Ribosomal Database Project (see http://rdp.life.uiuc.edu; Larsen *et al.*, 1993; Maidak *et al.*, 1997). Analysis of these molecules has been used to resolve the relationships among both prokaryotic and eukaryotic microorganisms. Since ribosomes play a central role in cellular information processing and interact specifically with many cellular components, they have coevolved over time with these components in each of the major cell lineages and are unlikely to have been transferred between distantly related organisms. Analysis of small subunit (SSU) rRNA has provided us with a view of life divided into three main groups, or domains (Woese, 1987; Woese *et al.*, 1990). Using distance, parsimony and maximum likelihood methods, the archaebacteria, eubacteria and eukaryotes appear as monophyletic groups. However, the relationship of these three groups to each other cannot be determined by RNA analysis alone because of the lack of a suitable outgroup in this data set when considering a universal phylogeny.

10.2 Duplicated genes and rooted phylogenies

The problem of rooting universal trees can be overcome by analysing proteins assumed to have arisen by ancient gene duplication. An ancient gene duplication would give rise within an organism to two diverging gene copies (genes A and B). If an analysis compares gene types A and B, the split between A and B reflects the gene duplication and not a speciation event; i.e., genes A and B are paralogous (Fitch, 1970) and not orthologous. Analysis of these duplicated genes can reach back to the time before the last common ancestor (e.g., Gogarten and Taiz, 1992). Using one of the paralogous genes as an outgroup for the other, analyses of duplicated genes allow the placement of the last common ancestor in the universal tree of life. Since the proposal of this method by Schwartz and Dayhoff (1978), several pairs of ancient duplicated genes, including elongation factors (1α/Tu and 2/G) and dehydrogenases (Iwabe *et al.*, 1989), aminoacyl-tRNA synthetases (Brown and Doolittle, 1995) and the catalytic and noncatalytic subunits of the ATP-hydrolysing head groups of H^+-ATPases (Gogarten *et al.*, 1989), have been used to create rooted, universal phylogenies.

10.3 Catalytic H^+-ATPase subunits

Since the vacuolar type ATPases (V-ATPases) of the eukaryotic endomembrane system are homologous to the bacterial coupling factor ATPases (Zimniak *et al.*, 1988; Harvey and Nelson, 1992), this molecular marker is present in all three cell lineages, i.e., in eubacteria, in archaea and in the nucleocytoplasmic component of eukaryotes. The ATP-hydrolysing headgroup of all three ATPase types (eubacterial or F-type, archaeal or A-type, and eukaryotic or V-type) consists of two homologous subunits, the so-called catalytic and noncatalytic subunits. These arose by a gene duplication which occurred before the three cellular lineages diverged from each other. Using this ancient gene duplication to root the universal tree of life, the last common ancestor is placed between eubacteria on one side and eukaryotes and archaea on the other (Gogarten *et al.*, 1989). Analysis of elongation factors, dehydrogenases, aminoacyl-tRNA synthetases, and tRNAs also support this placement of the root. This closer relationship between archaea and eukaryotes was also suggested by other characters that had not undergone an ancient gene duplication, e.g., RNA polymerases (Puhler *et al.*, 1989) and TATA binding proteins (Marsh *et al.*, 1994; Langer *et al.*, 1995). However, without the use of an outgroup, one cannot exclude the possibility that the characters shared between the archaea and eukaryotes reflect the state of the last common ancestor and that these characters were changed or abolished in the lineage leading to the eubacteria.

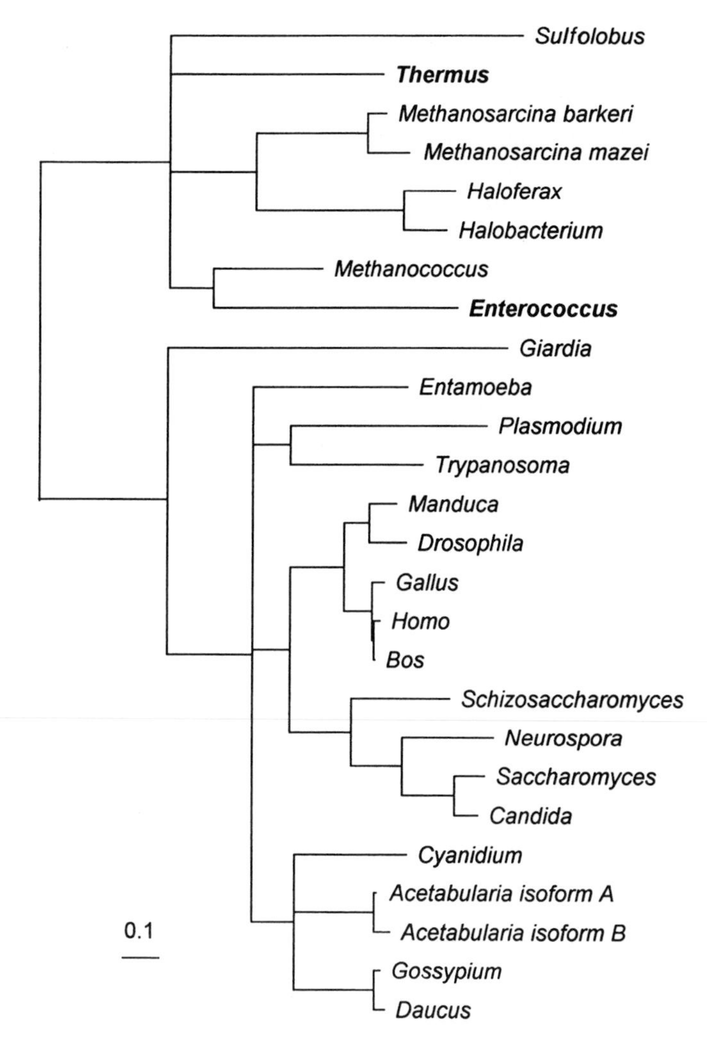

Figure 10.1 Molecular phylogeny of the catalytic subunits of vacuolar and archaeal type H^+-ATPases calculated by quartet puzzling as in PUZZLE 3.0 (Strimmer and von Haesler, 1996). Branches with insufficient support are collapsed into 'multi-furcations'. Branch lengths were calculated assuming a site-to site variation (STSV) described by a Gamma-distribution. The tree is scaled with respect to amino substitutions per site. The maximum likelihood estimate for the shape parameter (α) is 0.62 (Yang, 1994).

As with SSU rRNA trees, the ATPases clearly reflect the division of life into three distinct and separate cellular lineages or domains (Gogarten *et al.*, 1992). Within each of the three domains there is good agreement between the phylogenies calculated for 16S rRNAs and ATPases subunits (Gogarten *et al.*, 1996).

However, there are some important differences between the 16S rRNA and ATPase catalytic subunit phylogenies. Figure 10.1 shows a phylogenetic reconstruction of the evolution of the catalytic A subunit of the archaeal and vacuolar type ATPases, calculated using amino acid sequences. Based on rRNAs and other biochemical characters (cell wall, lipids, antibiotic resistance), the genus *Thermus* is classified as a eubacterium, with closest similarities to the genus *Deinococcus* (Woese, 1987; Hensel *et al.*, 1986). However,

Thermus thermophilus (also known as *T. aquaticus* HB8) does not have an F-ATPase like other eubacteria, but an A/V-ATPase (Yokoyama *et al.*, 1990; Tsutsumi *et al.*, 1991). Both major subunits of the *Thermus* ATPase are clearly archaeal in character (Gogarten *et al.*, 1992). A similar result is seen for the sodium-pumping ATPase from another eubacterium, *Enterococcus hirae*. This bacterium has a normal F-ATPase that groups as expected with other low-GC Gram-positives; in addition *E. hirae* has a sodium-pumping ATPase that, based on sequence, subunit composition and inhibitor sensitivity, is classified as a vacuolar or archaeal type ATPase (Takase *et al.*, 1994). Both of these archaeal type ATPases found in eubacteria group with the archaeal coupling factor ATPases (Hilario and Gogarten, 1993; Figure 10.1).

10.4 Conflicting molecular phylogenies

Conflicts in the topologies of trees constructed with different molecular markers can arise for a number of reasons. The molecule being analysed may contain insufficient data to resolve the phylogenetic relationships under study, for example if the molecule is too short or if regions within the molecule are saturated with substitutions or not variable enough. Over evolutionary time, additional substitutions can occur that obscure a phylogenetic signal and act as noise during tree construction. This would lead to incorrect resolution of deeper bifurcations. Statistical analyses, for example maximum likelihood methods or analyses of bootstrapped data sets (cf. Felsenstein, 1988) can be used to recognise some of these ill-resolved branching patterns. However, artefacts that are due to a systematic bias in the data that are not appropriately considered in the analysis (e.g., nucleotide bias; Lockhart *et al.*, 1992), or that are due to the attraction of long branches (e.g., Hillis *et al.*, 1994), are not always associated with low bootstrap values and thus may not be recognised.

Unrecognised paralogy can also lead to conflicting topologies. If a phylogeny mistakenly compares paralogous genes assumed to be orthologous, the resulting phylogenies reflect not only speciation but also a gene duplication event. This can lead to trees which differ greatly from the accepted species phylogeny. When comparing two conflicting rooted phylogenies, the assumption of paralogous genes explains the deeper of the conflicting bifurcations as a gene duplication event.

Horizontal gene transfer between organisms can explain conflicting topologies among or between well-resolved molecular phylogenies that are not comparing paralogous genes. Two types of horizontal gene transfer can be distinguished by the comparison of phylogenies constructed from different molecular markers. If only a single molecular marker yields a differing topology from the majority consensus of other molecular markers, this indicates that only one or a few genes have been transferred from one line of descent into another. The transfer of only one or a few molecules is known as xenology (Gray and Fitch, 1983). Synology, (Gogarten, 1994), however, can be recognised by the occurrence of two incompatible patterns, each of which is strongly supported by many molecular markers. Synology represents the fusion of formerly independent lineages to form a new line of descent. Both cases of horizontal gene transfer reconcile conflicting rooted phylogenies by explaining the more recent of the conflicting bifurcations as a case of horizontal transfer. The transfer of single genes or operons (xenology) leads to difficulties in finding the correct species tree; the occurrence of transfer events that involved substantial portions of the respective genomes (synology) points towards major events in the history of life.

10.5 Two distinct prokaryotic domains?

Although the positions of *Thermus* and *Enterococcus* in the vacuolar ATPase phylogeny have been proposed to be due to unrecognised paralogy (Forterre *et al.*, 1993), a more likely explanation is horizontal gene transfer from the archaeal lineage (in two separate events) to these eubacteria (Hilario and Gogarten, 1993; Gogarten *et al.*, 1996). Except for the position of these bacteria in the ATPase phylogeny, the phylogeny of H^+-ATPases, and those based on SSU rRNA and elongation factors (Cammarano *et al.*, 1992), support the division of prokaryotes into two distinct and separate groups, the archaea and the eubacteria. Other characters, including ribosomal proteins, promoter organisation, membrane lipids, cell wall structure, antibiotic sensitivity and flagellins, all suggest that the eubacteria and archaea are two distinct lineages (Zillig *et al.*, 1992). Recently, however, a growing list of molecular characters has been discovered that do not group the archaea as separate from the eubacteria. The most prominent of these are glutamine synthetases (Tiboni *et al.*, 1993; Kumada *et al.*, 1993; Brown *et al.*, 1994; Pesole *et al.*, 1995), homologues to the 70 kDa heat shock proteins (HSP70) (Gupta and Singh, 1992; Gupta and Golding, 1993; Gogarten *et al.*, 1996), and glutamate dehydrogenases (Benanchenhou-Lafha *et al.*, 1993; Hilario and Gogarten, 1993). These molecules have undergone ancient gene duplications, and in most cases the root appears to be placed between all of the prokaryotes on one side, and the eukaryotes on the other. In case of those molecular markers that allow a better resolution among the prokaryotes, it appears that the archaea group as a paraphyletic group within the eubacteria in a clade that is well supported by the respective data (e.g., Brown *et al.*, 1994; Gupta and Golding, 1993).

Another marker which does not group the archaea as separate from the eubacteria is the internal duplication found in carbamoyl-phosphate synthetase (Figure 10.2; Schofield, 1993; van den Hoff *et al.*, 1995; Olendzenski *et al.*, 1998; Lawson *et al.*, 1996; Lazcano, Puente, and Leguina, unpublished). In this case the first half of the molecule shares homology with the second half and each can be used as an outgroup of the other. One interesting feature of this analysis is that not all of the archaea are among the eubacteria; rather the molecule from *Methanococcus* groups among the eubacteria, while the *Sulfolobus* sequence groups with the eukaryotes.

The difference between those molecular markers that group the archaea within the bacterial domain and those that reflect two distinct prokaryotic domains does not appear to be due to ill-resolved branching patterns (Gupta and Golding, 1993; Brown *et al.*, 1994). If one wanted to reconcile the conflict assuming paralogous genes, one would have to postulate that the clear separation between eubacteria and archaea, which is found for many molecular markers, is due to so far unrecognised ancient gene duplication. Using paralogy as an explanation, the true phylogeny would then be reflected by HSP70 homologues, glutamate dehydrogenases, and glutamine synthetases. Since H^+-ATPases, elongation factors, RNA polymerases, aminoacyl-tRNA synthases, and ribosomes have retained comparable functions in all extant organisms, it seems unlikely that the last common ancestor would have had duplicate divergent copies for all of these activities. It also seems extremely unlikely that nearly identical sets of these characters were lost several times in the lineages leading to the different eubacterial groups, whereas exactly the complementary set would have had to have been lost multiple times in the lineages leading to present-day archaea and the eukaryotic nucleocytoplasm.

The third and remaining possibility is that the genes that show the close association between archaea and eubacteria were obtained by the archaea via horizontal gene transfer (Figure 10.3). Since a number of genes group the archaea together with the bacteria, this

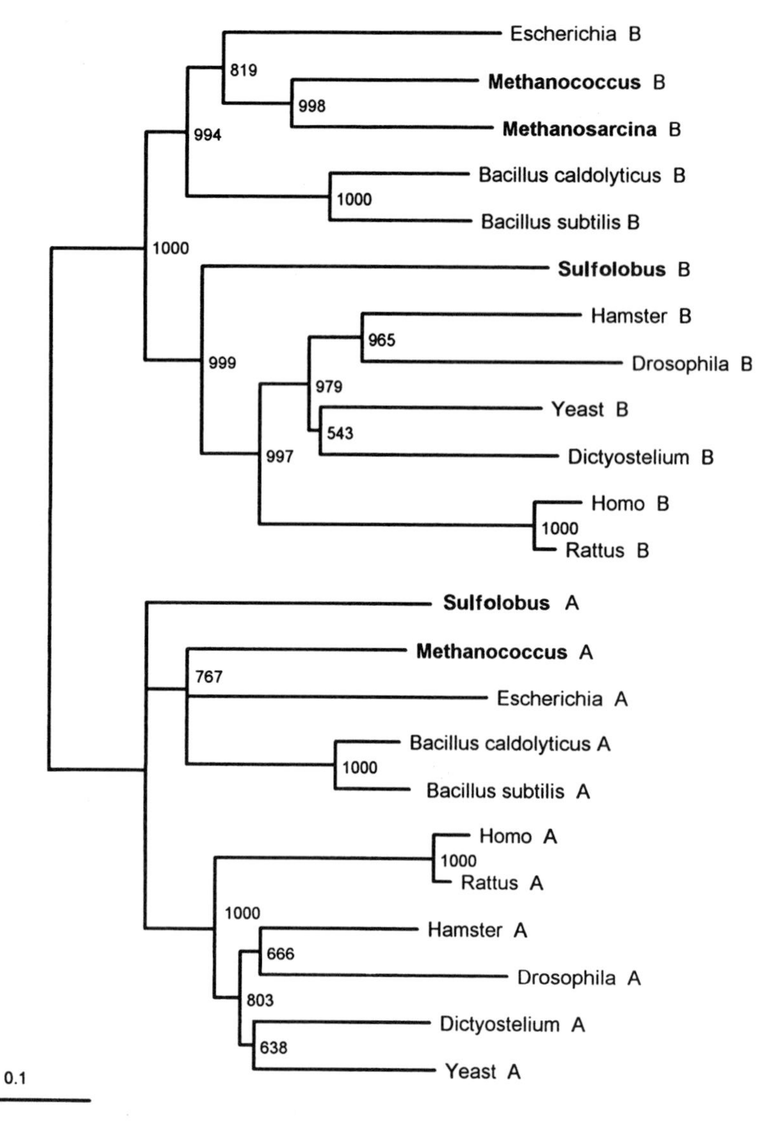

Figure 10.2 Bootstrapped distance analysis of carbamoyl-phosphate synthetase. The aligned data set consists of 273 amino acid positions within each half of the internal duplication in the large subunit of carbamoyl-phosphate synthetase (carB). The homologous regions in *Methanococcus janaschii* are found in two separate genes. Each half of the internal duplication was determined by dot-matrix plot analyses of the yeast carB gene. Positions corresponding to homologous domains in the amino-terminal half (A) and the carboxy-terminal half (B) were aligned using ClustalW v. 1.6 (Thompson *et al.*, 1994) and truncated to correspond to the homologous 273 amino acid positions of the available B fragment of *Methanosarcina barkeri* (complete sequence not available). One thousand bootstrapped replicates were performed using the phylogenetic tree option of ClustalW v. 1.6 with gap positions excluded. The tree was rooted by using all the amino-terminal (A) sequences as the outgroup. All branches supported in less than 50% of the bootstrapped samples were collapsed. Numbers give the percentage of bootstrapped samples that support the branch to the left of the number. Trees with the same toplogy among the carboxy-terminal segment (B) were obtained using parsimony (PROTPARS program of PHYLIP; Felsenstein, 1993) and maximum liklihood analyses (PUZZLE). Amino acid sequence alignments are available from the authors. The tree is scaled with respect to amino substitutions per site.

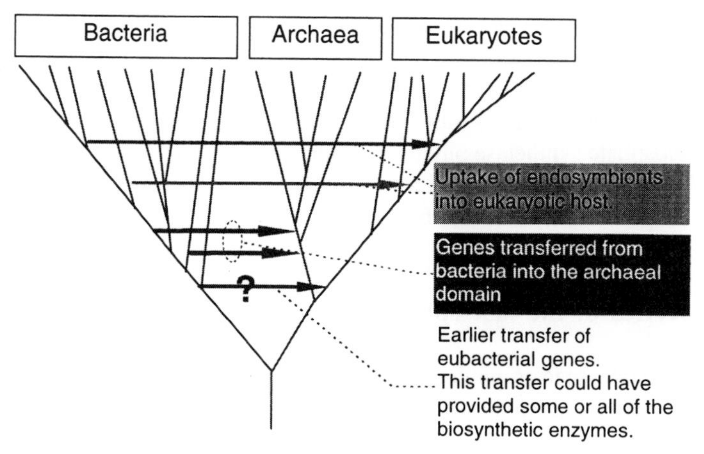

Figure 10.3 Proposed transfer events from eubacteria to the archaea and eukaryotes.

suggests that a significant portion of the genome was involved in this transfer. Many of the genes that suggest a closer association between eubacteria and archaea are involved in biosynthetic pathways (e.g., glutamine synthetase, carbamoyl-phosphate synthetase, glutamate dehydogenase). Some other well-studied enzymes, for example those involved in nitrogen fixation (Chien and Zinder, 1994; Souillard *et al.*, 1988), also suggest a close association between Gram-positives and at least some archaea. In addition, analysis of the *Methanococcus janaschii* genome shows that 44% of the proteins share closest similarity with eubacterial rather than eukaryotic genes (Koonin, 1997).

10.6 Open questions

These findings suggest that many enzymes involved in biosynthetic pathways were obtained by the archaea through horizontal transfer from bacteria. If this interpretation is correct, the following questions arise: Under what conditions did archaea obtain these genes and why was it a selective advantage for archaea to retain them? Did the archaeal ancestor lack these enzymes before the transfer took place? If these genes were already present in the last common ancestor, then they would have been present at the bifurcation leading to archaea and the eukaryotic nucleocytoplasm. Thus many of the genes transferred from the bacteria to the archaea would have replaced similar enzymes already present in this lineage.

Alternatively, there may have been an early eubacterial contribution to the eukaryotic nucleocytoplasm (Zillig *et al.*, 1992). This contribution, supported by glyceraldehyde-3-phosphate dehydrogenase (GAPDH) phylogeny (Hensel *et al.*, 1989; Markos *et al.*, 1993), predated the symbiosis that gave rise to mitochondria and plastids. This earlier eubacterial contribution could also have contributed the genes involved in the various anabolic pathways (Figure 10.3). According to this alternative scheme, these pathways would have been absent in the last common ancestor and evolved only in the eubacterial lineage. They would have been first transferred into the eukaryotic ancestor, and later transferred from the eubacteria to the archaea. This scenario bears some resemblance to Kandler's pre-cell proposal (Kandler, 1994) in that one lineage makes successive inventions that are subsequently passed on into the three domains; however, here the last common ancestor is hypothesised to have already had well-functioning genetic and bioenergetic machineries.

The currently available data argue against a single transfer event at the root of the archaeal domain: if there had been a single transfer event of a large number of genes, one would expect the archaea to form a single group within the bacteria in phylogenies based on the transferred genes. This is not observed for glutamine synthetase. Furthermore, the carbamoyl-phosphate synthetase phylogeny indicates that at least this gene was transfered into the archaeal gene after the split between cren- and euryarcheota had already occurred and that *Sulfolobus* never received the eubacterial gene.

10.7 Properties of the last common ancestor

Clearly, a reticulate species phylogeny greatly complicates the inference of characteristics present in the last common ancestor. The fact that a character is present in all three lineages is no longer sufficient to demonstrate that it was already present in the last common ancestor. To infer that a trait was present in the last common ancestor the phylogeny of this trait has to be compared and reconciled with the net-like species phylogeny.

In many instances a definitive conclusion remains elusive at present. However, the presently available data indicate that the last common ancestor already possessed effective transcription and translation machineries. H^+-ATPases present in the last common ancestor were likely to function as ATP synthases under physiological conditions (Gogarten and Taiz, 1992). The ancestor of eubacteria and archaea also had a cytochrome *c* oxidase homologue (Castresana *et al.*, 1995). At least two types are present in archaea: a halobacterial cytochrome oxidase which groups with the majority of all other eubacterial cytochrome oxidase homologues, while another type, found in *Sulfolobus* groups at the base of the eubacteria (Gogarten *et al.*, 1996). The cytochrome *c* oxidase homologues do not reflect a close association between archaea and Gram-positives, but seem to group at the base of the eubacteria, suggesting that this enzyme was not part of the horizontal transfer events discussed previously. However, the trees depicted by Castresana *et al.* (1994) are unrooted phylogenies and many of the deeper branches remain ill-resolved, so alternative interpretations cannot be excluded at present.

10.8 Conclusions

Ancient gene duplications provide a means to place the last common ancestor of all living organisms in the universal tree of life. Contradictions between well-resolved molecular phylogenies can be due to unrecognised gene duplication events or to horizontal gene transfer. Two examples that conflict with well-documented phylogenetic relationships are the presence of archaeal type H^+-ATPases in *Thermus* and *Enterococcus* and the close association between eubacteria and archaea found for many different genes. In both cases the best explanation for these conflicts is horizontal gene transfer. Assuming unrecognised paralogies (duplicated genes) as an explanation would necessitate many instances of convergent evolution and gene loss; furthermore this assumption would also result in species phylogenies that are at odds with two distinct prokaryotic domains (Archaea and Eubacteria). Horizontal gene transfer events contributed significantly to the evolution of the three cell lineages (Figure 10.4). Since a number of characters reflect the close association between archaea and eubacteria it seems likely that a substantial portion of the eubacterial genome was transferred to the archaea.

Eubacteria

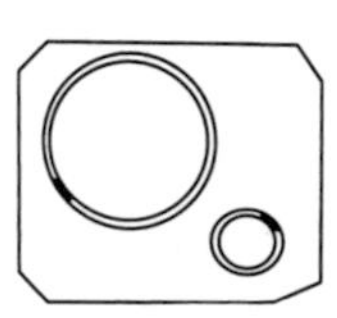

Few archaeal genes;
e.g.: A-ATPases in *Thermus* and *Enterococcus*

Archaea

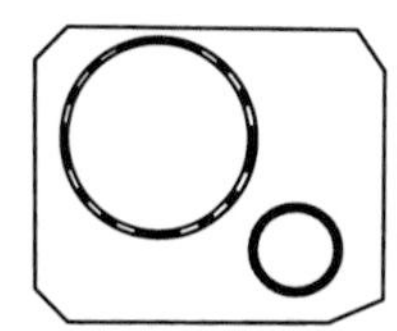

Several eubacterial genes, in particular genes involving biosynthesis, e.g.: glutamate dehydrogenase, glutamine synthase, carbamoyl-phosphate synthetase, also HSP70 homologues

Eukaryotes

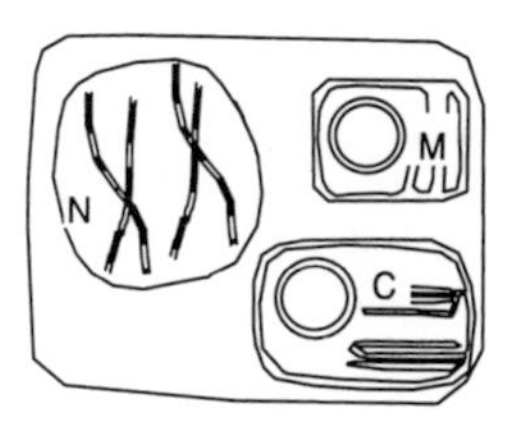

Many archaeal genes involved in transcription and translation, also V/A-ATPases;

Eubacterial genes in mitochondria and chloroplasts

?Other eubacterial genes, e.g.: cytoskeleton, GAPDH; many genes that encode biosynthetic enzymes reflect prokaryote/eukaryote dichotomy

Figure 10.4 Distribution of genes among eubacteria, archaea, and eukaryotes based on conflicting molecular phylogenies.

Horizontal transfer and the fusion of independent lineages transform the tree of life into a net of life complicating the inference of the properties of the last common ancestor. However, the organismal evolution is still visible as the majority consensus of different molecular phylogenies. The data strongly indicate that the last common ancestor was a cellular organism, with a DNA-based genome and a sophisticated transcription and translation machinery. Furthermore, the analysis of ATPase structure–function relationships and the evolution of cytochrome oxidase homologues suggest that the last common ancestor already used chemiosmotic coupling to synthesise ATP, and that electron transport chains could be used to energise the plasma membrane.

10.9 Summary

Molecular data found in extant organisms are useful in the study of early evolution. Ancient gene duplications allow placement of the last common ancestor in the universal tree of life. Comparison of phylogenies inferred from different molecular markers, however, often give conflicting topologies owing to inadequate resolution, unrecognised paralogy or horizontal gene transfer events. Horizontal transfer and the fusion of independent lineages point to major events in the history of life and transform the tree of life into a net of life.

Acknowledgements

Research in the authors' laboratory was supported by NSF grant BSR-9020868, and through the NASA-Exobiology programme. J.P.G. thanks Henrik Kibak and Lincoln Taiz for numerous stimulating discussions.

References

BENACHENHOU-LAFHA, N., FORTERRE, P. and LABEDAN, B. (1993) Evolution of glutamate dehydrogenase genes: evidence for two paralogous protein families and unusual branching patterns of the archaebacteria in the universal tree of life. *J. Mol. Evol.*, **36**, 335–346.

BROWN, J. R. and DOOLITTLE, W. F. (1995) Root of the universal tree of life based on ancient aminoacyl-tRNA synthetase. *Proc. Natl Acad. Sci. USA*, **92**, 2441–2445.

BROWN, J. R., MASUCHI, Y., ROBB, F. T. and DOOLITTLE, W. F. (1994) Evolutionary relationships of bacterial and archaeal glutamine synthetase genes. *J. Mol. Evol.*, **38**, 566–576.

CAMMARANO, P., PALM, P., CRETI, R., CECCARELLI, E., SANANGELANTONI, A. M. and TIBONI, O. (1992) Early evolutionary relationships among known life forms inferred from elongation factor ef-2 (ef-g) sequences. Phylogenetic coherence and structure of the archaeal domain. *J. Mol. Evol.*, **34**, 396–405.

CASTRESANA, J., LUBBEN, M., SARASTE, M. and HIGGINS, D. G. (1994) Evolution of cytochrome oxidase, an enzyme older than atmospheric oxygen. *Eur. Mol. Biol. Org. J.*, **13**, 2516–2525.

CASTRESANA, J., LUBBEN, M. and SARASTE, M. (1995) New archaebacterial genes coding for redox proteins: implications for the evolution of aerobic metabolism. *J. Mol. Biol.*, **250**, 202–210.

CHIEN, Y. T. and ZINDER, S. H. (1994) Cloning, DNA sequencing, and characterisation of a nifD-homologous gene from the archaeon *Methanosarcina barkeri* 227 which resembles nifDl from the eubacterium *Clostridium pasteurianum*. *J. Bacteriol.*, **176**, 6590–6598.

FELSENSTEIN, J. (1988) Phylogenies from molecular sequences: inference and reliability. *Annu. Rev. Genet.*, **22**, 521–565.

FELSENSTEIN, J. (1993) *Phylogeny Inference Package*, version 3.5c. Distributed by the author. Dept. of Genetics, Univ. of Washington, Seattle.

FITCH, W. S. (1970) Distinguishing homologous from analogous proteins. *System. Zool.*, **19**, 99–113.

FORTERRE, P., BENACHENHOU-LAFHA, N., CONFALONIERI, F., DUGUET, M., ELIE, C. and LABEDAN, B. (1993) The nature of the last universal ancestor and the root of the tree of life, still open questions. *BioSystems*, **28**, 15–32.

GOGARTEN, J. P. (1994) Which is the most conserved group of proteins? Homology-orthology, paralogy, xenology and the fusion of independent lineages. *J. Mol. Evol.*, **39**, 541–543.

GOGARTEN, J. P. (1995) The early evolution of cellular life. *Trends Ecol. Evol.*, **10**, 147–151.

GOGARTEN, J. P. and TAIZ, L. (1992) Evolution of proton pumping ATPases: rooting the tree of life. *Photosynth. Res.*, **33**, 137–146.

GOGARTEN, J. P., KIBAK, H., DITTRICH, P. *et al.* (1989) Evolution of the vacuolar H^+-ATPase: implications for the origin of eukaryotes. *Proc. Natl Acad. Sci. USA*, **86**, 6661–6665.

GOGARTEN, J. P., STARKE, T., KIBAK, H., FICHMANN, J. and TAIZ, L. (1992) Evolution and isoforms of V-ATPase subunits. *J. Exp. Biol.*, **172**, 137–147.

GOGARTEN, J. P., HILARIO, E. and OLENDZENSKI, L. (1996) Gene duplications and horizontal gene transfer during early evolution. In *Evolution of Microbial Life, Society for General Microbiology Symposium 54*, ed. D. McL. Roberts, P. Sharp, G. Alderson and M. Collins, pp. 267–292 (Cambridge: Cambridge University Press).

GRAY, G. S. and FITCH, W. S. (1983) Evolution of antibiotic resistance genes: the DNA sequence of a kanamycin resistance gene from *Staphylococcus aureus*. *Mol. Biol. Evol.*, **1**, 57–66.

GUPTA, R. S. and GOLDING, G. B. (1993) Evolution of HSP70 gene and its implications regarding relationships between archaebacteria, eubacteria and eukaryotes. *J. Mol. Evol.*, **37**, 573–582.

GUPTA, R. S. and SINGH, B. (1992) Cloning of the HSP70 gene from *Halobacterium marismortui*: relatedness of archaebacterial HSP70 to its eubacterial homologues and a model of the evolution of the HSP70 gene. *J. Bacteriol.*, **174**, 4594–4605.

HARVEY, W. R. and NELSON, N. (eds) (1992) V-ATPases. *J. Exp. Biol.*, **172**.

HENSEL, R., DEMHARTER, W., KANDLER, O., KROPPENSTEDT, M. and STACKEBRANDT, E. (1986) Chemotaxonomic and molecular-genetic studies of the genus *Thermus*: evidence for a phylogenetic relationship of *Thermus aquaticus* and *Thermus ruber* to the genus *Deinococcus*. *Int. J. System. Bacteriol.*, **36**, 444–453.

HENSEL, R., ZWICKL, P., FABRY, S., LANG, J. and PALM, P. (1989) Sequence comparison of glyceraldehyde-3-phosphate dehydrogenases from the three urkingdoms: evolutionary implication. *Can. J. Microbiol.*, **35**, 81–85.

HILARIO, E. and GOGARTEN, J. P. (1993) Horizontal transfer of ATPase genes – the tree of life becomes a net of life. *BioSystems*, **31**, 111–119.

HILLIS, D. M., HUELSENBECK, J. P. and SWOFFORD, D. L. (1994) Hobgoblin of phylogenetics? *Nature*, **3692**, 363–364.

IWABE, N., KUMA, K-I., HASEGAWA, M., OSAWA, S. and MIYATA, T. (1989) Evolutionary relationships of archaebacteria, eubacteria and eukaryotes inferred from phylogenetic trees of duplicated genes. *Proc. Natl Acad. Sci. USA*, **86**, 9355–9359.

KANDLER, O. (1994) The early diversification of life. In *Early Life on Earth*, Nobel Symposium No. 84, ed. S. Bengston, pp. 152–160 (New York: Columbia University Press).

KOONIN, E. V., MUSHEGIAN, A. R., GALPERIN, M. Y. and WALKER, D. R. (1997) Comparison of archaeal and bacterial genomes: computer analysis of protein sequences predicts novel functions and suggests a chimeric origin for the archaea. *Mol. Microbiol.*, **25**, 619–637.

KUMADA, Y., BENSON, D. R., HILLEMANN, D. *et al.* (1993) Evolution of the glutamine synthase gene, one of the oldest existing and functioning genes. *Proc. Natl Acad. Sci. USA*, **90**, 3009–3013.

LANGER, D., HAIN, J., THURIAUX, P. and ZILLIG, W. (1995) Transcription in archaea: similarity to that in eukarya. *Proc. Natl Acad. Sci. USA*, **92**, 5768–5772.

LARSEN, N., OLSEN, G. J., MAIDAK, B. L. *et al.* (1993) The ribosomal database project. *Nucleic Acids Res.*, **21**, 3021–3023.

LAWSON, F. S., CHARLEBOIS, R. L. and DILLON, J. A. (1996) Phylogenetic analysis of carbamoyl-phosphate synthetase genes: complex evolutionary history includes an internal duplication within a gene which can root the tree of life. *Mol. Biol. Evol.*, **13**, 970–977.

LOCKHART, P. J., BEANLAND, T. J., HOWE, C. J. and LARKUM, A. W. (1992) Sequence of *Prochloron didemni* atpBE and the inference of chloroplast origins. *Proc. Natl Acad. Sci. USA*, **89**, 2742–2746.

MAIDAK, B. L., OLSEN, G. J., LARSEN, N., OVERBEEK, R., MCCAUGHEY, M. J. and WOESE, C. R. (1997) The RDP (Ribosomal Database Project). *Nucleic Acids Res.*, **25**, 109–111.

MARGULIS, L. (1993) *Symbiosis in Cell Evolution*, 2nd edn (New York: W. H. Freeman).

MARKOS, A., MIRETSKY, A. and MULLER, M. (1993) A glyceraldehyde-3-phosphate dehydrogenase with eubacterial features in the amitochondriate eukaryote, *Trichomonas vaginalis*. *J. Mol. Evol.*, **37**, 631–643.

MARSH, T. L., REICH, C. I., WHITELOCK, R. B. and OLSEN, G. J. (1994) Transcription factor IID in the Archaea: sequences in the *Thermococcus celer* genome would encode a product closely related to the TATA-binding protein of eukaryotes. *Proc. Natl Acad. Sci. USA*, **91**, 4180–4184.

MOROWITZ, H. J. (1992) *Beginnings of Cellular Life: Metabolism Recapitulates Biogenesis* (New Haven: Yale University Press).

OLENDZENSKI, L., HILARIO, E. and GOGARTEN, J. P. (1998) Horizontal gene transfer and fusing lines of descent: the Archaebacteria – a chimera? In *Horizontal Gene Transfer*, ed. M. Syvanen and C. Kado (Chapman and Hall, London) (in press).

PESOLE, G., GISSI, C., LANAVE, C. and SACCONE, C. (1995) Glutamine synthetase gene evolution in bacteria. *Mol. Biol. Evol.*, **12**, 189–197.

PUHLER, G., LEFFERS, H., GROPP, F. *et al.* (1989) Archaebacterial DNA-dependent RNA polymerases testify to the evolution of the eukaryotic nuclear genome. *Proc. Natl Acad. Sci. USA*, **86**, 4569–4573.

SCHOFIELD, J. P. (1993) Molecular studies on an ancient gene encoding for carbamoyl-phosphate synthetase. *Clin. Sci.*, **84**, 119–128.

SCHOPF, J. W. (1993) Microfossils of the Early Archaean Apex chert: new evidence of the antiquity of life. *Science*, **260**, 640–646.

SCHOPF, J. W. and CORNELIUS, K. (eds) (1992) *The Proterozoic Biosphere: A Multidisciplinary Study* (Cambridge: Cambridge University Press).

SCHWARTZ, R. M. and DAYHOFF, M. O. (1978) Origins of prokaryotes, mitochondria, and chloroplasts. *Science*, **199**, 395–403.

SOUILLARD, N., MAGOT, M., POSSOT, O. and SIBOLD, L. (1988) Nucleotide sequence of regions homologous to nifH (nitrogenase Fe protein) from the nitrogen-fixing archaebacteria *Methanococcus thermolithotrophicus* and *Methanobacterium ivanovii*: evolutionary implications. *J. Mol. Evol.*, **27**, 65–76.

STRIMMER, K. and VON HAESELER, A. (1996) Quartet puzzling: a quartet maximum likelihood method for reconstructing tree topologies. *Mol. Biol. Evol.*, **13**, 964–969.

TAKASE, K., KAKINUMA, S., YAMATO, I., KONISHI, K., IGARASHI, K. and KAKINUMA, Y. (1994) Sequencing and characterisation of the ntp gene cluster for vacuolar-type Na(+)-translocating ATPase of *Enterococcus hirae*. *J. Biol. Chem.*, **269**, 11037–11044.

THOMPSON, J. D., HIGGINS, D. G. and GIBSON, T. J. (1994) CLUSTAL W: improving the sensitivity of progressive multiple sequence alignment through sequence weighting, positions-specific gap penalties and weight matrix choice. *Nucleic Acids Res.*, **22**, 4673–4680.

TIBONI, O., CAMMARANO, P. and SANANGELANTONI, A. M. (1993) Cloning and sequencing of the gene encoding glutamine synthase I from the Archaeum *Pyrococcus woesei*: anomalous phylogenies inferred from analysis of archaeal and bacterial glutamine synthase I sequences. *J. Bacteriol.*, **175**, 2961–2969.

TSUTSUMI, S., DENDA, K., YOKOYAMA, K., OSHIMA, T., DATE, T. and YOSHIDA, M. (1991) Molecular cloning of genes encoding major two subunits of a eubacterial V-Type ATPase from *Thermus thermophilus*. *Biochim. Biophys. Acta*, **1098**, 13–20.

VAN DEN HOFF, M. J. B., JONKER, A., BEINTEMA, J. J. and LAMERS, W. H. (1995) Evolutionary relationships of the carbamoylphosphate synthetase genes. *J. Mol. Evol.*, **41**, 813–832.

WÄCHTERSHÄUSER, G. (1992) Groundworks for an evolutionary biochemistry: the iron–sulfur world. *Prog. Biophys. Mol. Biol.*, **58**, 85–201.

WOESE, C. R. (1987) Bacterial evolution. *Microbiol. Rev.*, **51**, 221–271.

WOESE, C. R., KANDLER, O. and WHEELIS, M. L. (1990) Towards a natural system of organisms: proposal for the domains Archaea, Bacteria and Eukarya. *Proc. Natl Acad. Sci. USA*, **87**, 4576–4579.

YANG, Z. (1994) Maximum likelihood phylogenetic estimation from DNA sequences with variable rates over sites: approximate methods. *J. Mol. Evol.*, **39**, 306–314.

YOKOYAMA, K., OSHIMA, T. and YOSHIDA, M. (1990) *Thermus thermophilus* membrane-associated ATPase; indication of a eubacterial V-type ATPase. *J. Biol. Chem.*, **265**, 21946–21950.

ZILLIG, W., PALM, P. and KLENK, H-P. (1992) A model of the early evolution of organisms: the arisal of the three domains of life from the common ancestor. In *The Origin and Evolution of the Cell*, ed. H. Hartman and K. Matsuno, pp. 163–182 (Singapore: World Scientific).

ZIMNIAK, L., DITTRICH, P., GOGARTEN, J. P., KIBAK, H. and TAIZ, L. (1988) The cDNA sequence of the 69kDa subunit of the carrot vacuolar H^{+}-ATPase. Homology to the beta-chain of FoF1-ATPases. *J. Biol. Chem.*, **263**, 9102–9112.

11

Lateral Gene Exchange, an Evolutionary Mechanism for Extending the Upper or Lower Temperature Limits for Growth of Microorganisms? A Hypothesis

JUERGEN WIEGEL

Departments of Microbiology and Biochemistry & Molecular Biology, Center for Biological Resource Recovery, University of Georgia, Athens, Georgia, USA

11.1 Evolution and horizontal gene transfer

'In the beginning . . . the earth was without forms and void . . .' (*Genesis* 1) There is no quarrel with this picture. However, quite a variety of hypotheses and opinions exist about the temperature(s) at the time life started to evolve (Miller and Lazcano, 1995; Forterre, 1996; Lowe, 1994; Schwartzman *et al.*, 1993, and Chapter 3 of this book; Kasting, 1993; Baross, Chapter 1 of this book). Was the early earth hyperthermobiotic (i.e., above 85 °C), thermobiotic (between 50 °C and 85 °C), mesobiotic (20 °C and 50 °C), or fluctuating between the different scenarios in the environment(s) where life started to evolve? In other words, was the first progenote or rather – as the author believes – were the various progenotes from which the myriads of modern microorganisms evolved extreme thermophilic/hyperthermophilic or mesophilic entities or did several of them coexist? In regard to the hypothesis presented here, the question is whether the original entities had to adapt to an environment with more moderate temperatures or had to adapt to increasing temperatures. Regardless of how this question is answered, all the scenarios include the necessity for the evolving diversity of self-replicating and metabolic systems-containing forms of life to adapt to changing temperatures. This author favours heavily a model of multiple progenotes or mesogenotes as designated and developed by Otto Kandler (1994; and further elegantly presented in Chapter 2 of this book). In the author's opinion there is no need or no reason to believe that life had to originate from only one (type of) progenote and that only one of the emerging entities successfully survived and developed into the lines of modern life. The model of Kandler's 'evolutionary bush' for the origin of the three domains of life (according to Woese's scheme) or of any other number of basic units used in other models (Margulis, 1992; Margulis and Schwartz, 1997; Lake, 1995; Rivera and Lake, 1996) would easily allow for the occurrence of horizontal transfer of genes (Gogarten, 1995) or of gene precursors at very early stages of life. One can assume that, in places where early forms of life started, the conditions, especially the temperature, changed drastically, for example, through impacts of bolides (Sleep *et al.*, 1989; Gogarten-Boekels *et al.*, 1995). Thus, the author hypothesises that early thermophilic and more mesophilic entities coemerged and coevolved more or less simultaneously in these environments. Consequently, if these different 'temperature variants' were present in the same vicinity and were constantly disintegrating and reforming, one can propose that some entities took up genetic information which enabled them to cope better with the new existing – either hotter or cooler – temperatures. This would represent an early form of horizontal 'gene' transfer. Assuming that life evolved on surfaces probably of iron–sulfur compounds such as pyrite (Wächtershäuser 1988, 1992; Chapter 4 of this book; Russell *et al.* Chapter 6 of this book), and thus a close proximity of adsorbed entities, this scenario becomes very likely. Indications for the occurrence of horizontal gene transfer are increasing (Hilario and Gogarten, 1993; Gogarten, 1995; Gogarten *et al.*, 1996, and further discussed in Chapter 10 of this book; Hensel *et al.*, 1989). Thus, for the hypothesis put forward, the possibility that horizontal gene transfer can occur is taken as a given. The transfer could have occurred during the early stages of emerging life forms discussed above or at later stages of fully developed present-day genes. In the future, the hypothesis presented and the explanation of horizontal gene transfers may be substantiated or disproved as more genomes, including those from bacteria which exhibit biphasic Arrhenius graphs for growth, are fully sequenced and most of the genes in the genomes are functionally identified. At the present time only about half of the genes on studied genomes can be associated with known gene products. Thus far, with respect to the hypothesis presented here, the genome-level analysis is not particularly meaningful or conclusive.

11.2 Overview of bacteria with an extended temperature range for growth and a biphasic Arrhenius plot for growth

Generally, pure cultures of microorganisms grow over a temperature span of 25–30 °C. However, among the thermophiles are several bacteria and archaea which exhibit an extended temperature span for growth of more than 35 °C (Wiegel 1990, 1992). *Methanobacterium thermoautotrophicum* exhibits so far the largest span, with growth between 22 °C and 78 °C and methane formation between at least 15 °C and 78 °C (Wiegel, 1990). Interestingly, all the microorganisms exhibiting such an extended temperature span for growth – as far as examined – exhibit a biphasic relationship between growth temperature and growth rate. The biphasic curve is best illustrated using the Arrhenius plot. Although developed for chemical reactions and generally applied to enzyme kinetics, the Arrhenius plot can also be used to demonstrate elegantly the overall relation between growth temperatures and growth parameters such as doubling times. Usually, if only one reaction (or in a complex system, effectively a combination of overlapping reactions) is rate-limiting, a straight line is observed over the temperature span investigated. For complex systems like whole cells, however, no physical parameters (such as ΔH for a single chemical reaction) can be associated with the slopes of the graphs. However, on obtaining graphs with strong intermediate plateaus and in some cases a change in the slope of the graphs below and above the plateau, one can assume that this indicates a major change in the rate limitation for the growth of an organism. Examples of thermophilic microorganisms with an increased temperature span include a variety of different types and include bacteria such as the aerobic *Bacillus stearothermophilus* (heterotroph) and *B. schlegelii* (chemolithoautotrophic), and the anaerobic *Thermoanaerobacter ethanolicus* (glycolytic) and *Thermoanaerobacter kivui* (chemolithoautotrophic), but also archaea such as the anaerobic, thermophilic methanogen *Methanobacterium thermoautotrophicum* (chemolithoautotrophic) and the extremely thermophilic *Desulfurococcus mobilis* (sulfur-dependent metabolism) and methanogen *Methanonococcus jannaschii* (chemolithoautotrophic). These few examples represent a wide variety of physiologically and phylogenetically different microorganisms. Some other examples of archaea and bacteria for which a growth span exceeding 35 °C has been reported are (with T_{min}–T_{max} and the temperature for the intermediate plateau given in parentheses): *Thermodesulfobacterium commune* (45–85 °C, plateau at 38–43 °C); *Calorimator* (*Clostridium*) *fervidus* (about 40–80 °C); *Thermoanaerobacter brockii*, *T. thermohydrosulfuricus* and *T. 'sulfurigignens'* (35–77/8 °C, plateau at 55–60 °C); *Thermoanaerobacter* (*Acetogenium*) *kivui* (35–75 °C, *plateau* 57–60 °C); *Methanococcus thermolithotrophicus* (30 °C to above 70 °C); *Clostridium* (*sensu stricto*) *thermobutyricum* (26–61.5 °C, plateau around 35–38 °C with a drastic change of the slope above and below a relatively narrow plateau); *Thermoanaerobacterium desulfurigenes* (35–75 °C, plateau at 53–85 °C) (but not the closely related *Thermoanaerobacterium aotearoense* (34–66 °C), *T. thermosaccharolyticum*, or *T. saccharolyticum* (35–65 °C)). Owing to the research interest of the author, most of the examples analysed in more detail belong to the group of anaerobic eubacterial thermophiles. It is speculated that more examples of bacteria and archaea with the wide temperature span and the biphasic curve will be found in groups other than those mentioned above when their species are more closely investigated with respect to their temperature dependence of growth rates.

It is interesting to note that analogous biphasic pH-dependent curves are also found for bacteria with a rather wide pH range for growth, e.g., the above-mentioned *Thermoanaerobacter 'sulfurigignens'* (pH range from below 4.0 to above 8.0 with an optimum around 6.0 and 6.5 and the plateau around pH 5.0 and 5.7) or *Clostridium paradoxum* and *C. thermoalkaliphilum* with a temperature range for growth from 30 °C to 63 °C (the temperature plateau for this bacterium is very small), $pH^{25\,°C}$ range from 6.9 to 11.1, and pH optimum around 10.1 and plateau at 8.5–9.5. An explanation analogous to that for the existence of the temperature-dependent plateaux in the Arrhenius graphs as outlined in this chapter could also be made here for the biphasic growth response to the pH of the growth medium.

11.3 Hypothesis

Based on the above observations and some additional preliminary results, such as a comparison of two-dimensional protein gels from cells grown at temperatures above and below the intermediate plateau in the Arrhenius graph (Ljungdahl *et al.*, 1978), a working hypothesis had been postulated (Wiegel, 1990). Microorganisms able to grow in temperature spans of more than 33 °C should exhibit two-phased Arrhenius graphs (using a logarithmic term representing growth rates plotted versus the inverse growth temperatures in Kelvin). Thus, for these microorganisms the temperature curves can be divided into a lower and an upper temperature range for growth. Such curves could be explained by mixed cultures containing two organisms with optimal growth temperatures, e.g., one in the thermophilic and the other in the extremely thermophilic range. Such cultures would yield similar biphasic Arrhenius graphs. The farther apart the optimal growth temperatures of the two microorganism are, the stronger is the intermediate plateau. Subsequently, for cultures demonstrated to be pure (see below), the working hypothesis postulates that the two-phased temperature curves are due to the action of two different subsets of genes coding for otherwise growth-limiting gene products so that the microorganisms can grow in an extended temperature range. In other words, these organisms have one subset of genes allowing the organism to grow in the lower temperature range of the curve and a second subset of genes whose gene products allow the organisms to grow within the upper temperature range. For many catalytic and structural proteins, lipids, and other gene products, however, only one set of genes exists. These gene products are stable and functioning over the whole temperature span of both the lower and the upper temperature range.

It is further speculated that the organisms – at least in part – acquired the required information by horizontal gene transfer in order to supplement their existing gene pool and thus overcome the original limitation. If this is true, horizontal gene transfer can be regarded as an evolutionary mechanism in the struggle to survive in or to cope more successfully with a changing or a changed environment. Again, it must be stressed that this evolutionary process could have occurred at a very early stage when life evolved at the level of the mesogenotes (Kandler, 1994) or more recently after the presently recognised domains and the major phylogenetic branches had developed, or even could occur among the present-day microorganisms living in environments with drastically changing or fluctuating temperatures. It is possible that both scenarios occur(red) and thus in some instances the adaptation was from a (extreme) thermophile to a more mesophilic microorganism and in other instances from a thermophile to a more extreme thermophilic microorganism.

11.4 Examples of gene products which support the hypothesis of the two sets of genes

Special properties of organisms which grow over extended temperature spans include temperature-dependent formation of enzyme activities, temperature-dependent sensitivity to antibiotics, different protein patterns for the two temperature ranges, a change in sporulation frequency, and possibly a change in the archaea lipid pattern. These examples include the following gene products. Since they have been discussed in more detail by Wiegel (1990), not all examples and details are repeated here.

Bacillus stearothermophilus is able to grow anaerobically only above 50 °C (i.e., in the upper range of the two postulated temperature ranges) producing ethanol as one of its main products. Only in this upper temperature range does it produce a thermostable (4Fe–4S) ferredoxin and alcohol dehydrogenase. On the other hand, this bacterium also contains enzymes which are stable over both temperature ranges through conformational changes (Ljungdahl, 1979) and thus for these enzymes the bacterium does need the presence of an alternate enzyme for growth in an extended temperature range. Such temperature-dependent conformational changes have been investigated in detail and their existence was clearly demonstrated by Talsky (1971). Another strong argument for the validity of the working hypothesis is the example of temperature-dependent synthesis of two glucose-6-phosphate isomerases in *B. caldotenax*: the enzyme form which is produced in the range between 30 °C and 50 °C is not stable at temperatures above 55 °C, but a high-temperature-stable form is produced between 60 °C and 70 °C. *T. ethanolicus* and *T. thermohydrosulfuricus* have two ferredoxins. One is a thermostable 4Fe–4S cluster ferredoxin which is stable at 70 °C for several hours, and one is a thermolabile 8Fe–8S ferredoxin which is as thermolabile as ferredoxin isolated from anaerobic mesophiles such as *C. pasteurianum* (Wiegel *et al.*, 1979, and unpublished results). However, a strong temperature-dependent synthesis of neither ferredoxin was observed, nor have knockout mutants for one or the other ferredoxin been investigated to demonstrate a temperature-dependent requirement for these electron carriers.

The archaeon *M. thermoautotrophicum* exhibits different antibiotic resistance patterns for temperatures below and above the intermediate plateau. Noticeably 5′, 5′, 5′-trifluoroleucine, neomycin, kanamycin, and amphotericin B inhibit this methanogen only above 50 °C, the temperature at which the intermediate plateau starts in the Arrhenius plot. This difference in the temperature-dependent inhibition was used to isolate mutants with a different temperature range. The mutants grew only up to 69 °C instead of 77 °C with a change in T_{opt} from 69 °C to 50–55 °C. Most interestingly, these mutants did not show a biphasic Arrhenius plot. Several strains of *Methanobacterium thermoautotrophicum* which were isolated either in our (e.g., strain JW 510) or in D. Ward's laboratory (personal communication) from hot spring samples of Yellowstone National Park (Wyo) and identified by 16S rDNA sequence analysis (Bateson *et al.*, 1989) grow only up to 69 °C and they do not exhibit a biphasic Arrhenius plot (Wiegel, 1990). The analysed strain JW510, which was isolated from a hot spring in Yellowstone National Park, lacks one of the major lipids of *Methanobacterium thermoautotrophicum* ΔH, JW500 or JW501 (isolated from sewage sludge and mesobiotic sediments, respectively). The defect in the lipid synthesis is presently still under investigation. This strain could be a natural temperature mutant or the entity that existed before additional information was obtained by horizontal gene transfer. Again, for a more detailed discussion of other examples the reader is referred to Wiegel (1990).

11.5 Critical assessment of the hypothesis

The previously formulated concept of 'temperature tolerant' microorganisms (Wiegel, 1990) was based on observations of microorganisms able to grow over an extended temperature span for growth and exhibiting biphasic Arrhenius graphs. It was postulated that such microorganisms should be found not only among those growing in the thermophilic and the lower end of the extremely thermophilic temperature range, but also among those growing in the hyperthermophilic and the psychrophilic temperature range. However, more detailed analyses of these groups have not been done; thus at this time this working hypothesis has not been further verified.

In this chapter I have put forward the additional hypothesis that horizontal (interspecies) gene transfer – a mechanism for obtaining additional genetic information for extending the temperature range for growth – might be a more general mechanism in the evolution of microorganisms when changing temperatures impose an evolutionary pressure. Microorganisms which are able to adapt to grow over a wider temperature range clearly have an advantage in competing for nutrients or in survival.

The observed biphasic curves in the Arrhenius graphs could be simply due to the presence of mixed cultures of similar bacteria but with different temperature optima. For most of the given examples, careful analyses were done to ensure a pure culture: for example, besides the traditional criteria of repeatedly obtaining single cell colonies, these cultures were grown for many generations and subcultured separately at temperatures near T_{min} from the lower and near T_{max} from the upper temperature span, thus establishing a T_{min} and a T_{max} culture line, for which the growth rates and product formations were then compared for several temperatures over the whole temperature range. Both growth lines exhibited the same properties for each of the temperatures compared; thus, taken together with the traditional tests and the fact that during 16S rDNA sequence analysis no indication of a mixed culture was obtained, one can assume that at least these tested cultures are axenic.

As well as the explanation in terms of contaminated cultures, which has been ruled out in several instances (Wiegel, 1990, 1992; Wiegel *et al.*, 1989), there are other possibilities than the hypothesis presented above to explain the biphasic temperature curves. These include the occurrence of temperature-dependent phase transitions of lipids and/or temperature-dependent conformational changes of proteins, with possible concomitant changes in their catalytic activities or regulatory properties allowing them to be stable and active over an extended temperature span. The temperature range in which this transition occurs leads to a decrease of active and/or stable compounds and thus to a lower growth rate. At higher temperature when the transition to the other stable form is complete, an increased rate (although not necessarily the same as before) is again observed.

The presence of two forms of enzymes or other gene products of different temperature stability does not necessarily have to be due to the occurrence of horizontal gene transfer, but could also be due to gene doubling and subsequent modification through mutations, deletions, or combinations. Changes which conferred a different temperature dependence were then kept and further modified. However, in instances where the two genes have totally different sequences or the gene product is only synthesised in one of the temperature ranges, this explanation appears to be unlikely. Again, future comparative analysis of those genes will answer the question. It is speculated that probably both mechanisms are involved in the adaptation and evolution discussed.

Biphasic Arrhenius plots are found in examples of both (eu)bacteria and archaea (although less investigated in the latter). If one assumes that the above theory of horizontal gene transfer as a means of evolution is correct, this could mean that the horizontal gene transfer occurred during the time of the start of Kandler's 'evolutionary bush', that is, at the time of the 'mesogenotic' stage (Kandler, 1994, and Chapter 2 this book) and at a time when mesophilic 'pre-cells' had already evolved. It could also be the other way round, assuming that life evolved from mesophilic mesogenotes. On the other hand, it is likely that the horizontal gene transfer occurred as a secondary adaptation of already evolved mesophilic (eu)bacterial and archaeal lines to thermophilic growth conditions and thus was a rather recent event. Indications that it was rather recent include the fact that the gene products mentioned as indications for the validity of the working hypothesis are all different in the various microorganisms and do not – as so far seen – represent a unifying set of gene products. A unifying set of genes responsible for the extended temperature range would be an indication of an early event before differentiation into the diversity of present-day microorganisms. This is especially true for the comparison of the various bacteria belonging to the clostridia–bacillus subphylum branch. The assumption of a recent event could also be used to explain the scattered distribution of temperature-tolerant species among other present-day mesophiles. On the other hand, one could argue that most of the mesophiles which are closely related to the temperature-tolerant species have lost all or some important part of the required subset of genes enabling them to grow in the upper temperature range. The temperature-tolerant microorganisms kept this information. That this might have occurred is indicated by bacteria that have been called 'cryptic thermophiles' (Wiegel, 1990). These can be transformed from bacteria only able to grow at mesobiotic temperatures into bacteria able to grow at significantly more elevated temperatures like their closely related thermophilic species (e.g., *B. stearothermophilus*) by supplying one or two required enzymes (e.g., gyrase and topoisomerase; Lindsay and Creaser, 1975; Droffner and Yamamoto, 1985). Since growth under thermophilic conditions requires more than just a couple of enzymes conferring thermoresistance (Ljungdahl, 1979; Kogut, 1980; Amelunxen and Murdock, 1978; Wiegel, 1990), these microorganisms must have originally been thermophiles which had lost the capability of making functional copies of the enzymes reintroduced by the authors. The introduced enzymes could also have indirectly overcome the loss of other missing functions. Again, when more genomes of this type of microorganism have been sequenced, this question can, at least in part, be answered.

11.6 Conclusion

Horizontal gene transfer appears to be a possible explanation for the observation of the biphasic Arrhenius plots and extended temperature ranges exhibited by some microorganisms. Furthermore, horizontal gene transfer can be viewed as a means for microorganisms to extend their temperature range (and perhaps also pH range?) for growth and thus adapt successfully to environments with changed or frequently changing temperatures. Thus, horizontal gene transfer as discussed above could have been an avenue by which the originally thermophilic life forms adapted to lower temperatures as the earth cooled down. However, whether this process occurred only at the early mesogenotic stage or also at a much later state, or even occurs at the present time, cannot be answered from the present knowledge of the phenomenon. The hypothesis does not in particular require the assumption that life began under thermobiotic conditions and that this early life form had to adapt to more mesobiotic environments later. The author believes that various mesophilic to extremely thermophilic mesogenotes and 'temperature variants' of the early life forms coexisted in an environment with fluctuating or changing temperatures and that, thus, there is no single type of a progenote. Consequently, the working hypothesis for temperature-related adaptions could have gone in either directions, to higher or to lower temperatures.

11.7 Summary

Most microorganisms can grow only over a limited temperature span of less than 35 °C, in some cases even less than 20 °C. So far only a few thermophilic microorganisms have been identified that are able to grow over a temperature span of more than 35 °C, e.g., between 35 °C and above 75 °C. Interestingly, all these microorganisms – as far as examined – exhibit biphasic curves for their temperature dependence of growth. Arrhenius graphs, adapted for doubling times and growth temperatures, show this biphasic behaviour and the presence of an intermediate plateau. This chapter presents the hypothesis that microorganisms which can grow over a temperature span of more than 35 °C have acquired additional genetic information by horizontal gene transfer. The acquired information can be, for example, for enzymes, electron transfer compounds, cofactors, structural proteins and lipid components. Examples given to support the hypothesis include temperature-dependent synthesis of enzymes in *Bacillus stearothermophilus*, the presence of two ferredoxins differing in thermostability in *Thermoanaerobacter ethanolicus*, and temperature dependence of antibiotic resistance in and temperature-sensitive mutants of *Methanobacterium thermoautotrophicum*. These observations led to the hypothesis of horizontal gene transfer as a means for microorganisms to adapt to changing growth conditions and to evolve into microorganisms able to grow over extended temperature ranges in environments with strongly fluctuating temperatures or temperatures which changed to above or below the original growth temperatures. At this time, owing to lack of detailed analyses of the putatively involved genes, there is no indication whether these microorganisms have acquired genetic information to enable growth at the lower temperature (e.g., mesobiotic to moderate thermobiotic range) or at the higher temperature (e.g., thermobiotic to the lower end of the extreme thermobiotic range). Furthermore, the hypothesis does not address whether this occurred during the early evolutionary phase (e.g., Kandler's mesogenotic stage) or whether it is a secondary evolutionary event which has occurred over time and independently in different microorganisms and groups.

Acknowledgement

This research was supported by a grant from the US Department of Energy (DE-FG05-95ER-20199).

References

AMELUNXEN, R. E. and MURDOCK, A. L. (1978) Microbial life at high temperatures: mechanisms and molecular aspects. In *Microbial Life in Extreme Environments*, ed. D. Kushner, pp. 216–278 (London: Academic Press).

BATESON, M., WIEGEL, J. and WARD, D. M. (1989) Comparative analysis of 16S ribosomal RNA sequences of thermophilic fermentative bacteria isolated from hot springs in cyanobacterial mats. *System. Appl. Microbiol.*, **12**, 1–7.

DROFFNER, M. L. and YAMAMOTO, N. (1985) Isolation of thermophilic mutants of *Bacillus subtilis* and *Bacillus pumulis* and transformation of thermophilic trait to mesophilic strains. *J. Gen. Microbiol.*, **131**, 221–229.

FORTERRE, P. (1996) A hot topic: the origin of hyperthermophiles. *Cell*, **85**, 789–792.

GOGARTEN-BOEKELS, M., HILARIO, E. and GOGARTEN, J. P. (1995) The effects of heavy meteoritic bombardments of the early evolution – the emergence of the three domains of life. *Origins Life and Evolution of the Biosphere*, **25**, 251–264.

GOGARTEN, J. P. (1995) The early evolution of cellular life. *Trends Ecol. Evol.*, **10**, 147–151.

GOGARTEN, J. P., HILARIO, E. and OLENDZENSKI, L. (1996) Gene duplication and horizontal gene transfer during early evolution. In *Evolution of Microbial Life*, Society for General Microbiology Symposium 54, ed. D. McL Roberts, P. Sharp, G. Alderson and M. Collins, pp. 267–292 (Cambridge: Cambridge University Press).

HENSEL, R., ZWICKL, P., FABRY, S., LANG, J. and PALM, P. (1989) Sequence comparisons of glyceraldehyde-3-phosphate dehydrogenase from the three urkingdoms: evolutionary implication. *Can. J. Microbiol.*, **35**, 81–85.

HILARIO, E. and GOGARTEN, J. P. (1993) Horizontal transfer of ATP genes – the tree of life becomes a net of life. *Biosystems*, **31**, 111–119.

KANDLER, O. (1994) The early diversification of life. In *Early Life on Earth*, Nobel Symposium No. 84, ed. S. Bengtson, pp. 152–160 (NewYork: Columbia University Press).

KASTING, J. F. (1993) New spin on ancient climate. *Nature*, **364**, 759–760.

KOGUT, M. (1980) Are there strategies of microbial adaptation to extreme environments? *Trends Biochem. Sci.* **5**, 1–29.

LAKE, J. A. (1995) Calculating the probability of multitaxon evolutinary trees: bootstrappers gambit. *Proc. Natl Acad. Sci. USA*, **92**, 9662–9666.

LINDSAY, J. A. and CREASER, E. H. (1975) Enzyme thermostability is a transformable property betweeen *Bacillus* spp. *Nature* (*London*), **255**, 650–652.

LJUNGDAHL, L. G. (1979) Physiology of thermophilic bacteria. *Adv. Microbial Physiol.*, **19**, 149–243.

LJUNGDAHL, L. G., YANG, S. S., LIU, M. T., WIEGEL, J. and MAYER, F. (1978) On the physiology and thermophilic properties of *Clostridium thermoaceticum* and some other thermophilic anaerobes. In *Biochemistry of Thermophily*, ed. S. M. Friedmann, pp. 385–400 (New York: Academic Press).

LOWE, D. R. (1994) Early environments: constraints and opportunities for early evolution. In *Early Life on Earth*, Nobel Symposium No. 84, ed. S. Bengtson, pp. 24–35 (New York: Columbia University Press).

MARGULIS, L. (1992) Biodiversity molecular biological domains symbiosis and kingdom origins, *Biosystems*, **27**, 39–51.

MARGULIS, L. and SCHWARTZ, K. V. (1997) *Five Kingdoms: an Illustrated Guide to the Phyla of Life on Earth*, 3rd edn (San Francisco: W. H. Freeman).

MILLER, S. L. and LAZCANO, A. (1995) The origin of life at high temperature? *J. Mol. Evol.*, **41**, 689–692.

RIVERA, M. C. and LAKE, J. A. (1992) Evidence that eukaryotes and eocyte prokaryotes are immediate relatives. *Science*, **257**, 74–76.

RIVERA, M. C. and LAKE, J. A. (1996) The phylogeny of *Methanopyrus Kandlere. Int. J. System. Bacteriol.*, **46**, 348–351.

SCHWARTZMAN, D., MCMENAMIN, M. and VOLK, T. (1993) Did surface temperatures constrain microbial evolution? *BioScience*, **43**, 390–393.

SLEEP, N. H., ZAHNLE, K. J., KASTING, J. F. and MOROWITZ, H. J. (1989) Annihilitation of ecosystems by large asteroid impacts on the early earth. *Nature*, **342**, 139–142.

TALSKY, G. (1971) Zur anomalen Temperatureabhägignkeit enzymkatalysierter Reaktionen. *Angew. Chem.*, **83**, 553–594.

WÄCHTERSHÄUSER, G. (1988) Pyrite formation, the first energy source for life: a hypothesis. *System. Appl. Microbiol.*, **10**, 207–210.

WÄCHTERSHÄUSER, G. (1992) Groundworks for an evolutionary biochemistry: the iron–sulfur world. *Prog. Biochem. Mol. Biol.*, **58**, 85–201.

WIEGEL, J. (1990) Temperature spans for growth: hypothesis and discussion. *FEMS Microbiol. Rev.*, **75**, 155–170.

WIEGEL, J. (1992) The obligately anaerobic thermophilic bacteria. In *Thermophilic Bacteria*, ed. J. K. Kristjansson, chap. 6, pp. 105–184 (Boca Raton, FL: CRC Press).

WIEGEL, J., YANG, S. S. and LJUNGDAHL, L. G. (1979) Properties of ferredoxins and a rubredoxin from extreme thermophilic anaerobes, presented at the Annual Meeting of the American Society for Microbiology, Los Angeles, Abstr. K 63.

WIEGEL, J., KUK, S. and KOHRING, G. W. (1989) *Clostridium thermobutyricum*, spec. nov., a moderate thermophile isolated from a cellulolytic culture producing butyrate as a major product. *Int. J. System. Bacteriol.*, **39**, 199–204.

12

Evidence in Anaerobic Fungi of Transfer of Genes Between Them and from Aerobic Fungi, Bacteria and Animal Hosts

LARS G. LJUNGDAHL*, XIN-LIANG LI AND HUIZHONG CHEN

Center for Biological Resource Recovery and Department of Biochemistry & Molecular Biology, The University of Georgia, Athens, Georgia, USA

12.1 Introduction

Although this chapter does not deal with thermophiles, it deals with the concept of the workshop covered by this publication; that is, the origin of genes in microorganisms and the transfer of them between microorganisms and also higher eukaryotes. Specifically, evidence will be discussed indicating that anaerobic fungi present in the intestines of herbivorous animals contain genes apparently originating from aerobic bacteria and fungi, anaerobic bacteria, and their animal hosts. Gene transfers between the anaerobic fungi as well as gene duplications within them have also occurred. In addition, several hydrolytic enzymes with different catalytic properties involved in degradation of plant materials including cellulose and hemicellulose have very conserved domains such as dockerins and cohesins functioning in protein–protein interaction and in binding cellulose.

Anaerobic fungi have had a varied history with respect to their phylogenetic position and their classification is still inconsistent (Li *et al.*, 1993). Flagellated zoospores described by Liebetanz (1910) were considered protozoa. Orpin (1975) for the first time isolated from the rumen of a sheep the mycelium of a strictly anaerobic fungus and found that flagellates were released from thalli of the fungal rhizoids. He named the fungus *Neocallimastix frontalis*. Anaerobic fungi are now placed in the order Spizellomycetales and family Neocallimasticaceae (Barr 1988; Trinci *et al.*, 1994). Li *et al.* (1993) have suggested Neocallimasticales as a new order for the anaerobic fungi. Recent studies on the rRNA and other sequences (Li and Heath, 1992; Brownlee, 1994) confirmed the earlier classification that anaerobic fungi belong to the Chytridiomycota (Heath *et al.*, 1983).

To date, 17 morphologically distinct anaerobic fungi have been described. They have been isolated from the rumen and caecum of herbivorous animals as well as from ecosystems of animal pastures. The fungi have been placed into seven genera (Orpin, 1994). Furthermore, according to the number of sporangia developed from the thallus, they have bee divided into polycentric and monocentric fungi (Wubah *et al.*, 1991). Among the most-studied anaerobic fungal isolates, *Orpinomyces* species are polycentric whereas

Neocallimastix and *Piromyces* genera represent monocentric fungi. The morphology of sporangia and zoospores as well as the number of flagella of the zoospores of the fungi differ (Orpin, 1994). However, comparison of rRNA and internal transcribed spacer sequences between the species reveals that they are closely related to each other (Li and Heath, 1992).

12.2 Gene duplication

Gene duplication refers to the presence of multiple or similar sequences in the genome with identical or similar functions. Gene duplication occurs in several ways. Here we refer to genes coding for proteins that are duplicated and code for separate polypeptides or, alternatively, a single gene coding for a protein with multiple domains of similar functions.

Cloning and sequencing of genes of anaerobic fungi coding for plant cell wall-degrading enzymes have provided evidence that both types of gene duplication have occurred in these organisms. Examples of gene duplication within single genes are *xynA* encoding XYLA of *Neocallimastix patriciarum* (Gilbert *et al.*, 1992), and *xynA* encoding XYLA of a *Piromyces* strain (Fanutti *et al.*, 1995). The 1821 bp *xynA* of *N. patriciarum*, the first gene sequenced from an anaerobic fungus, encodes a polypeptide which can be dissected into four functional domains (Figure 12.1) consisting of an N-terminal leader or signal peptide, and two catalytic domains (CD), which are separated from a reiterated C-terminal sequence by a linker rich in Thr and Pro residues. The C-terminal sequence has now been identified as a protein-docking domain (see below) and below is referred to as a NCRPD (noncatalytic repeated peptide domain). The amino acid sequences of the two CDs display 92% identity and 96% similarity. The identity between the nucleotides of the two regions is 93% without any deletion or insertion between them. The two CDs have similar catalytic properties. The gene (*xyn3*) coding for a xylanase (XYN3) of a closely related fungus, *N. frontalis*, has an almost identical sequence to that of *xynA* of *N. patriciarum* (Durand *et al.*, 1996). These data, together with other information, suggest that the two fungi are phylogenetically clustered. Similar to the *Neocallimastix* xylanase genes, *Piromyces xynA* codes for a polypeptide with two similar CDs, but the domain organisation is different (Fanutti *et al.*, 1995). The two CDs, which exhibit 57% identity, are separated by the NCRPD (Figure 12.1). It is noteworthy that when the two CDs were expressed separately in *E. coli* only the CD of the 3′-coding end of *xynA* exhibited activity.

A duplication different from the above xylanase genes has been discovered in *xynB* coding for XYLB of *N. patriciarum* (Figure 12.1; Black *et al.*, 1994). This gene encodes a distinct xylanase with only one CD, which is located at the N-terminal region. It is followed by 12 tandem repeats containing TLPG, which in turn are followed by 45 repeats of the sequence XSKTLPGG. The function of these tandem repeats is presently unknown.

Several examples of gene duplications resulting in multiple genes coding for similar hydrolases have been found in anaerobic fungi. Six distinct cDNAs coding for cellulases have been isolated and sequenced from the polycentric fungus *Orpinomyces* PC-2 (Figure 12.1). Among them, CelA, CelC, CelD, and CelF consist of CD sequences, which places them into family 6 glycosyl hydrolases, whereas CelB and CelE are family 5 enzymes (Henrissat and Bairoch, 1993; Li *et al.*, 1997a,c; Chen *et al.*, 1998a,b). CD regions of the family 6 enzymes are between 60% and 80% identical to each other. However, CelA, CelC, and CelD have NCRPDs, which are placed N-terminally, whereas CelF has a cellulose-binding domain also at its N-terminal region (Figure 12.1). The family 5 enzymes, CelB and CelE, are 67% identical. Both have, in contrast to CelA, CelC, and

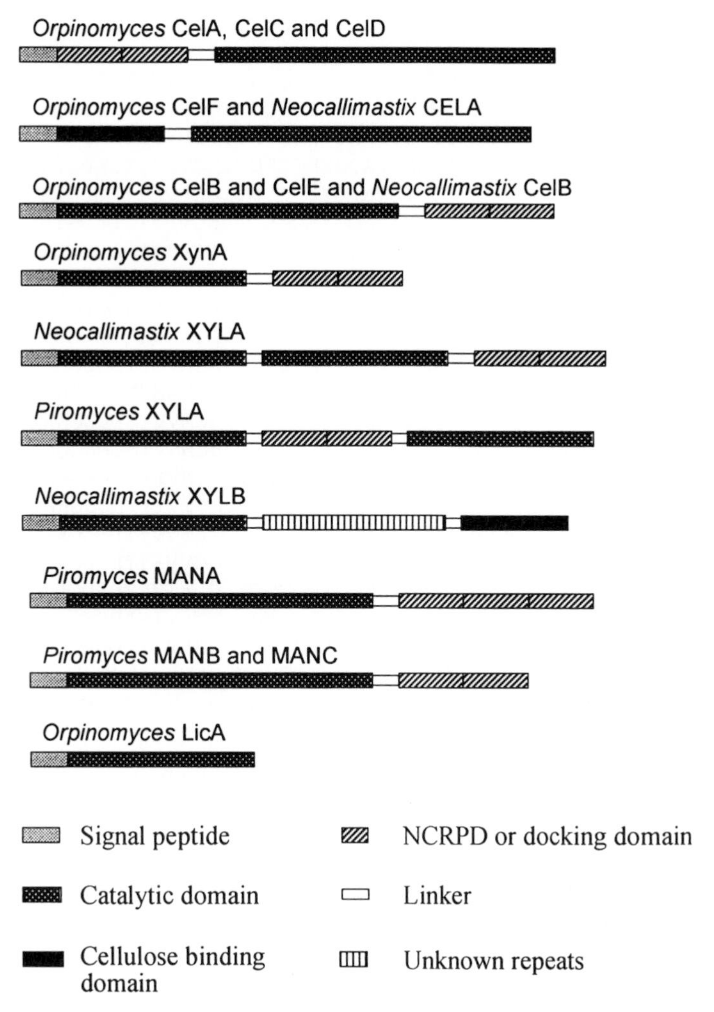

Figure 12.1 Domain organisations of glycosyl hydrolases of anaerobic fungi. Enzymes include *Orpinomyces* sp. strain PC-2 CelA, CelC (Li *et al.*, 1997c), CelB (Li *et al.*, 1997a), CelE (Chen *et al.*, 1998a), CelD (Chen *et al.*, unpublished data), CelF (Chen *et al.*, 1998b), XynA (Li *et al.*, 1997a), LicA (Chen *et al.*, 1997); *N. patriciarum* CELA (Denman *et al.*, 1996), CelB (Zhou *et al.*, 1994), XYLA (Gilbert *et al.*, 1992), XYLB (Black *et al.*, 1994); and *Piromyces* XYLA, MANA (Fanutti *et al.*, 1995), MANB, and MANC (Millward-Sadler *et al.*, 1996).

CelD, NCRPDs placed at the C-termini. The sequences of the NCRPDs are very similar, but apparently the placement in the enzyme is not critical. The regions coding for the CDs of the hydrolases of these two families clearly represent two distinct cellulase genes of different origins. One may speculate why *Orpinomyces* and also other cellulolytic microorganisms need to have many structurally similar cellulases. Comparison of biochemical properties revealed, however, that the highly similar cellulases indeed display distinct substrate and product specificities (Li *et al.*, 1997c; Chen *et al.*, 1998a,b), suggesting that the enzymes have been selected to hydrolyse substrates of different complexity. The family 5 enzymes are typical endoglucanases, whereas the family 6 enzymes show homology to cellobiohydrolases of aerobic fungi and endoglucanases of aerobic bacteria.

```
Orpinomyces CelA        (1)   20 CH.W...Q.YPCC.K..DCTVYYTDTEGKWGVLNNDWCMID
Orpinomyces CelB        (1)  390 C..FSTRLGYSCC.N..GFDVLYTDNDGQWGVENGNWCGIK
Orpinomyces CelC        (1)   20 CH.P....SYPCC.N..GCNVEYTDTEGNWGVENFDWCFID
Orpinomyces CelD        (1)   19 CH.P…...NYPCCQN..CGEVFYTDSDGQWGIENNDWCLIQ
Orpinomyces CelE        (1)  396 C..FSTRLGYSCC.N..GCDVFYTDNDGKWGVENGNWCGIK
Orpinomyces XynA        (1)  279 CSAKITAQGYKCCSDP.NCVVYYTDEDGTWGVENNQWCGCG
Neocallimastix CelB     (1)  392 C..FSVNLGYSCC.N..GCEVEYTDSDGEWGVENGNWCGIK
Neocallimastix XYLA     (1)  524 CSARITAQGYKCCSDP.NCVVYYTDEDGTWGVENNDWCGCG
Piromyces XYLA          (1)  286 CPSTITSQGYKCCSS..NCDIIYRDQSGDWGVENDEWCGCG
Piromyces MANA          (1)  492 CWS..INLGYPCCIG..DY.VVTTDENGDWGVENNEWCGIV
Piromyces MANB          (1)  492 CFS..IPLGYPCCKG..NT.VVYTDNDGDWGVENNEWCGIG
Piromyces MANC          (1)  490 CFS..IPLGYPCCKG..NT.VVYTDNDGDWGVENNEWCGIG
Orpinomyces CelA        (2)   63 CSSSITSQGYPCCSNN.NCKVEYTDNDGKWGVENNNWCGIS
Orpinomyces CelB        (2)  435 CWS..ERLGYPCCQY..TTNAEYTDNDGRWGVENGNWCGIY
Orpinomyces CelC        (2)   61 C..KFEALGYSCCK...GCEVVYSDEDGNWGVENQQWCGIR
Orpinomyces CelD        (2)   63 C..KFNALGYSCCS...HCNSVYSDNDGQWGIENGNWCGLK
Orpinomyces CelE        (2)  441 CWS..ERLGYPCCQY..TTNVEYTDNDGRWGVENGNWCGIY
Orpinomyces XynA        (2)  322 CSGKITAQGYKCCSDP.KCVVYYTDDDGKWGVENNEWCGCG
Neocallimastix CelB     (2)  437 CWS..EKLGYPCCQN..TSSVVYTDNDGKWGVENGNWCGIY
Neocallimastix XYLA     (2)  567 CSSKITSQGYKCCSDP.NCVVFYTDDDGKWGVENNDWCGCG
Piromyces XYLA          (2)  333 CPSSIKNQGYKCCSDS.C.EIVLTDSDGDWGIENDEWCGCG
Piromyces MANA          (2)  531 CWS..EPLGYPCCVG..NT.VISADESGDWGVENNEWCGIV
Piromyces MANB          (2)  535 CWS..EALGYPCCVS..SSDVYYTDNDGEWGVENGDWCGII
Piromyces MANC          (2)  533 CWS..EALGYPCCVSS.SDVYYTTDNDGEWGVENGDWCGII
Piromyces MANA          (3)  570 CWA..EFLGYPCCVG..NT.VISTDEFGDWGVENDDWCGIL
Consensus                        *       ** **           *  * * ***  ***
```

Figure 12.2 Amino acid sequence alignments of NCRPDs found in glycosyl hydrolases of anaerobic fungi. Enzyme abbreviations are the same as in Figure 12.1. Numbers in parentheses represent the position of the repeated peptides in the NCRPDs. The numbers in front of the first cysteine residues represents its position in polypeptide sequences of the enzymes. Residues highly conserved (over 90%) between the peptides are marked as consensus residues.

Duplications may have yielded three different genes, *manA, manB*, and *manC* (Figure 12.1), encoding mannanases in a *Piromyces* species (Fanutti *et al.*, 1995; Millward-Sadler *et al.*, 1996). The three mannanases are 80–85% identical and they have similar domain organisations. A major distinction is that the NCRPD of MANA consists of three repeated peptides (Fanutti *et al.*, 1995) in contrast to two as found for MANB and MANC (Millward-Sadler *et al.*, 1996).

Evidence has been provided that anaerobic fungi secrete multienzyme aggregates or complexes having molecular masses of over 2000 kDa, which efficiently degrade plant cell walls (Wilson and Wood, 1992; Ali *et al.*, 1995; Fanutti *et al.*, 1995; Li *et al.*, 1997b). These complexes appear to correspond to the cellulosomes of the anaerobic bacteria *Clostridium thermocellum* (Felix and Ljungdahl, 1993; Béguin and Lemaire, 1996) and *C. cellulovorans* (Doi *et al.*, 1994). The clostridial cellulosomes are composed of as many as 26 different polypeptides and include the scaffolding polypeptides, CipA and CbpA, and many catalytically active subunits. The binding between the scaffolding polypeptides and the catalytic subunits is mediated by internally repeated domains or cohesins of the scaffolding polypeptides and conserved duplicated regions or dockerins of the catalytic subunits (Bayer *et al.*, 1994; Choi and Ljungdahl, 1996). A majority of the sequenced hydrolases of the anaerobic fungi contain the NCRPD (Figures 12.1 and 12.2). This domain seems to function equivalently to that of the dockerins of cellulosomal subunits

(Fanutti *et al.*, 1995) and the consensus is that enzymes from anaerobic fungi with NCRPDs are subunits of cellulase/hemicellulase complexes produced by the fungi. However, there is no homology between sequences of the bacterial dockerins and the fungal NCRPDs, indicating that, even if they have the same function, they have developed or evolved independently. All the NCRPDs of the anaerobic fungal hydrolases contain two highly homologous peptides of 34–40 amino acid residues with the exception that the domain of MANA contains three peptides (Figure 12.2). Apparently, the sequences coding for the peptides in the fungi arose through gene duplications, transfers, and recombinations.

12.3 Gene transfer between anaerobic fungi

Analyses of rRNA sequences indicate that species of *Neocallimastix*, *Piromyces*, and *Orpinomyces* are closely related (Li and Heath, 1992) despite the fact that species of the first two genera are monocentric whereas species of *Orpinomyces* are polycentric. The comparison of CD sequences of xylanases from the anaerobic fungi strongly suggests that they have the same origin, and that the splitting of genes coding for XYLA in *Neocallimastix* (Gilbert *et al.*, 1992, Durand *et al.*, 1996) and for XynA of *Orpinomyces* (Li *et al.*, 1997a) occurred more recently than between those of *Neocallimastix* and *Piromyces* XYLA (Fanutti *et al.*, 1995). As discussed, there is only one CD in *Orpinomyces* XynA but two CDs in *Neocallimastix* XYLA and *Piromyces* XYLA (Figure 12.1). One possible explanation is that a horizontal gene transfer event occurred between *Neocallimastix* and *Orpinomyces* and that the *xynA* gene was subsequently duplicated in *Neocallimastix*.

Sequences coding for the NCRPDs are found in many of the genes encoding plant cell wall-degrading enzymes of anaerobic fungi. They are highly conserved among anaerobic fungal species (Figure 12.2), which indicates that an original NCRPD sequence was developed before the fungi evolved into distinct genera. Alternatively, it was first established in one gene of a single fungus and then finally spread into multiple genes and also by horizontal transfers into other fungi. The fact that sequence conservation between the xylanases and between CelBs of *Neocallimastix* and *Orpinomyces* is higher than between xylanases and CelBs within the fungi favours the latter possibility.

12.4 Genes of fungal origin

It would be too early to draw the conclusion that plant cell wall-degrading enzyme systems of anaerobic fungi resemble those of anaerobic bacteria and are rather distant from those of aerobic microorganisms. The distinction of the plant cell wall hydrolase systems between aerobic fungi and anaerobic bacteria centres on how the hydrolases are organised: for example, complex-associated as in the cellulosome of *C. thermocellum* versus free enzymes as in *Trichoderma reesei*. It seems that anaerobic fungi have acquired both strategies since they produce both types of enzymes. Thus CELA of *N. patriciarum* (Denman *et al.*, 1996) and CelF of *Orpinomyces* PC-2 (Chen *et al.*, 1998b) resemble cellobiohydrolase II (CBHII) of *T. reesei* and they have cellulose-binding domains (CBD). This is in contrast to CelA, CelB, CelC, CelD, CelE of *Orpinomyces* PC-2, and CelB of *N. patriciarum* (Zhou *et al.*, 1994), which all have NCRPDs (Figures 12.1 and 12.2) and are believed to be components of high-molecular-mass complexes (Li *et al.*, 1997b).

Sequences of genes encoding *T. reesei* CBH I-type cellulases have only been reported for aerobic organisms (Henrissat and Bairoch, 1993). CBH I is the most abundant extracellular protein when *T. reesei* is grown on cellulosic substrates. Using a cDNA library of

N. frontalis and a DNA probe encoding a part of the CBH I gene of *T. reesei*, a cDNA was found which was transcribed into a 2.1 kb RNA (Reymond *et al.*, 1991). This transcript was induced to high levels in cultures grown on cellulose. The sequence of the cDNA has not been reported and the presence of a CBD or NCRPD has not been established.

12.5 Genes of bacterial origin

Sequence analyses for genes and their encoded polypeptides of anaerobic fungal enzymes have provided extensive evidence that the fungi have, during the path of evolution, acquired genetic materials from prokaryotes and in particular from anaerobic bacteria found in gastrointestinal tracts of herbivores. An early observation was that the two catalytic domains of the XYLA, family 11 glycosyl hydrolases (Henrissat and Bairoch, 1993), of *N. patriciarum* are homologous to bacterial xylanases of the same family, particularly to a xylanase from the rumen bacterium *Ruminococcus flavefaciens* (Gilbert *et al.*, 1992). This indicated that the *Neocallimastix* and *Orpinomyces xynA*s were products of horizontal gene transfer from anaerobic bacteria. However, this became questionable after the *T. reesei xyn1* and *xyn2* sequences were made available (Törrönen *et al.*, 1992). They are also significantly homologous to the anaerobic fungal enzymes (Henrissat and Bairoch, 1993; Li *et al.*, 1997a). Whether the *Trichoderma* genes also originated from a bacterium through a recent transfer event needs further investigation. CelBs of both *N. patriciarum* (Zhou *et al.*, 1994) and *Orpinomyces* PC-2 (Li *et al.*, 1997a), on the other hand, have been shown to be homologous to endoglucanases found only in anaerobic bacteria, indicating the likelihood that the genes were derived from rumen bacteria. Again, the chance of the presence of such genes in aerobic fungi should not be ruled out.

An argument for gene transfer from bacteria to anaerobic fungi is in line with the presence of *licA* in *Orpinomyces* PC-2 coding for LicA (Figure 12.1), a β-1,3-1,4-glucanase (lichenase, EC 3.2.1.73), (Chen *et al.*, 1997). β-Glucan, a major hemicellulose in cereal cell walls, is a polymer consisting of alternating β-1,3 and β-1,4 glucoside units. Enzymes specific for β-glucans have been found in bacteria and plants, but the enzymes from these two sources are unrelated (Borriss *et al.*, 1990; Fincher *et al.*, 1986). Specific β-glucanases have not been found in aerobic fungi, although several cellulases from these fungi hydrolyse β-glucans. The *Orpinomyces* LicA is highly homologous to bacterial β-glucanases (Chen *et al.*, 1997), and particularly to that of the rumen bacterium *Streptococcus bovis*, with which it has sequence identity of 64% (Figure 12.3) (Ekinci *et al.*, 1997). These facts strongly support the idea that the *Orpinomyces* enzyme, those of *S. bovis*, and those of other bacteria (40–56%) have a common origin. It should be mentioned that PCR and zymogram analyses failed to detect a specific β-glucanase gene in *N. frontalis*.

The presence of introns disrupting the coding sequences has been used as a criterion for eukaryotic origin of genes, since the occurrence of introns in genes of prokaryotic organisms is rare. The lack of introns in *celB*, *celE*, and *licA* of *Orpinomyces* and *celB* of *N. patriciarum*, which are homologous to bacterial counterparts, support the view that they have prokaryotic origins.

Anaerobic fungi lack mitochondria. Instead they contain hydrogenosomes which are involved in energy generation (Yarlett *et al.*, 1986). Functions of hydrogenosomes involve the conversion of pyruvate and malate to hydrogen and carbon dioxide with the generation of ATP (Müller, 1993). The origins of hydrogenosomes in lower eukaryotes such as anaerobic protozoa, trichomonads, and ciliates as well as anaerobic fungi have been debated. In contrast to mitochondria, hydrogenosomes seem not to have an organelle

```
  1 MKSIISIAALSVLGLISKTMAAPAPAPVPGTAWNGSHDVMDFNYHESNRF 50
    |.. :|.  : :|::::   :..|...         |   ::||.::| :
  1 MWKKVSCLIVLLLAFFGFQQGVDAQSKY........HYSQELNYYNGNAM 42

 51 EMSNWPNGEMFNCRWTPNNDKFENGKLKLTIDRDG.SGYTCGEYRTKNYY 99
    |:.| .||:||||.:.|.|  |:|| :.|.||.|| :|||.||:|.|: :
 43 ELRNGSNGGMFNCNFVPGNVGFNNGLMSLKIDSDGRGGYTGGEWRSKERF 92

100 GYGMFQVNMKPIKNPGVVSSFFTYTGPSDGTKWDEIDIEFLGYDTTKVQF 149
    |||:|||||||||||||||||||||||||||||||||||||| |||||||
 93 GYGLFQVNMKPIKNPGVVSSFFTYTGPSDGTKWDEIDIEFLGKDTTKVQF 142

150 NYYTNGQGHHEHIHYLGFDASQGFHTYGFFWARNSITWYVDGTAVYTAYD 199
    ||||.|||:||.:. |||||||||||||| |. : ||||||| ||||||:
143 NYYTSGQGNHEYLYNLGFDASQGFHTYGFDWQADHITWYVDGRAVYTAYN 192

200 NIPDTPGKIMMNAWNGI.GVDDWLRPFNGRTNISAYYDWVSYDA 242
    |||.|||||||||| |. :||.|| ::|||| : |||||:|||.
193 NIPSTPGKIMMNAWPGTHEVDSWLGAYNGRTPLYAYYDWISYDQ 236
```

Figure 12.3 Comparison of the β-1,3-1,4-glucanases between the rumen anaerobic fungus *Orpinomyces* PC-2 (Chen *et al.*, 1997) and the rumen anaerobic bacterium *Streptococcus bovis* JB1 (Ekinci *et al.*, 1997). Amino acid residues with an identical match (|) and those with different degrees of conservation (: or .) are indicated.

genome. Recent determinations of nucleotide and amino acid sequences of the hydrogenosomal malic enzyme (van der Giezen *et al.*, 1997) and the β subunit of succinyl-CoA synthetase (Brondijk *et al.*, 1996) from *N. frontalis* strengthen the idea that hydrogenosomes are modified mitochondria. Both enzymes contain a mitochondrial-like targeting signal, which is cleaved when the proteins are transported into the hydrogenosomes. Further characterisation of the distinctions between mitochondria and hydrogenosomes will undoubtedly answer some basic biological questions.

12.6 Genes of animal origin

A cDNA for the cyclophilin (CyP) B sequence of *Orpinomyces* PC-2 has been sequenced (Chen *et al.*, 1995). CyPs display *in vitro* peptidylprolyl *cis–trans* isomerase activity and function in mammals as intracellular receptors of cyclosporin A, an immune-suppressive drug. CyP sequences are highly conserved between organisms. The *Orpinomyces* CyP B sequence is as high as 70% identical with human, bovine, and mouse sequences but less than 55% identical with *Neurospora* and *Saccharomyces* CyP Bs. Phylogenetic analysis using Protpars placed the *Orpinomyces* sequence closely clustered with those of animals but farther away from the fungal or bacterial sequences (Figure 12.4). Furthermore, deletions found in the fungal sequences in comparison to the animal CyP Bs are not found in *Orpinomyces* CyP B. Lastly, in contrast to genes for hydrolytic enzymes that are devoid of introns, the gene of the *cypB* is heavily interrupted by intron sequences (Figure 12.5), a characteristic of higher eukaryotic genes. An intron is also present in the enolase gene of *N. frontalis* (Durand *et al.*, 1995). These observations have led to the scenario that the *Orpinomyces cypB* has been introduced to *Orpinomyces* PC-2 probably by a recent transfer event from a higher eukaryotic source. The tight symbiotic relation between anaerobic fungi and herbivorous animals supports the view that this involves lateral transfer from

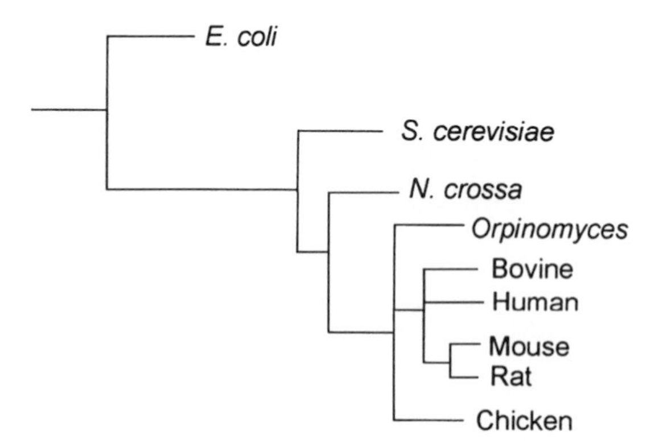

Figure 12.4 A unrooted phylogenetic tree based on the cyclophilin B sequences of various organisms using the Protpars program. The phylogenetic distances are measured by the lengths of the vertical lines.

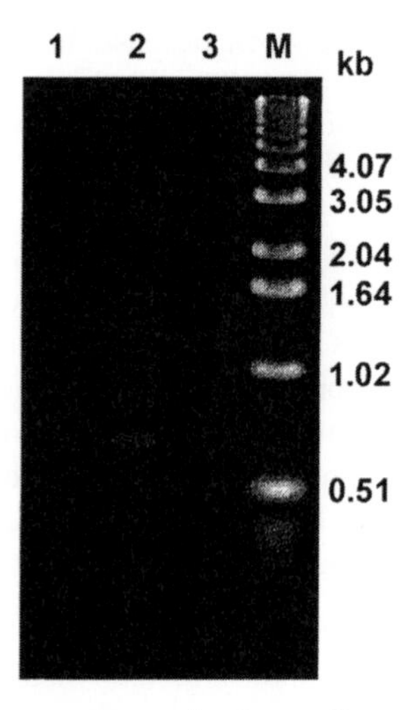

Figure 12.5 Detection of intron sequences in the coding region of *cypb* coding for CypB of *Orpinomyces* PC-2 using polymerase chain reactions. Forward and reverse primers corresponded to the 5′ and 3′ ends, respectively, of the *cypb* cDNA (Chen *et al.*, 1995). Templates were *Orpinomyces* genomic DNA (lane 1), cDNA library (lane 2), and negative control (lane 3).

its host, the cow, to the fungus. The nucleotide sequence of phosphoenolpyruvate carboxykinase of *N. frontalis* when compared with sequences of the enzyme from yeast and animals also showed the presence of highly conserved regions with the animal enzyme and not with the enzyme of the yeast (Reymond *et al.*, 1992).

12.7 Conclusions

Thermophilic microorganisms may well have been the first living cells capable of reproduction. Their enzymes function at very high temperatures but, as the earth's temperature decreased, mutations occurred allowing the production of altered enzymes functioning at lower temperatures. This led to the development of mesophilic organisms, which in turn, through additional mutations, allowed the development of bacteria and also higher organisms with special functions. Horizontal transfer of genes and development of short DNA segments encoding for domains giving enzymes special properties (e.g. cellulose-binding or protein-binding abilities), as discussed in this article, may then have led to a relatively fast evolution of fungi and other higher organisms.

12.8 Summary

Anaerobic fungi live in close contact with bacteria and other microorganisms in the rumen and caecum of herbivorous animals, where they digest ingested plant food. The fungi produce a number of hydrolytic enzymes including cellulases, xylanases, glucanases, and esterases. Genes encoding many of these and other enzymes have been sequenced. Comparison of the sequences has led to conclusions that some of the enzymes have origins similar to those of bacteria, whereas other enzymes appear to have aerobic fungal or even animal origins. These observations suggest that horizontal transfers of genes are common in ecosystems such as the rumen. In addition, gene duplication and insertions of various conserved domains with specific functions into genes are prevalent in the anaerobic fungi.

Acknowledgements

Work on anaerobic fungi was funded by grant from the US Department of Energy (DE-FG05-93ER20127). Support by a Georgia Power Distinguished Professorship in Biotechnology (L.G.L.) is also gratefully acknowledged.

References

Ali, B. R. S., Zhou, L., Graves, F. M. *et al.* (1995) Cellulases and hemicellulases of the anaerobic fungus *Piromyces* constitute a multiprotein cellulose-binding complex and are encoded by multigene families. *FEMS Microbiol Lett.*, **125**, 15–22.

Barr, D. J. S. (1988) How modern systematics relates to the rumen fungi. *BioSystems*, **21**, 351–356.

Bayer, E. A., Morag, E. and Lamed, R. (1994) The cellulosome – a treasure-trove for biotechnology. *Trends Biotech.* (*TIBTECH*), **12**, 379–386.

Béguin, P. and Lemaire, M. (1996) The cellulosome: an exocellular, multiprotein complex specialised in cellulose degradation. *Crit. Rev. Biochem. Mol. Biol.*, **31**, 201–236.

Black, G. W., Halewood, G. P., Xue, G.-P., Orpin, C. G. and Gilbert, H. J. (1994) Xylanase B from *Neocallimastix patriciarum* contains a non-catalytic 455-residue linker sequence comprised of 57 repeats of an octapeptide. *Biochem. J.*, **299**, 381–387.

Borriss, R., Buettner, K. and Maentsaelae, P. (1990) Structure of the β-1,3-1,4-glucanase gene of *Bacillus macerans*: homologies to other β-glucanases. *Mol. Gen. Genet.*, **222**, 278–283.

Brondijk, T. H. C., Durand, R., van der Giezen, M., Gottschal, J. C., Prins, R. A. and Fèvre, M. (1996) *scsB*, A cDNA encoding the hydrogenosomal β subunit of succingl-CoA synthetase from the anaerobic fungus *Neocallimastix frontalis. Mol. Gen. Genet.*, **253**, 315–323.

Brownlee, A. G. (1994) The nucleic acids of anaerobic fungi. In *Anaerobic Fungi*, ed. D. O. Mountfort and C. G. Orpin, pp. 241–256 (New York: Marcel Dekker).

CHEN, H., LI, X.-L. and LJUNGDAHL, L. G. (1995) A cyclophilin from the polycentric anaerobic fungus *Orpinomyces* sp. strain PC-2 is highly homologous to vertebrate cyclophilin B. *Proc. Natl Acad. Sci. USA*, **92**, 2587–2591.

CHEN, H., LI, X.-L. and LJUNGDAHL, L. G. (1997) Sequencing of a 1,3-1, 4-β-D-glucanase (lichenase) from the anaerobic fungus *Orpinomyces* strain PC-2: Properties of the enzyme expressed in *Escherichia coli* and evidence that the gene has a bacterial origin. *J. Bacteriol.*, **179**, 6028–6034.

CHEN, H., LI, X.-L., BLUM, D. L. and LJUNGDAHL, L. G. (1998a) Duplication of genes encoding endoglucanases in the anaerobic fungus *Orpinomyces* sp. strain PC-2. *FEMS Microbiol. Lett.*, **159**, 63–68.

CHEN, H., LI, X.-L., BLUM, D. L., XIMENES, E. A. and LJUNGDAHL, L. G. (1998b) A cellulase gene, *celF*, from the anaerobic fungus *Orpinomyces* sp. strain PC-2 contains an intron in its open reading frame region (manuscript to be submitted).

CHOI, S. K. and LJUNGDAHL, L. G. (1996) Structural role of calcium for the organisation of the cellulosome of *Clostridium thermocellum*. *Biochemistry*, **35**, 4906–4910.

DENMAN, S., XUE, G.-P. and PATEL, B. (1996) Characterisation of a *Neocallimastix patriciarum* cellulase cDNA (*celA*) homologous to *Trichoderma reesei* cellobiohydrolase II. *Appl. Environ. Microbiol.*, **62**, 1889–1896.

DOI, R. H., GOLDSTEIN, M., HASHIDA, S., PARK, J.-S. and TAKAGI, M. (1994) The *Clostridium cellulovorans* cellulosome. *Crit. Rev. Microbiol.*, **20**, 87–93.

DURAND, R., FISCHER, M., RASCLE, C. and FÈVRE, M. (1995) *Neocallimastix frontalis* enolase gene, *enol*: first report of an intron in an anaerobic fungus. *Microbiology*, **141**, 1301–1308.

DURAND, R., RASCLE, C. and FÈVRE, M. (1996) Molecular characterisation of *xyn3*, a member of the endoxylanase multigene family of the rumen anaerobic fungus *Neocallimastix frontalis*. *Curr. Genet.*, **30**, 531–540.

EKINCI, M. S., MCCRAE, S. I. and FLINT, H. J. (1997) Isolation and overexpression of a gene encoding an extracellular β-(1,3-1,4)-glucanase from *Streptococcus bovis* JB1. *Appl. Environ. Microbiol.*, **63**, 3752–3756.

FANUTTI, C., PONYI, T., BLACK, G. W., HAZLEWOOD, G. P. and GILBERT, H. J. (1995) The conserved noncatalytic 40-residue sequence in cellulases and hemicellulases from anaerobic fungi functions as a protein docking domain. *J. Biol. Chem.*, **270**, 29314–29322.

FELIX, C. R. and LJUNGDAHL, L. G. (1993) The cellulosome: the exocellular organelle in *Clostridium thermocellum*. *Annu. Rev. Microbiol.*, **47**, 791–819.

FINCHER, G. B., LOCK, P. A., MORGAN, M. M. (1986) Primary structure of the (1-3, 1-4)-β-D-glucan 4-glucohydrolase from barley aleurone. *Proc. Natl Acad. Sci. USA*, **83**, 2081–2085.

GILBERT, H. J., HAZLEWOOD, G. P., LAURIE, J. I., ORPIN, C. G. and XUE, G. P. (1992) Homologous catalytic domains in a rumen fungal xylanase: evidence for gene duplication and prokaryotic origin. *Mol. Microbiol.*, **6**, 2065–2072.

HEATH, I. B., BAUCHOP, T. and SKIPP, R. A. (1983) Assignment of the rumen anaerobe *Neocallimastix frontalis* to the Spizellomycetales (Chytridiomycetes) on the basis of its polyflagellate zoospore ultrastructure. *Can. J. Bot.*, **61**, 295–307.

HENRISSAT, B. and BAIROCH, A. (1993) New families in the classification of glycosyl hydrolases based on amino acid sequence similarities. *Biochem. J.*, **293**, 781–788.

LI, J. and HEATH, I. B. (1992) The phylogenetic relationships of the anaerobic chytridiomycetous fungi (Neocallimasticaceae) and the Chytridiomycota, I. Cladistic analysis of rRNA sequences. *Can. J. Bot.*, **70**, 1738–1746.

LI, J., HEATH, I. B. and PACKER, L. (1993) The phylogenetic relationships of the anaerobic chytridiomycetous gut fungi (Neocallimasticaceae) and the Chytridiomycota. II. Cladistic analysis of structural data and description of Neocallimasticales ord. nov. *Can. J. Bot.*, **71**, 393–407.

LI, X.-L., CHEN, H. and LJUNGDAHL, L. G. (1997a) Monocentric and polycentric anaerobic fungi produce structurally related cellulases and xylanases. *Appl. Environ. Microbiol.*, **63**, 628–635.

LI, X.-L., CHEN, H. Z., HE, Y., BLUM, D. L. and LJUNGDAHL, L. G. (1997b) High molecular weight cellulase/hemicellulase complexes of anaerobic fungi. *ASM 97th Gen. Meeting*, p. 424, abstr. O-31.

LI, X.-L., CHEN, H. and LJUNGDAHL, L. G. (1997c) Two cellulases, CelA and CelC, from the polycentric anaerobic fungus strain PC-2 cotain N-terminal docking domains for a cellulase/hemicellulase complex. *Appl. Environ. Microbiol.*, **63** (in press).

LIEBETANZ, E. (1910) Die parasitischen Protozoen, des Wiederkauermagens. *Arch. Protistenkd.*, **19**, 19–80.

MILLWARD-SADLER, S. J., HALL, J., BLACK, G. W., HAZLEWOOD, G. P. and GILBERT, H. J. (1996) Evidence that the *Piromyces* gene family encoding endo-1, 4-mannanases arose through gene duplication. *FEMS Microbiol Lett.*, **141**, 183–188.

MÜLLER. M. (1993) The hydrogenosome. *J. Gen. Microbiol.*, **139**, 2879–2889.

ORPIN, C. G. (1975) Studies of the rumen flagellate *Neocallimastix frontalis. J. Gen. Microbiol.*, **91**, 270–280.

ORPIN, C. G. (1994) Anaerobic fungi: taxonomy, biology and distribution in nature. In *Anaerobic Fungi*, ed. D. O. Mountfort and C. G. Orpin, pp. 1–45 (New York: Marcel Dekker).

REYMOND, P., DURAND, R., HEBRAUD, M. and FÈVRE, M. (1991) Molecular cloning of genes from the rumen anaerobic fungus *Neocallimastix frontalis*: expression during hydrolase induction. *FEMS Microbiol. Lett.*, **77**, 107–112.

REYMOND, P., GEOURJON, C., ROUX, B., DURAND, R. and FÈVRE, M. (1992) Sequence of the phosphoenolpyruvate carbokinase-encoding cDNA from the rumen anaerobic fungus *Neocallimastix frontalis*: comparison of the amino acid sequence with animals and yeast. *Gene*, **110**, 57–63.

TÖRRÖNEN, A., MACH, R. L., MESSNER, R. *et al.* (1992) The two major xylanases from *Trichoderma reesei*: characterisation of both enzymes and genes. *Biotechnology*, **10**, 1461–1465.

TRINCI, A. P. J., DAVIES, D. R., GULL, K. *et al.* (1994) Anaerobic fungi in herbivorous animals. *Mycol. Res.*, **96**, 129–152.

VAN DER GIEZEN, M., RECHINGER, K. B., SVENDSEN, I. *et al.* (1997) A mitochondrial-like targeting signal on the hydrogenosomal malic enzyme from the anaerobic fungus *Neocallimastix frontalis*: support for the hypothesis that hydrogenosomes are modified mitochondria. *Mol. Microbiol.*, **23**, 11–21.

WILSON, C. A. and WOOD, T. M. (1992) Studies on the cellulase of the rumen anaerobic fungus *Neocallimastix frontalis*, with special reference to the capacity of the enzyme to degrade crystalline cellulose. *Enzyme Microbiol. Technol.*, **14**, 258–264.

WUBAH, D. A., FULLER, M. S. and AKIN, D. E. (1991) Isolation of monocentric and polycentric fungi from the rumen and feces of cows in Georgia. *Can. J. Bot.*, **69**, 1232–1236.

YARLETT, N., ORPIN, C. G., MUNN, E. A., YARLETT, N. and GREENWOOD, A. C. (1986) Hydrogenosomes in the rumen fungus *Neocallimastix patriciarum. Biochem. J.*, **236**, 729–739.

ZHOU, L., XUE, G.-P., ORPIN, C. G., BLACK, G. W., GILBERT, H. J. and HAZLEWOOD, G. P. (1994) Intronless *celB* from the anaerobic fungus *Neocallimastix patriciarum* encodes a modular family A endoglucanase. *Biochem. J.*, **297**, 359–364.

PART FIVE

Enzyme-based Phylogenies

13

DNA Topoisomerases, Temperature Adaptation, and Early Diversification of Life

PURIFICACIÓN LÓPEZ-GARCÍA

Institut de Génétique et Microbiologie, Université Paris-Sud, Orsay, France

13.1 Introduction

In recent times, two sets of convergent arguments have been used to favour the hypothesis of a hot autotrophic origin of life over the classic heterotrophic origin proposed to occur in a colder prebiotic soup (Baross and Hoffman, 1985; Woese, 1987; Wächtershäuser, 1988; Woese *et al.*, 1990; Pace, 1991). Geological and astronomical data suggested that conditions governing the Earth at the time when life arose (3.8–4.2 Ga) consisted of more extensive volcanism and higher temperatures. These conditions resembled those of hyperthermophilic biotopes today (Baross and Hoffman, 1985; Nisbet, 1985; Shock, 1996). 16S rRNA sequence-based phylogenetic analyses, on the other hand, revealed that present-day hyperthermophiles are the deepest-branching organisms (Woese *et al.*, 1990; Stetter, 1996). The proposed bacterial rooting of the tree of life has further supported this view (Woese *et al.*, 1990; Brown and Doolittle, 1995; Baldauf *et al.*, 1996). However, the position of the root and the hot origin of life are themselves controversial (Doolittle and Brown, 1994; Forterre, 1996; Lazcano and Miller, 1996). In this regard, even the possibility that the last common ancestor (or cenancestor) was the only survivor of the last large meteorite bombardment has been formulated in order to make its putative hyperthermophilic nature compatible with a cold origin of life (Miller and Lazcano, 1995).

More generally accepted is the concept of an already complex cenancestor endowed with a DNA genome and transcriptional and translational apparatus (Zillig, 1991; Forterre *et al.*, 1993). Comparative analyses of archaeal, bacterial and eukaryal genome sequences seemed to confirm this assumption (Edgell and Doolittle, 1997; Olsen and Woese, 1997). The cenancestor should have thus experienced a previous evolutionary pathway leading from obscure beginnings, through a hypothetical RNA world, to the establishment of our present-day DNA-based world. During this process, the evolution of systems assuring the integrity and functionality of DNA molecules must have occurred accordingly. As essential components of these systems, DNA topoisomerases have evolved to solve problems of disentangling DNA strands or duplexes. These universal enzymes are thus involved in a variety of cellular functions ranging from transcription, recombination, and replication (including decatenation and segregation of chromosomes), to the regulation of DNA

Table 13.1 Classification and main features of DNA topoisomerases[a]

Mechanistic type	Main features	Phylogenetic subfamilies	Main features	Subtypes	Representative/first described example(s)	Reference
I	Catalyse the passage of one DNA strand or helix through ssDNA breaks ATP independent[b] Monomeric[c]	IA	ssDNA preferential activity	RG-like	*Sulfolobus acidocaldarius*	Kikuchi and Asai (1984), Forterre *et al.* (1985), for review see Duguet (1995)
			Relax only negative supercoils	I-like	Topo I *Escherichia coli*	Wang (1971), for review see Wang (1996)
			Covalently linked to 5′ end of transient cleavage	III-like	Topo III *E. coli*	Srivenugopal *et al.* (1984), for review see Wang (1996)
		IB	dsDNA preferential binding Relax negative and positive supercoils Covalently linked to 3′ end of cleavage		Eukaryotic Topo I (*Mus musculus*) Archaeal Topo V (*Methanopyrus kandleri*)	Champoux and Dulbecco (1972) Kozyavkin *et al.* (1994)
II	Catalyse the passage of a DNA helix through dsDNA breaks ATP-dependent Homodimer (eukaryotes) or heterotetramer (prokaryotes)	IIA (Classic)	Relax positive and negative supercoils	G-like	Topo II *E. coli* (gyrase, GyrA+GyrB)	Gellert *et al.* (1976), for review see Wang (1996)
			Catenation/decatenation 5′ end cleavage binding	Non-G	Topo IV *E. coli* (ParC+ParE)	Kato *et al.* (1990), for review see Wang (1996)
		IIB	Relax positive and negative supercoils Catenation/decatenation 5′ end cleavage binding		Topo VI *Sulfolobus shibatae*	Bergerat *et al.* (1997)

[a] Topo, topoisomerase; G, gyrase; RG, reverse gyrase.
[b] Except reverse gyrase, with an ATP-binding site in the helicase-like domain.
[c] Except *M. kandlerii* RG, which is split in the middle of the topoisomerase domain (Krah *et al.*, 1996).

supercoiling and maintenance of genome stability (Wang *et al.*, 1990; Drlica, 1992; Luttinger, 1995; Wang, 1996). Depending on their ability to cut one or both DNA strands, they have been classified into two mechanistic types, each one of which is subdivided into two phylogenetic families (Table 13.1). In Archaea, Bacteria and Eukarya, type I and II enzymes coexist in the cell, suggesting that both mechanistic types evolved before domain diversification.

Reverse gyrase (RG), a unique type I topoisomerase able to introduce positive supercoils in DNA molecules, constitutes a hallmark of hyperthermophily. So far, it has been found in all bacteria and archaea growing optimally above 80 °C (Duguet, 1995). Functional questions arise, such as the role of positive supercoiling in DNA stability at high temperatures, as well as evolutionary ones. If the cenancestor was a hyperthermophile, it should have had RG. But the RG gene seems to be the result of a fusion event (topoisomerase plus a putative helicase) (Confalonieri *et al.*, 1993), and some authors have suggested that RG is a derived trait that appeared in the course of adaptation to hyperthermophily from mesophily (Forterre, 1995; Forterre *et al.*, 1995). Alternatively, RG could have evolved on the way to a cenancestor that was already quite complex. But then, were its predecessors, which were devoid of RG, hyperthermophilic? Does this imply that the evolution of early life was necessarily mesophilic?

As discussed in this chapter, a study of the phylogenetic distribution of DNA topoisomerases as essential regulators of a biologically appropriate DNA state suggests that adaptation to mesophily may have appeared later in evolution. I speculate the possibility of a thermophilic ancestor (60–80 °C) whose DNA had spontaneously an optimal structure for function. Subsequently, systems involving specialisation of DNA topoisomerases evolved to keep homeostatic control of that configuration when organisms adapted to different temperature niches.

13.2 Homeostasis of DNA geometry

13.2.1 DNA supercoiling, DNA topology and topoisomerase function

The shape of DNA in space, i.e. its geometry, determines its biological function. Relative distances in the molecule must be recognised by proteins processing DNA information. Geometrical parameters are generally simplified into two main components, the twist (Tw), representing the wrapping of one strand around the other, and directly related to the path of the helix, and the writhe (Wr), or supercoiling, related by the equation $\Delta Lk = \Delta Tw + \Delta Wr$ (for review see Bates and Maxwell, 1993). As depicted schematically in Figure 13.1, a variety of factors, as different as protein tracking in different DNA-derived processes and cytoplasmic salt concentrations, can affect DNA geometry. Cells can actively control DNA geometry, changing the topological number of links between the DNA strands (Lk), by using DNA topoisomerases. They release locally induced tensions, most probably their primary function in early living systems, and they control global supercoiling directly or in combination with DNA-binding proteins (see below).

In general, relative variations of Lk by comparison to the relaxed state (superhelical density, specific linking difference) are used as estimates of DNA geometry, which otherwise is very difficult to measure *in vivo*. Since the twist can change only locally in active DNA (if not, it would be denatured or inactive), DNA supercoiling and topology are concepts sometimes employed indistinctly. In mesophilic archaea, bacteria and eukaryotes,

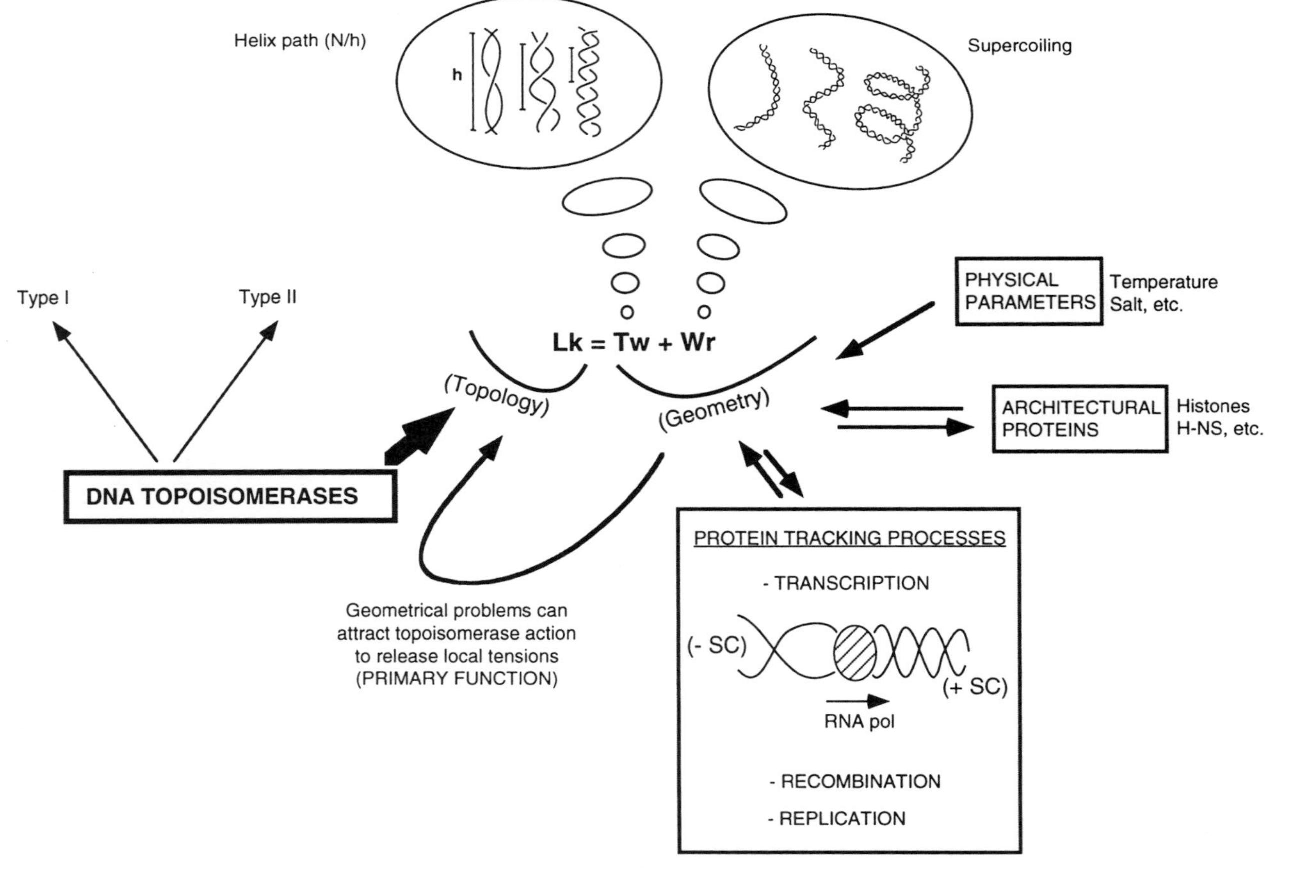

Figure 13.1 Factors affecting DNA geometrical and topological parameters, related by the central equation: Lk, linking number; Tw, DNA twist; Wr, DNA writhe; SC, supercoiling.

DNA is negatively supercoiled, with specific linking differences (σ) *in vitro* between −0.07 and −0.05 (Worcel and Burgi, 1972; Germond *et al.*, 1975; Wang, 1987; Charbonnier and Forterre, 1994). In contrast, in hyperthermophilic archaea, plasmid DNA exists, in general, from relaxed to positively supercoiled at their optimal growth temperatures, with σ from −0.006 to +0.03, exhibiting a linking excess by comparison to a mesophile's DNA (Charbonnier and Forterre, 1994; López-García and Forterre, 1997). Although protein-tracking effects may cause more significant topological changes in small DNAs, at least in mesophilic bacteria supercoiling levels in chromosomes and plasmids match reasonably well (Lukomski and Wells, 1994).

13.2.2 *The hypothesis of an optimal DNA geometry*

The differences in DNA topology between mesophiles and hyperthermophiles, and their possible compensatory effects for temperature, suggest the existence of a precise overall geometry of DNA molecules, in which the relative distances are maintained for correct function (Forterre *et al.*, 1996) (see Figure 13.2). This optimal spatial structure, keeping a balance between integrity and functionality, must have been conserved throughout evolution.

The ideas that follow are based in the above hypothesis. However, it must be interpreted in relative terms, since active DNA molecules are dynamic entities, continuously changing their local configurations during replication, transcription, etc. Indeed, these might constitute a regulatory device by affecting nearby supercoiling-sensitive promoters (Wang and Lynch, 1993). Moreover, at global levels, changes of supercoiling vary for any given organism living at any temperature within certain limits kept under homeostatic control. In mesophilic bacteria, supercoiling levels change depending on growth conditions (Balke and Gralla, 1987; Jensen *et al.*, 1995), or as response to environmental shifts of osmolarity (Hsieh *et al.*, 1991b), oxygen concentration (Hsieh *et al.*, 1991a), pH (Karem and Foster, 1993), or temperature (Goldstein and Drlica, 1984). For some pathogens, environmentally induced superhelical variations could even have a role in overall gene expression regulation (Higgins *et al.*, 1990; Dorman, 1995). In hyperthermophilic archaea, plasmid DNA supercoiling also varies during growth and temperature shifts within an apparently regulated range (López-García and Forterre, 1997). In spite of these variations, and favouring the hypothesis of a global DNA homeostasis, chromosomal *E. coli* domains are not believed to differ in functional superhelical density (Miller and Simons, 1993).

13.2.3 *Strategies for homeostatic control of DNA supercoiling*

In mesophiles, at least two different strategies to control DNA supercoiling levels exist (Figure 13.2), which could suggest that adaptation to mesophily has occurred independently in different lineages. In bacteria, supercoiling levels are mainly controlled by the antagonistic action of gyrase, the only topoisomerase introducing specifically negative superturns in the molecule, and Topo I relaxing them (Table 13.1) (Drlica, 1992; Wang, 1996). In eukaryotes, carrying only relaxing DNA topoisomerases (Table 13.1), negative supercoiling is generated by DNA wrapping around the histone core and subsequent relaxation of the internucleosomal tension by these enzymes (Saavedra and Huberman, 1986). In mesophilic archaea, we lack studies about the regulation of DNA supercoiling. In halophiles, at the end of the euryarchaeal branch, a bacterial-like mechanism could be

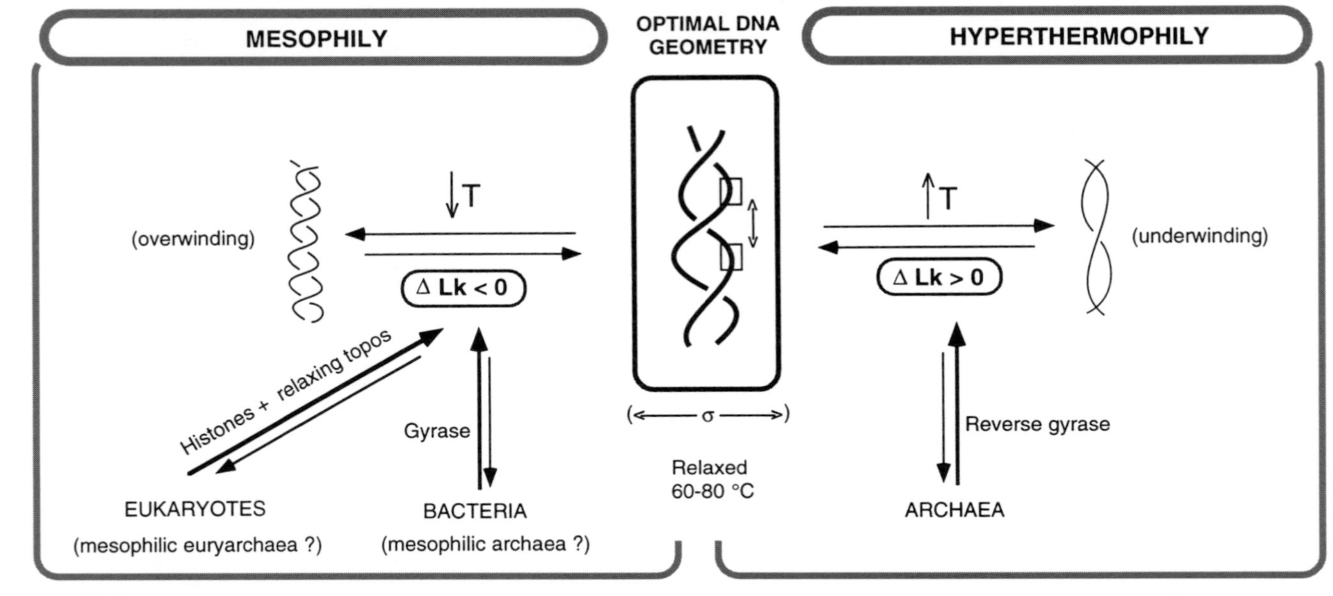

Figure 13.2 Hypothesis of an optimal DNA geometry under homeostatic control. Living organisms have developed different strategies for modifying DNA topology (ΔLk) in order to compensate for the physical effect of temperature (T) shifts on DNA. Boxes in the DNA molecule indicate recognition sequences. σ, a relative measure of ΔLk, may vary within a certain range.

used. DNA in these organisms is highly negatively supercoiled (Mojica *et al.*, 1994; López-García *et al.*, 1994), and they probably contain a bacterial-like gyrase. The presence of bacterial *gyr*-like genes has been reported (Holmes and Dyall-Smith, 1991), and treatment at low doses with specific gyrase inhibitors induces DNA relaxation, suggesting a similar enzymatic activity (Sioud *et al.*, 1988).

In hyperthermophilic archaea, control of DNA supercoiling most likely involves the antagonistic action of reverse gyrase (RG) and one relaxing activity (Forterre and Elie, 1993; Forterre *et al.*, 1996) (Figure 13.2). Among the relaxing topoisomerase candidates that are widespread in both archaeal kingdoms (see Figure 13.3), only a type II enzyme exists, Topo VI (Table 13.1). Topo VI, first purified from *Sulfolobus shibatae* (Bergerat *et al.*, 1994) is the prototype of a new phylogenetic family with archaeal and eukaryal representatives (Bergerat *et al.*, 1997). Regarding type I topoisomerases with exclusively relaxing activities, Topo III-like genes have been detected in the genomes of *Methanococcus jannaschii* (Bult *et al.*, 1996), *Pyrobaculum aerophilum* (Fitz-Gibbon *et al.*, 1997), and *Pyrococcus furiosus* (F. Robb, personal communication) (Figure 13.3), and the enzymatic activity may have been previously detected in archaea (Slesarev *et al.*, 1991). Also Topo V, a biochemical member of the 'eukaryotic' IB subfamily, has been described in *Methanopyrus kandlerii* (Slesarev *et al.*, 1993) (Table 13.1), but whether its presence is widespread in archaea is not known.

In hyperthermophilic bacteria, *Thermotoga maritima* possesses both RG and gyrase (Guipaud *et al.*, 1997). While the presence of RG argues in favour of the need for a mechanism introducing linking excess for survival at high temperatures, the existence of a gyrase raises several questions. However, in *T. maritima*, gyrase seems to be the only type II topoisomerase in the cell (Guipaud *et al.*, 1996; `http://www.tigr.org`) and, in contrast to the *Escherichia coli* gyrase – devoted essentially to the gyration activity (decatenation is carried out by Topo IV) – it decatenates very efficiently. Since chromosome decatenation and segregation are essential for the cell, and are carried out by type II enzymes (Wang, 1996), this could be the primary function of the *Thermotoga* gyrase (see below).

Interestingly, euryarchaea also possess histones homologous to eukaryotic ones (especially to the highly conserved H4 family), and form nucleosomal structures with canonical histone folds (Grayling *et al.*, 1996; Starich *et al.*, 1996). DNA wraps around the histone core in a positive sense, but may also wrap in a negative sense depending on conditions such as protein:DNA ratio (Musgrave *et al.*, 1991; Reeve *et al.*, 1997). Although the role of archaeal nucleosomes remains unknown, it is tempting to speculate on their involvement in the regulation of DNA supercoiling by helping to generate linking deficit or excess in mesophilic or hyperthermophilic archaea, respectively.

13.3 Gyrase and reverse gyrase: did early prokaryote diversification correlate with adaptation to distinct temperature niches?

13.3.1 Evolution of DNA topoisomerase function

This chapter is not intended to be an exhaustive review on topoisomerases or topoisomerase function. Rather, I will try to provide some clues about the functional evolution of type I and type II enzymes from their distribution in living beings, which can be relevant in understanding temperature adaptation. Although far from complete, a general picture of topoisomerase distribution in present-day organisms is displayed in Figure 13.3.

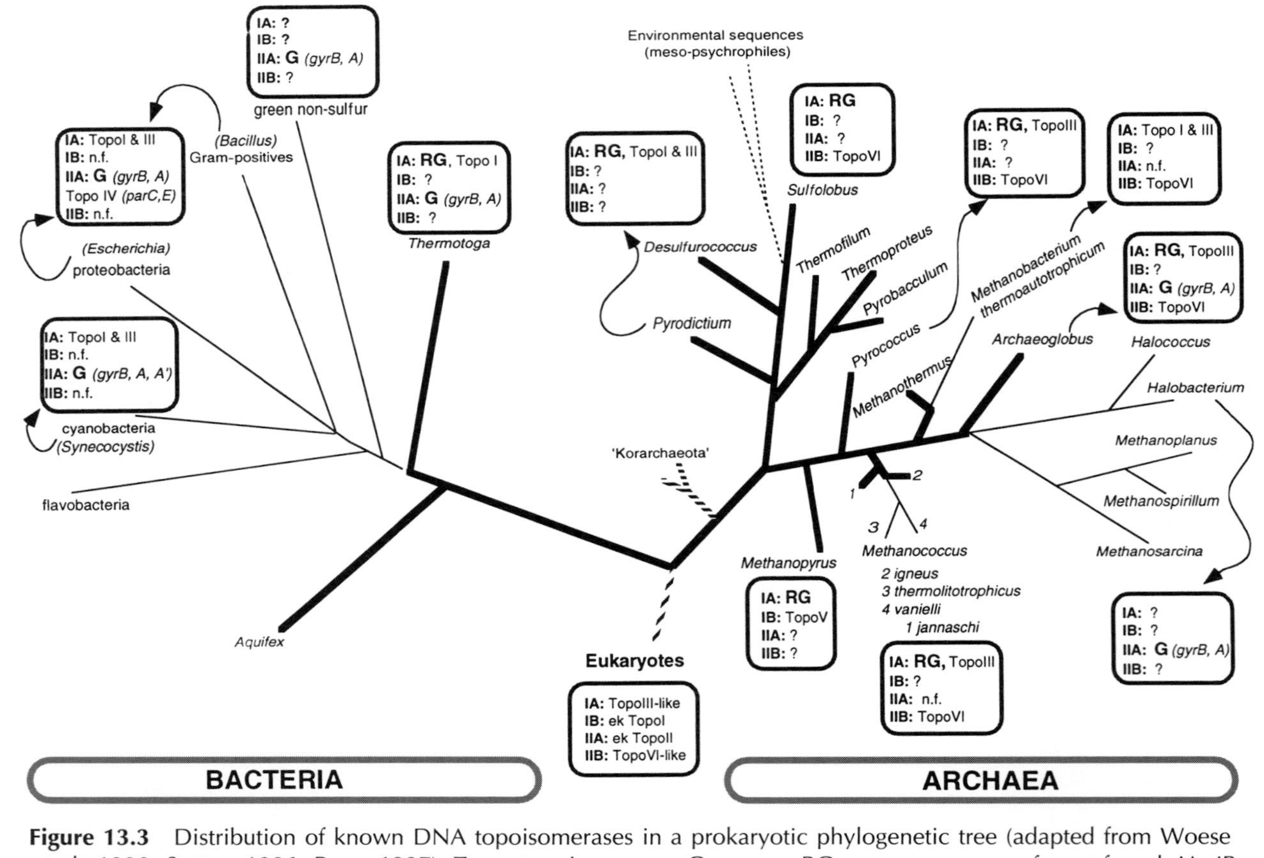

Figure 13.3 Distribution of known DNA topoisomerases in a prokaryotic phylogenetic tree (adapted from Woese *et al.*, 1990; Stetter, 1996; Pace, 1997). Topo, topoisomerase; G, gyrase; RG, reverse gyrase; n.f., not found; IA, IB, IIA, IIB, distinct topoisomerase phylogenetic subfamilies. Bold lines represent hyperthermophilic lineages, and broken lines environmental phylotypes.

Type I topoisomerases

Topoisomerases evolved to resolve topological problems in nucleic acid molecules. Early replicative genomes might have been composed of small linear nucleic acids before the evolution of any topoisomerase activity. Locally generated tensions might have been released spontaneously (see Figure 13.4, bottom). In this context, type I topoisomerases, simpler and ATP-independent, likely evolved first, allowing size increase of small replicons. This might have occurred before the DNA era, during a putative RNA–protein world. Moreover, type I enzymes might have played an essential role facilitating the transition between RNA-based and DNA-based worlds, particularly Topo III-like enzymes. Not only are these ubiquitous in the three domains (Figure 13.3), but also the *E. coli* Topo III has been shown to be a true RNA topoisomerase (DiGate and Marians, 1992; Wang *et al.*, 1996). Their function would thus be related to processes involving RNA, such as transcription and RNA replication.

Once this elementary relaxing function was in place, duplication and recruitment of a new function could give birth to RG, and allow hyperthermophily (by the creation of positive DNA supercoils in association with a putative helicase). Since RG is found in both hyperthermophilic bacteria and archaea, it may have been present in their common ancestor. However, the possibility remains for a horizontal transfer from archaea to bacteria.

Type II topoisomerases

A priori, type II topoisomerases are more complicated enzymes, since they are generally composed of two different kinds of subunits and are ATP-dependent (Table 13.1). All type II enzymes endowed with exclusively relaxing activities (all except gyrase), act on positive and negative supercoils, being specifically involved in chromosome segregation and decatenation upon replication or recombination (Luttinger, 1995; Wang, 1996). This is an essential function in increasingly larger genomes, and although type I enzymes are able *in vitro* to decatenate, in particular Topo III (Wang, 1996), they must not be very efficient. Type II topoisomerases probably evolved to perform this task, although they could also provide better efficiency for the removal of supercoils.

As in the case of type I enzymes, one may hypothesise that once the primary function (chromosome decatenation/segregation) was in place, duplication and specialisation in a second function, such as the active introduction of negative supercoils (gyrase) was allowed. Indeed, that is the case in most bacteria, possessing two sets of type II enzymes, gyrase and Topo IV-like (Table 13.1, Figure 13.3) (Huang, 1996). However, the analysis of type II topoisomerase distribution raises two fundamental problems. At present, as mentioned above, the early-branching *T. maritima* has only one type II topoisomerase, which is a gyrase. Moreover, Huang failed to amplify additional *gyrA*-like sequences in the radioresistant micrococci and *Thermus*, green sulfur, green non-sulfur and *Planctomyces* (Huang, 1996). Only one set of genes can be retrieved from the genome sequence of *Deinococcus radiodurans* `http://www/tigr.org`. These organisms appear to branch slightly earlier in the bacterial bush (Pace, 1997). Also, it is interesting to note than in the recently sequenced genome of the cyanobacterium *Synechocystis* sp., two *gyrA* sequences (one more similar to its Topo IV homologue *parC*) but only one *gyrB* can be detected. The C-terminal end of GyrA is responsible for the gyrase activity, and its removal converts the gyrase in a relaxing Topo IV-like enzyme (Kampranis and Maxwell, 1996). This could suggest that a duplication event took place just before, or during, the differentiation of cyanobacteria, and that in these organisms GyrB could interact with both GyrA-like and ParC-like proteins to perform two already individualised functions (GyrB + GyrA, a gyrase to control overall levels of supercoiling, and GyrB + ParC-like, to decatenate chromosomes). Later in the bacterial branch, a second duplication event involving *gyrB* could have led to the complete independence of both enzymatic functions and to a higher versatility for the colonisation of mesophilic and psychrophilic environments. Interestingly, many of the bacteria with only one putative complete set of type II topoisomerases, can be quite thermophilic. *Thermus* spp. can grow up to 85 °C, and cyanobacteria and photosynthetic bacteria up to 73 °C (Brock, 1986).

Second, the only type II topoisomerases present in early branching bacteria and archaea belong to different phylogenetic families. However, the respective B subunits of Topo VI and gyrase still exhibit some homology (Bergerat *et al.*, 1997). This could indicate that at the time of the prokaryotic ancestor a primitive type II topoisomerase activity existed, involving the interaction of a proto-B subunit with other DNA-interacting proteins able to cut and/or ligate DNA. The speciation of Archaea and Bacteria implied also the speciation of both topoisomerase families (IIA and IIB). This speculation may make sense in the context of a primitive DNA replication system at the time of the cenancestor (Edgell and Doolittle, 1997; Olsen and Woese, 1997).

In the archaeal branch, halophiles have most probably imported a gyrase from bacteria (likely Gram-positive). The genes branch in the middle of bacterial ones (Forterre *et al.*, 1994), and haloarchaea seem to have acquired other bacterial genes in this way (Horne *et al.*, 1988; Altekar and Rajopalan, 1990). Indeed, horizontal import of genes may have had a profound impact in archaeal evolution (Hilario and Gogarten, 1993; Gogarten 1994). Gyrase import might have taken place earlier in euryarchaea, since *Archaeoglobus fulgidus* already contains *gyr* homologues, although the high similarity with the *Thermotoga* gyrase would rather suggest an independent horizontal transfer. Indeed, both genera occupy identical biotopes, being frequently co-isolated (L'Haridon *et al.*, 1995).

13.3.2 A thermophilic ancestor

The generation of overall linking excess or deficit in DNA genomes may be essential for temperature adaptation. Hyperthermophiles utilise reverse gyrase, and mesophiles use gyrase or histone-mediated mechanisms. Mesophily appears to be an adaptation at least as complicated as hyperthermophily in terms of topoisomerase requirements. Reverse gyrase likely evolved earlier than (or at least at the same time as) gyrase or the 'eukaryotic' mechanisms to control overall supercoiling. One appealing possibility is that the ancestor was a thermophile, still devoid of supercoil-introducing activities, thriving between 60 °C and 80 °C, in which the optimal global DNA structure was generated spontaneously. This configuration would be the most energetically favoured at that temperature, corresponding to a compromise between stability and melting capability.

Two theoretical possibilities can be envisioned for the further evolution of this thermophile (Figure 13.4). Either RG evolved in the archaeal lineage leading to hyperthermophily, and was subsequently horizontally transferred to bacteria thriving in similar or nearby environments, or RG evolved prior to the diversification of both domains, implying an already hyperthermophilic ancestor. In the latter case, the bacterial RG would be a remnant of early prokaryotic history.

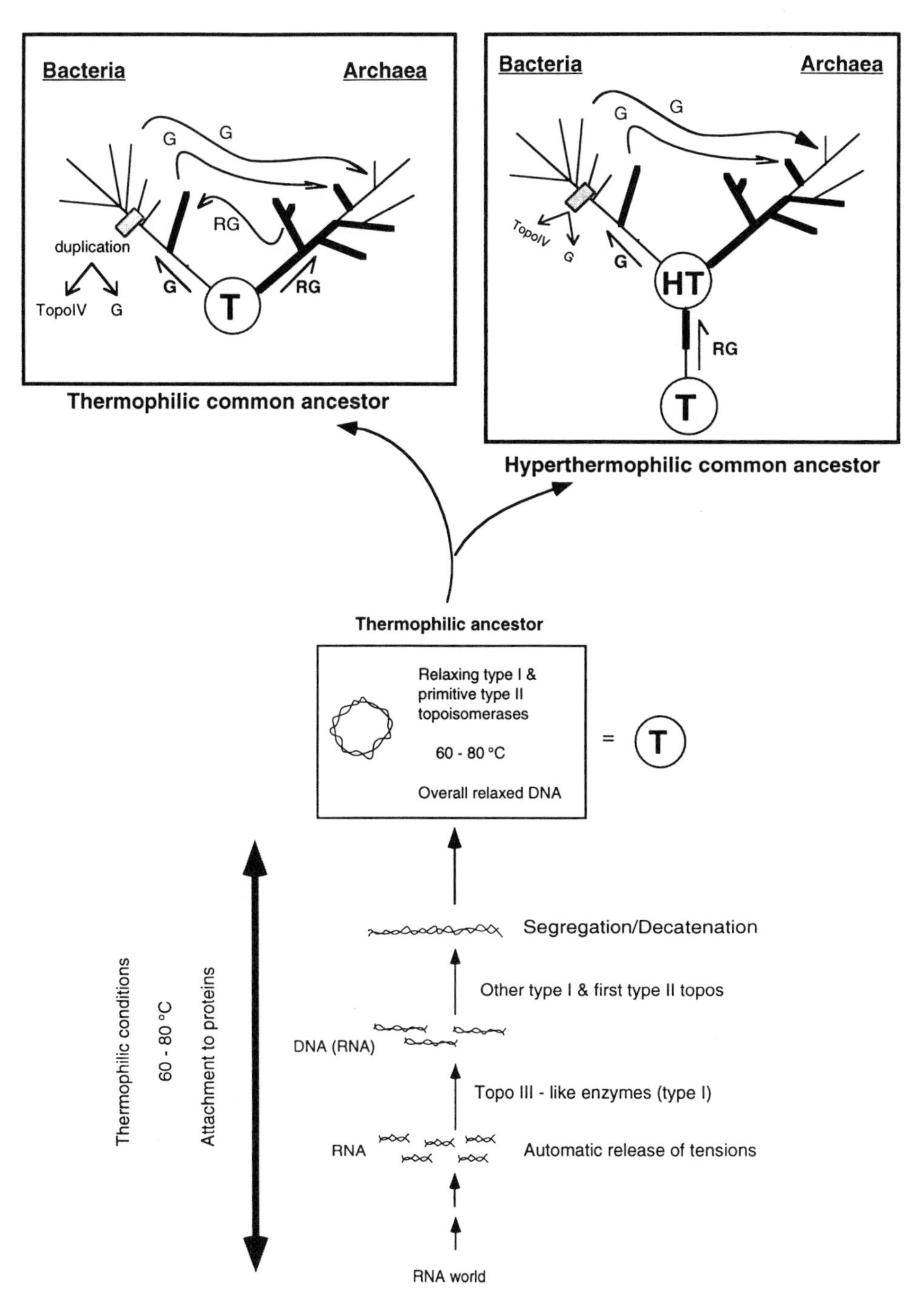

Figure 13.4 Possibilities of prokaryotic divergence and DNA topoisomerase evolution from a thermophilic ancestor. The common prokaryotic ancestor was either a thermophile (T, 60–80 °C), or a hyperthermophile (HT) endowed with a reverse gyrase (RG). G, gyrase. Sinuous arrows indicate horizontal transfer events. Bold lines correspond to hyperthermophilic lineages in idealised phylogenetic trees. At the bottom are shown putative coevolutionary steps of genomes and topoisomerase function.

13.4 Conclusions

I have speculated about the idea of an optimal global DNA configuration, energetically favoured and spontaneously generated under thermophilic (60–80 °C) conditions, that has been maintained throughout evolution. The development of mechanisms for its homeostatic control in hyperthermophiles and mesophiles, as a way to compensate for the physical effects of thermal fluctuations, would therefore have been necessary for evolution from a thermophilic ancestor. The evolution of supercoil-introducing DNA topoisomerases, namely reverse gyrase and gyrase (especially after its duplication in the bacterial line), would have played a central role in the adaptation to hyperthermophily and mesophily, respectively, and their distributions in the phylogenetic tree provide valuable information on early prokaryotic diversification.

Molecular ecology, allowing the identification of many unculturable bacteria and archaea, in combination with comparative genomics should allow us to complete the picture of topoisomerase occurrence to test these ideas. Comparing *Thermotoga* and other putative bacterial RG sequences with their archaeal counterparts might help to discriminate between a horizontal transfer between domains or a common hyperthermophilic ancestor. The analysis of the topoisomerase complement of psychrophilic crenarchaea to see whether they have acquired a gyrase by horizontal transfer, or developed alternative mechanisms to generate DNA linking deficit, or the study of the hyperthermophilic 'Korarchaeota' (Pace, 1997), would be most helpful in answering these 'hot' questions in early evolution.

13.5 Summary

Since a specific DNA geometry is essential for biological function, mechanisms involved in structural DNA homeostasis must have evolved simultaneously with DNA-based genomes. Cells can modulate DNA structure by adjusting DNA topology thanks to DNA topoisomerases, which are able to cut and religate nucleic acid strands, thus solving topological problems. Valuable information in early evolution could therefore be retrieved from studies of the function and phylogenetic distribution of these universal enzymes (four different families arranged into mechanistic types I and II). While in extant organisms DNA is negatively supercoiled in mesophiles belonging to the three domains of life, Archaea, Bacteria and Eukarya, plasmid DNA in hyperthermophilic (optimal growth from 80 °C up) archaea appears to vary from relaxed to positively supercoiled. This seems to be linked to the specific presence of reverse gyrase (type I) in hyperthermophiles, which introduces positive supercoils into DNA. In contrast, two different mechanisms are known to generate global levels of DNA negative supercoiling, suggesting that adaptation to mesophily took place at least twice in evolution. Universal to Bacteria, DNA gyrase (type II) introduces negative supercoiling, although most probably in early-branching bacteria gyrase is mainly devoted to chromosome decatenation upon replication, being the only type II topoisomerase present in the cell. Later in evolution, a second type II enzyme appeared by duplication in the bacterial branch, allowing functional speciation and, possibly, an improved adaptability to mesophily. I speculate that before domain diversification a thermophilic ancestor (60–80 °C) existed which was endowed only with relaxing topoisomerases and whose optimal DNA structure was generated spontaneously at that temperature. Evolution of DNA supercoil-introducing activities, namely reverse gyrase and gyrase, would have facilitated adaptation to hyperthermophily and mesophily in Archaea and Bacteria, respectively. If reverse gyrase evolved before domain diversification (if it is not a horizontally acquired character in hyperthermophilic bacteria), the last common ancestor or cenancestor would have already been a hyperthermophile.

Acknowledgements

I am most grateful to O. Guipaud, D. Moreira, D. Musgrave and P. Forterre for helpful discussions and critiques. The *Archaeoglobus fulgidus*, *Thermotoga maritima*, *Deinococcus radiodurans* and *Methanobacterium thermoautotrophicum* genome sequences were available from The Institute of Genomic Research (TIGR), and Genome Therapeutics Corporation (GTC), in the frame of the US DOE Microbial Genome Project.

References

ALTEKAR, W. and RAJAGOPALAN, R. (1990) Ribulose bisphosphate carboxylase activity in halophilic Archaebacteria. *Arch. Microbiol.*, **153**, 169–174.

BALDAUF, S. L., PALMER, J. P. and DOOLITTLE, W. F. (1996) The root of the universal tree and the origin of eukaryotes based on elongation factor phylogeny. *Proc. Natl Acad. Sci. USA*, **93**, 7749–7754.

BALKE, V. L. and GRALLA, J. D. (1987) Changes in linking number of supercoiled DNA accompany growth transitions in *Escherichia coli*. *J. Bacteriol.*, **169**, 4499–4506.

BAROSS, J. A. and HOFFMAN, S. E. (1985) Submarine hydrothermal vents and associated gradient environments as sites for the origin and evolution of life. *Origins of Life*, **15**, 327–45.

BATES, A. D. and MAXWELL, A. (1993) DNA supercoiling. In *DNA Topology*, ed. D. Rickwood, 'In focus' series, pp. 17–55 (Oxford: IRL Press).

BERGERAT, A., GADELLE, D. and FORTERRE, P. (1994) Purification of a DNA topoisomerase II from the hyperthermophilic archaeon *Sulfolobus shibatae*. *J. Biol. Chem.*, **269**, 27663–27669.

BERGERAT, A., DE MASSY, B., GADELLE, D., VAROUTAS, P. C., NICOLAS, A. and FORTERRE, P. (1997) An atypical topoisomerase II from archaea with implications for meiotic recombination. *Nature*, **386**, 414–417.

BROCK, T. D. (1986) Introduction: an overview of the thermophiles. In *Thermophiles*, ed. T. D. Brock, pp. 1–16 (New York: Wiley-Interscience).

BROWN, J. R. and DOOLITTLE, W. F. (1995) Root of the universal tree of life based on ancient aminoacyl-tRNA synthetase gene. *Proc. Natl Acad. Sci. USA*, **92**, 2441–2445.

BULT, C. J., WHITE, O., OLSEN, G. J. *et al.* (1996) Complete genome sequence of the methanogenic archaeon, *Methanococcus jannaschii*. *Science*, **273**, 1058–1073.

CHAMPOUX, J. J. and DULBECCO, R. (1972) An activity from mammalian cells that untwists superhelical DNA – a possible swivel for DNA replication (polyoma–ethydium bromide–mouse-embryo cells dye binding assay). *Proc. Natl Acad. Sci. USA*, **69**, 143–146.

CHARBONNIER, F. and FORTERRE, P. (1994) Comparison of plasmid DNA topology among mesophilic and thermophilic eubacteria and archaeobacteria. *J. Bacteriol.*, **176**, 1251–1259.

CONFALONIERI, F., ELIE, C., NADAL, M., BOUTHIER DE LA TOUR, C., FORTERRE, P. and DUGUET, M. (1993) Reverse gyrase: a helicase-like domain and a type I topoisomerase in the same polypeptide. *Proc. Natl Acad. Sci. USA*, **90**, 4753–4757.

DIGATE, R. J. and MARIANS, K. J. (1992) *Escherichia coli* topoisomerase III-catalysed cleavage of RNA. *J. Biol. Chem.*, **267**, 20532–20535.

DOOLITTLE, W. F. and BROWN, J. R. (1994) Tempo, mode, the progenote and the universal root, *Proc. Natl Acad. Sci. USA*, **91**, 6721–6728.

DORMAN, C. J. (1995) DNA topology and the global control of bacterial gene expression: implications for the regulation of virulence gene expression. *Microbiology*, **141**, 1271–1280.

DRLICA, K. (1992) Control of bacterial DNA supercoiling. *Mol. Microbiol.*, **6**, 425–433.

DUGUET, M. (1995) Reverse gyrase. In *Nucleic Acids and Molecular Biology*, ed. F. Eckstein and D. M. J. Lilley, vol. 9, pp. 84–114 (Berlin: Springer-Verlag).

EDGELL, D. R. and DOOLITTLE, W. F. (1997) Archaea and the origin(s) of DNA replication proteins. *Cell*, **89**, 995–998.

FITZ-GIBBON, S., CHOI, A. J., MILLER, J. M. *et al.* (1997) A fosmid-based genomic map and identification of 474 genes of the hyperthermophilic archaeon *Pyrobaculum aerophylum. Extremophiles*, **1**, 36–51.

FORTERRE, P. (1995) Thermoreduction, a hypothesis for the origin of prokaryotes. *C. R. Acad. Sci. Paris*, **318**, 415–422.

FORTERRE, P. (1996) A hot topic: the origin of hyperthermophiles. *Cell*, **85**, 789–792.

FORTERRE, P. and ELIE, C. (1993) Chromosome structure, DNA topoisomerases, and DNA polymerases in archaebacteria (archaea). In *The Biochemistry of Archaea*, ed. M. Kates, D. Kushner and A. T. Matheson, pp. 325–366 (Amsterdam: Elsevier).

FORTERRE, P., MIRAMBEAU, G., JAXEL, C., NADAL, M. and DUGUET, M. (1985) High positive supercoiling *in vitro* catalysed by an ATP and polyethylene glycol-stimulated topoisomerase from *Sulfolobus acidocaldarius. EMBO J.*, **4**, 2123–2128.

FORTERRE, P., BENACHENOU-LAHFA, N., CONFALONIERI, F., DUGUET, M., ELIE, C. and LABEDAN, B. (1993) The nature of the last universal ancestor and the root of the tree of life, still open questions. *Biosystems*, **28**, 15–32.

FORTERRE, P., BERGERAT, A., GADELLE, D. *et al.* (1994) Evolution of DNA topoisomerases and DNA polymerases: a perspective from Archaea. *System. Appl. Microbiol.*, **16**, 746–758.

FORTERRE, P., CONFALONIERI, F., CHARBONNIER, F. and DUGUET, M. (1995) Speculations on the origin of life and thermophily: review of available information on reverse gyrase suggests that hyperthermophilic procaryotes are not so primitive. *Origins of Life and Evolution of the Biosphere*, **25**, 235–249.

FORTERRE, P., BERGERAT, A. and LÓPEZ-GARCÍA, P. (1996) The unique DNA topology and DNA topoisomerases of hyperthermophilic archaea. FEMS *Microbiol. Rev.*, **18**, 237–248.

GELLERT, M., MIZUUCHI, K., O'DEA, M. H. and NASH, H. A. (1976) DNA gyrase: an enzyme that introduces superhelical turns into DNA. *Proc. Natl Acad. Sci. USA*, **73**, 3872–3876.

GERMOND, J. E., HIRT, B., OUDET, P., GROSS-BELLARD, M. and CHAMBON, P. (1975) Folding of the DNA double helix in chromatin-like structures from simian virus 40. *Proc. Natl Acad. Sci. USA*, **72**, 1843–1846.

GOGARTEN, P. (1994) Which is the most conserved group of proteins? Homology-orthology, paralogy, xenology, and the fusion of independent lineages. *J. Mol. Evol.*, **39**, 541–543.

GOLDSTEIN, E. and DRLICA, K. (1984) Regulation of bacterial DNA supercoiling: plasmid linking numbers vary with growth temperature. *Proc. Natl Acad. Sci. USA*, **81**, 4046–4050.

GRAYLING, R. A., SANDMAN, K. and REEVE, J. N. (1996) Histones and chromatin structure in hyperthermophilic Archaea. *FEMS Microbiol. Rev.*, **18**, 203–213.

GUIPAUD, O., LABEDAN, B. and FORTERRE, P. (1996) A *gyr*B-like gene from the hyperthermophilic bacterion *Thermotoga maritima. Gene*, **174**, 121–28.

GUIPAUD, O., MARGUET, E., KNOLL, K. M., BOUTHIER DE LA TOUR, C. and FORTERRE, P. (1997) Both DNA gyrase and reverse gyrase are present in the hyperthermophilic bacterium *Thermotoga maritima. Proc. Natl Acad. Sci. USA*, **94**, 10606–10611.

HIGGINS, C. F., DORMAN, C. J. and NI BHRIAN, N. (1990) Environmental influences on DNA supercoiling: a novel mechanism for the regulation of gene expression. In *The Bacterial Chromosome*, ed. K. Drlica and M. Riley, pp. 421–432 (Washington DC: ASM Press).

HILARIO, E. and GOGARTEN, J. P. (1993) Horizontal transfer of ATPase genes – the tree of life becomes a net of life. *Biosystems*, **31**, 111–119.

HSIEH, L. S., BURGER, M. and DRLICA, K. (1991a) Bacterial DNA supercoiling and [ATP]/[ADP] ratio: changes associated with a transition to anaerobic growth. *J. Mol. Biol.*, **219**, 443–450.

HSIEH, L. S., ROUVIÈRE-YANIV, J. and DRLICA, K. (1991b) Bacterial DNA supercoiling and [ATP]/[ADP] ratio: changes associated with salt shock. *J. Bacteriol.*, **173**, 3914–3917.

HOLMES, M. L. and DYALL-SMITH, M. L. (1991) Mutations in DNA gyrase result in novobiocin resistance in halophilic archaebacteria. *J. Bacteriol.*, **173**, 642–648.

HORN, M., ENGLERT, C. and PFEIFER, F. (1988) Two genes encoding gar vacuole proteins in *Halobacterium halobium*. *Mol. Gen. Genet.*, **213**, 459–464.

HUANG, W. M. (1996) Bacterial diversity based on type II DNA topoisomerase genes. *Annu. Rev. Genet.*, **30**, 79–107.

JENSEN, P. R., LOMAN, L., PETRA, B., VAN DER WEIJDEN, C. and WESTERHOFF, H. (1995) Energy buffering of DNA structure fails when *Escherichia coli* runs out of substrate. *J. Bacteriol.*, **177**, 3420–3426.

KAMPRANIS, S. C. and MAXWELL, A. (1996) Conversion of DNA gyrase into a conventional type II topoisomerase. *Proc. Natl Acad. Sci. USA*, **93**, 14416–14421.

KAREM, K. and FOSTER, J. W. (1993) The influence of DNA topology on the environmental regulation of a pH-regulated locus in *Salmonella typhimurium*. *Mol. Microbiol.*, **10**, 75–86.

KATO, J., NISHIMURA, Y., IMAMURA, R., NIKI, H., HIRAGA, S. and SUZUKI, H. (1990) New topoisomerase essential for chromosome segregation in *E. coli*. *Cell*, **63**, 393–404.

KIKUCHI, A. and ASAI, K. (1984) Reverse gyrase – a topoisomerase which introduces positive superhelical turns into DNA. *Nature*, **309**, 667–681.

KOZYAVKIN, A., KRAH, R., GELLERT, M., LAKE, J. A., STETTER, K. O. and SLESAREV, A. I. (1994) A reverse gyrase with an unusual structure. A type-I DNA topoisomerase from the hyperthermophile *Methanopyrus kandleri* is a 2-subunit protein. *J. Biol. Chem.*, **269**, 11081–11089.

KRAH, R., KOZYAVKIN, S. A., SLESAREV, A. I. and GELLERT, M. A. (1996) Two-subunit type I topoisomerase (reverse gyrase) from a hyperthermophilic methanogen. *Proc. Natl Acad. Sci. USA*, **93**. 106–110.

LAZCANO, A. and MILLER, S. L. (1996) The origin and early evolution of life: prebiotic chemistry, the pre-RNA world, and time. *Cell*, **85**, 793–798.

L'HARIDON, S., REYSENBACH, A.-L., GLÉNAT, P., PRIEUR, D. and JEANTHON, C. (1995) Hot subterranean biosphere in a continental oil reservoir. *Nature*, **377**, 223–224.

LÓPEZ-GARCÍA, P. and FORTERRE, P. (1997) DNA topology in hyperthermophilic archaea: reference states and their variation with growth phase, growth temperature, and temperature stresses. *Mol. Microbiol.*, **23**, 1267–1279.

LÓPEZ-GARCÍA, P., ANTÓN, J., ABAD, J. P. and AMILS, R. (1994) Halobacterial megaplasmids are negatively supercoiled. *Mol. Microbiol.*, **11**, 421–427.

LUKOMSKI, S. and WELLS, R. D. (1994) Left-handed Z-DNA and *in vivo* supercoil density in the *Escherichia coli* chromosome. *Proc. Natl Acad. Sci. USA*, **91**, 9980–9984.

LUTTINGER, A. (1995) The twisted 'life' of DNA in the cell: bacterial topoisomerases. *Mol. Microbiol.*, **15**, 601–606.

MILLER, S. L. and LAZCANO, A. (1995) The origin of life – did it occur at high temperatures? *J. Mol. Evol.*, **41**, 689–692.

MILLER, W. G. and SIMONS, R. W. (1993) Chromosomal supercoiling in *Escherichia coli*. *Mol. Microbiol.*, **10**, 675–684.

MOJICA, F. J. M., CHARBONNIER, F., JUEZ, G., RODRÍGUEZ-VALERA, F. and FORTERRE, P. (1994) Effects of salt and temperature on plasmid topology in the halophilic archaeon *Haloferax volcanii*. *J. Bacteriol.*, **176**, 4966–4973.

MUSGRAVE, D. R., SANDMAN, K. M. and REEVE, J. N. (1991) DNA binding by the archaeal histone HMf results in positive supercoiling. *Proc. Natl Acad. Sci. USA*, **88**, 10397–10401.

NISBET, E. G. (1985) The geological setting of the earliest life forms. *J. Mol. Evol.*, **21**, 289.

OLSEN, G. J. and WOESE, C. R. (1997) Archaeal genomics: an overview. *Cell*, **89**, 991–994.

PACE, N. R. (1991) Origin of life – facing up to the physical setting. *Cell*, **65**, 531–533.

(1997) A molecular view of microbial diversity and the biosphere. *Science*, **276**, 734–740.

REEVE, J. N., SANDMAN, K. and DANIELS, C. (1997) Archaeal histones, nucleosomes, and transcription initiation. *Cell*, **89**, 999–1002.

SAAVEDRA, R. A. and HUBERMAN, J. A. (1986) Both DNA topoisomerases I and II relax 2 μm plasmid DNA in living yeast cells. *Cell*, **45**, 65–70.

SHOCK, E. L. (1996) Hydrothermal systems as environments for the emergence of life. In *Evolution of Hydrothermal Systems on Earth (and Mars?)*, Ciba Foundation Symposium 202, pp. 40–60 (Chichester: Wiley).

SIOUD, M., POSSOT, O., ELIE, C., SIBOLD, L. and FORTERRE, P. (1988) Coumarin and quinolone action in archaebacteria: evidence for the presence of a DNA gyrase-like enzyme. *J. Bacteriol.*, **170**, 946–953.

SLESAREV, A. I., ZAITZEV, D. A., KOPYLOV, V. M., STETTER, K. O. and KOZYAVKIN, S. A. (1991) DNA topoisomerase III from extremely thermophilic archaebacteria: ATP-independent type I topoisomerase from *Desulfurococcus amylolyticus* drives extensive unwinding of closed circular DNA at high temperature. *J. Biol. Chem.*, **266**, 12321–12328.

SLESAREV, A., STETTER, K., LAKE, J., GELLERT, M., KRAH, R. and KOZYAVKIN, S. (1993) DNA topoisomerase V is a relative of eukaryotic topoisomerase I from a hyperthermophilic prokaryote. *Nature*, **364**, 735–737.

SRIVENUGOPAL, K. S., LOCKSHON, D. and MORRIS, D. R. (1984) *Escherichia coli* DNA topoisomerase III: purification and characterization of a new type I enzyme. *Biochemistry*, **23**, 1899–1906.

STARICH, M. R., SANDMAN, K., REEVE, J. N. and SUMMERS, M. F. (1996) NMR structure of HMf from the hyperthermophile *Methanothermus fervidus* confirms that this archaeal protein is a histone. *J. Mol. Biol.*, **255**, 187–203.

STETTER, K. O. (1996) Hyperthermophilic prokaryotes. *FEMS Microbiol. Rev.*, **18**, 149–58.

WÄCHTERSHÄUSER, G. (1988) Pyrite formation, the first energy source for life: a hypothesis, *System. Appl. Microbiol.*, **10**, 207–210.

WANG, J. C. (1971) Interaction between DNA and an *Escherichia coli* protein omega. *J. Mol. Biol.*, **55**, 523–533.

WANG, J. C. (1987) Recent studies of DNA topoisomerases. *Biochim. Biophys. Acta*, **909**, 1–9.

WANG, J. C. (1996) DNA topoisomerases. *Annu. Rev. Biochem.*, **65**, 635–692.

WANG, J. C. and LYNCH, A. S. (1993) Transcription and DNA supercoiling. *Curr. Opin. Genet. Dev.*, **3**, 764–768.

WANG, J. C., CARON, P. R. and KIM, R. A. (1990) The role of DNA topoisomerases in recombination and genome stability: a double-edged sword? *Cell*, **62**, 403–406.

WANG, H., DIGATE, R. J. and SEEMAN, N. C. (1996) An RNA topoisomerase. *Proc. Natl Acad. Sci. USA*, **93**, 9477–9482.

WOESE, C. R. (1987) Bacterial evolution. *Microbiol. Rev.*, **51**, 221–271.

WOESE, C. R., KANDLER, O. and WHEELIS, M. L. (1990) Towards a natural system of organisms: proposal for the domains Archaea, Bacteria and Eucarya. *Proc. Natl Acad. Sci. USA*, **87**, 4576–4579.

WORCEL, A. and BURGI, E. (1972) On the structure of the folded chromosome of *Escherichia coli*. *J. Mol. Biol.*, **71**, 127–147.

ZILLIG, W. (1991) Comparative biochemistry of Archaea and Bacteria. *Curr. Opin. Genet. Dev.*, **1**, 544–551.

14

Aminoacyl-tRNA Synthetases: Evolution of a Troubled Family

JAMES R. BROWN

Department of Bioinformatics, SmithKline Beecham Pharmaceuticals, Collegeville, Pennsylvania, USA

The evolution of aminoacyl-tRNA synthetases is particularly interesting for two reasons. First, aminoacyl-tRNA synthetases function in translation by attaching an amino acid to its cognate tRNA. Therefore, the development of the genetic code and the evolution of aminoacyl-tRNA synthetases are thought to be strongly linked. Second, aminoacyl-tRNA synthetases provide one of the best examples of a large multiple protein family whose radiation likely occurred before the emergence of the first single-cell organisms. An understanding of the evolutionary tempo and mode of this protein family is important to a general understanding of how genomes evolve. This latter aspect of aminoacyl-tRNA synthetase evolution will be the focus of this chapter.

14.1 Aminoacyl-tRNA synthetase structure and function

Aminoacyl-tRNA synthetases have been studied intensively with respect to their mode of tRNA recognition and esterification reactions. Crystallographic structures have been determined for several synthetases. These topics have been reviewed extensively elsewhere (Carter, 1993; Cavarelli and Moras, 1993; Moras, 1992) and will only be briefly touched upon here.

Aminoacyl-tRNA synthetases catalyse the esterification or charging of a tRNA molecule with a specific amino acid via a two-step reaction. First, the amino acid is activated in the presence of ATP and Mg^{2+}, which results in the attachment of an aminoacyl-adenylate (amino acid-AMP) to the synthetase and the release of PPi. Second, the synthetase aminoacylates the tRNA then releases the amino acid–tRNA complex and AMP (Eriani *et al.*, 1995).

For each of the 20 amino acids there exists a specific aminoacyl-tRNA synthetase, although there are significant exceptions which will be discussed later. The 20 different synthetases are evenly split among two families called class I and class II which differ in terms of primary sequences, crystallographic structures, and tRNA charging sites (Table 14.1; Eriani *et al.*, 1990). Class I synthetases share the amino acid motifs HIGH

Table 14.1 Classes of aminoacyl-tRNA synthetases and some general characteristics

Features	Class I	Class II
Amino acid specificity	Arginine Cysteine Glutamate Glutamine Isoleucine Leucine Methionine Tryptophan Tyrosine Valine	Alanine Asparagine Aspartate Glycine Histidine Lysine Phenylalanine Proline Serine Threonine
Generali Motifs	HIGH KMSKS	1) +G(F/Y)xx(V/L/I)xxPϕϕ 2) +ϕϕxϕxxxFRxE 3) ϕGϕGϕGϕERϕϕϕϕ
Structure of binding site	alternating α-helices and β-sheets (Rossmann fold)	6 antiparallel β-sheets with 3 flanking α-helices
tRNA charging sites	2′-OH	3′-OH (Phensaminoacylates 2′-OH

In the motif row symbols are used to represent variable (x), positively charged (+) and hydrophobic (ϕ) residues.

and KMSKS which are located in a nucleotide-binding fold of alternating α-helices and parallel β-sheets called the Rossmann fold (Moras, 1992; Eriani *et al.*, 1995). Class II synthetases lack these class I motifs and, instead have three different amino acid motifs which are more loosely conserved. The class II binding site is also different, being comprised of 6 antiparallel β-sheets and two long α-helices. The two classes of synthetases acylate the tRNA molecule at different sites; class I attaches the activated amino acid to the 2′-OH, while class II charges the 3′-OH (except for phenylalanyl-tRNA synthetase which aminoacylates the 2′-OH).

Class I and II synthetases are widely considered to be evolutionarily distinct and unrelated, although an attempt was made to show that the enzymes from the two classes might have originated from complementary strands of the same primordial nucleic acid sequence (Rodin and Ohno, 1995). The two aminoacyl-tRNA synthetase subdivisions are not concordant with either amino acid biochemistry, anti-codon assignments in the genetic code, or tRNA sequences (Nicholas and McClain, 1995).

However, within each class, specific aminoacyl-tRNA synthetases appear to be evolutionarily related. Nagel and Doolittle (1991, 1995) showed that, for most aminoacyl-tRNA synthetases, bacterial versions were more closely related to the same type of synthetase from a eukaryote than to any other type of synthetase from the same bacterium. Thus, tyrosyl-tRNA synthetase, or TyrRS (specific synthetases will be abbreviated as *aaRS* where *aa* is the three-letter amino acid code), of *Escherichia coli* is more similar to yeast TyrRS than to TrpRS from *E. coli* (Figure 14.1). On this basis, Nagel and Doolittle (1991) concluded that the divergence of class I and class II type synthetases was more ancient than that separating prokaryotes and eukaryotes.

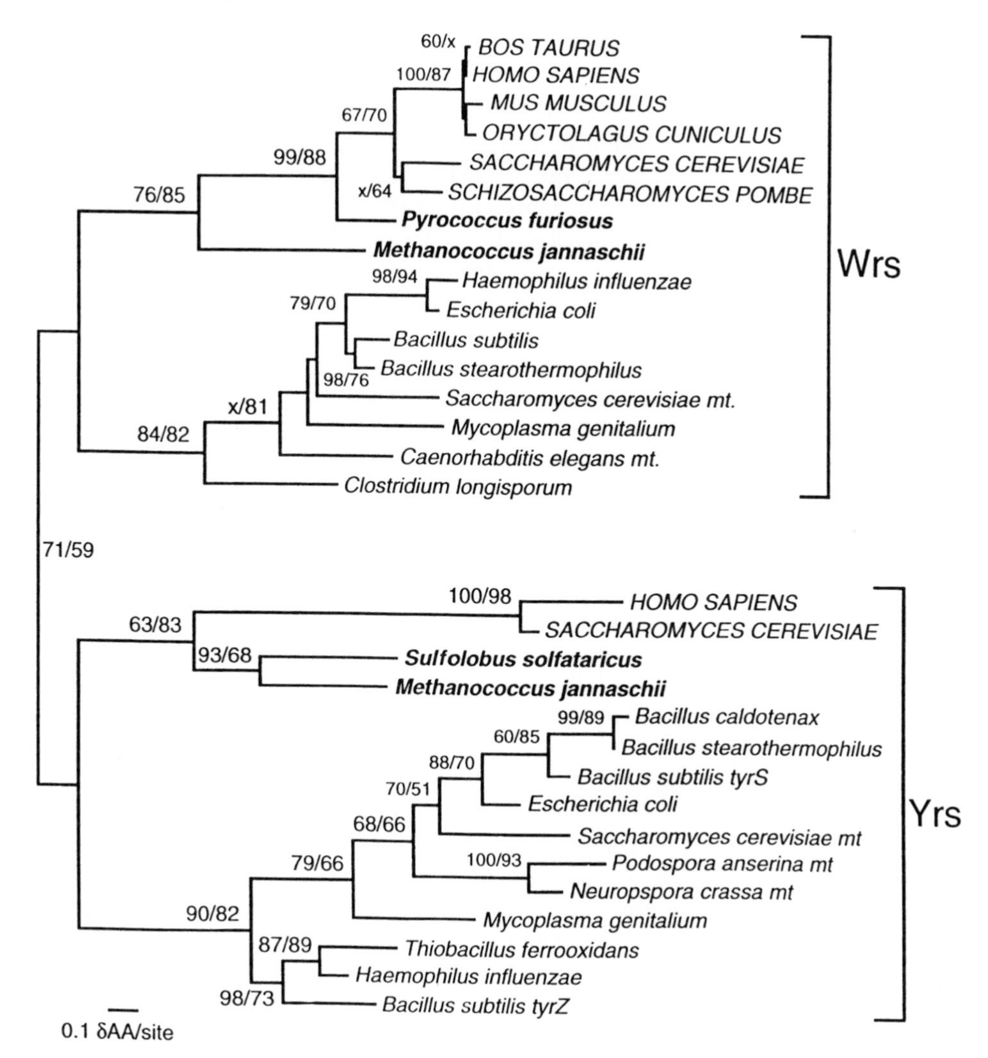

Figure 14.1 Phylogeny of tryptophanyl- and tyrosyl-tRNA synthetases (Wrs and Yrs, respectively). Unless noted otherwise, all phylogenies were constructed using the neighbour-joining method as implemented by the program NEIGHBOR of the PHYLIP 3.57c package (Felsenstein, 1993). The scale bar represents 0.1 expected amino acid replacements per site as estimated by the program PROTDIST using the Dayhoff option. In this figure, the numbers are the percentage occurrence of nodes in 500 bootstrap re-samplings of maximum-parsimony and neighbour-joining analyses, respectively (Brown *et al.*, 1997). In all subsequent figures, numbers represent the percentage occurrence of nodes in 1000 bootstrap re-samplings of neigbour-joining analyses. Values less than 50% are either not shown or reported as 'x'. Species names of Bacteria are in plain type, those of Eukarya in uppercase, and those of Archaea in bold type. Also indicated are eukaryotic nuclear-encoded mitochondria (mt. or mito.) or chloroplast (chlo.) targeted isoforms.

14.2 Rooting the universal tree

Suggestions that aminoacyl-tRNA synthetases are more ancient that their host genomes occurred at a time when the actual definitions *prokaryote* and *eukaryote* were being hotly debated. The impetus behind these arguments was the taxonomic status of the archaebacteria. In the late 1970s, it became apparent from rRNA phylogenies that there was a deep split among the prokaryotes (Fox *et al.*, 1977; Woese and Fox, 1977). The existence of two prokaryotic subdivisions, eubacteria and archaebacteria, was further supported by discoveries of the special nature of the archaebacteria such as their unique membrane structure, and important genetic characteristics they shared with eukaryotes but not eubacteria (reviewed in Brown and Doolittle, 1997). In 1990, Woese, Kandler and Wheelis proposed that the categories of prokaryotes and eukaryotes be replaced by a tripartite domain classification scheme in which there were the Bacteria (eubacteria), the Archaea (archaebacteria) and the Eukarya (eukaryotes). While there was wide acceptance of their re-classification by most archaebacteriologists and many molecular evolutionists, other evolutionary biologists challenged the elevation of archaebacteria (and hence eubacteria) to a taxonomic rank equal that of eukaryotes (reviewed in Brown and Doolittle, 1997). Furthermore, there was the important and perplexing issue of trying to root the universal tree of life.

There are three possible topologies for any universal tree: (1) the first divergence separated Bacteria from a line which was to produce Archaea and eukaryotes; (2) instead, a proto-eukaryotic lineage from a fully prokaryotic (Bacteria and Archaea) clade tree; or (3) the Archaea on the one hand from eukaryotes and Bacteria on the other. However, on the basis of a solitary gene, it is impossible to derive an objective rooting for the universal tree. Typically, the rooting for a particular organism tree, for example all mammalian species, would be determined by including sequence data from a known outgroup species, such as some cold-blooded vertebrates. However, outgroup species are not available for a gene tree consisting of all living organisms unless specific assumptions are made such as the progression of life from a prokaryotic to a eukaryotic cell. Therefore, the branching order of the three domains emerging from their last common ancestor – which Fitch and Upper (1987) called the cenancestor – can only be established by some method unrelated to either outgroup organisms or theories about primitive and advanced states.

In 1989, a solution to this problem using ancient duplicated genes was proposed simultaneously in separate papers by Gogarten and colleagues and Iwabe and colleagues. Their collective reasoning was as follows; although there can be no organism which is an outgroup for a tree relating all organisms, one could root a tree based on the sequences of outgroup genes produced by an early gene duplication. Iwabe *et al.* (1989) applied this concept using elongation factors, a paralogous family of GTP-binding proteins which facilitate the binding of aminoacylated tRNA molecules to the ribosome (EF-Tu in Bacteria and EF-1α in eukaryotes and Archaea) and the translocation of peptidyl-tRNA (EF-G in Bacteria and EF-2 in eukaryotes and Archaea). Gogarten *et al.* (1989) rooted the universal tree using catalytic subunits of V-type (found in Archaea and eukaryotes) and F-type (found in Bacteria) ATPase subunits. Both paralogous gene trees found agreement in a rooting which placed Archaea and eukaryotes as sister groups, hence Bacteria was the outgroup. Woese, Kandler and Wheelis (1990) incorporated these protein rootings in their formulation of three domains of life.

However, the rooting of the universal was still highly controversial. Some analyses of rRNA and elongation factors suggested that the Archaea were not monophyletic, rather one archaeal kingdom, the Crenarchaeota, was more closely related to the Eukarya than the other archaeal kingdom, the Euryarchaeota (Lake, 1988; Rivera and Lake, 1992). Since both Iwabe *et al.* (1989) and Gogarten *et al.* (1989) included only a single archaeal protein homologue in their respective phylogenetic analyses, concerns about the integrity of the domain Archaea could not be addressed. Other criticisms of the rooting focused on the low proportion of homologous amino acid positions between the two groups of elongation factors and the subsequent discovery of bacteria with archaea-like V type ATPases (reviewed in Brown and Doolittle, 1997).

Thus other rootings of the universal tree were needed, and the duplicated nature of the aminoacyl-tRNA synthetases made them obvious candidates. In 1995, Brown and Doolittle derived a universal tree based on archaeal, bacterial and eukaryotic IleRS sequences rooted with bacterial and eukaryotic ValRS and LeuRS sequences (no archaeal versions of the latter two synthetases were available). The most conserved region of IleRS, that between the HIGH and KMSKS motifs, was sequenced from two species of Archaea as well as a lower eukaryote and two deeply branching thermophilic bacteria. Phylogenetic analyses showed that (1) IleRS, LeuRS, and ValRS were monophyletic groups, which partially confirmed the study of Nagel and Doolittle (1991); (2) IleRS sequences from the Archaea, Bacteria, and Eukarya formed separate monophyletic groups as depicted by rRNA trees; and (3) the two closest domains were the Archaea and Eukarya, which agreed with earlier rootings based on different molecules and fewer species. Thus, a canonical view of aminoacyl-tRNA synthetase evolution emerged which was of an ancient family of proteins that diverged before the appearance of contemporary cellular lineages, and that the evolution of this protein family tracked that of the conventional universal tree of life.

In a contrary finding, Ribas de Pouplana *et al.* (1996) proposed that bacterial TrpRS and TyrRS were more similar to each other than to eukaryotic TrpRS and TyrRS. However, subsequent analyses which included new archaeal sequences showed that TrpRS and TyrRS were separate monophyletic groups. Furthermore, within each synthetase cluster, Archaea and eukaryotes were sister groups and the Bacteria were the outgroup which provided further confirmation of earlier rootings (Brown *et al.*, 1997).

14.3 Charging for glutamine, asparagine and lysine are the exceptions

Early biochemical studies had shown that the synthesis of Gln-tRNA^{Gln} and Glu-tRNA^{Glu} could occur by either of two distinct pathways. 'Crown eukaryotes' (animals, plants, and fungi) and *Escherichia coli* have specific GlnRS and GluRS enzymes (Lazzarini and Mehler, 1964; Ravel *et al.*, 1965). GlnRS has been biochemically demonstrated not to exist in Gram-positive bacteria (Lapointe *et al.*, 1986; Breton *et al.*, 1990), cyanobacteria, an α-subdivision proteobacterium *Rhizobium meliloti* (Gagnon *et al.*, 1996), chloroplasts, and mitochondria (Schön *et al.*, 1988), with the exception of *Leishmania tarentolae* mitochondria (Nabholz *et al.*, 1997). In these organisms, Gln-tRNA^{Gln} is synthesised via a two-step process in which tRNA^{Gln} is first misacylated with glutamate by GluRS, then glutamate is converted to glutamine by a specific amidotransferase. Although not confirmed experimentally, this pathway can be inferred to exist in many other bacterial species based on overall sequence similarities among bacterial GluRS gene sequences, the apparent absence of GlnRS gene homologues and, more recently, the simultaneous presence of gene homologues to the Glu-tRNA^{Gln} amidotransferase operon (*gatCAB*) of Gram-positive *Bacillus subtilis* (Curnow *et al.*, 1997). Similarly, GlnRS gene homologues have been found in at least two other bacteria which belong to the γ-proteobacteria subdivision along with *E. coli* (Figure 14.2). Since, the same GluRS charges either tRNA^{Glu} or tRNA^{Gln} with glutamate, the activity of this synthetase has been called nondiscriminatory

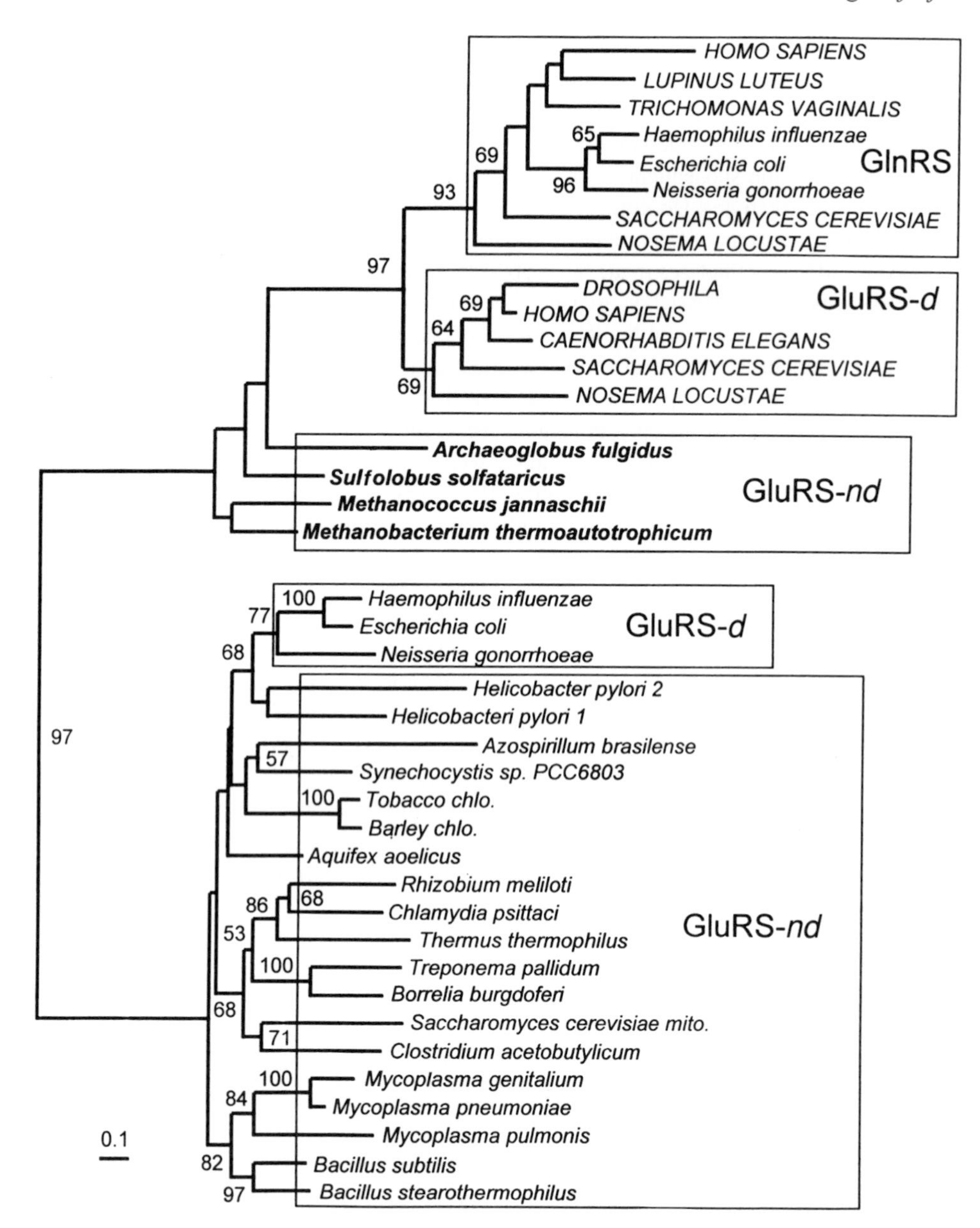

Figure 14.2 Phylogeny of glutaminyl- and glutamyl-tRNA synthetases (GlnRS and GluRS, respectively). Clusters of glutamyl-tRNA synthetases proposed either to discriminate between $tRNA^{Gln}$ and $tRNA^{Glu}$ (GluRS-*d*) or to not discriminate (GluRS-*nd*) are indicated.

(Lamour *et al.*, 1994), or GluRS-*nd*. Conversely, GluRS of eukaryotes and γ-proteobacteria is discriminatory (GluRS-*d*) in that it will aminoacylate $tRNA^{Glu}$ but not $tRNA^{Gln}$.

Eukaryotic GlnRS and GluRS-*d* amino acid sequences are more similar to each other than they are to any bacterial GluRS-*nd*, while γ-proteobacteria GlnRS are most similar to eukaryotic GlnRS. In the absence of archaeal sequences, Lamour *et al.* (1994) and Gagnon *et al.* (1996) suggested that the gene for GlnRS arose from a duplication of a eukaryotic GluRS gene and was subsequently transferred to the γ-proteobacteria. A more complete phylogenetic analysis, which included not only archaeal GluRS sequences but GlnRS and GluRS sequences from two lower eukaryotes as well, suggested that these two events – the duplication leading to separate GluRS and GlnRS genes, and the subsequent transfer of a GlnRS gene from a eukaryote to a γ-proteobacterium – occurred near simultaneously in early eukaryotic evolution (Brown and Doolittle, 1998). Archaea have a GluRS-like synthetase but not GlnRS. However, archaeal GluRS appears to be ancestral to the eukaryotic GlnRS and GluRS-*d* cluster.

Interestingly, *Helicobacter pylori*, which is the closest bacterial outgroup to the γ-proteobacteria, has two highly similar GluRS genes, which suggests a recent gene duplication. Whether both genes are actively transcribed and have GluRS-*nd* activity or whether one copy shows a novel GlnRS activity is unknown. Other duplicated aminoacyl-tRNA synthetase genes include HisRS in *Bacillus subtilis*, *Synechocystis*, and *Aquifex aeolicus*; TyrRS and ThrRS in *B. subtilis*; and SerRS in *Clostridium actebutylicum*. Furthermore, most eukaryotic synthetases are encoded by two separate genes, with one isoform targeted to the mitochondria while the other functions in the cytoplasm.

In a halophilic archaeon, recent evidence was found for the synthesis of Asn-tRNA^{Asn} via transamidation of Asp-tRNA^{Asn} formed by AspRS (Curnow *et al.*, 1996). This alternative pathway might also exist in certain bacteria, namely *H. pylori* (Tomb *et al.*, 1997) and *Aquifex aoelicus* (Deckert *et al.*, 1998), since their completely sequenced genomes also lack a recognisable AsnRS gene.

The genes of aminoacyl-tRNA synthetases charging for lysine and cysteine, as well as asparagine and glutamine were not found in the completely sequenced genomes of the archaea *Methanococcus jannaschii* (Bult *et al.*, 1996), *Methanobacterium thermoautotrophicum* (Smith *et al.*, 1997), and *Archaeoglobus fulgidus* (Klenk *et al.*, 1997). Since the proteins of these organisms clearly use the amino acids charged by these 'missing synthetases', either there is an alternative pathway where one synthetase cannot discriminate between two different tRNA species and a secondary modification of the misacylated amino acid-tRNA complex occurs, or there exists a novel aminoacyl-tRNA synthetase which is unrecognisable at the amino acid sequence level. The latter instance was found to be true for archaeal LysRS. As a member of the class II family, LysRS has been found in all eukaryotes and bacteria, and in the archaeon *Sulfolobus solfataricus* which belongs to the kingdom Crenarchaeota. However, a novel class I-like synthetase was found to catalyse the formation of Lys-tRNA^{Lys} in four species from the other archaeal kingdom, the Euryarchaeota (Ibba *et al.*, 1997), and one group of bacteria, the spirochaetes (Ibba *et al.*, 1997). It was suggested that this class I LysRS originated from the Archaea and that spirochaetes obtained their copy through horizontal gene transfer.

Figure 14.3 shows the phylogeny of class II LysRS, AspRS, and AsnRS based on a multiple alignment of 87 amino acids. The sequence length is short because only residues surrounding the three amino acid motifs characteristic of class II aminoacyl-tRNA synthetases were alignable. Thus many nodes are not well resolved according to bootstrap replicate values. However, four clusters of sequences are strongly supported – LysRS, AsnRS, bacterial AspRS, and archaeal-eukaryotic AspRS. Surprisingly, AspRS is not one cluster, rather bacterial AspRS sequences are equally distant from archaeal-eukaryotic AspRS as they are from LysRS and AsnRS sequences. Moreover, in the LysRS portion of the tree the archaeon *Sulfolobus solfataricus*, is more closely related to bacteria than to eukaryotes. Either the various mechanisms for charging tRNA with Lys, Asn, and Asp were transferred among certain lineages after their descent from the cenancestor or the cenancestor was a complex organism in which multiple pathways and enzymes coexisted only to be later refined or eliminated in specific descendants.

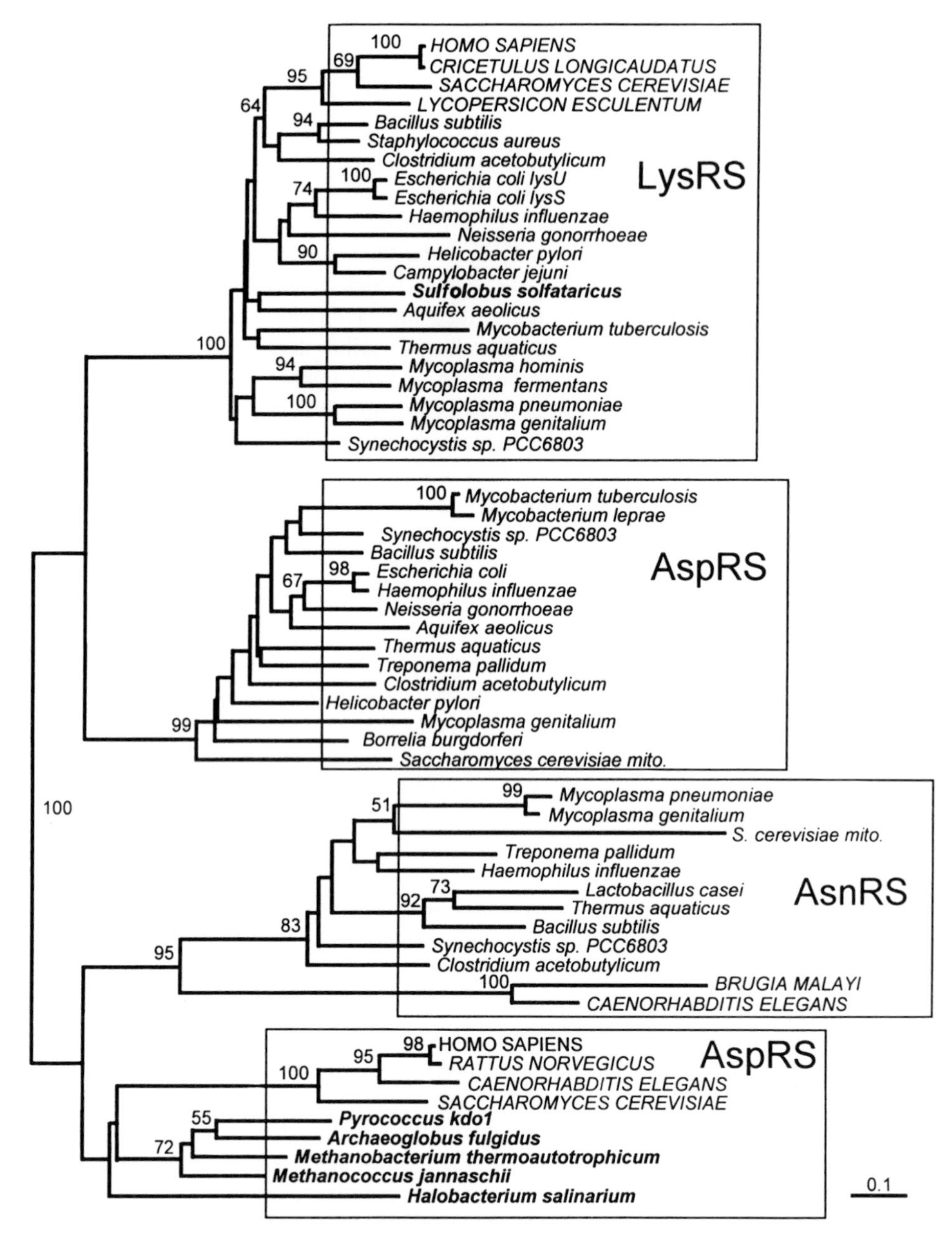

Figure 14.3 Phylogeny of lysyl-, aspartyl-, and asparaginyl-tRNA synthetases (LysRS, AspRS, and AsnRS, respectively).

14.4 Loss and gain of aminoacyl-tRNA synthetases

Is there any discernible pattern in the loss or gain of aminoacyl-tRNA synthetase genes in the context of the universal tree? Furthermore, can the aminoacyl-tRNA synthetase gene complement of the cenancestor genome be reconstructed? In addition to the above-mentioned variations in aminoacyl-tRNA synthetase occurrences, glycyl-tRNA synthetase

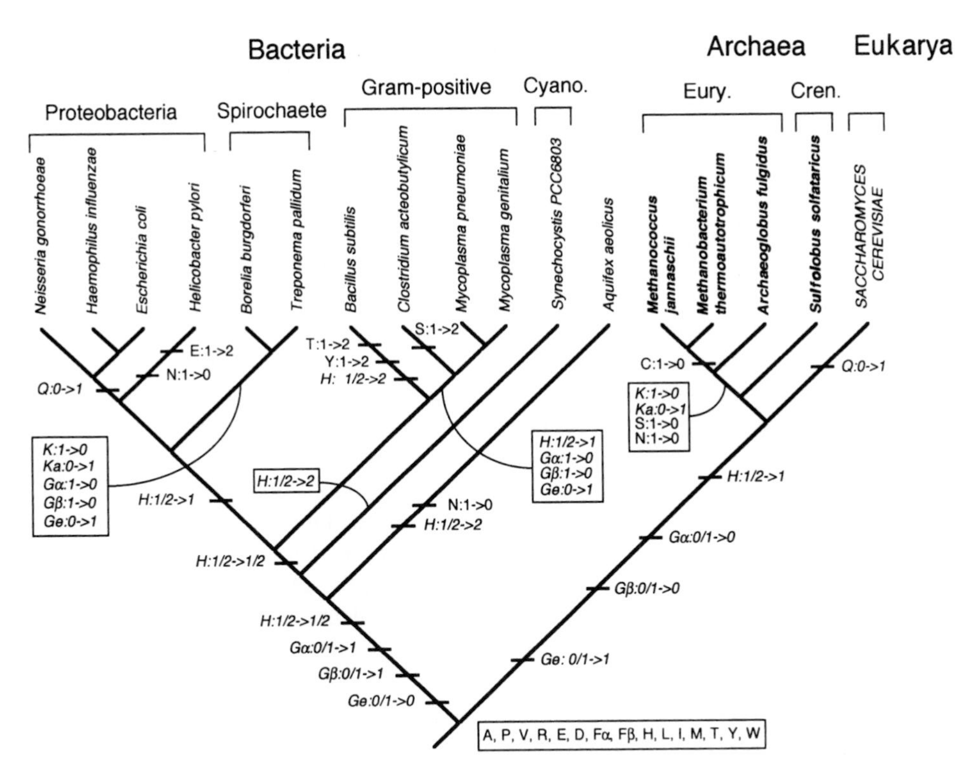

Figure 14.4 Proposed losses (0), gains (1), and duplications (2) of aminoacyl-tRNA synthetase genes overlaid on the canonical universal tree. The tree topology is a composite based on phylogenies derived from duplicated protein families (Gogarten *et al.*, 1989; Iwabe *et al.*, 1989) and rRNA (Olsen *et al.*, 1994). Only shown are those organisms for which complete or nearly complete genomic sequences are known. The presence or absence of a particular aminoacyl-tRNA synthetase was determined by searching each genome with the complete complement of aminoacyl-tRNA synthetases from *Escherichia coli* and *Saccharomyces cerevisiae* using the program BLAST 2.0 (Altschul *et al.*, 1990). Subject open reading frames were considered to be homologous to the query aminoacyl-tRNA synthetase if the *p*(*N*) value was less than 10^{e-10}. Using the program MACCLADE (Madison and Madison, 1992) the presence and absence of aminoacyl-tRNA synthetases were coded as character states with values of 1 and 0, respectively, and forced upon the universal tree topology. Estimated changes in character states and their direction are denoted on the branches. Aminoacyl-tRNA synthetases found in all genomes are listed at the base of the tree. Specific aminoacyl-tRNA synthetases are denoted by the single-letter amino acid code except where Ka is a novel class I LysRS found in some archaea and spirochaetes, and Gα and Gβ are dimeric subunits of GlyRS while Ge is the tetrameric form typical of eukaryotes. Abbreviations are used for cyanobacteria (Cyano.), Euryarchaeota (Eury.) and Crenarchaeota (Cren.).

exhibits either a tetrameric ($\alpha_2\beta_2$) or dimeric (α_2) structure in different species (reviewed in Friest *et al.*, 1996). Dimeric GlyRS occurs in a few Gram-positive (*Mycoplasma* spp., *Clostridium acteobutylicum*, and *Staphylococcus aureus*), and Gram-negative (*Thermus thermophilus* and spirochaetes) bacteria, and all archaea and eukaryotes, while the majority of Gram-positive and Gram-negative bacteria have tetrameric GlyRS.

When overlaid on the conventional universal tree, particular synthetase genes appear to be either lost or gained multiple times in different, dispersed lineages (Figure 14.4). It is somewhat unsatisfactory from the cladist perspective to resort repeatedly to either horizontal gene transfer or, worse, convergent evolution as an explanation for this aminoacyl-tRNA synthetase distribution. At present, no other recourse than horizontal transfer is available to rationalise the appearance of GlnRS in eukaryotes and γ-proteobacteria, GlyRS α_2 in spirochaetes, *Mycoplasma*, *Clostridium*, the Archaea, and Eukarya, and a class I LysRS in the Euryarchaeota and spirochaetes.

Thus far, eukaryotes appear to show the greatest solidarity with respect to aminoacyl-tRNA synthetase complement, with the exception of mitochondrial targeted isoforms which likely originated from an α-proteobacterial endosymbiont. However, the report of a mitochondria-targeted GlnRS from the protist *Leishmania tarentolae* (Nabholz *et al.*, 1997) perhaps suggests that more eukaryotic surprises are to be expected.

However, it is remarkable that so few of the 20 aminoacyl-tRNA synthetases are involved in playing genomic musical chairs. Duplicate HisRS genes are found in a Gram-positive bacterium (*B. subtilis*), a cyanobacterium (*Synechococcus* sp.), and an extreme thermophile (*A. aeolicus*). Discounting eukaryotic mitochondria and cytoplasmic-targeted isoforms, no other synthetase is so often found in duplicate. Other synthetase duplications appear to be lineage specific, such GluRS in *Helicobacter* or TyrRS and ThrRS in *B. subtilis*. Were duplicated HisRS genes the ancestral state in the cenancestor, with one copy lost independently in different organisms, or are independent gene duplications of HisRS favoured because this synthetase is particularly useful for other cellular processes? Do the life cycles of these organisms require two differently regulated HisRS genes?

The aminoacyl-tRNA synthetases that can be eliminated from the genome over long evolutionary time periods without adversely affecting protein synthesis are relatively few, namely GlnRS and AsnRS, with SerRS and CysRS as other possibilities. Again, widely dispersed lineages have similar deficiencies in their aminoacyl-tRNA synthetase complement, which argues for independent losses. However, before any synthetase can discarded, the machinery for the compensatory aminoacylation pathway must be in place and functioning. The evolutionary relationships of known enzymes in these compensatory pathways, such as the genes of the Glu-tRNA^{Gln} amidotransferase operon, have not been studied in depth, yet such information might reveal the mode of acquisition of these alternative aminoacylation mechanisms.

14.5 Up-rooting the universal tree

When Brown and Doolittle (1995) used IleRS, ValRS, and LeuRS sequences to derive a third, independent rooting of the universal tree of life, there were no archaeal sequence data available for the latter two synthetases. In their rooted tree, the branch leading to the LeuRS and IleRS clusters bisected ValRS sequences into two groups, one of eukaryotes and *E. coli* while the other had several Gram-positive bacteria. Mitochondrial and cytoplasmic ValRS are unusual in that both isoforms encoded by the same gene (Martindale *et al.*, 1989). The phyletic position of eukaryotic ValRS sequences, which included one from an amitochondriate protist, with *E. coli* suggests that the nuclear gene for ValRS originated from a bacterial endosymbiont.

Reconstruction of the ValRS phylogenetic tree using sequences from a wider range of species, including the Archaea, support the notion that eukaryotic ValRS originated from bacteria (Figure 14.5). When rooted by IleRS, there are two major ValRS clusters; an archaeal group and a bacterial-eukaryotic group. Although eukaryotes are monophyletic, their clade is imbedded in a larger bacterial cluster (Hashimoto *et al.*, 1998). The relative

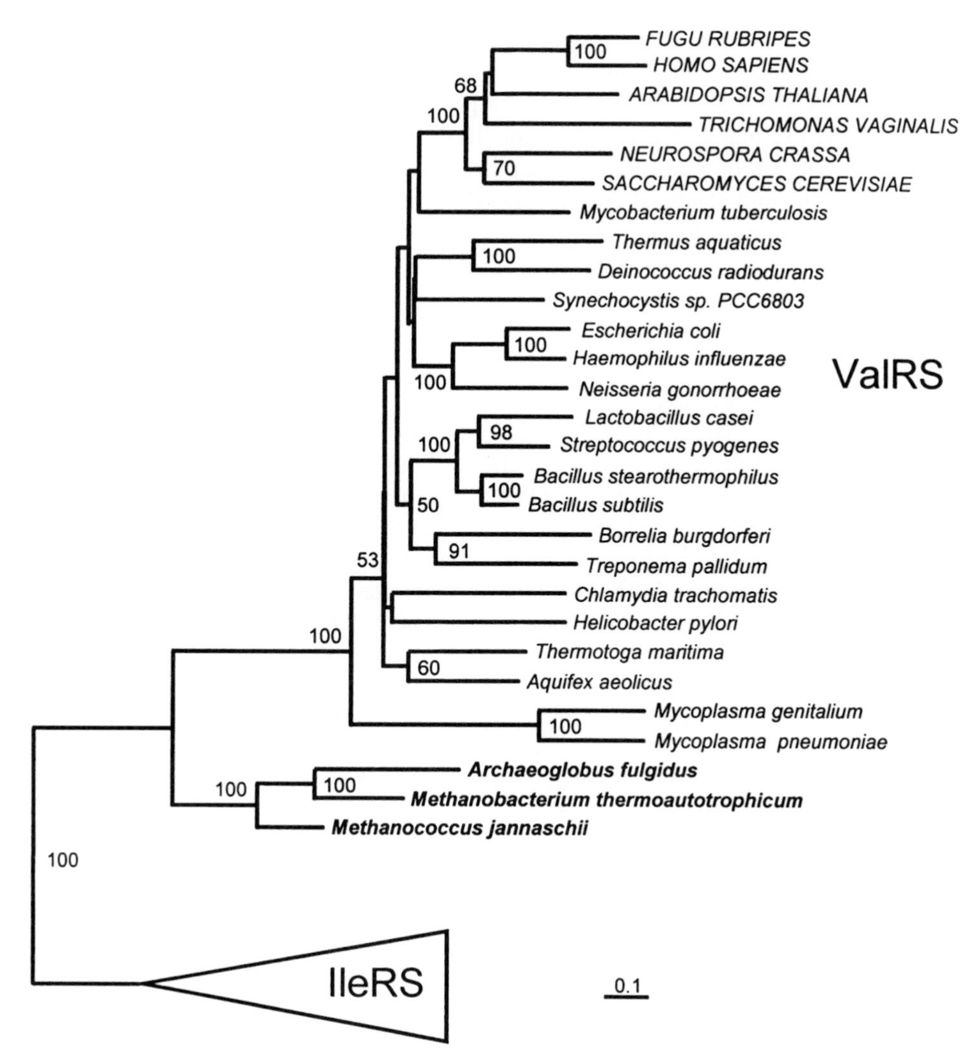

Figure 14.5 Phylogeny of valyl-tRNA synthetases (ValRS) rooted by 10 isoleucyl-tRNA synthetases (IleRS).

branching orders among the Bacteria are not well resolved and no α-proteobacterial ValRS sequences are available, so it is not possible to confirm whether eukaryotic ValRS specifically originated from the mitochondrial endosymbiont.

As sequence databases grow, it becomes possible to derive new reciprocally rooted universal trees as well as to revisit past phylogenetic analyses with a wider range of species. Baldauf *et al.* (1996) recently completed a phylogenetic analyses of an expanded elongation factor dataset and still found strong support for the rooting proposed by Iwabe *et al.* (1989). Hilario and Gogarten (1993) suggested that the lateral gene transfer of an ATPase subunit gene between two species of Archaea and Bacteria was a rare event which has been supported by their negative search results for other bacterial V-type ATPases (Gogarten *et al.*, 1996). Interestingly, new bacterial sequence data suggest that eukaryotes and some bacteria have exchanged IleRS genes – a transfer that was possibly mediated by selection for antibiotic resistance (Brown *et al.*, 1998).

Only a few years ago, the existence of two classes of aminoacyl-tRNA synthetases was recognised and the evolutionary relationships within each class were demonstrated. As hypotheses, these findings are now being put to the test by new sequence data from evolutionarily exciting groups of organisms. In the end, what we should hope for is not only a re-evaluation of the evolution of aminoacyl-tRNA synthetases but a major reconsideration of the importance of radical genetic processes, such as horizontal gene transfers, in the evolution of genomes.

14.6 Summary

Aminoacyl-tRNA synthetases serve the critical role in protein synthesis of catalysing the esterification or charging of an amino acid to its cognate tRNA molecule. Therefore, the development of the genetic code and the evolution of aminoacyl-tRNA synthetases are thought to be strongly linked. In addition, aminoacyl-tRNA synthetases provide one of the best examples of a large multiple protein family whose radiation likely occurred before the emergence of the first single-cell organisms. An understanding of the evolutionary tempo and mode of this protein family is important to a general understanding of how genomes evolve. The canonical view of aminoacyl-tRNA synthetase evolution is of two distinct protein families, each consisting of 10 different synthetases which duplicated prior to the divergence of prokaryotes and eukaryotes. However, complete genomic sequences from species of Archaea and Bacteria have now seriously challenged earlier views of aminoacyl-tRNA synthetase evolution. Complex evolutionary processes such as gene loss, duplication and transfer must be considered in light of contemporary genome distributions of aminoacyl-tRNA synthetase genes.

Acknowledgements

I thank M. Ibba and D. Söll for an early preview of their manuscript. This work was supported by the Department of Bioinformatics and the Anti-Infectives Research Division of SmithKline Beecham.

References

ALTSCHUL, S. F., GISH, W., MILLER, W., MYERS, E. W. and LIPMAN, D. J. (1990) Basic local alignment search tool. *J. Mol. Biol.*, **251**, 403–410.

BALDAUF, S. L., PALMER, J. D. and DOOLITTLE, W. F. (1996) The root of the universal tree and the origin of eukaryotes based on elongation factor phylogeny. *Proc. Natl Acad. Sci. USA*, **93**, 7749–7754.

BRETON, R., WATSON, D., YAGUCHI, M. and LAPOINTE, J. (1990) Glutamyl-tRNA synthetases of *Bacillus subtilis* 168T and of *Bacillus stearothermophilus*. *J. Biol. Chem.*, **265**, 18248–18255.

BROWN, J. R. and DOOLITTLE, W. F. (1995) Root of the universal tree of life based on ancient aminoacyl-tRNA synthetase gene duplications. *Proc. Natl Acad. Sci. USA*, **92**, 2441–2445.

BROWN, J. R. and DOOLITTLE, W. F. (1997) Archaea and the prokaryote to eukaryotes transition. *Microbiol. Mol. Biol. Rev.*, **61**, 456–502.

BROWN, J. R. and DOOLITTLE, W. F. (1998) Gene descent, duplication and horizontal transfer in the evolution of glutamyl- and glutaminyl-tRNA synthetases. *J. Mol. Evol.* (in press).

BROWN, J. R., ROBB, F. T., WEISS, R. and DOOLITTLE, W. F. (1997) Evidence for the early divergence of tryptophanyl- and tyrosyl-tRNA synthetases. *J. Mol. Evol.*, **45**, 9–16.

BROWN, J. R., ZHANG, J. and HODGSON, J. E. (1998) A bacterial antibiotic resistance gene with eukaryotic origins. *Current Biology*, **8**, R365–R367.

BULT, C. J., WHITE, O., OLSEN, G. J. *et al.* (1996) Complete genome sequence of the methanogenic archaeon, *Methanococcus jannaschii. Science*, **273**, 1058–1073.

CARTER, C. W. (1993) Cognition, mechanism, and evolutionary relationships in aminoacyl-tRNA synthetases. *Ann. Rev. Biochem.*, **62**, 715–748.

CAVARELLI, J. and MORAS, D. (1993) Recognition of tRNAs by aminoacyl-tRNA synthetases. *FASEB J.*, **7**, 79–86.

CURNOW, A. W., IBBA, M. and SÖLL, D. (1996) tRNA-dependent asparagine formation. *Nature*, **382**, 589–590.

CURNOW, A. W., HONG, K.-W., YUAN, R. *et al.* (1997) Glu-tRNAGln amidotransferase: a novel heterotrimeric enzyme required for correct decoding of glutamine codons during translation. *Proc. Natl Acad. Sci. USA*, **94**, 11819–11826.

DECKART G., WARREN, P. V., GAASTERLAND, T. *et al.* (1998) The complete genome of the hyperthermophilic bacterium *Aquifex aeolicus* genome. *Nature*, **392**, 353–358.

ERIANI, G., DELARUE, M., POCH, O., GANGLOFF, J. and MORAS, D. (1990) Partition of tRNA synthetases into two classes based on mutually exclusive sets of sequence motifs. *Nature*, **347**, 203–206.

ERIANI, G., CAVARELLI, J., MARTIN, F. *et al.* (1995) The class II aminoacyl-tRNA synthetases and their active site: evolutionary conservation of an ATP binding site. *J. Mol. Evol.*, **40**, 499–508.

FELSENSTEIN, J. (1993) *PHYLIP* (*Phylogeny Inference Package*) version 3.57c, distributed by the author: http://evolution.genetics.washington.edu/phylip.html, Department of Genetics, University of Washington, Seattle.

FITCH, W. M. and UPPER, K. (1987) The phylogeny of tRNA sequences provides evidence for ambiguity reduction in the origin of the genetic code. *Cold Spring Harbour Symp. Quant. Biol.*, **52**, 759–767.

FOX, G. E., MAGRUM, L. J., BALCH, W. E., WOLFE, R. S. and WOESE, C. R. (1977) Classification of methanogenic Bacteria by 16S ribosomal RNA characterisation. *Proc. Natl Acad. Sci. USA*, **74**, 4537–4541.

FRIEST, W., LOGAN, D. T. and GAUSS, D. H. (1996) Glycyl-tRNA synthetases. *Biol. Chem. Hoppe-Seyler*, **377**, 343–356.

GAGNON, Y., LACOSTE, L., CHAMPAGNE, N. and LAPOINTE, J. (1996) Widespread use of the Glu-$tRNA^{Gln}$ transamidation pathway among Bacteria. A member of the alpha purple Bacteria lacks glutaminyl-tRNA synthetase. *J. Biol. Chem.*, **271**, 4856–4863.

GOGARTEN, J. P., KIBAK, H., DITTRICH, P. *et al.* (1989) Evolution of the vacuolar H^+-ATPase: Implications for the origin of eukaryotes. *Proc. Natl Acad. Sci. USA*, **86**, 6661–6665.

GOGARTEN, J. P., HILARIO, E. and OLENDZENSKI, L. (1996) Gene duplications and horizontal transfer during early evolution. In *Evolution of Microbial Life*, SGM 54, ed. D. McL. Roberts, P. Sharp, G. Alderson and M. Collins, pp. 267–292 (Cambridge: Cambridge University Press).

HASHIMOTO, T., SÁNCHEZ, L. B., SHIRAKURQ, T., MÜLLER, M. and HASEGAWA, M. (1998) Secondary absence of mitochondria in *Giardia lamblia* and *Trichomonas vaginalis* revealed by valyl-tRNA synthetase phylogeny. *Proc. Natl. Acad. Sci. USA*, **95**, 6860–6865.

HILARIO, E. and J. P. GOGARTEN (1993) Horizontal transfer of ATPase genes – the tree of life becomes the net of life. *BioSystems*, **31**, 111–119.

IBBA, M., MORGAN, S., CURNOW, A. W. (1997) A euryarchaeal lysyl-tRNA synthetase: resemblence to class I synthetases. *Science*, **278**, 1119–1122.

IBBA, M., BONO, J. L., ROSA, P. A. and SÖLL, D. (1998) Archaeal-type lysyl-tRNA synthetase in the lyme disease spirochaete *Borrelia burgdorferi. Proc. Natl Acad. Sci. USA*, **94**, 14383–14388.

Iwabe, N., Kuma, K.-I., Hasegawa, M., Osawa, S. and Miyata, T. (1989) Evolutionary relationship of Archaea, Bacteria, and eukaryotes inferred from phylogenetic trees of duplicated genes. *Proc. Natl Acad. Sci. USA*, **86**, 9355–9359.

Klenk, H.-P., Clayton, R. A., Tomb, J.-F. *et al.* (1997) The complete genome sequence of the hyperthermophilic, sulfate-reducing archaeon *Archaeoglobus fulgidus*. *Nature*, **390**, 364–370.

Lake, J. A. (1988) Origin of the eukaryotic nucleus determined by rate-invariant analysis of rRNA sequences. *Nature*, **331**, 184–186.

Lamour, V., Quevillon, S., Diriong, S., N'Guyen, V. C., Lipinski, M. and Mirande, M. (1994) Evolution of the Glx-tRNA synthetase family: The glutaminyl eznyme as a case of horizontal gene transfer. *Proc. Natl Acad. Sci. USA*, **91**, 8670–8674.

Lapointe, J., Duplain, L. and Proulx, M. (1986) A single glutamyl-tRNA synthetase aminoacylates $tRNA^{Glu}$ and $tRNA^{Gln}$ in *Bacillus subtilis* and efficiently misacylates *Escherichia coli* $tRNA^{Gln}$ *in vitro*. *J. Bacteriol.*, **165**, 88–93.

Lazzarini, R. A. and Mehler, A. H. (1964) Separation of specific glutamate- and glutamine-activating enzymes from *Escherichia coli*. *Biochemistry*, **3**, 1445–1449.

Maddison, W. P. and Maddison, D. R. (1992) *MacClade Version 3* (Sunderland: Sinauer Associates).

Martindale, D. W., Gu, Z. M. and Csank, C. (1989) Isolation and complete sequence of the yeast isoleucyl-tRNA synthetase gene (ILS1). *Curr. Genet.*, **15**, 99–106.

Moras, D. (1992) Structural and functional relationships between aminoacyl-tRNA synthetases. *Trends Biochem. Sci.*, **17**, 159–164.

Nabholz, C. E., Hauser, R. and Schneider, A. (1997) *Leishmania tarentolae* contains distinct cytosolic and mitochondrial glutaminyl-tRNA synthetase activities. *Proc. Natl Acad. Sci. USA*, **94**, 7903–7908.

Nagel, G. M. and Doolittle, R. F. (1991) Evolution and relatedness in two aminoacyl-tRNA synthetase families. *Proc. Natl Acad. Sci. USA*, **88**, 8121–8125.

Nagel, G. M. and Doolittle, R. F. (1995) Phylogenetic analysis of the aminoacyl-tRNA synthetases. *J. Mol. Evol.*, **40**, 487–498.

Nicholas, H. B. Jr. and McClain, W. H. (1995) Searching tRNA sequences for relatedness to aminoacyl-tRNA synthetase families. *J. Mol. Evol.*, **40**, 482–486.

Olsen, G. J., Woese, C. R. and Overbeek, R. (1994) The winds of (evolutionary) change: breathing new life into microbiology. *J. Bacteriol.*, **176**, 1–6.

Ribas de Pouplana, L., Furgier, M., Quinn, C. L. and Schimmel, P. (1996) Evidence that two present-day components needed for the genetic code appeared after nucleated cells separated from eubacteria. *Proc. Natl Acad. Sci. USA*, **93**, 166–170.

Rivera, M. C. and Lake, J. A. (1992) Evidence that eukaryotes and eocyte prokaryotes are immediate relatives. *Science*, **257**, 74–76.

Rodin, S. and Ohno, S. (1995) Two types of aminoacyl-tRNA synthetases could be originally encoded by complementary strands of the same nucleic acid. *Origins of Life and Evolution of the Biosphere*, **25**, 565–589.

Schön, A., Kannangara, C. G., Gough, S. and Söll, D. (1988) Protein biosynthesis in organelles requires misaminoacylation of tRNA. *Nature*, **331**, 187–190.

Smith, D. S., Doucette-Stamm, L. A., Deloughery, C. *et al.* (1997) Complete genome sequence of *Methanobacterium thermoautotrophicum* ΔH: Functional analysis and comparative genomics. *J. Bacteriol.*, **179**, 7135–7155.

Tomb, J. F., White, O., Kerlavage, A. R. *et al.* (1997) The complete genome sequence of the gastric pathogen *Helicobacter pylori*. *Nature*, **388**, 539–547.

Woese, C. R. and Fox, G. E. (1977) Phylogenetic structure of the prokaryotic domain: the primary kingdoms. *Proc. Natl Acad. Sci. USA*, **51**, 221–271.

Woese, C. R., Kandler, O. and Wheelis, M. L. (1990) Towards a natural system of organisms: proposal for the domains Archaea, Bacteria and Eukarya. *Proc. Natl Acad. Sci. USA*, **87**, 4576–4579.

15

The Evolutionary History of Carbamoyltransferases: Insights on the Early Evolution of the Last Universal Common Ancestor

BERNARD LABEDAN[1] **AND ANNE BOYEN**[2]

[1] *Institut de Génétique et de Microbiologie, CNRS URA 1354, Université Paris-Sud, Orsay, France*
[2] *Microbiologie, Vrije Universiteit Brussel, and Vlaams Interuniversitair Instituut voor Biotechnologie, Brussels, Belgium*

15.1 Introduction

Evolutionarily related (homologous) proteins are very useful in trying to retrace both protein evolution and gene ancestry. Orthologous proteins, which are encoded by genes descending from a unique ancestor, can be used to reconstruct phylogenetic trees of organisms. Such a reconstruction may go as far as the last universal common ancestor, if homologues have been detected in the three domains of life (Woese *et al.*, 1990). Paralogous proteins which are encoded by genes descending from copies of an ancestral duplication allow us to progress even further in understanding protein evolution. Since paralogy events occur independently of speciation and may be very ancient, they may help to describe the set of genes carried by ancestral organisms. In this chapter, we retrace the history of present-day carbamoyltransferases which appear to be encoded by paralogous genes, descending from ancestral genes which duplicated at least twice during the early evolution of the last universal common ancestor. This analysis helps to understand how early proteins have evolved and gives us some insights into the complexity of the gene set present in this universal ancestor.

15.2 Ornithine carbamoyltransferases and aspartate carbamoyltransferases are evolutionarily related proteins

Ornithine carbamoyltransferase (OTCase; EC 2.1.3.3) and aspartate carbamoyltransferase (ATCase; EC 2.1.3.3) catalyse analogous reactions in two different metabolic pathways. ATCase catalyses the first committed step of pyrimidine biosynthesis, the reaction between carbamoyl phosphate and aspartate, whereas OTCase is implicated in arginine biosynthesis where it assures the transfer of the carbamoyl moiety of carbamoyl phosphate to the 5-amino group of ornithine, thereby forming citrulline. In those organisms which display

Table 15.1 Sequences of extremophilic carbamoyltransferases recently determined in our laboratories

Enzyme	Species	Domain[a]	Source	Accession number
Ornithine carbamoyltransferase (OTCase)	*Pyrococcus furiosus*[b]	A	Roovers *et al.* (1997)	X99225
	Thermotoga maritima[b]	B	M. Van de Casteele	Y10661
	Thermus thermophilus[d]	B	Sanchez *et al.* (1997)	Y10266
	Vibrio spp. 2693[e]	B	Z. Liang	Y11033
Aspartate carbamoyltransferase (ATCase)	*Pyrococcus abyssi*[b]	A	Purcarea *et al.* (1997)	U61765
	Sulfolobus solfataricus[c]	A	V. Durbecq, D. Charlier	X99872
	Thermotoga maritima[b]	B	C. Pingguo, M. Van de Casteele	Y10300
	Thermus aquaticus[d]	B	Van de Casteele *et al.* (1997)	Y09536
	Vibrio spp. 2693[e]	B	M. Van de Casteele, Z. Yuan-Fu, Y. Xu, Z. Liang	Y09786

[a] A = Archaea, B = Bacteria.
[b] Hyperthermophile.
[c] Extreme thermophile.
[d] Thermophile.
[e] Psychrophile.

the deiminase pathway for arginine degradation, a second, catabolic, OTCase mediates the thermodynamically unfavourable reverse reaction, the phosphorolysis of citrulline into ornithine and carbamoyl phosphate (Cunin *et al.*, 1986). These enzymes share a common functional domain, the carbamoyl phosphate-binding domain (Wild and Wales, 1990).

The functional relationship between OTCases and ATCases seems to be parallelled by structural similarities. Indeed, the amino acid sequences of the OTCase subunit and of the chain composing the catalytic subunit of ATCase display significant amounts of identical and similar residues, especially in the N-terminal moiety, the carbamoyl phosphate-binding domain. Although the primary sequences of the C-terminal parts, responsible for the binding of the respective amino acid substrates, are more divergent, significant conservation at the level of secondary and tertiary structures was revealed by the three-dimensional analysis of the catabolic OTCase of *Pseudomonas aeruginosa* (gene *arcB*) and the comparison of its globular structure with that *of the Escherichia coli* ATCase (Villeret *et al.*, 1995).

These functional and structural relationships suggest that OTCases and ATCases are paralogous (Fitch, 1970) proteins, i.e. that they are encoded by genes resulting from the duplication of a single ancestral gene followed by divergence of the resulting copies. The recent characterisation in our laboratory of several ATCases and OTCases from different thermophilic and one psychrophilic prokaryotes (Bacteria and Archaea; see Table 15.1) allowed us to extend the comparison between both enzymes over the three domains of life (Woese *et al.*, 1990) and to retrace at least part of their evolutionary history, using a molecular phylogeny approach.

15.3 Phylogenetic study of carbamoyltransferases

15.3.1 Multiple alignment and phylogenetic tree

In a first step the predicted amino acid sequences of OTCases and ATCases from different extremophilic prokaryotes were used to perform a homology search in the protein databases. This provided us with a body of 64 sequences of carbamoyltransferases: 33 OTCases and 31 ATCases, including our newly determined sequences (Table 15.1). A multiple alignment of these sequences was constructed using different automatic procedures (Gonnet *et al.*, 1992; Thompson *et al.*, 1994). As expected, all these sequences show significant sequence similarities, with some of the similarity blocks specific for either OTC or ATCases but many others common to both groups of enzymes, a feature that helped to anchor the multiple alignment.

As could be anticipated, the obtained alignment was satisfactory in the carbamoyl phosphate-binding domain (N-terminal moiety) but rather ambiguous in the substrate-binding C-terminal part where residues were allowed to diverge further. Fortunately, the experimentally determined, well conserved, secondary structures allowed us to resolve the majority of the ambiguities on visual inspection, thus improving the alignment by hand.

If the mere existence of this multiple alignment with conserved similarity blocks throughout the whole polypeptide sequences reinforces the suspicion of ancestral parology, it becomes very clear when the multiple alignment of all 64 sequences of OTCases and ATCases is used to reconstruct a phylogenetic tree (Felsenstein, 1989; Gonnet *et al.*, 1992; Swofford, 1993). Indeed, as shown in Figure 15.1, the sequences resolve in two well-separated trees, each specific for either OTCases or ATCases, which are interconnected. The same tree topology was obtained using entirely different methods (maximum parsimony, distance, and approximation to maximum likelihood) and did not change

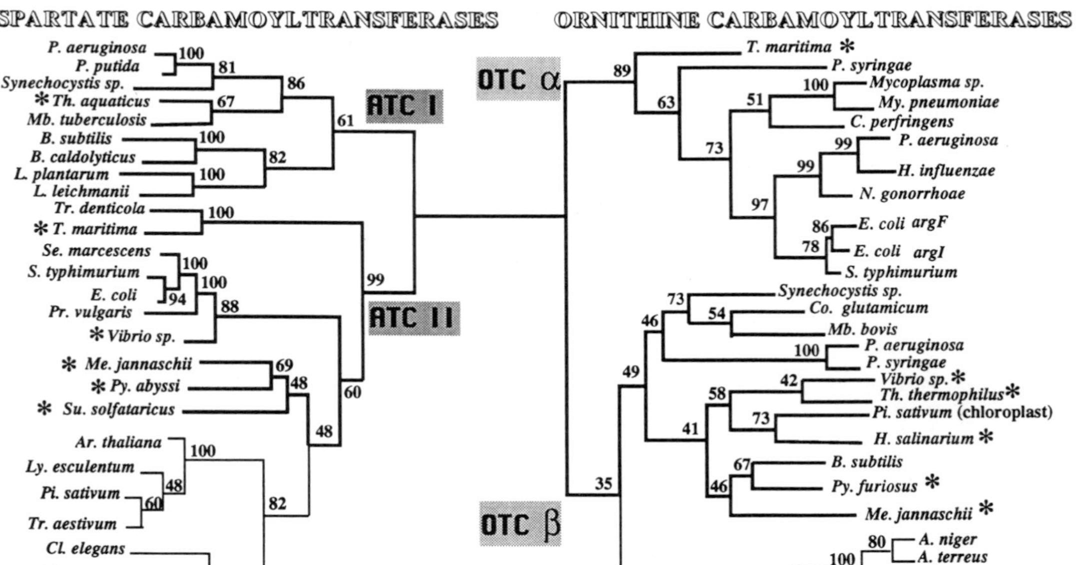

Figure 15.1 A composite phylogenetic tree for ornithine and aspartate carbamoyltransferases. The multiple alignment was used to reconstruct a tree using the maximum parsimony method. Each enzyme tree was rooted by its paralogous one. The branch lengths (computed using the PAUP program) are drawn to scale and indicated thick lines for prokaryotes and thin lines for eukaryotes, respectively. Extremophiles are indicated by a star (*) next to the species name. Each node was tested using the bootstrap approach and its strength given as percentage of its presence in the output of 100 random trees is shown as bold figures next to each internal node. The bootstrap values were determined using the maximum parsimony (PROTPARS) method. Slightly different values were obtained when using a distance method (e.g. NEIGHBOR; may be obtained by electronic mail from labedan@igmors.u-psud.fr).

when various sequences were left out. Therefore, it seems rather robust in spite of the rather low bootstrap values obtained at various nodes. Moreover, using only the fairly well conserved N-terminal carbamoyl phosphate-binding domain in the tree reconstruction instead of the complete sequence (including the more divergent substrate-binding domains) still resulted in the same topology, a feature that strongly favours the notion that the present-day genes encoding OTCases and ATCases derive from ancestral ones which had already the actual size, rather than from the fusion of independent modules resulting in specialised carbamoyltransferases.

As the three domains of life are represented in the OTCase as well as in the ATCase tree, it would appear reasonable to root each of them by the outgroup formed by the paralogous carbamoyltransferase and to use it as a species tree in challenging the topology of the universal tree of life. Indeed, this approach has been used repeatedly for other couples of paralogous proteins (Gogarten *et al.*, 1989; Iwabe *et al.*, 1989; Benachenhou-Lahfa *et al.*, 1993; Brown *et al.*, 1994; Brown and Doolittle, 1995). Unfortunately, it shows, upon closer inspection, that both trees exhibit a polyphyly of Bacteria (and in the case of the OTCase tree also of the Archaea) and can therefore not be used as phylogenetic probes to infer organismal trees.

15.3.2 Identification of subfamilies

Two putative subfamilies could be identified for both OTCases and ATCases by searching for unique signature sequences (Labedan *et al.*, 1998), which are distributed over the organisms under investigation as follows.

The ATCase subfamilies

The ATCase I subfamily corresponds to the proteins from the genera *Bacillus*, *Lactobacillus*, and *Pseudomonas*, and from the species *Mycobacterium tuberculosis*, *Synechocystis* sp. and *Thermus aquaticus* ZO5 and thus until now includes bacteria only. In the ATCase II subfamily, the bacteria remain polyphyletic with a paraphyletic grouping of the evolutionarily distantly related bacteria *Treponema denticola* and *Thermotoga maritima*. This unexpected grouping appears to be borne out by the tertiary structure of the two enzymes: in both organisms the gene encoding ATCase has been found to be a fusion product consisting of a proximal part homologous to the gene encoding the basic catalytic monomer (*pyrB* in *E. coli*) and a distal part very similar to the gene encoding the regulatory subunit (*pyrI*) found in enterobacteria and also present in the archaea *Sulfolobus solfataricus*, *Pyrococcus abyssi* and *Methanococcus jannaschii* (Van de Casteele *et al.*, 1994, 1997). The archaea form a monophyletic group and share a common node with the whole of the eukaryotes; the cluster formed by archaea and eukaryotes is found branching with the enterobacteria. If this situation is reminiscent of the canonical topology (Woese *et al.*, 1990), it must be remarked that the node common to archaea and eukaryotes is not supported by a high bootstrap value.

The OTCase subfamilies

The OTC α subfamily includes several of the bacterial proteins, for instance those from *E. coli* and *T. maritima*, whereas others belong to the OTC β subfamily. The bacterial sequences in the OTC β group are polyphyletic and interspersed with the archaeal sequences, which are polyphyletic themselves. Indeed, the hyperthermophylic *Py. furiosus* shares a common node with *B. subtilis*, but this pair is paraphyletic to *M. jannaschii*. Remarkably, the OTCase from the extreme halophile *Halobacterium salinarium* is unexpectedly close to the *Pisum sativum* chloroplast sequence and this pair shares a common ancestor with the enzymes of two extremophilic bacteria, the psychrophilic *Vibrio* sp. and the thermophile *Thermus thermophilus*. It must however be stressed that the nodes of this particular cluster have low bootstrap values and that, moreover, the OTC β subfamily is the one part of the tree whose topology is affected by discarding the C-terminal part of the sequences in the tree reconstruction process. If these unexpected groupings allow the addition of new sequences to the already considerable sample used here, several of them might be explained by repeated events of horizontal transfer. In the case of *E. coli* K 12, such an event appears to have occurred fairly recently (Van Vliet *et al.*, 1988). Most probably a mechanism of horizontal transfer is also responsible for the unusual association of the OTCase of the halophile *H. salinarium* with that of the *P. sativum* chloroplast (which is presumably of bacterial descent) as well as for the mixture of anabolic and catabolic OTCases, a feature that is sustained by immunological data (Falmagne *et al.*, 1985). It has been shown that the conversion of regulated catabolic OTCases into Michaelian anabolic ones, or vice versa, would require only very few mutations (Baur *et al.*, 1990; Kuo *et al.*, 1989). Horizontal transfer followed by divergence might therefore have occurred rather recently and the functional difference between catabolic and anabolic OTCases may have been acquired independently in several organisms, well after the divergence of the new species. Moreover, for those catabolic OTCases that are not allosteric, the metabolic specialisation is set at the level of transcriptional regulation, not at the structural level.

Species sampling in the subfamilies

The relative positions of the prokaryotes in the composite tree are most unexpected. When considering the bacteria only, ATC I and OTC β subfamilies seem to be coupled (for example in *B. subtilis*, *Pseudomonas*, *Mycobacterium* and *Thermus* species), and ATC II and OTC α likewise (in *T. maritima* and the enterobacteria). It appears puzzling therefore that the archaea take opposite places in the two protein trees. Indeed, whereas they group with the enterobacteria (ATC II) and appear very distant from *B. subtilis* in the ATCase tree, in the OTCase tree they cluster with *B. subtilis* (OTC β) and appear very distant from the enterobacteria. These data strongly suggest that successive duplications have occurred and that the ancestors of the subfamilies kept only one copy of the duplicated genes.

15.4 A possible scenario for the evolutionary history of carbamoyltransferases

The present sampling of carbamoyltransferases in the organisms studied up to now can be accounted for if it is accepted that the common ancestor to all extant life (the so-called last universal common ancestor) contained already (at least) two copies of an ancestral ATCase and two or more of an ancestral OTCase as well (Figure 15.2). In the course of further evolution these respective copies may have been either selectively maintained or lost at some step of the speciation process leading to the present-day organisms. This hypothesis looks the simplest way of explaining the polyphyly shown by Bacteria in both trees, as well as the striking differences in relative positions of the prokaryotes in both trees.

Indeed, if we call the ancestral genes *atc1*, *atc2*, *otcα* and *otcβ* respectively, it appears that one of the last ancestors to Archaea and Eukaryotes would have kept only the copies

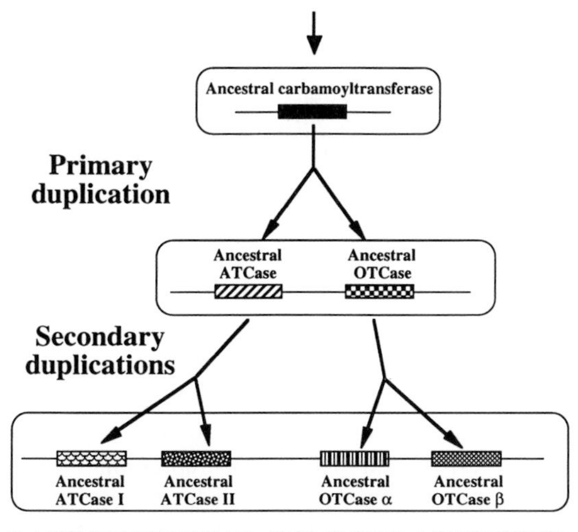

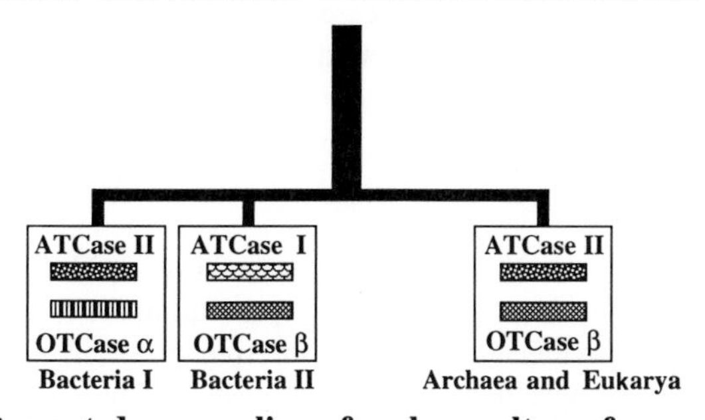

Figure 15.2 A possible scenario for the evolutionary history of carbamoyltransferases. The putative successive events of gene duplication leading to the presence of two ancestral ATCases and two ancestral OTCases in the last universal common ancestor are schematised (top) as the simplest explanation for the sampling of both carbamoyltransferases in present-day species (bottom).

atc2 and *otcβ*, since as yet, all Archaea and all Eukaryotes contain exclusively the subfamilies ATCII and OTC α. The loss of the ATCI and OTC α copies would have occurred somewhere after the emergence of the bacterial domain. This sequence of events corresponds to current views (Woese *et al.*, 1990) in which Archaea and Eukaryotes would have shared a lapse of time in their evolutionary history. In the case of Bacteria, two categories have developped: one harbouring the pair *atc2* and *otcα* (for instance *E. coli*), the other the pair *atc1* and *otcβ* (for instance *B. subtilis*). Such a phenomenon of secondary parology followed by the specific loss of one copy at different moments of their evolutionary history could explain at least some of the unexpected associations of unrelated prokaryotic species.

If the last common ancestor already contained two copies of each ancestral carbamoyltransferase, the gene duplications from which they arose must have occurred either in this last common ancestor or in an even earlier organism (Figure 15.2). Even if we do not know how specialised these ancestral enzymes were, this conclusion is in support of the hypothesis (see for example, Forterre *et al.*, 1993) that this universal ancestor was a rather sophisticated organism already, possibly endowed with a fair amount of genetic redundancy.

The duplications responsible for the emergence of *atc1* and *atc2*, and *otcα* and *otcβ*, were probably not contemporaneous and must have been preceded by an even earlier gene duplication of a very ancient carbamoyltransferase displaying substrate ambiguity (Jensen, 1976) from which evolved the ancestral differentiated ornithine and aspartate carbamoyltransferases. Thus, our data suggest that this initial duplication was already an ancient event in the evolution of the last universal common ancestor.

15.5 Conclusions

This work shows that tracing back the history of two paralogous proteins present in the three domains of life could be a successful approach to progressively disclosing the gene equipment of the last universal common ancestor. Contrary to what was recently believed, and taking the history of carbamoyltransferases as representative of many other proteins, it appears clear that this ancient organism must have been a very sophisticated organism: its enzymes would have been already specialised and its genes were the result of an already long and complex history. This peculiar case of the carbamoyltransferases history could be a model for future studies with other ubiquitous paralogous proteins.

15.6 Summary

Thirty-three sequences of ornithine carbamoyltransferases (OTCases) and 31 sequences of aspartate carbamoyltransferases (ATCases) representing the three domains of life were multiply aligned and a phylogenetic tree was inferred from this multiple alignment. The global topology of the composite rooted tree suggests that present-day genes are derived from paralogous ancestral genes which were already of the same size and argues against a mechanism of fusion of independent modules. The polyphyly for Bacteria in both enzyme trees and for the Archaea in the OTCase tree makes this composite tree unsuitable for assessing the actual order of organismal descent. This complexity may be explained by assuming the occurrence of two subfamilies in the OTCase tree (OTC α and OTC β) and two others in the ATCase tree (ATC I and ATC II). These subfamilies could have arisen from duplication and selective losses of some differentiated copies during the successive speciations. We suggest that Archaea and Eukaryotes share a common ancestor in which the ancestral copies giving the present-day ATC II/OTC β combinations were present, whereas Bacteria comprise two classes: one containing the ATC II/OTC α combination; the other harbouring the ATC I/OTC β combination. Moreover, multiple horizontal gene transfers could have occurred rather recently amongst prokaryotes. Whichever the actual history of carbamoyltransferases, our data suggest that the last common ancestor to all extant life possessed differentiated copies of genes coding for both carbamoyltransferases, indicating it as a rather sophisticated organism.

References

Baur, H., Tricot, C., Stalon, V. and Haas, D. (1990) Converting catabolic ornithine carbamoyltransferase to an anabolic enzyme. *J. Biol. Chem.*, **265**, 14728–14731.

Benachenhou-Lahfa, N., Forterre, P. and Labedan, B. (1993) Evolution of glutamate dehydrogenase genes: evidence for two paralogous protein families and unusual branching patterns of the archaebacteria in the universal tree of life. *J. Mol. Evol.*, **36**, 335–346.

BROWN, J. R. and DOOLITTLE, W. F. (1995) Root of the universal tree of life based on ancient aminoacyl-tRNA synthetase gene duplication. *Proc. Natl Acad. Sci. USA*, **92**, 2441–2445.

BROWN, J. R., MASUCHI, Y., ROBB, F. T. and DOOLITTLE, W. F. (1994) Evolutionary relationships of bacterial and archaeal glutamine synthetase genes. *J. Mol. Evol.*, **38**, 566–576.

CUNIN, R., GLANSDORFF, N., PIÉRARD, A. and STALON, V. (1986) Biosynthesis and metabolism of arginine in bacteria. *Microbiol. Rev.*, **50**, 314–352.

FALMAGNE, P., PORTETELLE, D. and STALON, V. (1985) Immunological and structural relatedness of catabolic ornithne carbamoyltransferase and the anabolic enzymes of enterobacteria. *J. Bacteriol.*, **161**, 714–719.

FELSENSTEIN, J. (1989) PHYLIP – a phylogeny inference package (version 3.2). *Cladistics*, **5**, 164–166.

FITCH, W. D. (1970) Distinguishing homologous from analogous proteins. *System. Zool.*, **19**, 99–113.

FORTERRE, P., BENACHENHOU, N., CONFALONIERI, F., DUGUET, M., ELIE, C. and LABEDAN, B. (1993) The nature of the last universal ancestor and the root of the tree of life, still open questions. *BioSystems*, **28**, 15–32.

GOGARTEN, J. P., KIBAK, H., DITTRICH, P. *et al.* (1989) Evolution of the vacuolar H^+-ATPase: implications for the origin of eukaryotes. *Proc. Natl Acad. Sci. USA*, **86**, 6661–6665.

GONNET, G. H., COHEN, M. A. and BENNER, S. A. (1992) Exhaustive matching of the entire protein sequence database. *Science*, **256**, 1443–1445.

IWABE, N., KUMA, K., HASEGAWA, M., OSAWA, S. and MIYATA, T. (1989) Evolutionary relationship of archaebacteria, eubacteria, and eukaryotes inferred from phylogenetic trees of duplicated genes. *Proc. Natl Acad. Sci. USA*, **86**, 9355–9359.

JENSEN, R. (1976) Enzyme recruitment in evolution of new function. *Ann. Revi. Microbiol.*, **30**, 409–425.

KUO, L. C., ZAMBIDIS, I. and CARON, C. (1989) Triggering of allostery in an enzyme by a point mutation, ornithine transcarbamoylase. *Science*, **245**, 522–524.

LABEDAN, B., BOYEN, A., BAETENS, M. *et al.* (1998) The evolutionary history of carbamoyltransferases: a complex set of paralogous genes was already present in the last universal common ancestor. *J. Mol. Evol.* (in press).

PURCAREA, C., HERVÉ, G., LADJIMI, M. M. and CUNIN, R. (1997) Aspartate transcarbamylase from the deep-sea hyperthermophilic archaeon *Pyrococcus abyssi*: genetic organisation, structure and expression in *Escherichia coli*. *J. Bacteriol.*, **179**, 4134–4157.

ROOVERS, M., HETHKE, C., LEGRAIN, C., THOMM, M. and GLANSDORFF, N. (1997). Isolation of the gene encoding *Pyrococcus furiosus* ornithine carbamoyltransferase and study of its expression profile *in vivo* and *in vitro*. *Eur. J. Biochem.*, **247**, 1038–1045.

SANCHEZ, R., BAETENS, M., VAN DE CASTEELE, M., ROOVERS, M., LEGRAIN, C. and GLANSDORFF, N. (1997) Ornithine carbamoyltransferase from the extreme thermophile *Thermus thermophilus*: analysis of the gene and characterisation of the protein. *Eur. J. Biochem.*, **248**, 466–474.

SWOFFORD, D. L. (1993) *PAUP: Phylogenetic Analysis Using Parsimony*, Version 3.1.1 (Washington, DC: Smithsonian Institute).

THOMPSON, J. D., HIGGINS, D. G. and GIBSON, T. J. (1994) CLUSTAL W: improving the sensitivity of progressive multiple sequence alignment through sequence weighting, position-specific gap penalties and weight matrix choice. *Nucleic Acids Res.*, **22**, 4673–4680.

VAN DE CASTEELE, M., DEMAREZ, M., LEGRAIN, C. *et al.* (1994) Genes encoding thermophilic aspartate carbamoyltransferases of *Thermus aquaticus* ZO5 and *Thermotoga maritima* MSB8: modes of expression in *E. coli* and properties of their products. *Biocatalysis*, **11**, 165–179.

VAN DE CASTEELE, M., CHEN, P. G., ROOVERS, M., LEGRAIN, C. and GLANSDORFF, N. (1997) Structure and expression of a pyrimidine gene cluster from the extreme thermophile *Thermus* strain ZO5. *J. Bacteriol.*, **179**, 3470–3481.

VAN VLIET, F., BOYEN, A. and GLANSDORFF, N. (1988) On interspecies gene transfer: the case of *argF* gene of *Escherichia coli*. *Ann. Inst. Pasteur/Microbiol.*, **139**, 493–496.

VILLERET, V., TRICOT, C., STALON, V. and DIDEBERG, O. (1995) Crystal structure of *Pseudomonas aeruginosa* catabolic ornithine transcarbamoylase at 3.0-Å resolution: a different oligomeric organisation in the transcarbamoylase family. *Proc. Natl Acad. Sci. USA*, **92**, 10762–10766.

WILD, J. and WALES, M. (1990) Molecular evolution and genetic engineering of protein domains involving aspartate carbamoylase. *Ann. Rev. Microbiol.*, **44**, 193–218.

WOESE, C. R., KANDLER, O. and WHEELIS, M. L. (1990) Towards a natural system of organisms: proposal for the domains Archaea, Bacteria, and Eukarya. *Proc. Natl Acad. Sci. USA*, **87**, 4576–4579.

XU, Y., ZHANG, Y. F., LIANG, Z. Y., VAN DE CASTEELE, M., LEGRAIN, C. and GLANSDORFF, N. (1998) Aspartate carbamoyltransferase from a psychrophilic deep-sea bacterium, *Vibrio* strain 2693. Properties of the enzyme, genetic organisation and synthesis in *Escherichia coli*. *Microbiology*, **144**, 1435–1441.

PART SIX

Enzyme Evolution

16

Evolution of the Histone Fold

KATHLEEN SANDMAN[1], WENLIAN ZHU[2], MICHAEL F. SUMMERS[2] AND JOHN N. REEVE[1]

[1] *Department of Microbiology, Ohio State University, Columbus, Ohio, USA 43210*
[2] *Howard Hughes Medical Institute, University of Maryland Baltimore County, Baltimore, Maryland, USA 21228*

16.1 Introduction

Histones are small, basic, abundant DNA-binding proteins that form the protein core around which chromosomal DNA is wrapped, by electrostatic interactions, to generate nucleosomes (Wolffe, 1992). Histone-containing nucleoprotein complexes are present in both Archaea (Pereira *et al.*, 1997) and Eukarya, consistent with their divergence from a common histone-containing ancestral structure that existed before the separation of the archaeal and eukaryal lineages. Histones may have originated in hyperthermophiles as a means to protect their chromosomes from thermal denaturation. All histones contain the histone fold structural motif, namely three α-helices separated by β-strand regions that are stabilised by protein dimerisation. Archaeal histones are essentially dimers of histone fold motifs; eukaryal histones are dimers of histone folds with short N-terminal and C-terminal amino acid extensions; and histone folds also exist as domains within more complex proteins. In this chapter we review the structure of the histone fold, and use a structure-based alignment to discuss the evolution and divergence of histone fold-containing proteins. Bacteria also contain several different abundant, small, positively charged proteins, often termed 'histone-like' proteins (Schmid, 1990). These proteins do not, however, have primary sequences in common with histones, they do not contain histone fold motifs, and the complexes that they form with DNA are unrelated to nucleosomes.

16.2 Archaeal histones

The nucleosome core is a histone octamer that contains two copies of four genetically distinct histone polypeptides, assembled as two (H3+H4) dimers flanked by (H2A+H2B) dimers. These histones, especially H3 and H4, have highly conserved primary sequences, consistent with the structural and functional constraints in packaging and accessing nuclear DNA being conserved in all Eukarya. The discovery of histones in Archaea (Sandman

```
                     |-helix I-|            |---------helix II---------|        |-helix III-|
HMfB                 MELPIAPIGRIIKDAG--AERVSDDARITLAKILEEMGRDIASEAIKLARHAGRKTIKAEDIELAVRRFKK*
                     1       10          20        30        40        50        60       69

                            |   ||         |  |   |   || ||  ||  ||  ||          |    |  ||
H2A          (22)    LQFPVGRVHRLLRKGNY-AERVGAGAPVYLAAVLEYLTAEILELAGNAARDNKKTRIIPRHLQLAIRNDEE (37)
H2B          (33)    KESYSIYVYKVLKQVHP-DTGISSKAMGIMNSFVNDIFERIAGEASRLAHYNKRSTITSREIQTAVRLLLP (22)

H3           (63)    KLPFQRLVREIAQDFKT-DLRFQSSAVMALQEASEAYLVGLFEDTNLCAIHAKRVTIMPKDIQLARRIRGE (2)
H4           (26)    QGITKPAIRRLARRGG--VKRISGLIYEETRGVLKVFLENVIRDAVTYTEHAKRKTVTAMDVVYALKRQGR (7)

HAP3         (39)    RWLPINNVARLMKNTLPPSAKVSKDAKECMQECVSELISFVTSEASDRCAADKRKTINGEDILISLHALGF (34)
HAP5        (157)    HSLPFARIRKVMKTDED-VKMISAEAPIIFAKACEIFITELTMRAWCVAERNKRRTLQKADIAEALQKSDM (15)

dTAF42       (17)    VPKDAQVIMSILKELN--VQEYEPRVVNQLLEFTFRYVTCILDDAKVYANHARKKTIDLDDVRLATEVTLD (192)
dTAF62        (3)    GSSISAESMKVIAESIG-VGSLSDDAAKELAEDVSIKLKRIVQDAAKFMNHAKRQKLSVRDIDMSLKVRNV (519)

dTAF28       (89)    PMLTKPRLTELVREVDT-TTQLDEDVEELLLQIIDDFVRDTVKSTSAFAKHRKSNKIEVRDVQLHFERKYN (37)

NCB1         (56)    THFPPAKVKKIMQTDED-IGKVSQATPVIAGRSLEFFIALLVKKSGEMARGQGTKRITAEILKKTILNDEK (22)
NCB2          (7)    VSLPKATVQKMISEILDQDLMFTKDAREIIINSGIEFIMILSSMASEMADNEAKKTIAPEHVIKALEELEY (68)
                            |   ||         |  |   |   || ||  ||  ||  ||          |    |  ||
```

Figure 16.1 Alignment of histone fold domains. The top line shows the sequence of HMfB with the α-helical regions identified. Vertical lines identify the hydrophobic residues which participate in the hydrophobic core of the histone fold. Each eukaryal sequence is paired with its heterodimer partner, except for dTAF28, which is thought to form homodimers (Hoffman *et al.*, 1996). The number of N- and C-terminal amino acid residues in the polypeptide that are not shown in the alignment are indicated by the figures in parentheses preceding and following the sequence. The H2A, H2B, H3, and H4 sequences are from chicken (Wells and McBride, 1989), HAP from *S. cerevisiae* (McNabb *et al.*, 1995), dTAF from *Drosophila* (Kokubo *et al.*, 1994), and NCB from *S. cerevisiae* (Gadbois *et al.*, 1997).

et al., 1990; Starich *et al.*, 1996) added support to the prediction of the small subunit ribosomal RNA (ssu-rRNA) tree that the Archaea and Eukarya are more closely related to each other than either is to the Bacteria (Iwabe *et al.*, 1989), and extended the use of histones for chromosomal packaging to include euryarchaeotal species, including hyperthermophiles, that position near the root of the evolutionary tree. The most-studied archaeal histones, HMfA and HMfB (histones A and B from *Methanothermus fervidus*) have molecular masses of ~7.5 kDa and sequences that are very similar to each other and homologous to the globular histone fold domains of the four eukaryal core histones. When all core histone sequences are aligned, based on their homologies to the HMf sequences, the common ancestry of histones is clearly revealed (Figure 16.1). Each of the eukaryal histones has a sequence that is more similar to the archaeal HMf sequences than to the sequences of the other eukaryal nucleosome core histones. The common ancestor probably therefore had a sequence similar to a consensus sequence for contemporary archaeal histones. Consistent with a common and presumably ancient mode of DNA interaction, archaeal histones wrap DNA into structures that resemble the structure formed at the centre of the eukaryal nucleosome by the $(H3+H4)_2$ tetramer (Pereira *et al.*, 1997; Reeve *et al.*, 1997).

Eighteen archaeal histone gene sequences are currently available (Figure 16.2), from hyperthermophiles, moderate thermophiles, and mesophiles, and the biochemical properties and DNA interactions of archaeal histones have been reviewed recently (Grayling *et al.*, 1996a,b; Reeve *et al.*, 1997). The two most divergent archaeal histone sequences still exhibit 60% similarity, there are no gaps in the alignment of the 18 sequences, and there are only one to three residue length variations at the N- and C-termini. Because of their small sizes (66 to 69 amino acid residues) and extensive sequence conservation, the phylogenetic relationships of these archaeal histones cannot be defined precisely by primary sequence comparisons, although some conclusions can be drawn. All of the histones in one archaeon are generally more similar to each other than to histones in other archaea, and there are different numbers of histone genes in different archaea (Reeve *et al.*, 1997). A variable number of histone gene duplications have therefore probably occurred in the different archaeal lineages after their separation from a common ancestor that had one histone gene. It is, however, also possible that the ancestral histone gene duplicated before the archaeal lineages diverged, consistent with a minimum of two histone genes in all extant archaea, and this was then followed by different numbers of gene duplications and/or by lateral transfers of histone genes.

16.2.1 The hydrophobic core of a histone dimer is the monomer–monomer interface

Archaeal and eukaryal histones exist only as dimers in solution, and oligomerise into higher-order structures when complexed with DNA (Grayling *et al.*, 1996a; Wolffe, 1992). An extensive network of hydrophobic contacts occurs between the two monomers within a rHMfB dimer (Figures 16.3 and 16.4; Starich *et al.*, 1996). Each monomer contains a central 27-residue α-helix (helix II) flanked by two shorter α-helices, helices I and III, that contain 10 and 12 residues, respectively. Separating the α-helices are short regions of coil and β-strand structure (Figure 16.3). All three α-helices are amphipathic, with charged faces on the dimer surface and hydrophobic faces interacting within the dimer's hydrophobic core. The hydrophobic contacts formed in a dimer between two long α-helix IIs are shown in Figure 16.4A, and the positions of additional hydrophobic residues that

```
                |-helix I-|          |---------helix II---------|       |-helix III-|
HMfB       M-ELPIAPIGRIIKDAGAERVSDDARITLAKILEEMGRDIASEAIKLARHAGRKTIKAEDIELAVRRFKK*
           1        10        20        30        40        50        60      69

                    |  ||     |  |   |   || ||  ||  ||  ||            |    |  ||
HMfA       MGELPIAPIGRIIKNAGAERVSDDARIALAKVLEEMGEEIASEAVKLAKHAGRKTIKAEDIELARKMFK*
HMtA1      MAELPIAPVGRIIKNAGAQRISDDAREALAKILEEKGEEIAKEAVKLAKHAGRKTVKASDIELAAKKL*
HMtA2      MAELPIAPVGRIIKNAGAQRISDDAKEALAKALEEMGEEISRKAVELAKHAGRKTVKATDIEMAAKQL*
HMtB       M-ELPIAPIGRIIKNAGAEIVSDDAREALAKVLEAKGEEIAENAVKLAKHAGRKTVKASDIELAVKRM*
HFoA1      MAELPIAPVGRIIKNAGAPRVSDDARDALAKVLEEMGEGIAAEAVKLAKHAGRKTVKASDIEMAVKAA*
HFoA2      MAELPIAPVGRIIKNAGAQRISDDAKEALAKALEENGEELAKKAVELAKHAGRKTVKAEDIEMAVKSA*
HFoB       M-ELPIAPIGRIIKNAGAERVSDDAREALAKALEEKGETIATEAVKLAKHAGRKTVKASDVELAVKRL*
HTz        MAELPIAPIDRLIRKAGAERVSEDAAKALAEYLEEYAIEVGKKATEFARHAGRKTVKAEDVRLAVKA*
HPyA1      MGELPIAPVDRLIRKAGAERVSEEAAKILAEYLEEYAIEVSKKAVEFARHAGRKTVKAEDIKLAIKS*
HPyA2      MAELPIAPVDRLIRKAGAQRVSEQAAKLLAEHLEEKALEIARKAVDLAKHAGRKTVKAEDIKLAIRS*
HAfA       MAELPMAPVDRLIRKAGAERVSADAVEKMVEVLEDYAITVAKKAVEIAKHSGRKTVTADDIKLALSM*
HAfB       M-ELPLAPVERLLRKAGASRVSEDAKVELAKAIEEYAMQIGKKAAELAKHAGRKTVKVDDIKLALREL*
MJ0168     MAELPVAPFERILKKAGAERVSEAAAEYLAEAVEEIALEIAKEAVELAKHAKRKTVKVEDIKLALKK*
MJ0932     MAELPVAPFERILKKAGAERVSRAAAEYLAEAVEEIALEIAKEAVELAKHAKRKTVKVEDIKLALKQ*
MJ1258     MAELPVAPCVRILKKAGAQRVSEAAGKYFAEALEEIALEIARKSVDLAKHAKRKTVKVEDVKAALRG*
MJECL17    MAELPVAPFVRILKKDGAERVSRAAAEYFAEAIEDLALEIAKEAVDLAKHAKRKTVKVEDVKLALKK*
MJECL29    MTELPVAPFERILKKVGAERVSRAAAEYLAEAFEEIALEIAKEAVDLAKHAKRKTVKVEDIKLALKK*
                    |  ||     |  |   |   || ||  ||  ||  ||            |    |  ||

            1       10        20        30        40        50        60     67
HMvA        MIPKGTVKRIMKDNTEMYVSTESVVAL-DILQEMIVTTTKIAEENAAKDKRKTIKARDIEECDAERLK
MJ1647      MLPKATVKRIMKQHTDFNISAEAVDELCNMLEEIIKITTEVAEQNARKEGRKTIKARDIKQCDDERLK

            70        80        90      98
HMvA       EKILQVSERTEKVNMLANEILHVIASELERY*
MJ1647     RKIMELSERTDKMPILIKEMLNVITSEL*
```

Figure 16.2 Alignment of archaeal histone sequences. The structure of the HMfB histone, top line, has been solved by NMR spectroscopy and the positions of the three α-helices are indicated above the amino acid sequence. The asterisks indicate the locations of stop codons and the hyphen indicates a gap introduced to improve the alignment. *HmvA* and MJ1647 are archaeal genes in the genomes of *M. voltae* (Agha-Amiri and Klein, 1993) and *M. jannaschii* (Bult *et al.*, 1996), respectively, that encode proteins with N-terminal domains that align with histones. HMf sequences are from *M. fervidus* (Sandman *et al.*, 1990; Tabassum *et al.*, 1992), HMt from *Methanobacterium thermoautotrophicum* strain ΔH (Tabassum *et al.*, 1992; Smith *et al.*, 1997), HFo from *Methanobacterium formicicum* (Darcy *et al.*, 1995), HTz from *Thermococcus zilligii* (Ronimus and Musgrave, 1996), HPy from *Pyrococcus* strain GB-3a (Sandman *et al.*, 1994b), HAf from *Archaeoglobus fulgidus* (http://www.ncbi.nlm.gov/BLAST/tigr_db.html), and MJ from *Methanococcus jannaschii* (Bult *et al.*, 1996).

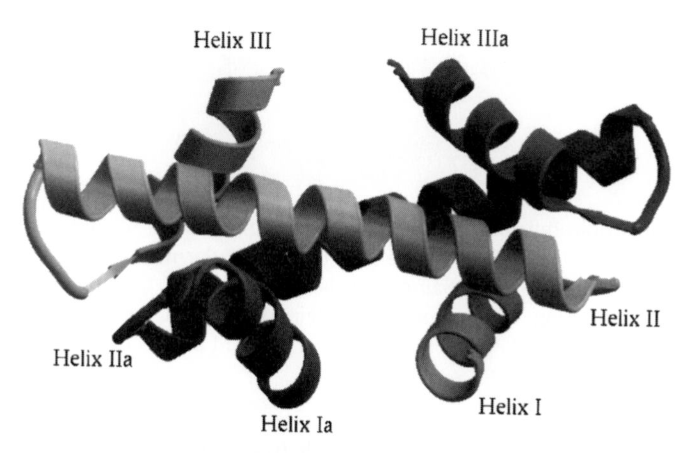

Figure 16.3 Structure of a $(rHMfB)_2$ homodimer (Starich *et al.*, 1996). Helices I, II and III comprise one monomer while helices Ia, IIa and IIIa comprise the other.

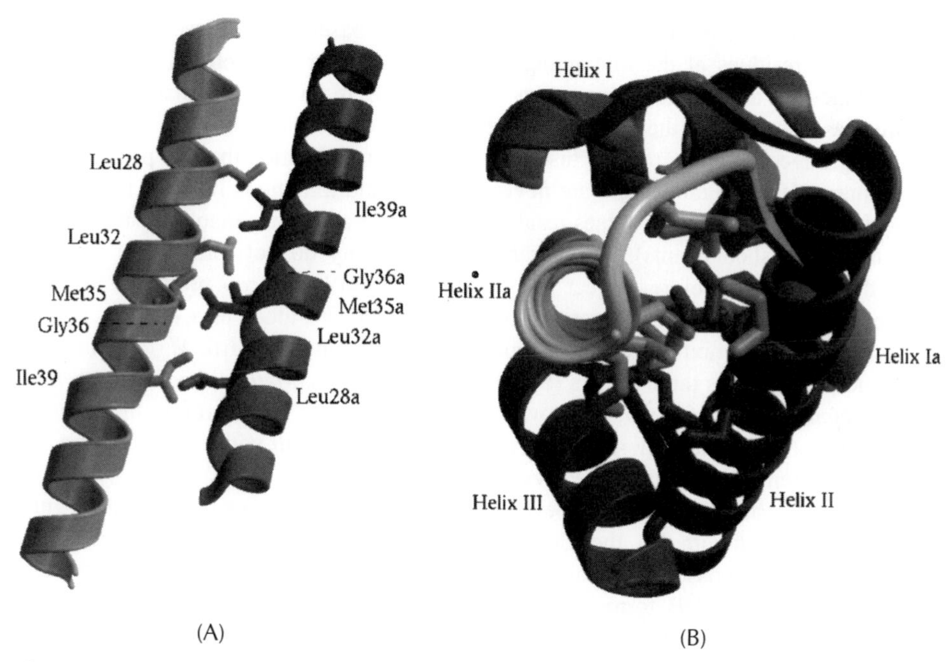

Figure 16.4 (A) Helices II and IIa are shown in their antiparallel orientation. Multiple hydrophobic residues on each helix comprise a network of hydrophobic contacts between the helices. (B) The $(rHMfB)_2$ dimer, looking down the axis of helix IIa (centre, left). Helix IIIa has been deleted from the foreground to provide a view of the side-chains comprising the hydrophobic core.

contribute to the dimer core, side-chains from α-helices I and III and from the β-strand regions, are identified in Figure 16.4B. As demonstrated by the alignment of primary sequences (Figure 16.2), the hydrophobic residues that facilitate dimer formation are highly conserved in all archaeal histones, predicting the formation of dimers with structures very similar to that of $(rHMfB)_2$. Preliminary NMR studies have confirmed this prediction for $(rHMfA)_2$, $(rHFoB)_2$, and $(rHPyA1)_2$ (unpublished results).

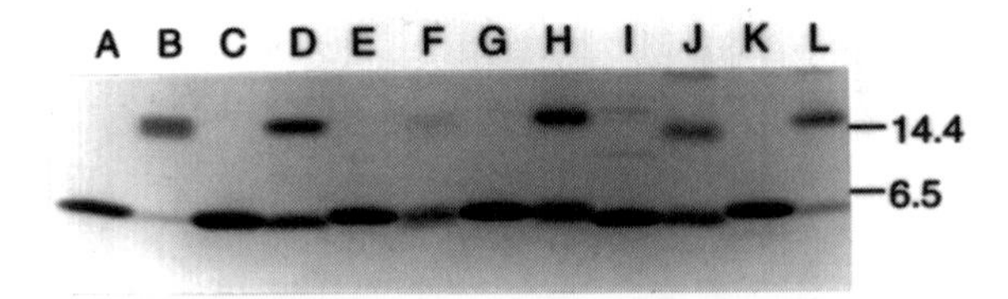

Figure 16.5 Electrophoretic separation of recombinant archaeal histones cross-linked with formaldehyde through a SDS–tricine gel. The molecular masses of the size standards, in kDa, are indicated to the right. Samples in lanes B, D, F, H, J, L were exposed to formaldehyde cross-linking. Lanes A and B contained rHMfB; C and D contained rHMfA; E and F contained rHFoA1; G and H contained rHFoA2; I and J contained rHFoB; K and L contained rHPyA1.

16.2.2 Archaeal histones can form both homodimers and heterodimers

Both $(HMfA)_2$ and $(HMfB)_2$ homodimers, and (HMfA+HMfB) heterodimers are formed *in vivo* and *in vitro* (Sandman *et al.*, 1994a). Essentially identical monomer–monomer interface contacts are predicted within the heterodimer, based on the presence of identical hydrophobic residues at the interacting positions. Many of the archaeal histones have been synthesised as recombinant proteins in *E. coli* and, in all cases, formaldehyde cross-linking has shown that they form homodimers in solution (Figure 16.5). Success in producing a recombinant archaeal histone is, in fact, an indication that the protein forms homodimers. Histone monomers are never detected, and mutations in *hmfB* that change the hydrophobicity of core residues result in rHMfB variants that cannot be purified from *E. coli* (unpublished results). These variants presumably do not fold and/or dimerise correctly, and are therefore subject to rapid degradation by *E. coli* proteases. Archaea with multiple histones with conserved hydrophobic core residues have the potential to synthesise complex mixtures of homodimers and heterodimers *in vivo*. *M. fervidus*, with 2 histones, synthesises three dimers (Sandman *et al.*, 1994a). *Methanobacterium thermoautotrophicum* and *M. formicicum*, however, have 3 histones and therefore the potential to synthesise 6 different dimers, and *Methanococcus jannaschii* with five histones could form 15 different histone dimers. Provided with this flexibility, Archaea may synthesise some histones that have functions beyond genome packaging, possibly in gene regulation, and, consistent with this, the HMfA to HMfB ratio is not constant in *M. fervidus* but varies with growth phase (Sandman *et al.*, 1994a). The complexes formed by these two proteins do differ; HMfB forms more compact structures with DNA than HMfA, and HMfB is more abundant in stationary phase cells, presumably when minimal access to genomic information is required. A family of eukaryal sequence-specific transcription factors, the bHLH proteins, similarly dimerise through the association of hydrophobic faces of amphipathic helices (Ma *et al.*, 1994), and different dimers have different DNA-sequence recognition properties.

Homodimers and heterodimers are formed either to recognise different sequences, or to sequester particular monomers into nonbinding dimers (Maleki *et al.*, 1997). The archaeal histones do not bind to specific DNA sequences, but they do bind preferentially to curved DNAs (Howard *et al.*, 1994; Grayling *et al.*, 1997), and different archaeal histone dimers may have different affinities for different DNA shapes. HMvA and MJ1647 (Agha-Amiri and Klein, 1993; Bult *et al.*, 1996) encode archaeal proteins with sequences similar to archaeal histone sequences, but with C-terminal extensions (Figure 16.2). They have hydrophobic residues at the appropriate positions for dimer formation, although the identities of these residues are quite divergent. They may therefore only form homodimers, but potentially could also heterodimerise with the histone-fold containing archaeal histones. HMvA and MJ1647 homologues are not present in the genomes of *M. thermoautotrophicum* (Smith *et al.*, 1997) and *Archaeoglobus fulgidus* (http://www.ncbi.nlm.nih.gov/BLAST/tigr_db.html).

16.3 Eukaryal histones

16.3.1 The hydrophobic character of residues at the monomer–monomer interface is conserved between archaeal and eukaryal histones

The structure of rHMfB, three α-helices separated by two β-strands (Figure 16.3), can be superimposed almost exactly on the histone fold structure of the globular domains of the eukaryal nucleosome core histones (Starich *et al.*, 1996). In Figure 16.1, the archaeal and eukaryal histone sequences have been aligned so that the hydrophobic residues that form the hydrophobic core and stabilise the $(rHMfB)_2$ structure are aligned with hydrophobic residues in every sequence. Previous histone alignments maximised the alignment of identical residues, which required the introduction of gaps (Sandman *et al.*, 1990) whereas, in this new alignment, the boundaries of α-helices coincide precisely with only the introduction of a single amino acid gap between helix I and strand I. The eukaryal core histones form only (H2A+H2B) and (H3+H4) heterodimers, and the residues that form the dimer interfaces vary considerably. It therefore seems likely that the hydrophobic surface of each histone monomer has a different shape which directs the association of correct partners, and precludes incorrect dimer formation.

16.3.2 Scenarios for evolution of eukaryal histones from archaeal-like ancestral sequences

How could the two different and exclusive pairs of eukaryal heterodimers have evolved from one ancestral histone? An initial gene duplication, followed by sequence divergence, must have generated two genes that encoded histones capable of forming both homodimers and heterodimers, as now exist in *M. fervidus*. As these genes diverged further, the shape of the hydrophobic monomer–monomer interface must have changed to the extent that homodimerisation eventually was prevented and heterodimer formation mandated. The second pair of eukaryal histone genes could have been generated by a repetition of this scenario, or by a duplication of the two genes that already encoded a heterodimer pair. As all eukarya now appear to have H2A, H2B, H3 and H4, and to assemble these four histones into structurally identical nucleosomes, the two heterodimer pairs were apparently established before the divergence of the Eukarya. Finding a species that has only one fixed heterodimer pair would add an important evolutionary link, and if such a species is found, the prediction is that it will be a prokaryote.

16.4 Histone fold motifs in eukaryal proteins

Histone fold-encoding genes have apparently continued to duplicate, diverge, and combine with other sequences, as histone fold motifs are now found in a variety of proteins. Several eukaryal transcription factors have sequence consistent with the presence of histone fold domains, and in one case this has been confirmed by structural studies (Xie *et al.*, 1996). The pattern of hydrophobic residues that constitutes the hydrophobic core of the histone fold is conserved in all of these proteins, although the specific amino acids vary. A variety of hydrophobic surfaces can apparently be generated in a histone fold motif that can direct the association of monomers to form specific heterodimers. Lesk and Chothia (1980) similarly found that the structure of globins was conserved, despite primary sequence variations, by maintaining hydrophobic residues at the sites that determined the position and orientations of α-helices and of the haem group.

Histones do not bind DNA in a sequence-specific manner (Grayling *et al.*, 1997; Baldi *et al.*, 1996), but sequence-specific DNA binding ability has been conferred on histone fold-containing proteins by association with additional polypeptides. For example, the CCAAT-box binding factor, a general eukaryal transcription factor designated HAP in yeast and CBF in mammalian systems, contains three polypeptides, two of which (HAP3 and HAP5) associate via histone fold domains, while the third (HAP2) provides the sequence-specific recognition function (Kim *et al.*, 1996). Similarly, the Dr1-DRAP repressor complex (NCB1 and NCB2 in yeast) heterodimerises through histone fold domains but binds DNA via the TATA-box binding protein (TBP; Mermelstein *et al.*, 1996). In the eukaryal transcription pre-initiation complex, some TAFII components of transcription factor TFIID form a histone octamer-like structure (Xie *et al.*, 1996) and the crystal structure of the (dTAF42+dTAF62)$_2$ tetramer bears a striking resemblance to that of the (H3+H4)$_2$ tetramer at the centre of the nucleosome core. Most likely, two (dTAF28)$_2$ homodimers bind to the (dTAF42+dTAF62)$_2$ tetramer (Hoffman *et al.*, 1996), in a manner similar to (H2A+H2B) dimer binding to the (H3+H4)$_2$ tetramer, to complete a TAF octamer. This octamer requires association with TBP for sequence recognition. TBP binding to the TATA box introduces a substantial bend into the DNA (Kim *et al.*, 1993), which could help recruit the TAFII components or the Dr1-DRAP repressor to the promoter, assuming that these proteins also bind preferentially to curved DNA.

16.5 Conclusions

Clearly the histone fold is an ancient structural motif that existed in a common ancestor of the Archaea and Eukarya, possibly in a hyperthermophile that employed histones to stabilise its genome at high temperatures. Today, the histone fold persists in its simplest form in archaeal histones that compact and organise archaeal genomes. In eukaryal histones, the histone fold serves a similar function, but these polypeptides also have N- and C-terminal extensions that participate in gene regulation, and in eukaryal transcription factors histone fold domains are buried within much larger proteins and combined with sequence-recognition motifs to facilitate gene regulation. Arents and Moudrianakis (1993) predicted a path for the DNA on the surface of the nucleosome based on surface charges, and suggested that the paired N-terminal ends of α-helix Is, and paired β-strand regions between the α-helices should contact the DNA; however, the precise molecular details of histone fold–DNA contacts remain unknown. Aligning histone fold domains offers few primary sequence clues, although basic residues are always present at positions 10 and 13 in α-helix I, and at positions 51, 52, and 53 in β-strand 2. The archaeal histones bind and compact DNA readily *in vitro*, and therefore provide a simple and tractable system for structure–function studies. Site-directed mutagenesis of *hmfB* and subsequent synthesis of rHMfB variants in *E. coli* has already identified several residues that are involved in DNA binding and compaction. T54, for example, a residue that is conserved in the second β-strand residue in all archaeal histones, is essential for DNA binding (unpublished results).

16.6 Summary

The evolution and divergence of the histone fold, an ancient protein structural motif that existed in a common ancestor of the Archaea and the Eukarya, is discussed in light of structural studies of archaeal histones, which are dimers of the histone fold. The monomer–monomer interface is the hydrophobic core of the dimer, and the specific identities of the amino acid residues that comprise this surface determine the dimerisation properties. Archaeal histones have the greatest plasticity, forming both homodimers and heterodimers, and this multiplicity in DNA wrapping proteins may play a role in gene regulation for these organisms. Eukaryal histones and more complex eukaryal proteins containing the histone fold motif are differentiated histone folds that clearly have a common ancestor as the archaeal histones, but they are limited to heterodimerisation with a specific partner and may require association with sequence-specific DNA-binding motifs to carry out their roles in the regulation of gene expression.

Acknowledgements

Archaeal histone research at The Ohio State University is supported by NIH grant GM53185.

References

AGHA-AMIRI, K. and KLEIN, A. (1993) Nucleotide sequence of a gene encoding a histone-like protein in the archaeon *Methanococcus voltae*. *Nucleic Acids Res.*, **21**, 1491.

ARENTS, G. and MOUDRIANAKIS, E. N. (1993) Topography of the histone octamer surface: repeating structural motifs utilised in the docking of nucleosomal DNA. *Proc. Natl Acad. Sci. USA*, **90**, 10489–10493.

BACON, D. J. and ANDERSON, W. F. (1988) A fast algorithm for rendering space-filling molecule pictures. *J. Mol. Graphics*, **6**, 219–220.

BALDI, P., BRUNAK, S., CHAUVIN, Y. and KROGH, A. (1996) Naturally occurring nucleosome positioning signals in human exons and introns. *J. Mol. Biol.*, **263**, 503–510.

BULT, C. J., WHITE, O., OLSEN, G. J. *et al.* (1997) Complete genome sequence of the methanogenic archaeon, *Methanococcus jannaschii*. *Science*, **273**, 1058–1073.

DARCY, T. J., SANDMAN, K. and REEVE, J. N. (1995) *Methanobacterium formicicum*, a mesophilic methanogen, contains three HFo histones. *J. Bacteriol.*, **177**, 858–860.

GADBOIS, E. L., CHAO, D. M., REESE, J. C., GREEN, M. R. and YOUNG, R. A. (1997) Functional antagonism between RNA polymerase II holoenzyme and global negative regulator NC2 *in vivo*. *Proc. Natl Acad. Sci. USA*, **94**, 3145–3150.

GRAYLING, R. A., SANDMAN, K. and REEVE, J. N. (1996a) DNA stability and DNA binding proteins. *Adv. Protein Chem.*, **48**, 437–467.

GRAYLING, R. A., SANDMAN, K. and REEVE, J. N. (1996b) Histones and chromatin structure in hyperthermophilic Archaea. *FEMS Microbiol. Rev.*, **18**, 203–213.

GRAYLING, R. A., BAILEY, K. A. and REEVE, J. N. (1997) DNA binding and nuclease protection by the HMf histones from the hyperthermophilic archaeon *Methanothermus fervidus*. *Extremophiles*, **1**, 79–88.

HOFFMAN, A., CHIANG, C.-M., OELGESCHLAGER, T. *et al.* (1996) A histone octamer-like structure within TFIID. *Nature*, **380**, 356–359.

HOWARD, M. T., SANDMAN, K., REEVE, J. N. and GRIFFITH, J. D. (1992) HMf, a histone-related protein from the hyperthermophilic archaeon *Methanothermus fervidus*, binds preferentially to DNA containing phased tracts of adenines. *J. Bacteriol.*, **174**, 7864–7867.

IWABE, N., KUMA, K., HASEGAWA, M., OSAWA, S. and MIYATA, T. (1989) Evolutionary relationship of archaebacteria, eubacteria, and eukaryotes inferred from phylogenetic trees of duplicated genes. *Proc. Natl Acad. Sci. USA*, **86**, 9355–9359.

KIM, I.-S., SINHA, S., DECROMBRUGGHE, B. and MAITY, S. N. (1996) Determination of functional domains in the C subunit of the CCAAT-binding factor (CBF) necessary for formation of a CBF–DNA complex: CBF-B interacts simultaneously with both the CBF-A and CBF-C subunits to form a heterotrimeric CBF molecule. *Mol. Cell. Biol.*, **16**, 4003–4013.

KIM, J. L., NIKOLOV, D. B. and BURLEY, S. K. (1993) Cocrystal structure of TBP recognising the minor groove of a TATA element. *Nature*, **365**, 520–527.

KOKUBO, T., GONG, D.-W., WOOTTON, J. C., HORIKOSHI, M., ROEDER, R. G. and NAKATANI, Y. (1994) Molecular cloning of *Drosophila* TFIID subunits. *Nature*, **367**, 484–487.

KRAULIS, P. J. (1991) MOLSCRIPT: a program to produce both detail and schematic plots of protein structures. *J. Appl. Crystallogr.*, **24**, 946–950.

MA, P. C. M., ROULD, M. A., WEINTRAUB, H. and PABO, C. O. (1994) Crystal structure of MyoC bHLH domain-DNA complex: perspectives on DNA recognition and implications for transcriptional activation. *Cell*, **77**, 451–459.

MALEKI, S. J., ROYER, C. A. and HURLBUT, B. K. (1997) MyoD-E12 heterodimers and MyoD-MyoD homodimers are equally stable. *Biochemistry*, **36**, 6762–6767.

MCNABB, D. S., XING, Y. and GUARENTE, L. (1995) Cloning of yeast HAP5: a novel subunit of a heterotrimeric complex required for CCAAT binding. *Genes Dev.*, **9**, 47–58.

MERMELSTEIN, F., YEUNG, K., CAO, J. (1996) Requirement of a corepressor for Dr1-mediated repression of transcription. *Genes Dev.*, **10**, 1033–1048.

PEREIRA, S. L., GRAYLING, R. A., LURZ, R. and REEVE, J. N. (1997) Archaeal nucleosomes, *Proc. Natl Acad. Sci. USA*, **94**, 12633–12637.

REEVE, J. N., SANDMAN, K. and DANIELS, C. J. (1997) Archaeal histones, nucleosomes, and transcription initiation. *Cell*, **89**, 999–1002.

RONIMUS, R. and MUSGRAVE, D. (1996) A gene, *han1A*, encoding an archaeal histone-like protein from the *Thermococcus* species AN1: homology with eukaryal histone consensus sequences and the implications for delineation of the histone fold, *Biochim. Biophys. Acta*, **1307**, 1–7.

SANDMAN, K., KRZYCKI, J. A., DOBRINSKI, B., LURZ, R. and REEVE, J. N. (1990) HMf, a DNA-binding protein isolated from the hyperthermophilic archaeon *Methanothermus fervidus*, is most closely related to histones. *Proc. Natl Acad. Sci. USA*, **87**, 5788–5791.

SANDMAN, K., GRAYLING, R. A., DOBRINSKI, B., LURZ, R. and REEVE, J. N. (1994a) Growth-phase-dependent synthesis of histones in the archaeon *Methanothermus fervidus. Proc. Natl Acad. Sci. USA*, **91**, 12624–12628.

SANDMAN, K., PERLER, F. B. and REEVE, J. N. (1994b) Histone-encoding genes from *Pyrococcus*: evidence for members of the HMf family of archaeal histones in a non-methanogenic *Archaeon. Gene*, **150**, 207–208.

SCHMID, M. (1990) More than just 'histone-like' proteins. *Cell*, **63**, 451–453.

SMITH, D. R., DOUCETTE-STAMM, L. A., DELOUGHERY, C. *et al.* (1997) The complete genome sequence of *Methanobacterium thermoautotrophicum* strain ΔH: functional analysis and comparative genomics, *J. Bacteriol.*, **179**, 7135–7155.

STARICH, M. R., SANDMAN, K., REEVE, J. N. and SUMMERS, M. F. (1996) NMR structure of HMfB from the hyperthermophile, *Methanothermus fervidus*, confirms that this archaeal protein is a histone. *J. Mol. Biol.*, **255**, 187–203.

TABASSUM, R., SANDMAN, K. M. and REEVE, J. N. (1992) HMt, a histone-related protein from *Methanobacterium thermoautotrophicum* ΔH. *J. Bacteriol.*, **174**, 7890–7895.

Wells, D. and McBride, C. (1989) A comprehensive compilation and alignment of histones and histone genes. *Nucleic Acids Res.*, **17**, r311–r346.

Woese, C. R., Kandler, O. and Wheelis, M. L. (1990) Towards a natural system of organisms: proposal for the domains Archaea, Bacteria, and Eukarya. *Proc. Natl Acad. Sci. USA*, **87**, 4576–4579.

Wolffe, A. (1992) *Chromatin: Structure and Function* (London: Academic Press).

Xie, X., Kokubo, T., Cohen, S. L. *et al.* (1996) Structural similarity between TAFs and the heterotetrameric core of the histone octamer. *Nature*, **380**, 316–322.

17

Comparative Enzymology as an Aid to Understanding Evolution

MICHAEL J. DANSON, RUPERT J. M. RUSSELL, DAVID W. HOUGH AND GARRY L. TAYLOR

Centre for Extremophile Research, Department of Biology & Biochemistry, University of Bath, Bath, UK

17.1 Introduction and aims

Virtually all chemical reactions within a cell are catalysed by enzymes. The sum total of these reactions is known as metabolism, which has two principal components: catabolism and anabolism. Catabolic pathways serve to degrade nutrients to produce energy and the precursors of cellular components, whereas anabolic routes use these precursors in biosynthetic, energy-requiring pathways. The metabolic link between them is provided by central metabolism, and so fundamental are these central pathways that they are found in all living organisms. A comparative analysis of central metabolism, of both the pathways and the component enzymes, can therefore yield invaluable information on the evolution of organisms and the adaptation to their environments.

These comparisons have greater impact when correlated with phylogenetic studies. Using rRNA sequence analyses, it has been proposed that all organisms fall within one of three evolutionarily distinct domains: the Bacteria, the Archaea and the Eukarya (Woese *et al.*, 1990). Moreover, with additional data from related genes whose common ancestor duplicated in the ancestral lineage, it is thought that the Archaea and Eukarya share a common stem on the tree, with the root lying between them and the Bacteria (Doolittle and Brown, 1994). From this universal phylogenetic tree, the hyperthermophilic Archaea, and a few thermophilic Bacteria, represent the shortest lineages and are close to the root, indicating that they have undergone the least change and therefore that they may be the most primitive of extant organisms. There is still controversy as to whether members of the Archaea form a monophyletic lineage, although there is a general agreement on the primitive nature of the hyperthermophiles, consistent with the consensual, but not universal, view that life arose under conditions of high temperatures.

From such considerations, it is thought that the study of archaeal metabolism is an important aspect of understanding evolution and the origin of life on earth. The purpose of this brief review is therefore to assess our current knowledge of the central metabolic pathways of the Archaea (thermophiles, halophiles and methanogens), highlighting areas of particular interest and activity, and then to progress to what factors are emerging from recent studies on the structures of the enzymes catalysing these reactions. For a general review of archaeal central metabolism the reader is referred to Danson (1993). For detailed descriptions of metabolism within the various archaeal phenotypes, the following reviews are recommended: Kelly and Adams (1994), Schönheit and Schäfer (1995) and Kengen *et al.* (1996) (hyperthermophiles); Blaut (1994) (methanogens); Rawal *et al.* (1988) (halophiles). The reader is also referred to these reviews for the majority of the original literature on the studies reported below.

17.2 Central metabolism within the Archaea

As the range of Archaea studied becomes wider, it becomes clearer that the pathways of central metabolism were established before the divergence of the three evolutionary domains. However, patterns and features characteristic of archaeal metabolism can be observed and, if the Archaea are truly primitive, they can give clues to how metabolism has evolved.

17.2.1 The interconversion of hexose sugars and pyruvate

The Embden–Meyerhof glycolytic pathway, or modifications thereof, has now been found throughout the Archaea. In both methanogens and halophiles, the gluconeogenic pathway seems predominant, methanogens growing autotrophically and fixing carbon into acetyl-CoA, and halophiles being able to grow well on amino acids that feed predominantly into the citric acid cycle. In those thermophilic Archaea that are able to grow on hexose sugars, it was initially thought that the glycolytic pathway could not be used catabolically as ATP-linked hexokinase and phosphofructokinase activities could not be detected. However, it is now known that alternative enzymes are present: a pyrophosphate-linked phosphofructokinase has been found in *Thermoproteus tenax* and ADP-linked hexokinase and phosphofructokinase are present in *Pyrococcus furiosus* and *Thermococcus* species. In the latter organisms, a further modification of the normal glycolytic pathway is seen in the conversion of glyceraldehyde 3-phosphate to 3-phosphoglycerate via a ferredoxin oxidoreductase, without the intermediate 1,3-diphosphoglycerate that would otherwise lead to the formation of ATP. This oxidoreductase is also found in *Desulfurococcus amylolyticus*, although this organism possesses ATP-linked hexokinase and phosphofructokinase (Selig *et al.*, 1997). These pathways are summarised in Figure 17.1.

The unusual ADP-kinases have not been looked for in other thermophiles (e.g. *Thermoplasma* or *Sulfolobus*), or in the halophiles or methanogens, although the remaining glycolytic enzymes are known to be present in these organisms.

Surprisingly, catabolism of hexoses via Entner–Doudoroff-like pathways (Figure 17.2) is also found in many Archaea. The halophiles possess a semi-phosphorylated Entner–Doudoroff pathway, whereas nonphosphorylated versions are found in the thermophilic genera of *Thermoplasma*, *Sulfolobus*, *Thermoproteus* and *Pyrococcus*.

In a recent *in vivo* and *in vitro* study of the catabolism of glucose in the hyperthermophilic Archaea, Selig *et al.* (1997) found that *Sulfolobus* exclusively uses the nonphosphorylated Entner–Doudoroff pathway; *Thermococcus, Desulfurococcus* and *Pyrococcus* use only the Embden–Meyerhof route; and *Thermoproteus* uses both pathways simultaneously, with the Embden–Meyerhof pathway being the major (85%) route. Why *Pyrococcus* possesses both Embden–Meyerhof and Entner–Doudoroff pathways is not known.

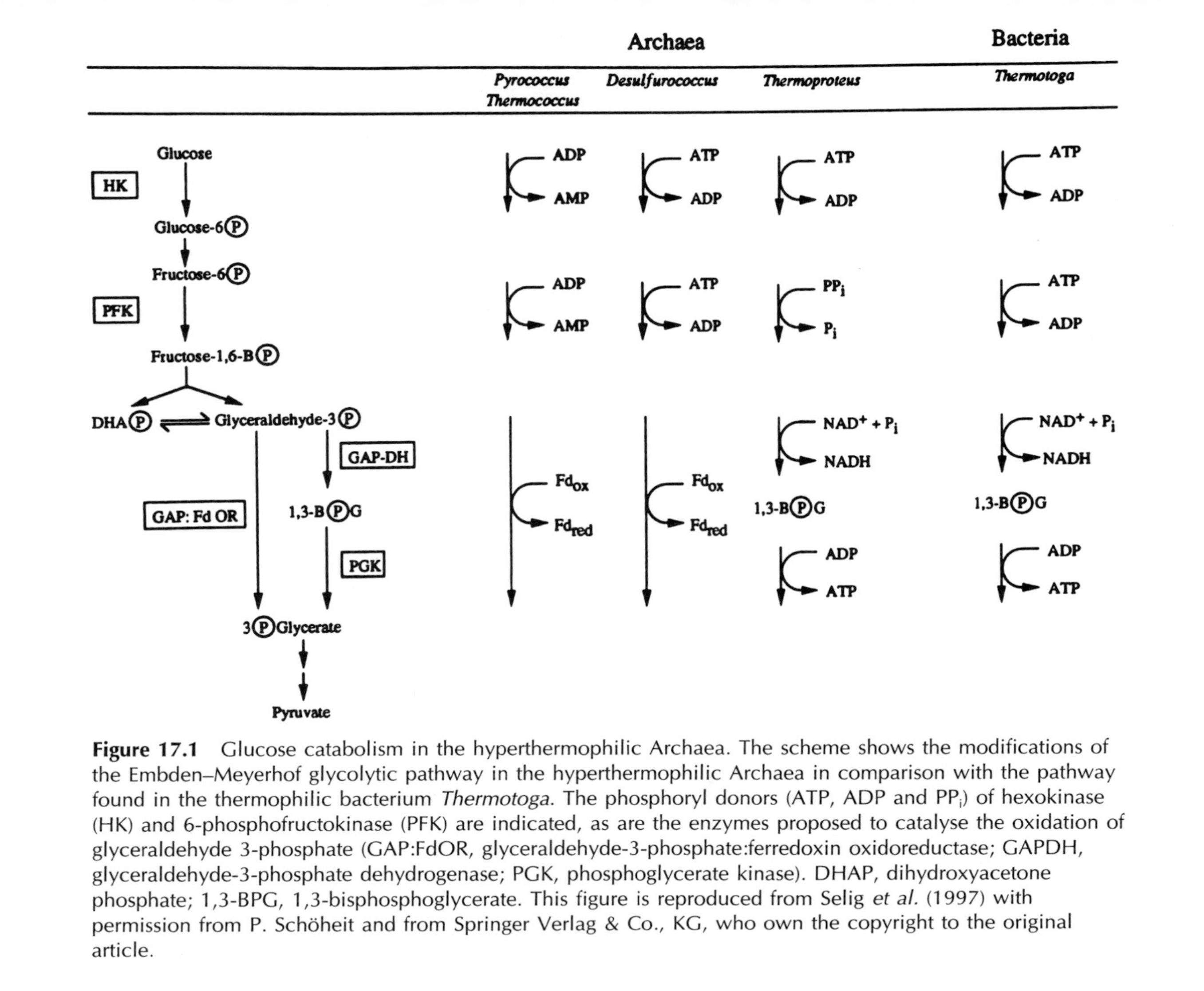

Figure 17.1 Glucose catabolism in the hyperthermophilic Archaea. The scheme shows the modifications of the Embden–Meyerhof glycolytic pathway in the hyperthermophilic Archaea in comparison with the pathway found in the thermophilic bacterium *Thermotoga*. The phosphoryl donors (ATP, ADP and PP_i) of hexokinase (HK) and 6-phosphofructokinase (PFK) are indicated, as are the enzymes proposed to catalyse the oxidation of glyceraldehyde 3-phosphate (GAP:FdOR, glyceraldehyde-3-phosphate:ferredoxin oxidoreductase; GAPDH, glyceraldehyde-3-phosphate dehydrogenase; PGK, phosphoglycerate kinase). DHAP, dihydroxyacetone phosphate; 1,3-BPG, 1,3-bisphosphoglycerate. This figure is reproduced from Selig *et al.* (1997) with permission from P. Schöheit and from Springer Verlag & Co., KG, who own the copyright to the original article.

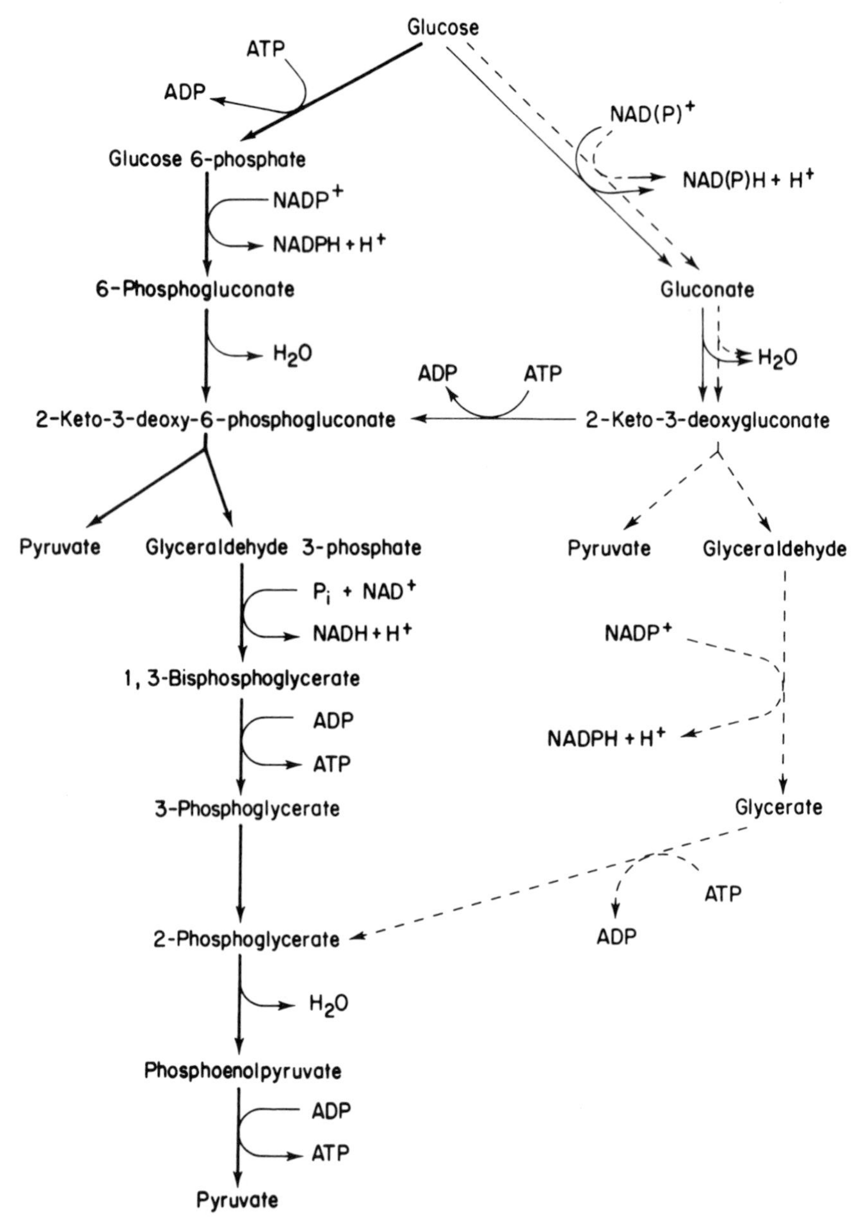

Figure 17.2 The modified Entner–Doudoroff pathway of thermophilic and halophilic Archaea. The semi-phosphorylated Entner–Doudoroff pathway of halophilic Archaea (solid lines) and the nonphosphorylated Entner–Doudoroff pathway of the thermophiles *Sulfolobus* and *Thermoplasma* (broken lines) are shown in comparison with the classical Entner–Doudoroff pathway of Bacteria (bold lines). The nonphosphorylated Entner–Doudoroff pathway in *Pyrococcus* is also shown by the broken lines, except that the oxidations of glucose and glyceraldehyde are catalysed by ferredoxin-linked oxidoreductases and not by NAD(P)-dehydrogenases. In the nonphosphorylated Entner–Doudoroff pathway of *Thermoproteus*, the NAD(P)-linked glucose dehydrogenase has been found (Siebers and Hensel, 1993), but the nature of the other enzymes has not been defined. This figure is reproduced from Danson (1993) with permission from Elsevier Science Publishers B.V., The Netherlands, who own the copyright to the original article.

17.2.2 Archaeal features of hexose catabolism

The energy yields from these archaeal pathways are low. The Embden–Meyerhof pathway in *Thermococcus, Desulfurococcus* and *Pyrococcus*, and the nonphosphorylated Entner–Doudoroff route do not generate a net yield of ATP. In *T. tenax*, which possesses a glyceraldehyde-3-phosphate dehydrogenase and a 3-phosphoglycerate kinase, one ATP per glucose oxidised to pyruvate will be produced, as in the case of the semiphosphorylated Entner–Doudoroff pathway in halophilic Archaea.

The use of ADP-linked kinases may reflect the greater thermostability of ADP compared with ATP, although Kengen *et al.* (1996) doubt the real advantage of this even in a hyperthermophile. The pyrophosphate-linked phosphofructokinase in *T. tenax* is common to the most deeply branching eukaryotes, such as *Entamoeba histolytica* and *Giardia lamblia*, and may reflect the proposed common lineage of Archaea and Eukarya.

Finally, we have noted the use of non-haem iron proteins in place of NAD(P) in the Archaea and have suggested that this may be a primitive feature related to the thermolability of the nicotinamide cofactors and the possible origin and early evolution of life at high temperatures (Daniel and Danson, 1995).

17.2.3 Conversion of pyruvate to acetyl-CoA

Until recently, it was thought that the three domains were distinctive in their enzymes that catalyse this step of central metabolism. However, new results concerning the enzymes of pyruvate metabolism in the Archaea indicate again that there may be common genes that are used to varying extents, or in different metabolic circumstances.

As a general feature, eukaryotes and aerobic and facultatively anaerobic Bacteria convert pyruvate to acetyl-CoA using the pyruvate dehydrogenase multienzyme complex. However, in facultative or strictly anaerobic Bacteria, ferredoxin/flavodoxin pyruvate oxidoreductases or pyruvate formate lyase are operative. Indeed, *Escherichia coli* has been found to possess all three enzyme systems, although in markedly different catalytic amounts.

In contrast, pyruvate dehydrogenase complex enzymic activity has not been detected in any Archaeon to date. Rather, throughout the Archaea, whether aerobes or anaerobes, pyruvate oxidoreductase is present, and it has been suggested that this and other 2-oxoacid oxidoreductases existed before the divergence of Archaea, Eukarya and Bacteria, and that the 2-oxoacid dehydrogenase complexes evolved after the development of oxidative phosphorylation.

However, we have discovered the presence of dihydrolipoamide dehydrogenase activity in the halophilic Archaea, and in *Tp. acidophilum*. In Bacteria and Eukarya, this enzyme is the third component of the 2-oxoacid dehydrogenase complexes; its role within those systems is its only known function, and therefore its discovery in the Archaea in the absence of its normal complexes was unexpected. The gene for dihydrolipoamide dehydrogenase from *Haloferax volcanii* has been cloned and sequenced, and from sequence comparisons and its catalytic properties it is clearly homologous to the complexed enzyme in non-archaeal organisms.

We have recently sequenced the upstream region of the *Hf. volcanii* dihydrolipoamide dehydrogenase gene (K. A. Jolley, D. G. Maddocks, M. L. Dyall-Smith, D. W. Hough and M. J. Danson, unpublished data) and unexpectedly found open reading frames homologous to the other component polypeptides of the eukaryal and bacterial 2-oxoacid dehydrogenase complexes. Analysis of the sequences, and the possible expression of the genes, are currently in progress, but the data clearly indicate their presence in the halophilic Archaea and therefore in all three evolutionary domains. The genes could have originated in the halophiles by horizontal gene transfer, but this might not be a recent event given that dihydrolipoamide dehydrogenase is found throughout the halophilic Archaea and in *Thermoplasma*. Moreover, a gene for the enzyme has recently been identified in the genome of *Methanococcus jannaschii* (Bult *et al.,* 1996), although we have doubts about its identity since its sequence similarity to the *Haloferax* enzyme is significantly less than that between the halophile and the Bacteria and Eukarya. Also, the methanogenic gene is considerably shorter than all other dihydrolipoamide dehydrogenases. Finally, a gene showing similarity to dihydrolipoamide dehydrogenase has been identified in the genome of *Pyrobaculum aerophilum* (Fitz-Gibbon *et al.*, 1997).

Clearly, the evolution of this step in central metabolism still holds a considerable number of secrets.

17.2.4 The production of acetate

It has been noted above that the Embden–Meyerhof pathway in *P. furiosus* and the non-phosphorylated Entner–Doudoroff route do not generate a net yield of ATP. However, many thermophilic, hyperthermophilic and halophilic Archaea possess the capacity to produce ATP via substrate-level phosphorylation through an ADP-linked acetyl-CoA synthetase:

$$\text{Acetyl-CoA} + \text{ADP} + \text{P}_i \rightarrow \text{acetate} + \text{CoA} + \text{ATP}$$

This mechanism is not found in the Bacteria, but was in fact first detected in the anaerobic eukaryotes *Entamoeba histolytica* and *Giardia lamblia* (Schönheit and Schäfer, 1995).

17.2.5 The citric acid cycle

Of all the pathways of central metabolism, the citric acid cycle appears to be the most invariant in terms of the chemical transformations and therefore of the enzymes used, the only exception being the use of the 2-oxoglutarate oxidoreductase for the interconversion of 2-oxoglutarate and succinyl-CoA in the Archaea, whereas the aerobic Bacteria and Eukarya use the NAD-linked 2-oxoacid dehydrogenase complexes. This is an analogous situation to the conversion of pyruvate to acetyl-CoA, and is a further example of the use of non-haem iron proteins in place of NAD(P) in the Archaea.

However, within the Archaea there are variations in the direction of carbon flow, with oxidative, reductive and partial cycles being found. These have been extensively reviewed by Danson (1993) and Schönheit and Schäfer (1995), the broad conclusions being that the complete oxidative cycle has been found in the halophiles and the thermophiles *Thermoplasma*, *Sulfolobus*, *Thermoproteus tenax* and *Pyrobaculum*. The reverse reductive cycle for CO_2 fixation has been demonstrated in *Thermoproteus neutrophilus*, as it has in the Bacteria *Hydrogenobacter thermophilus* (Shiba *et al.*, 1985), *Desulfobacter hydrogenophilus* (Schauder *et al.*, 1987) and *Aquifex pyrophilus*. The Archaeon *Acidianus brierleyi* (originally named *Sulfolobus brierleyi*) can grow heterotrophically or autotrophically, and in the latter mode was thought to utilise a reverse citric acid cycle; however, doubt has now been thrown on that suggestion in the absence of detectable ATP-citrate lyase activity, and evidence has been provided for a modified 3-hydroxypropionate cycle (Ishii *et al.*, 1997).

Partial citric acid cycles, leading to 2-oxoglutarate either oxidatively via citrate or reductively via succinate, have been found in the methanogens. However, the extent of incompleteness of these partial cycles must be treated with caution. For example, *M. jannaschii* uses the partial reductive cycle, synthesising 2-oxoglutarate via succinate, yet from its genome sequence there appear to be genes encoding both isocitrate dehydrogenase and aconitase (Bult *et al.*, 1996). Whether or not they are expressed into functional enzymes is at present unknown, although there is no evidence from the sequence alignments for the presence of a citrate lyase to complete the reductive cycle.

The essentially invariant chemistry of the citric acid cycle across all organisms supports the proposal from Wächtershäuser (1990) that it was one of the first pathways to have evolved. Moreover, the reductive cycle may be one of the most primitive central metabolic pathways, perhaps serving as a CO_2-fixing route, with the oxidative, bioenergetic cycle awaiting the availability of oxygen as a suitable electron acceptor. An alternative scenario, where the cycle evolved in two halves as a 'horseshoe' structure for amino acid biosynthesis, has been proposed by Meléndez-Hevia *et al.* (1996).

17.3 The enzymes of central metabolism

Comparative enzymology at the structural level is a powerful tool, where, in the case of archaeal proteins, it can provide invaluable information on the evolution of metabolism and on the structural basis of enzyme hyperstability. The enzymes of central metabolism are especially amenable to such investigations, as these pathways are essentially common to all three evolutionary domains (Danson, 1993). Consequently there are numerous reports on structural studies of archaeal central metabolic enzymes, the most detailed data being available for glucose dehydrogenase (John *et al.*, 1994), the aldehyde ferredoxin oxidoreductases (Chan *et al.*, 1995), citrate synthase (Russell *et al.*, 1994, 1997), glutamate dehydrogenase (Yip *et al.*, 1995; Rice *et al.*, 1996) and malate dehydrogenase (Dym *et al.*, 1995). It is beyond the scope of this review to report in detail on these studies. Rather, we have chosen to summarise our work on citrate synthase and, from the structural principles deduced, to suggest a hypothesis on the nature of archaeal enzyme structures.

17.3.1 *Citrate synthase*

We have chosen citrate synthase, which catalyses the condensation of oxaloacetate and acetyl-CoA at the start of the citric acid cycle, as a model protein for structural and evolutionary studies (Muir *et al.*, 1994). We have cloned, sequenced and expressed in *E. coli* the genes encoding citrate synthase from the Archaea *Pyrococcus furiosus* (optimum growth at 100 °C) (Muir *et al.*, 1995), *Sulfolobus solfataricus* (80 °C) (Connaris *et al.*, 1997), *Thermoplasma acidophilum* (55 °C) (Sutherland *et al.*, 1990, 1991), and a psychrophilic bacterium (10 °C) (Gerike *et al.*, 1997). With the crystal structure of the eukaryal citrate synthase from pig already available (Remington *et al.*, 1982), we have determined the 3-D structures of the recombinant enzyme from *Tp. acidophilum* and *P. furiosus* the psychrophile (Russell *et al.*, 1994, 1997, 1998), and from *S. solfataricus* and the psychrophile (unpublished data from our laboratory).

Consistent with the evidence from the nature of the pathways, the evidence from the citrate synthase structures also suggests the establishment of the central metabolic enzymatic machinery before the divergence of Archaea, Bacteria and Eukarya. Thus, the crystallography data show that these dimeric citrate synthases share a high degree of structural homology, despite the sequence identities being as low as 25% between the three domains, and the catalytic mechanism has been conserved. Comparisons of the two thermophilic archaeal citrate synthase structures with that from pig have revealed a number of structural trends that correlate with increasing thermostability (Russell *et al.*, 1997):

- An increase in the compactness of the subunits, achieved by shortening of loops, an increase in the number of atoms buried, optimised packing of side-chains in the interior of the protein, and a reduction in the number and size of internal cavities.
- A more intimate association of the subunits of the dimer, achieved by a greater complementarity of the monomers and, in the *Pyrococcus* enzyme, by the C-terminal region of each monomer folding over the surface of the other monomer.
- An increase in complex ion-pair networks, especially at the subunit interface.
- A reduction in the content of the thermolabile residues asn, gln and cys, and of aspartate, which can undergo chain-cleavage reactions at elevated temperatures.

Not all these features are common to other thermophilic enzymes studied from the Archaea, although the presence of ion-pair networks is being recognised as a consistent theme in hyperthermophilic proteins.

However, thermal stability is only one factor in the function of enzymes at high temperatures. With the range of recombinant citrate synthases that we have generated, we are now in a position to explore how the flexibility necessary for enzymic catalysis is achieved in these hyperstable enzymes, and whether structural rigidity to maintain conformational stability imposes limits on the catalytic efficiency that can be achieved. Furthermore, we have recently raised the question of significant changes in pK_a values of ionisable amino acid residues with increasing temperature, a factor that may be important in active-site histidine residues, for example, that act as proton donors and acceptors during catalysis, as they do in the citrate synthase mechanism (Danson *et al.*, 1996).

We have recently cloned and sequenced the citrate synthase gene from the Archaeon *Haloferax volcanii* (D. G. Maddocks, D. W. Hough and M. J. Danson, unpublished data) and so we may soon be in a position to investigate the structural basis of halophilicity as well as thermophilicity.

17.4 Do archaeal enzymes represent the minimal functional unit?

The differences in the homologous structures of the citrate synthases, and of other archaeal enzymes that have been studied at the atomic level, tend to be such that the archaeal enzymes are more compact structures than their non-archaeal counterparts. These observations have led us tentatively to suggest that archaeal enzymes might represent the *minimal functional unit*, defined as the minimal structural framework required to maintain structural integrity and yet retain specific catalytic activity. Furthermore, if the Archaea are the most primitive of extant organisms, then evolution might have involved additions to this minimum structural framework in solvent-exposed regions as environmental constraints relaxed. The maintenance of this minimal structure in many archaeal proteins may be due to the constraints imposed on these organisms and their constituents by the extreme environments in which they continue to live.

What is the evidence for this minimal functional unit? In the case of the citrate synthases, which are predominantly α-helical proteins, the archaeal enzymes show both a dramatic reduction in the sizes of many of the loops connecting the α-helices and an absence of the 35-residue N-terminal region, when compared with the pig enzyme. Both features result in a more compact archaeal protein structure. From sequence alignments of the *Sulfolobus* and *Haloferax* citrate synthases, loop truncations and the absence of the N-terminal regions are again predicted.

Similar features can be seen in archaeal dehydrogenases. The nucleotide-binding domain is a conserved structural feature present in many dehydrogenases. In the glucose dehydrogenase from *Tp. acidophilum*, this domain has fewer secondary structural elements than the homologous domain in non-archaeal dehydrogenases (John *et al.*, 1994). Furthermore, the malate dehydrogenase from *Haloarcula marismortui* (Dym *et al.*, 1995) has a smaller nucleotide-binding domain than that of the thermophilic bacterium *Thermus flavus* (Kelly *et al.*, 1993), resulting from a truncation of a number of solvent-exposed loops. In comparison to the dogfish lactate dehydrogenase, the nucleotide-binding domains are of similar size, but the catalytic domain is smaller in the archaeal enzyme. However, there are no significant differences in the sizes of the loops or solvent-exposed regions of the *P. furiosus* glutamate dehydrogenase compared with that from *Clostridium symbiosum* (Yip *et al.*, 1995; Rice *et al.*, 1996).

We have also found support for the hypothesis from sequence alignments where at least one representative crystal structure of the non-archaeal protein has been determined. At the time of writing, these include glyceraldehyde-3-phosphate dehydrogenase, triose-phosphate isomerase, superoxide dismutase, elongation factor-2, hsp70, enolase and aspartate aminotransferase, with both thermophilic and halophilic archaeal examples.

As whole genomes are being sequenced, and more enzyme crystal structures are being determined, many further comparisons will be possible in the near future, and the hypothesis that we have proposed will be either confirmed or refuted. Even at this point, it must be noted that not all archaeal proteins show significant regions that are absent when compared with the non-archaeal sequence, although we know of very few examples where the archaeal sequence is the significantly larger of the two. More crystal structures are clearly needed, especially of proteins from the nonthermophilic Archaea and the thermophilic Bacteria.

17.5 Does metabolite stability control the upper temperature limit of life?

Much of the work to date on archaeal enzymes has concentrated on the structural basis of protein stability to temperatures in excess of 100 °C. This has largely been driven by potential biotechnological applications of these enzymes as well as by a desire to understand how such enzymes can operate at the physiological temperatures of the hyperthermophilic Archaea. It is clear that, through a combination of intrinsic and extrinsic factors, enzyme stability at temperatures in excess of 120 °C is achievable. This may not be the case for all enzymes, and the maximum growth temperature will be determined by the most labile component. However, there is the general feeling that achievement of protein stability is not the overriding factor that determines the possible upper temperature limit.

On the other hand, metabolite stability may be the crucial determinant of the upper temperature at which any organism can live. Oxaloacetate, one of the substrates for citrate synthase, has a half-life of <1 s at 100 °C, and metabolites such as carbamoyl phosphate are equally unstable. If the metabolite(s) to or from which such compounds are converted are stable entities, then by definition the equilibrium position will lie in favour of the more stable component. For example, the equilibrium constant at pH 7 between malate and oxaloacetate is of the order of 10^{11} and cellular oxaloacetate will essentially be stored as the thermostable malate, thereby reducing its rate of thermal destruction.

However, unlike the high intracellular concentrations of unusual compounds that can stabilise proteins in hyperthermophilic Archaea (Hensel, 1993), there is no known analogous way of stabilising thermolabile metabolites. One possible enzymic mechanism might be substrate channelling, where the product of one enzyme would be handed on directly to the next enzyme in the pathway, and not be released into the bulk solution. If the metabolite were stable when protein-bound, this would be a mechanism for its protection. Evidence for the stabilisation of carbamoyl phosphate by this mechanism has been provided (Legrain *et al.*, 1995).

While these mechanisms may be effective for metabolites in a single metabolic reaction, it is unlikely to be possible for compounds participating throughout an organism's metabolism; e.g. NAD(P)H and ATP. It might then be the rate of hydrolysis of these metabolites that sets the upper temperature limit of life, and this was an important consideration that led us to propose that non-haem iron proteins may have been the earliest redox systems (Daniel and Danson, 1995).

17.6 Conclusions and future trends in archaeal enzymology

The study of archaeal enzymology is being driven by the fascination of the origin and evolution of life, the desire to understand the fundamentals of protein hyperstability, and the potential rewards that biotechnological applications are offering.

There are still only a few atomic structures of archaeal enzymes in the database, and it is clear that many more are required before we can begin to have confidence that the observed structural trends do indeed contribute to their intrinsic stabilities. Even then, these features will have to be supported by site-directed mutagenesis experiments and by thermodynamic analyses of unfolding processes, of which very few archaeal examples have been reported to date. In the absence of such data, the ability to engineer stability into a mesophilic enzyme of choice seems a long way off. Consequently, there is now a growing opinion that the quickest and most direct route to obtaining a specific and stable catalyst is not to engineer a mesophilic version but to isolate the required enzyme from an extremophile. Failing that, the best alternative might be to find an extremophilic enzyme that carries out the desired chemistry and then manipulate its substrate specificity to what is desired.

Finally, the speed of genomic sequencing has provided a new and additional role for the archaeal enzymologist. Taking the case of *Methanococcus jannaschii* as an example (Bult *et al.*, 1996), approximately 50% of the open-reading frames remain unidentified with respect to matches in the existing sequencing databases, and even those that do show homology can only be tentatively assigned on this basis. Therefore, a major task is now to express these genes and identify the function of the protein/enzyme product. This will be no mean feat – the easy task has been accomplished; the enzymologist now faces the more difficult challenge.

17.7 Summary

Central metabolism provides a valuable source of enzymes for comparative structural analyses and evolutionary studies. These metabolic pathways in the Archaea are briefly described, and factors specific to this evolutionary domain are highlighted. Our studies on the citric acid cycle enzyme, citrate synthase, are reviewed in the context of understanding the structural basis of archaeal protein hyperstability and catalytic function at high temperatures. The hypothesis is forwarded that archaeal enzymes may represent the minimal functional unit, and additions to this structural framework are considered with respect to the adaptation of life from thermophilic to mesophilic environments.

Acknowledgements

We thank the Biotechnology and Biological Sciences Research Council, UK, for financial support. We are also indebted to Dr M. L. Dyall-Smith (University of Melbourne, Australia) and to Dr J. Rossjohn (St Vincent's Medical School, Melbourne, Australia) for many helpful discussions. Finally, we thank Professor Peter Schönheit for sharing with us his insights into archaeal metabolism, and for allowing us to reproduce his data in Figure 17.1. Through limitations of space, it has not been possible to cite all original papers but merely to refer to reviews in which those references are quoted. We apologise to all authors concerned.

References

BLAUT, M. (1994) Metabolism of methanogens. *Antonie van Leeuwenhoek*, **66**, 187–208.

BULT, C. J., WHITE, O., OLSEN, G. J. *et al.* (1996) Complete genome sequence of the methanogenic Archaeon *Methanococcus jannaschii*. *Science*, **273**, 1058–1073.

CHAN, M. K., MUKUND, S., KLETZIN, A., ADAMS, M. W. W. and REES, D. C. (1995) Structure of a hyperthermophilic tungstopterin enzyme, aldehyde ferredoxin oxidoreductase. *Science*, **267**, 1463–1469.

CONNARIS, H., WEST, S. M., HOUGH, D. W. and DANSON, M. J. (1997) Cloning and overexpression in *Escherichia coli* of the gene encoding citrate synthase from the hyperthermophilic Archaeon *Sulfolobus solfataricus*. *Extremophiles*, **2**, 61–66.

DANIEL, R. M. and DANSON, M. J. (1995) Did primitive microorganisms use non-haem iron proteins in place of NAD(P)? *J. Mol. Evol.*, **40**, 559–563.

DANSON, M. J. (1993) Central metabolism of the Archaea. *New Comp. Biochem. (The Biochemistry of Archaea)*, **26**, 1–24.

DANSON, M. J., HOUGH, D. W., RUSSELL, R. J. M., TAYLOR, G. L. and PEARL, L. (1996) Enzyme thermostability and thermoactivity. *Protein Eng.*, **9**, 629–630.

DOOLITTLE, W. F. and BROWN, J. R. (1994) Tempo, mode, the progenote and the universal root. *Proc. Natl Acad. Sci.*, **91**, 6721–6728.

DYM, O., MEVARECH, M. and SUSSMAN, J. L. (1995) Structural features that stabilise halophilic malate dehydrogenase from an archaebacterium. *Science*, **267**, 1344–1346.

FITZ-GIBBON, S., CHOI, A. J., MILLER, J. H. *et al.* (1997) A fosmid-based genomic map and identification of 474 genes of the hyperthermophilic Archaeon *Pyrobaculum aerophilum*. *Extremophiles*, **1**, 36–51.

GERIKE, U., DANSON, M. J., RUSSELL, N. J. and HOUGH, D. W. (1997) Sequencing and expression of the gene encoding a cold-active citrate synthase from an Antarctic Bacterium, Strain DS2-3R. *Eur. J. Biochem.*, **248**, 49–57.

HENSEL, R. (1993) Proteins of extreme thermophiles. *New Comp. Biochemistry (The Biochemistry of Archaea)*, **26**, 209–221.

ISHII, M., MIYAKE, T., SATOH, T. (1997) Autotrophic carbon dioxide fixation in *Acidianus brierleyi*. *Arch. Microbiol.*, **166**, 368–371.

JOHN, J., CRENNELL, S. J., HOUGH, D. W., DANSON, M. J. and TAYLOR, G. L. (1994) The crystal structure of glucose dehydrogenase from *Thermoplasma acidophilum*. *Structure*, **2**, 385–393.

KELLY, C. A., NISHIYAMA, M., OHNISHI, Y., BEPPU, T. and BIRKTOFT, J. J. (1993) Determinants of protein thermostability observed in the 1.9 Å crystal structure of malate dehydrogenase from the thermophilic bacterium *Thermus flavus*. *Biochemistry*, **32**, 3913–3922.

KELLY, R. M. and ADAMS, M. W. W. (1994) Metabolism in hyperthermophilic microorganisms. *Antonie van Leeuwenhoek*, **66**, 247–270.

KENGEN, S. W. M., STAMS, A. J. M. and DE VOS, W. M. (1996) Sugar metabolism in hyperthermophiles. *FEMS Micro. Rev.*, **18**, 119–137.

LEGRAIN, C., DEMAREZ, M., GLANSDORFF, N. and PIERARD, A. (1995) Ammonia-dependent synthesis and metabolic channeling of carbamoyl-phosphate in the hyperthermophilic archaeon *Pyrococcus furiosus*. *Microbiology*, **141**, 1093–1099.

MELÉNDEZ-HEVIA, E., WADDELL, T. G. and CASCANTE, M. (1996) The puzzle of the Krebs citric acid cycle: assembling the pieces of chemically feasible reactions, and opportunism in the design of metabolic pathways during evolution. *J. Mol. Evol.*, **43**, 293–303.

MUIR, J. M., HOUGH, D. W. and DANSON, M. J. (1994) Citrate synthases from the Archaea. *System. Appl. Microbiol.*, **16**, 528–533.

MUIR, J. M., RUSSELL, R. J. M., HOUGH, D. W. and DANSON, M. J. (1995) Citrate synthase from the hyperthermophilic Archaeon *Pyrococcus furiosus*. *Protein Eng.*, **8**, 583–592.

RAWAL, N., KELKAR, S. M. and ALTEKAR, W. (1988) Alternative routes of carbohydrate metabolism in halophilic archaebacteria. *Ind. J. Biochem. Biophys.*, **25**, 674–686.

REMINGTON, S., WIEGAND, G. and HUBER, R. (1982) Crystallographic refinement and atomic models of two different forms of citrate synthase at 2.7 and 1.7 Å resolution. *J. Mol. Biol.*, **158**, 111–152.

RICE, D. W., YIP, K. S. P., STILLMAN, T. J. *et al.* (1996) Insights into the molecular basis of thermal stability from the structure determination of *Pyrococcus furiosus* glutamate dehydrogenase. *FEMS Microbiol. Rev.*, **18**, 105–117.

RUSSELL, R. J. M., HOUGH, D. W., DANSON, M. J. and TAYLOR, G. L. (1994) The crystal structure of citrate synthase from the thermophilic Archaeon *Thermoplasma acidophilum*. *Structure*, **2**, 1157–1167.

RUSSELL, R. J. M., CAMPBELL-FERGUSON, J. M., HOUGH, D. W., DANSON, M. J. and TAYLOR, G. L. (1997) Determinants of protein hyperthermostability revealed by the crystal structure of *Pyrococcus furiosus* citrate synthase. *Biochemistry, USA*, **36**, 9983–9994.

RUSSELL, R. J. M., GERIKE, U., DANSON, M. J., HOUGH, D. W. and TAYLOR, G. L. (1998) Structural adaptations of the cold-active citrate synthase from an Antarctic bacterium. *Structure*, **6**, 351–361.

SCHAUDER, R., WIDDEL, F. and FUCHS, G. (1987) Carbon assimilation pathways in the sulfate-reducing bacteria. II. Enzymes of a reductive citric acid cycle in the autotrophic *Desulfobacter hydrogenophilus*. *Arch. Microbiol.*, **148**, 218–225.

SCHÖNHEIT, P. and SCHÄFER, T. (1995) Metabolism of hyperthermophiles. *World J. Microbiol. Biotechnol.*, **11**, 26–57.

SELIG, M., XAVIER, K. B., SANTOS, H. and SCHÖNHEIT, P. (1997) Comparative analysis of Embden–Meyerhof and Entner–Doudoroff glycolytic pathways in hyperthermophilic archaea and the bacterium *Thermotoga*. *Arch. Microbiol.*, **167**, 217–232.

SHIBA, H., KAWASUMI, T., KODAMA, T. and MINODA, Y. (1985) CO_2 assimilation via the reductive tricarboxylic acid cycle in an obligately autotrophic hydrogen-oxidising bacterium, *Hydrogenobacter thermophilus*. *Arch. Microbiol.*, **141**, 189–203.

SIEBERS, B. and HENSEL, R. (1993) Glucose catabolism of the hyperthermophilic archaeon *Thermoproteus tenax*. *FEMS Microbiol. Lett.*, **111**, 1–8.

SUTHERLAND, K. J., HENNEKE, C. M., TOWNER, P., HOUGH, D. W. and DANSON, M. J. (1990) Citrate synthase from the thermophilic archaebacterium *Thermoplasma acidophilum*: cloning and sequencing of the gene. *Eur. J. Biochem.*, **194**, 839–844.

SUTHERLAND, K. J., DANSON, M. J., HOUGH, D. W. and TOWNER, P. (1991) Expression and purification of plasmid-encoded *Thermoplasma acidophilum* citrate synthase from *Escherichia coli*. *FEBS Lett.*, **282**, 132–134.

WÄCHTERSHÄUSER, G. (1990) Evolution of the first metabolic cycles. *Proc. Natl Acad. Sci. USA*, **87**, 200–204.

WOESE, C. R., KANDLER, O. and WHEELIS, M. L. (1990) Towards a natural system of organisms: proposals for the domains Archaea, Bacteria and Eukarya. *Proc. Natl Acad. Sci. USA*, **87**, 4576–4579.

YIP, K. S. P., STILLMAN, T. J. and BRITTON, K. L. (1995) The structure of *Pyrococcus furiosus* glutamate dehydrogenase reveals a key role for ion-pair networks in maintaining stability at extreme temperatures. *Structure*, **3**, 1147–1158.

18

Pyrophosphate-dependent Phosphofructokinases in Thermophilic and Nonthermophilic Microorganisms

HUGH W. MORGAN AND RON S. RONIMUS
Thermophile Research Unit, University of Waikato, Hamilton, New Zealand

18.1 Phosphoryl donors of phosphofructokinase and their distribution

Glycolysis is a central pathway of carbohydrate catabolism which is present in the three domains of life. This remarkable conservation reflects its importance throughout evolution. Phosphofructokinase (PFK) is regarded as one of the key enzymes in glycolysis because it usually catalyses an important irreversible reaction and is a major control point for allosteric effectors of the pathway. The phosphoryl donor for this reaction is conventionally adenosine triphosphate (ATP) and the enzyme ATP:D-fructose-6-phosphate 1-phosphotransferase (ATP-PFK, EC 2.7.1.11) catalyses the reaction

$$\text{ATP} + \text{fructose-6-P} \rightarrow \text{fructose-1,6-P}_2 + \text{ADP} \ (\Delta G^{\circ\prime} = -18.45 \text{ kJ/mol})$$

Other forms of phosphoryl donor for this reaction are known, of which the most common is the substitution of pyrophosphate (PP_i) for ATP (Mertens, 1991). This PP_i-PFK (inorganic pyrophosphate:D-fructose-6-phosphate 1-phosphotransferase, EC 2.7.1.90) is also found in organisms from all three domains of life and catalyses the reversible reaction

$$\text{PP}_i + \text{fructose-6-P} \quad \text{fructose-1,6-P}_2 + \text{P}_i \ (\Delta G^{\circ\prime} = -8.70 \text{ kJ/mol})$$

Higher plants and the photosynthetic protists, e.g. *Euglena gracilis*, contain cytoplasmic PP_i-PFKs as well as ATP-PFKs. The PP_i-PFKs from these organisms are heterotetrameric proteins which are strongly activated by fructose 2,6-bisphosphate. They appear to be involved in the regulation of gluconeogenesis/glycolysis in the photosynthetic tissues only, and are quite distinct from the dimeric, nonallosteric PP_i-PFKs characterised thus far in bacteria. These allosterically controlled PP_i-PFKs from higher plants and *Euglena* will not be discussed further in this review.

The thermophilic archaeon *Pyrococcus furiosus* has a completely unique phosphoryl donor for the PFK enzyme (Kengen *et al.*, 1996) and utilises ADP. This enzyme (ADP-PFK, ADP:D-fructose-6-phosphate 1-phosphotransferase) catalyses the reaction

$$\text{ADP} + \text{fructose-6-P} \rightarrow \text{fructose-1,6-P}_2 + \text{AMP}$$

Table 18.1 Summary of properties and distribution of pyrophosphate[a], ADP- and some ATP-dependent phosphofructokinases (PFK)

Genus/species	Molecular mass[b] (kDa)	Structure	K_m (F-6-P) (μmol/l)	K_m (P-donor) (μmol/l)	Reference
PP_i-dependent PFK: eubacterial					
Spirochaeta thermophila	57	Monomer	240	45	This work
Propionibacterium freudenreichii	43	Homodimer	100	69	O'Brien *et al.* (1975), Ladror *et al.* (1991)
Amycolatopsis methanolica	36	Tetramer	400	200	Alves *et al.* (1996)
Achloeplasma laidlawii	37	Dimer	650	110	De Santis *et al.* (1989)
Alcaligenes	—	—	—	—	Sawyer *et al.* (1977)
Pseudomonas marina	—	—	—	—	Sawyer *et al.* (1977)
Anaeroplasma intermedium	—	—	—	—	Petzel *et al.* (1989)
Streptococcus pleomorphus	—	—	—	—	Petzel *et al.* (1989)
Erysipelothrix rhusiopathiae	—	—	—	—	Petzel *et al.* (1989)
Clostridium inocuum	—	—	—	—	Petzel *et al.* (1989)
Bacteroides fragilis	—	—	—	—	Roberton and Glucina (1982)
Rhodospirillum rubrum	95	—	380	25	Pfleiderer and Klemme (1980)
Actinoplanes sp.	—	—	—	—	Alves *et al.* (1996)
Actinomyces naeslundii	—	—	—	—	Takahashi *et al.* (1995)
PP_i-dependent PFK: archaeal					
Thermoproteus tenax	—	—	235	50	Siebers and Hensel (1993)
PP_i-dependent PFK: primitive eukaryal					
Toxoplasma gondii	45	Dimer	270	33	Peng *et al.* (1992)
Giardia lamblia	64	Monomer	250	39	Li and Phillips (1995)
Trichomonas vaginalis	48	Dimer	80	12	Mertens *et al.* (1989)
Tritrichomonas foetus	—	—	120	14	Mertens *et al.* (1989)
Isotricha prostoma	48	Dimer	80	12	Mertens *et al.* (1989)
Entamoeba histolytica	41.5	Dimer	38	14	Huang *et al.* (1995)
Eimeria tenella	—	—	77	—	Denton *et al.* (1994)
Euglena gracilis	110	Monomer	300	100	Enomoto *et al.* (1988)
Naegleria fowleri	48	Tetramer	10	15	Mertens *et al.* (1993)
Cryptosporidium parvum	—	—	320	—	Denton *et al.* (1996)
ADP-dependent PFK					
Pyrococcus furiosus	—	—	—	—	Kengen *et al.* (1994)
Thermococcus celer	—	—	—	—	Selig *et al.* (1997)
Thermococcus litoralis	—	—	—	—	Selig *et al.* (1997)
Thermococcus zilligii	—	—	—	—	This work
ATP-dependent PFK: thermophilic eubacterial					
Thermotoga maritima	—	—	—	—	Selig *et al.* (1997)
Thermus thermophilus	36.5	Tetramer	60–100	20–30	Xu *et al.* (1990)
Thermus strain X-1	33	Tetramer	90	130	Cass and Stellwagen (1975)
Bacillus stearothermophilus	33.9	Tetramer	30	100	Byrnes *et al.* (1994)
ATP-dependent PFK: archaeal					
Desulfurococcus amylolyticus	—	—	—	—	Selig *et al.* (1997)

[a] PP_i-dependent PFK activity has also been identified in other species of *Amycolatopsis* (Alves *et al.*, 1994), *Alcaligenes* (Sawyer *et al.*, 1977), *Acholeplasma* (Pollack and Williams, 1986), *Naegleria* (Mertens *et al.*, 1993), *Rhodopseudomonas* (Pfleiderer and Klemme, 1980) and *Spirochaeta* (unpublished results).

[b] Data derived from either SDS-PAGE as is the case for *Spirochaeta thermophila*, or from gene sequence data where possible.

No $\Delta G^{o\prime}$ value has been given for the ADP-driven reaction but presumably it lies between the values of the ATP and PP_i reactions. The low $\Delta G^{o\prime}$ value for the PP_i-PFK reaction, which could be close to zero *in vivo* (Mertens, 1991), makes the reaction more easily reversible, whereas the ATP-dependent (and presumably also the ADP-dependent) reaction is regarded as irreversible under normal physiological conditions. Each of these three enzymes is nearly absolute in its requirement for the specified phosphoryl donor, with commonly less than 2% of activity being exhibited with the alternative donors.

The distribution of these different forms of PFK and some of the biochemical properties of the enzymes are summarised in Table 18.1. This table contains a current listing of all references to the PP_i- and ADP-dependent enzymes, but only illustrates the thermophilic examples of the ATP-dependent PFKs found in eubacteria. The assumption is made that ATP-dependent PFKs are widespread among all higher eukaryotic and most aerobic organisms.

18.2 Phylogeny and evolution of the different types of PFK

The occurrence and distribution of the various forms of PFK within the phylogenetic tree is not readily explained. The near universal distribution of ATP-PFKs has led to the belief in the primacy of this phosphoryl donor in the evolution of the enzyme (and of the glycolytic pathway). Mertens (1991) proposed that 'The distribution of PP_i-PFK in the living world appears to correlate better with metabolic characteristics (anaerobiosis) than with phylogenetic relationships. An interesting possibility is, therefore, that PP_i-PFK derived from ATP-PFK, probably not once but on several independent occasions'. The reverse argument, that PP_i-PFK preceded the evolution of ATP-PFK, has been made by Baltscheffsky (1996) largely on geochemical grounds and by Dijkhuizen (1996) on phylogenetic evidence.

The low $\Delta G^{o\prime}$ value of the PP_i-PFK enzyme facilitates its involvement in gluconeogenesis by allowing the reversal of the 'normal' glycolytic pathway. Wächtershäuser (1994) has proposed that the earliest form of metabolism was a reductive citric acid cycle and that related pathways, e.g. glycolysis and the Entner–Doudoroff pathway, arose from this later by evolving in the anabolic direction. Fothergill-Gilmore and Michels (1993) have suggested a convincing scenario in which glycolysis might have evolved in such a manner, and in which the requirement for a PP_i-PFK is a prerequisite.

The nonallosteric PP_i-PFKs are predominantly found in anaerobic eubacteria and primitive eukaryotes. These organisms use glycolysis as their major ATP-generating pathway, so the necessity for a reversible phosphofructokinase cannot be related to a gluconeogenic role when the overwhelming substrate flow is in the catabolic direction. The current hypotheses to account for the distribution of PP_i-PFKs in different organisms make quite different assumptions about the phylogeny of the enzymes. Mertens (1991) interprets the occurrence of the enzyme in quite diverse and unrelated phylogenetic lineages as evidence of its spontaneous derivation from an ATP-PFK on many separate occasions; a likely reason for this is adaptation to an anaerobic mode of existence. The rationale for this argument is that ATP yield by glycolytic anaerobes is barely sufficient for growth, and that the use of a PP_i-PFK provides an energetic advantage by scavenging the PP_i released as a by-product of other biosynthetic reactions (formation of nucleic acids, proteins and polysaccharides) which would otherwise be hydrolysed to inorganic phosphate by pyrophosphatase. Theoretically this could improve the ATP yield of glycolytic organisms by 50%, but for organisms with oxidative phosphorylation the yield gain would be far less significant. Mertens (1991) pointed out the prevalence of PP_i-PFKs in obligate anaerobes (including the anaerobic protists *Giardia* and *Entamoeba*) and its absence from aerobic bacteria as support for this contention. However, there are aerobic bacteria which possess a PP_i-PFK (Alves *et al.*, 1994) and the majority of glycolytic anaerobes utilise an ATP-PFK.

In Baltscheffsky's view (1996) the use of PP_i as phosphoryl donor reflects its use in energy conversion leading to its role in the origin and early evolution of life. The presence of PP_i in prebiotic environments could have provided both the phosphate and energy source for evolution of early metabolic pathways. Simulation of volcanic activity by heating of orthophosphate rocks under anhydrous conditions has demonstrated the continuous formation of pyrophosphate and tripolyphosphate, and these same products have been found in condensates from volcanic magma (Yamagata *et al.*, 1991) and submarine springs, albeit at low (0.37–0.45 μmol/l) concentrations (Karl, 1995). If evolution of metabolic pathways preceded the evolution of the progenote (as seems probable), then the use of PP_i as an energy source is much more likely than that of ATP. In this scenario, the use of PP_i precedes the use of ATP in all forms of metabolism, with the latter only being incorporated as the complexity and control of metabolic reactions evolved. In this case, the distribution of present-day PP_i-PFKs in diverse phylogenetic lineages presumably reflects the remnants of a once universally distributed enzyme. The reason for retention of the PP_i-PFK in different organisms might be unique to each particular case, but it would be expected that those organisms which have the deepest and shortest phylogenetic lineages would be more likely to have retained a PP_i-PFK, reflecting their close relationship with the common ancestor.

Since PP_i formation is a feature of hydrothermal and volcanic systems, there is the possibility that thermophilic bacteria might have retained the use of this energy source to a greater degree than other bacteria, because of its ready availability in their environment. Although some suggestions have been made favouring the use of PP_i by thermophiles, based on its greater thermal stability compared to ATP (Kengen *et al.*, 1994), this is not strongly supported by empirical data. Under typical physiological conditions (temperatures of 90 °C at pH 7.0) ATP has a reported half-life of 115 min whereas that of PP_i is 750 min (Tetas and Lowenstein, 1963), but these values are likely to be orders of magnitude greater than the half-lives of these molecules in metabolism, so that their use as phosphoryl donors is unlikely to be precluded on thermal stability considerations alone. Polyphosphate can essentially be regarded as fully stable under the same conditions.

Among the Archaea, the pathway of ATP generation has only been investigated in a small number of species. It is notable that in two species that carry out glycolysis the PFK of *Thermoproteus tenax* uses PP_i as the phosphoryl donor, and *Pyrococcus furiosus* has a PFK which appears to be completely unique to members of the *Thermococcales* in that ADP is used. Selig *et al.* (1997) have recently demonstrated the presence of an ADP-dependent PFK in *Thermococcus celer* and *T. litoralis*, both members of the *Thermococcales*, and we have shown that *Thermococcus zilligii* (previously *Thermococcus* strain AN1; Ronimus *et al.*, 1997) also possesses this enzyme activity. Since the *Thermococcales* represent one of the deepest branches of the universal phylogenetic tree, this enzyme might then reflect the most ancestral pathway of sugar fermentation. However, as yet no sequence information is available on ADP-dependent PFKs, so that similarity between this enzyme and those using other phosphoryl donors cannot be assessed. *Desulfurococcus amylolyticus* is the only sugar-fermenting archaeaon to posses a typical ATP-dependent PFK; *Sulfolobus acidocaldarius* oxidises sugars by a nonphosphorylated form of the Entner–Doudoroff pathway (Selig *et al.*, 1997). Thus the anaerobic hyperthermophilic archaea have representatives which possess all known forms of PFK with respect to phosphoryl donor. Sequence analysis of the genes for these enzymes will provide an interesting insight into the phylogeny of this enzyme and the glycolytic pathway.

Among the hyperthermophilic eubacteria, the anaerobic *Thermotogales*, e.g. *Thermotoga maritima*, represents the deepest-rooted branch in the eubacterial tree with the capacity to ferment sugars, and has been reported to possess an ATP-dependent PFK (Schröder *et al.*, 1994). Janssen and Morgan (1992) demonstrated the presence of a PP_i-PFK in a novel isolate of thermophilic (T_{opt} = 68 °C) spirochaete, and subsequently some moderately thermophilic (T_{opt} = 55 °C) spirochaetes were also shown to possess PP_i-PFK activity (unpublished results).

Several other thermophilic species including the aerobic *Thermus thermophilus* and *Bacillus stearothermophilus* also possess ATP-dependent PFKs. The most deeply rooted of the eubacterial lineages is that represented by *Aquifex*, but relatively little is known about the enzymology of this genus. Its ability to oxidise hydrogen and grow autotrophically might favour the reversible form of the phosphofructokinase reaction in order to facilitate glucose synthesis from the reductive TCA cycle. Mesophilic eubacteria possessing PP_i-dependent PFK are listed in Table 18.1. Comparisons of partial amino acid sequences of regions encoding the binding sites for fructose-6-phosphate and the phosphoryl donor of PFK enzymes from various organisms showed that the PP_i- and ATP-dependent enzymes formed a monophyletic group (Alves *et al.*, 1996). This strongly suggests that both types of PFK enzyme evolved from a common ancestor, and Alves *et al.* (1996) proposed that the ancestral PFK more likely resembled a simpler form of this protein: a PP_i-dependent homodimeric enzyme lacking allosteric control.

18.3 Phosphofructokinase activity in spirochaetes

We have further investigated the PFK activity in members of the *Spirochaetales*. The spirochaete lineage is not particularly deep-rooted but, importantly, it is conserved and contains thermophilic and mesophilic members. All species so far investigated possess only PP_i-dependent PFK activity (Table 18.2) so that thermophily is not the sole determinant of selection for the PP_i-based activity in this lineage.

Table 18.2 Specific activities of phosphofructokinase in crude cell extracts of spirochaetes

Species/strain	Optimum growth temperature (°C)	Specific activity (μmol/min per mg) with ATP	Specific activity (μmol/min per mg) with PP_i	Ratio of PP_i to ATP activity of cell extracts
S. thermophila				
Strain RI 19.B1	68	0.03	4.63	154
S. thermophila				
Strain Z-1203	70	0.01	0.30	300
Strain GAB 73	60	0.07	5.98	85
Strain Rt 118.B2	50	0.04	2.38	59
Strain Wai 21.B2	50	0.09	5.41	60
S. stenostrepta	25	0.01	0.08	8

Table 18.3 Substrate specificity of purified phosphofructokinase from *Spirochaeta thermophila* RI 19.B1

Substrate	Percentage activity, pH 6.8, 3.5 mmol/l $MgCl_2$	Percentage activity, pH 6.8, 30 mmol/l $MgCl_2$	Percentage activity, pH 6.0, 3.5 mmol/l $MgCl_2$
PP_i	100	100	100
PPP_i	73	95	100
Polyphosphate	0	56	nd
Adenosine-P_4	13	nd	6
ATP	3	nd	3
ADP	2	nd	nd
AMP	0	nd	nd
Acetyl phosphate	13	nd	26

PPP_i = tripolyphosphate. nd = not determined. No activity was recorded with either 2-phosphoglycerate, phosphoenolpyruvate or carbamoyl phosphate. Adenosine-P_4 = adenosine tetraphosphate.

The PFK from *Spirochaeta thermophila* has been purified to homogeneity. Unusually, it has a monomeric protein structure, a relatively high K_m for the fructose 6-phosphate substrate and is nonallosteric. The latter two characteristics are regarded by Alves *et al.* (1996) as likely properties of an ancestral enzyme. Although the N-terminal amino acid sequence of this enzyme did not show high homology to other bacterial PFKs, the validation of this claim must await the full amino acid sequence.

If phosphoanhydride bonds of geothermal origin were energy sources for the evolution of metabolic pathways prior to the first progenote (Baltscheffsky, 1996), then forms other than PP_i (for example, polyphosphates) might also be utilised. Polyphosphates serve a variety of metabolic functions and are ubiquitous in all life forms to the extent that they have been termed a 'molecular fossil' (Kornberg, 1995). Interestingly, polyphosphates functioned as phosphoryl donor for the purified PP_i-dependent PFK from *Spirochaeta thermophila.* The enzyme activity is influenced by pH and polyphosphate chain length; the latter is partly a reflection of the available magnesium concentration, which is a requirement for PFK activity. Polyphosphates chelate magnesium ions so that activity is only evident when an excess of magnesium is provided (Table 18.3). Polyphosphate activation of PFK enzymes has not previously been recorded, although polyphosphate substitution for ATP by glucokinase has been reported in several species of eubacteria (Hsieh *et al.*, 1993).

The phosphoryl substrate range for the purified PFK from *Spirochaeta thermophila* (Table 18.3) shows substantial activity against acetyl phosphate. This is the first report of a PP_i-dependent PFK exhibiting activity on such a substrate, and is of interest since Baltscheffsky (1996) has proposed that acetyl phosphate is a key intermediate in the evolution of energy flow in a prenucleotide world. In this scenario, reactions driven by the use of thioesters, which presupposes a heterotrophic origin of life, evolved to use PP_i via the intermediary of acetyl phosphate as energy carrier. The residual activity shown by the PP_i-dependent PFK might represent a molecular relic of that evolution. This must be viewed with some caution. First, there are strong arguments to support the hypothesis that the prenucleotide world originated autotrophically by using the reductive force of pyrite formation (Keller *et al.*, 1994); and second, the use of acetyl phosphate by a larger number of PP_i-dependent PFK enzymes needs to be ascertained.

18.4 Conclusions

The earlier hypothesis by Mertens (1991) has not been substantiated by recent findings. The presence of PP_i-dependent PFK activity in aerobic bacteria and the unique occurrence of an ADP-dependent PFK in species of *Pyrococcus* and *Thermococcus* seem at variance with a simple explanation based on energetics. A prenucleotide role for PP_i (and polyphosphate) in energising reactions prior to the evolution of the progenote might be more favourably considered, and the low $\Delta G^{\circ\prime}$ of the reaction is compatible with a heterotrophic or an autotrophic origin of life. Only the Archaea possess all three types of PFK but this might simply be a reflection of the small number of species, particularly thermophilic, which have been screened for this enzyme activity. Without a greater appreciation of the distribution of the three distinct forms of PFK, their unique biochemistry and their phylogeny through sequence comparisons, interpretation of current information must be regarded as speculation. However, we expect that further investigations into the species distribution, biochemistry and structure of phosphofructokinase genes will provide greater insights into life's origins.

18.5 Summary

Phosphofructokinases (PFKs) are key enzymes in glycolysis and their activity might predate the progenote. Most highly evolved organisms use ATP as an energy source for PFK activity, but among anaerobes and thermophiles other phosphoryl donors such as pyrophosphate (PP_i) or ADP are not uncommon. The apparently random distribution of these unusual phosphoryl donors within the universal phylogenetic tree cannot adequately be explained. Earlier hypotheses relating to energetic efficiency under anaerobic conditions are not substantiated. The occurrence and stability of pyrophosphate and polyphosphates in hydrothermal and volcanic waters has led them to be considered as feasible energy sources to drive metabolic reactions in the prenucleotide world. The PP_i-dependent PFK from *Spirochaeta thermophila* utilised both polyphosphates and acetyl phosphate as phosphoryl donor; the contention that PP_i-dependent PFK enzymes represent the ancestral form cannot be discounted.

References

Alves, A. M. C. R., Euverink, G. J. W., Hektor, H. J. *et al.* (1994) Enzymes of glucose and methanol metabolism in the Actinomycete *Amyycolatopsis methanolica. J. Bacteriol.*, **176**, 6827–6835.

Alves, A. M. C. R., Meijer, W. G., Vrijbloed, J. W. and Dijkhuizen, L. (1996) Characterisation and phylogeny of the *pfp* gene of *Amycolatopsis methanolica* encoding PP_i-dependent phosphofructokinase. *J. Bacteriol.*, **178**, 149–155.

Baltscheffsky, H. (1996) Energy conversion leading to the origin and early evolution of life: did inorganic pyrophosphate precede adenosine triphosphate? In *Origin and Evolution of Biological Energy Conversion*, ed. H. Baltscheffsky, pp. 1–9 (New York: VCH Publishers).

Byrnes, W. M., Zhu, X., Younathan, E. S. and Chang, S. H. (1994) Kinetic characteristics of phosphofructokinase from *Bacillus stearothermophilus*: MgATP nonallosterically inhibits the enzyme. *Biochemistry*, **33**, 3424–3431.

Cass, H. K. and Stellwagen, E. (1975) A thermostable phosphofructokinase from the extreme thermophile *Thermus* X-1. *Arch. Biochem. Biophys.*, **171**, 682–694.

DeSantis, D., Tryon, V. V. and Pollack, J. D. (1989) Metabolism of Mollicutes: the Emden–Meyerhof–Parnas pathway and the hexose monophosphate shunt. *J. Gen. Microbiol.*, **135**, 683–691.

Denton, H., Thong, K.-W. and Coombs, G. H. (1994) *Eimeria tenella* contains a pyrophosphate-dependent phosphofructokinase and a pyruvate kinase with unusual allosteric regulators. *FEMS Microbiol. Lett.*, **115**, 87–92.

Denton, H., Roberts, C. W., Alexander, J., Thong, K.-W. and Coombs, G. H. (1996) Enzymes of energy metabolism in the bradyzoites and tachyzoites of *Toxoplasma gondii*. *FEMS Microbiol. Lett.*, **137**, 103–108.

Dijkhuizen, L. (1996) Evolution of metabolic pathways. In *Evolution of Microbial Life*, ed. D. Mcl. Roberts, P. Sharp, G. Alderson and M. A. Collins, pp. 243–266 (Cambridge: Cambridge University Press).

Enomoto, T., Miyatake, K. and Kitaoka, S. (1988) Purification and immunological properties of fructose-2,6-biphosphate-sensitive pyrophosphate: D-fructose 6-phosphate 1-phosphotransferase from the protist *Euglena gracilis*. *Comptes Biochem. Physiol.*, **90B**, 897–902.

Fothergill-Gilmore, L. A. and Michels, P. A. M. (1993) Evolution of glycolysis. *Prog. Biophys. Mol. Biol.*, **59**, 105–236.

Hsieh, P.-C., Shenoy, B. C., Jentoft, J. E. and Phillips, N. F. B. (1993) Purification of polyphophate and ATP glucose phosphotransferase from *Mycobacterium tuberculosis* $H_{37}Ra$: evidence that poly(p) and ATP glucokinase activities are catalysed by the same enzyme. *Protein Express. Purific.*, **4**, 76–84.

Huang, M., Albach, R. A., Chang, K.-P., Tripathi, R. L. and Kemp, R. G. (1995) Cloning and sequencing a putative pyrophosphate-dependent phosphofructokinase gene from *Entamoeba histolytica*. *Biochim. Biophys. Acta*, **1260**, 215–217.

Janssen, P. H. and Morgan, H. W. (1992) Glucose catabolism by *Spirochaeta thermophila* RI 19.B1. *J. Bacteriol.*, **174**, 2449–2453.

Karl, D. M. (1995) Ecology of free-living hydrothermal vent microbial communities. In *The Microbiology of Deep-sea Hydrothermal Vents*, ed. David M. Karl, pp. 36–124 (Boca Raton FL: CRC Press).

Keller, M., Blochl, E., Wächtershäuser, G. and Stetter, K. O. (1994) Formation of amide bonds without a condensation agent and implications for the origin of life. *Nature*, **368**, 836–838.

Kengen, S. W. M., de Bok, F. A. M., van Loo, N.-D., Dijkema, C., Stams, A. J. M. and de Vos, W. (1994) Evidence for the operation of a novel Embden–Meyerhof pathway that involves ADP-dependent kinases during sugar fermentation by *Pyrococcus furiosus*. *J. Biol. Chem.*, **269**, 17537–17541.

Kengen, S. W. M., Stams, A. J. M. and de Vos, W. M. (1996) Sugar metabolism of hyperthermophiles. *FEMS Microbiol. Rev.*, **18**, 199–237.

Kornberg, A. (1995) Inorganic polyphosphate: toward making a forgotten polymer unforgettable. *J. Bacteriol.*, **177**, 491–496.

Ladror, U. S., Gollapudi, L., Tripathi, R. L., Latshaw, S. P. and Kemp, R. G. (1991) Cloning, sequencing and expression of pyrophosphate-dependent phosphofructokinase from *Propionibacterium freundenreichii*. *J. Biol. Chem.*, **266**, 16550–16555.

Li, Z. and Phillips, N. F. B. (1995) Pyrophosphate-dependent phosphofructokinase from *Giardia lamblia*: purification and characterisation. *Protein Express. Purific.*, **6**, 319–328.

Mertens, E. (1991) Pyrophosphate-dependent phosphofructokinase, an anaerobic glycolytic enzyme? *Fed. Eur. Biochem. Soc.*, **285**, 1–5.

Mertens, E., Van Schaftingen, E. and Müller, M. (1989) Presence of a fructose-2, 6-biphosphate-insensitive pyrophosphate: fructose-6-phosphate phosphotransferase in the anaerobic protozoa *Tritrichomonas foetus*, *Trichomonas vaginalis* and *Isotricha prostoma*. *Mol. Biochem. Parasitol.*, **37**, 183–190.

Mertens, E., de Jonckheere, J. and van Schaftingen, E. (1993) Pyrophosphate-dependent phosphofructokinase from the amoeba *Naegleria fowleri*, an AMP-sensitive enzyme. *Biochem. J.*, **292**, 797–803.

O'BRIEN, W. E., BOWIEN, S. and WOOD, H. G. (1975) Isolation and characterisation of a pyrophosphate-dependent phosphofructokinase from *Propionibacterium shermanii*. *J. Biol. Chem.*, **250**, 8690–8695.

PENG, Z.-Y., MANSOUR, J. M., ARAUJO, F., JU, J.-Y., MCKENNA, C. E. and MANSOUR, T. E. (1992) Some phosphonic acid analogues as inhibitors of pyrophosphate-dependent phosphofructokinase, a novel target in *Toxoplasma gondii*. *Biochem. Pharmacol.*, **49**, 105–113.

PETZEL, J. P., HARTMAN, P. A. and ALLISON, M. J. (1989) Pyrophosphate-dependent enzymes in walled bacteria phylogenetically related to the wall-less bacteria of the class *Mollicutes*. *Int. J. of System. Bacteriol.*, **39**, 413–419.

PFLEIDERER, C. and KLEMME, J.-H. (1980) Pyrophosphate-dependent D-fructose–6-phosphate-phosphotransferase in *Rhodospirillaceae*. *Z. Naturforschung*, **35c**, 229–238.

POLLACK, J. D. and WILLIAMS, M. V. (1986) PPi-dependent phosphotransferase (phosphofructokinase) activity in the Mollicutes (Mycoplasma) *Acholeplasma laidlawii*. *J. Bacteriol.*, **165**, 53–60.

ROBERTON, A. M. and GLUCINA, P. J. (1982) Fructose-6-phosphate phosphorylation in *Bacteroides* species. *J. Bacteriol.*, **15**, 1056–1060.

RONIMUS, R. S., REYSENBACH, A.-L., MUSGRAVE, D. R. and MORGAN, H. W. (1997) The phylogenetic position of the *Thermococcus* isolate AN1 based on 16S rRNA gene sequence analysis: a proposal that AN1 represents a new species, *Thermococcus zilligii* sp. nov. *Arch. Microbiol.*, **168**, 245–248.

SAWYER, M. H., BAUMANN, P. and BAUMANN, L. (1977) Pathways of D-fructose and D-glucose catabolism in marine species of *Alcaligenes*, *Pseudomonas marina* and *Alteromonas communis*. *Arch. Microbiol.*, **112**, 169–172.

SCHRÖDER, C., SELIG, M. and SCHÖNHEIT, P. (1994) Glucose fermentation to acetate, CO_2, and H_2 in the anaerobic hyperthermophilic eubacterium *Thermotoga maritima*: involvement of the Embden–Meyerhof pathway. *Arch. Microbiol.*, **161**, 460–470.

SELIG, M., XAVIER, K. B., SANTOS, H. and SCHÖNHEIT, P. (1997) Comparative analysis of Embden–Meyerhof and Entner–Doudoroff glycolytic pathways in the hyperthermophilic archaea and the bacterium *Thermotoga*. *Arch. Microbiol.*, **167**, 217–232.

SIEBERS, B. and HENSEL, R. (1993) Glucose metabolism of the hyperthermophilic archaeum *Thermoproteus tenax*. *FEMS Microbiol. Lett.*, **111**, 1–8.

TAKAHASHI, N., KALFAS, S. and YAMADA, T. (1995) Phosphorylating enzymes involved in glucose fermentation of *Actinomyces naeslundii*. *J. Bacteriol.*, **177**, 5806–5811.

TETAS, M. and LOWENSTEIN, J. M. (1963) The effect of metal ions on the hydrolysis of adenosine di- and triphosphate. *Biochemistry*, **2**, 350–357.

WÄCHTERSHÄUSER, G. (1994) Life in a ligand sphere. *Proc. Natl Acad. Sci. USA*, **91**, 4283–4287.

XU, J., OSHIMA, T. and YOSHIDA, M. (1990) Tetramer-dimer conversion of phosphofructokinase from *Thermus thermophilus* induced by its allosteric effectors. *J. Mol. Biol.*, **215**, 597–606.

YAMAGATA, Y., WATANABE, H., SAITOH, M. and NAMBA, T. (1991) Volcanic production of polyphosphates and its relevance to prebiotic evolution. *Nature*, **352**, 516–519.

PART SEVEN

Membrane Evolution

19

sn-Glycerol-1-phosphate Dehydrogenase: A Key Enzyme in the Biosynthesis of Ether Phospholipids in Archaea

MASATERU NISHIHARA[1], TAKAYUKI KYURAGI[2], NOBUHITO SONE[2] AND YOSUKE KOGA[1]

[1] *Department of Chemistry, University of Occupational and Environmental Health, Kitakyushu, Japan*
[2] *Department of Biochemical Engineering and Science, Kyushu Institute of Technology, Kawazu, Iizuka, Japan*

19.1 Introduction

Membrane structure is absolutely indispensable for a cell because the cell must be separated from the surrounding environment by a membrane. We do not know when and how the first membrane was established during evolution, but all living organisms now have membranes constructed with polar lipids. One of the distinctive molecular features common to all Archaea that distinguishes them from Bacteria and Eukarya and unites their disparate phenotypes is the nature of their membrane polar lipids, which consist of di- and tetraethers of glycerol with isoprenoid alcohols bound at the *sn*-2 and *sn*-3 positions. That is, the archaeal polar lipids have the enantiomeric configuration of their glycerophosphate (GP) backbone (*sn*-glycerol 1-phosphate, G-1-P) that has the mirror image structure of the bacterial or eukaryal counterpart (*sn*-glycerol 3-phosphate, G-3-P). This means that the absolute stereochemistry of the glycerol moiety of polar lipid in Archaea is the opposite of that of glycerol ester membrane lipids in Bacteria and Eukarya. No exception in this difference has been found so far, suggesting that the difference in the backbone structures between members of Archaea and Bacteria might be a reflection of an irreversible evolutionary step at the early stage of the development of the common ancestor.

Methanobacterium thermoautotrophicum is an autotrophic archaeon which synthesises sugars from CO_2 via dihydroxyacetone phosphate (DHAP) and D-glyceraldehyde. Zhang and coworkers have shown that G-1-P is the direct precursor of ether lipid synthesis in this microorganism (Zhang *et al.*, 1990; Zhang and Poulter, 1993). The other water-soluble compound – DHAP, glycerol, or G-3-P – does not serve a substrate for the enzyme reaction. The most efficient hydrocarbon substrate for ether lipid formation is geranylgeranyl pyrophosphate. These results indicate that the enzyme involved in G-1-P formation must be a key enzyme in the biosynthesis of enantiomeric polar lipid structures in this organism.

We have characterised for the first time the enzyme that catalyses G-1-P formation from DHAP in cell-free extracts in an archaeon. It was identified in *M. thermoautotrophicum* and termed *sn*-glycerol-1-phosphate:NAD(P)$^+$ oxidoreductase (G-1-P dehydrogenase) (Nishihara and Koga, 1995).

19.2 *M. thermoautotrophicum* G-1-P dehydrogenase

G-1-P dehydrogenase from *M. thermoautotrophicum* ΔH has a molecular mass of approximately 302 kDa and consists of a single subunit of 38 kDa. The enzyme appears to be a homooctamer. The purified enzyme was active for both DHAP and G-1-P. NADPH as well as NADH was active as a hydrogen donor. No activity was observed for G-3-P, D-glyceraldehyde, glycerol, or dihydroxyacetone. Kinetic constants were determined in the both directions of DHAP reduction and G-1-P oxidation. K_m for DHAP (2.17 mmol/l) was 7.5 times smaller than that for G-1-P (16.3 mmol/l), indicating that the formation of G-1-P is the natural direction catalysed by the enzyme in the cell. The presence of α-GP (a racemic mixture of G-1-P and G-3-P) in the assay mixture did not inhibit the enzyme activity up to 500 mmol/l, indicating that neither G-1-P nor G-3-P inhibits G-1-P synthesis.

The gene encoding G-1-P dehydrogenase (*egsA*) in *M. thermoautotrophicum* was cloned and sequenced. The deduced amino acid sequence indicated that G-1-P dehydrogenase consisted of 347 amino acids. The calculated molecular mass was 39 963 Da which corresponded well with that estimated by SDS-PAGE. A putative NAD-binding site containing the motif composed of several glycine residues and aspartic acid situated on a top of barrel structure was identified. This is a common structure for enzymes of the NAD-dependent dehydrogenase superfamily (Wierenga *et al.*, 1986). If the comparison is restricted to the vicinity of the NAD-binding region, this enzyme shows similarity to many NAD-dependent dehydrogenases, such as *B. stearothermophilus* glycerol dehydrogenase (Mallinder *et al.*, 1992), yeast alcohol dehydrogenase type IV (Williamson and Paquin, 1987), and lobster glyceraldehyde-3-phosphate dehydrogenase (Davidson *et al.*, 1967).

The corresponding GP-forming enzyme for lipid biosynthesis in Bacteria is the biosynthetic NAD-linked G-3-P dehydrogenase (EC 1.1.1.94) found in *Escherichia coli*, which catalyses the same reaction with DHAP and NADH as that of the methanogen except for the stereo configuration of the product (G-3-P) (Kito and Pizer, 1969; Edgar and Bell, 1978). The gene of *E. coli* G-3-P dehydrogenase (*gpsA*) has been cloned by Sofia *et al.* (1994) (76.0–81.5 minutes of *E. coli* gene, NIH database accession no. U00039). G-1-P dehydrogenase did not share any sequence homology with that of the corresponding biosynthetic G-3-P dehydrogenase in *E. coli*, although both enzymes are NAD-linked alcohol dehydrogenases with similar polypeptide lengths. It is thus clear that *M. thermoautotrophicum* G-1-P dehydrogenase has no close relationship with bacterial G-3-P dehydrogenase, although *M. thermoautotrophicum* G-1-P dehydrogenase is a member of NAD-dependent alcohol dehydrogenase family, and has a rather close relationship with glycerol dehydrogenase. These results indicate that *M. thermoautotrophicum* G-1-P dehydrogenase and *E. coli* G-3-P dehydrogenase might have originated from different ancestral enzymes.

19.3 GP dehydrogenase and glycerol kinase in other Archaea

In all organisms studied so far, α-GP arises from phosphorylation of glycerol by glycerol kinase or by the reduction of DHAP by GP dehydrogenase. G-1-P dehydrogenase is the enzyme responsible for the formation of G-1-P in Archaea. To investigate whether this reaction (the reduction of DHAP to G-1-P) was ubiquitous in *M. thermoautotrophicum*, we measured GP-forming activity in *Halobacterium cutirubrum* and *Pyrococcus* sp. KS8-1 and both activities, glycerol kinase and GP dehydrogenase, were found in these organisms. Glycerol kinase from both *H. cutirubrum* and *Pyrococcus* sp. KS8-1 was specific for the formation of G-3-P. G-1-P dehydrogenase was detected in both *Pyrococcus* sp. KS8-1 and *H. cutirubrum*, while G-3-P dehydrogenase was also detected in *H. cutirubrum*. These results indicate that G-1-P is formed by the stereospecific G-1-P dehydrogenase in both halophilic (*Halobacterium*) and heterotrophic, hyperthermophilic (*Pyrococcus*) Archaea, as well as in methanogens (*Methanobacterium*).

19.4 Conclusions

Considering the indispensability of the phospholipid membrane for the existence of a cell and the differences in the stereochemical structure of polar lipids between Archaea and Bacteria, it is likely that the membrane played an important role for the differentiation of the common ancestor in the early stage of evolution. The polar lipids of Archaea show a wide variety in their core lipids, sugars, and polar head groups (Kates, 1993; Koga *et al.*, 1993b). The diversity is not random but is dependent on the phylogenetic relationship as deduced by 16S rRNA sequences (Koga *et al.*, 1993a). It is concluded that the diversity of lipids has been determined by the phylogeny of organisms. The highest rank of classification of living organisms is the domain (Archaea, Bacteria, and Eukarya), and one the most distinctive features that discriminate Archaea and Bacteria is the enantiomeric structure of GP. The fact that the enzymes responsible for the formation of G-1-P and G-3-P are quite different in their amino acid sequences means that these enzymes must have originated from different ancestral enzymes, and that their stereospecificities could not have mutated to the opposite ones. Thus, the established stereospecificities of the enzymes and consequently the stereostructures of GP in Archaea and Bacteria are assumed to have been conserved since the time when Archaea and Bacteria diverged. A membrane made of phospholipids with either enantiomer of GP might separately insulate intracellular processes (metabolism) and it should have established a cell, which would be an ancestor of either domain of life (Archaea and Bacteria). Substantial metabolism without cells has been suggested to be present in forms such as surface metabolism on pyrite under anaerobic and thermophilic conditions, as proposed by Wächtershäuser (1988, 1992), before cells were enclosed by lipid membranes, since most basic biochemical features were shared by Archaea and Bacteria (i.e., coenzymes, central metabolic pathways, and genetic machineries). However, it should be noted that only a few enzymes involved in the initial stages of polar lipid biosynthesis, including G-1-P dehydrogenase and enzymes for glycerol ether bond formation, have been characterised so far. Investigations of the pathways of polar lipid biosynthesis and the characterisation of the enzymes involved would give us more important information about the evolution and differentiation of Archaea and Bacteria, and the deepest branching points of the phylogenetic tree of life.

19.5 Summary

One of the most characteristic features of archaebacterial ether phospholipid is the enantiomeric configuration of their glycerophosphate backbone (*sn*-glycerol 1-phosphate), which is the mirror image of the structure of the eubacterial or eukaryotic counterpart. The enzyme that forms glycerophosphate of this configuration was found for the first time in a cell-free extract of the methanogen *Methanobacterium thermoautotrophicum*, and has been purified. This enzyme was identified as *sn*-glycerol-1-phosphate:NAD$^+$ oxido-reductase (*sn*-glycerol-1-phosphate dehydrogenase). This enzyme formed *sn*-glycerol 1-phosphate from dihydroxyacetone phosphate with either NADH or NADPH as an electron donor. Estimation of molecular mass using gel chromatography and SDS-PAGE suggested that the enzyme is present as an octamer. Optimum temperature was 75 °C, and optimum pH was 6.7–7.4. The enzyme activity was stimulated by the presence of potassium ions. The gene encoding this enzyme was cloned and sequenced. The gene of this enzyme (*ORF1*) encodes 347 amino acids (calculated molecular mass 36 963 Da) with the charactaristic NADH-binding motif. The overall amino acid sequence did not show similarity to any other proteins found in electronic databases except for the glycerol dehydrogenases of *Bacillus stearothermophilus* (21.6%), *Escherichia coli* (21.5%), and *Citrobacter freundii* (18.4%). Because *sn*-glycerol-1-phosphate dehydrogenase and *sn*-glycerol-3-phosphate dehydrogenase are originated from different ancestor enzymes, and it would be almost impossible to interchange the stereospecificity of the enzymes, it seems likely that the stereostructure of membrane phospholipids of a cell must be maintained from the time of birth of the first cell. The hypothesis we propose here is that Archaea and Bacteria are differentiated by the occurrence of cells enclosed by membranes of phospholipids with *sn*-glycerol-1 phosphate and *sn*-glycerol 3-phosphate as their respective backbones.

References

DAVIDSON, B. E., SAJGO, M., NOLLER, H. F. and HARRIS, J. I. (1967) Amino-acid sequence of glyceraldehyde 3-phosphate dehydrogenase from lobster muscle. *Nature*, **216**, 1181–1185.

EDGAR, J. R. and BELL, R. M. (1978) Biosynthesis in *Escherichia coli* of *sn*-glycerol 3-phosphate, a precursor of phospholipid. Purification and physical characterisation of wild type and feedback-resistant forms of the biosynthetic *sn*-glycerol-3-phosphate dehydrogenase. *J. Biol. Chem.*, **253**, 6348–6353.

KATES, M. (1993) Membrane lipids in Archaea. In *The Biochemistry of Archaea (Archaebacteria)* ed. M. Kates, D. J. Kushner and A. T. Matheson, pp. 261–295 (Amsterdam: Elsevier Science).

KITO, M. and PIZER, L. I. (1969) Purification and regulatory properties of the biosynthetic L-glycerol-3-phosphate dehydrogenase from *Escherichia coli. J. Biol. Chem.*, **244**, 3316–3323.

KOGA, Y., AKAGAWA-MATSUSHITA, M., OHGA, M. and NISHIHARA, M. (1993a) Taxonomic significance of the distribution of the component parts of polar ether lipids in methanogens. *System. Appl. Microbiol.*, **16**, 342–351.

KOGA, Y., NISHIHARA, M., MORII, H. and AKAGAWA-MATSUSHITA, M. (1993b) The ether polar lipids of methanogenic bacteria. Structures, comparative aspects, and biosynthesis. *Microbiol. Rev.*, **57**, 164–182.

MALLINDER, P. R., PRITCHARD, A. and MOIR, A. (1992) Cloning and characterisation of a gene from *Bacillus stearothermophilus* var. non-diasticus encoding a glycerol dehydrogenase. *Gene*, **110**, 9–16.

NISHIHARA, M. and KOGA, Y. (1995) *sn*-Glycerol-1-phosphate dehydrogenase in *Methanobacterium thermoautotrophicum*: key enzyme in biosynthesis of the enantiomeric GP backbone of ether phospholipids of archaebacteria. *J. Biochem.*, **117**; 933–935.

SOFIA, H. J., BURLAND, V., DANIELS, D. L., PLUNKETT, G., III and BLATTNER, F. R. (1994) Analysis of the *Escherichia coli* genome V. DNA sequence of the region from 76.0 to 81.5 minutes. *Nucleic Acids Res.*, **22**, 2576–2586.

WÄCHTERSHÄUSER, G. (1988) Before enzymes and templates: theory of surface metabolism. *Microbiol. Rev.*, **52**, 452–484.

WÄCHTERSHÄUSER, G. (1992) Ground works for an evolutionary biochemistry. The iron–sulfur world. *Prog. Biophys. Mol. Biol.*, **58**, 85–201.

WIERENGA, R. K., TERPSTRA, P. and HOL, W. G. J. (1986) Prediction of the occurrence of the ADP-binding βαβ-fold in proteins, using an amino acid sequence fingerprint. *J. Mol. Evol.*, **187**, 101–107.

WILLIAMSON, V. M. and PAQUIN, C. E. (1987) Homology of *Saccharomyces cerevisiae* ADH4 to an iron-activated alcohol dehydrogenase from *Zymomonas mobilis*. *Mol. Gen. Genet.*, **209**, 374–381.

ZHANG, D. and POULTER, C. D. (1993) Biosynthesis of archaebacterial ether lipid. Formation of ether linkages by prenyltransferases. *J. Am. Chem. Soc.*, **115**, 1270–1277.

ZHANG, D. L., DANIELS, L. and POULTER, C. D. (1990) Biosynthesis of archaebacterial membranes. Formation of isoprene ethers by a prenyl transfer reaction. *J. Am. Chem. Soc.*, **112**, 1264–1265.

20

From the Common Ancestor of all Living Organisms to Protoeukaryotic Cell

AKIHIKO YAMAGISHI[1], **TAKAHIDE KON**[1], **GEN TAKAHASHI**[2] **AND TAIRO OSHIMA**[1]

[1] *Department of Molecular Biology, Tokyo University of Pharmacy and Life Science, Tokyo, Japan*
[2] *School of Medicine, Hirosaki University, Hirosaki, Japan*

20.1 Introduction

Figure 20.1 shows the phylogenetic tree constructed from the small subunit rRNA genes of living organisms reported by Woese *et al.* (1990), in which the root was placed on the basis of the results presented by Gogarten *et al.* (1989) and Iwabe *et al.* (1989). The species, which is a group of living organisms with sufficient gene interaction to constitute a single gene-pool, has divided into two species at the point Commonote. The species can be called the last and the latest common ancestor of all the living organisms on the earth. We have suggested that the last common ancestor must have already had a rigid genetic system with circular chromosomal DNA, based on the circularity of chromosomal DNAs of eubacteria and archaebacteria (Kondo *et al.*, 1993; Yamagishi and Oshima, 1995). In the first half of the chapter we will discuss the characteristics of the last common ancestor.

The Archaeal line splits into Archaebacteria and Urkaryotes, which is defined by Woese *et al.* (1977) as occurring at the point P. Urkaryotes represent the genes or genetic components of eukaryotes. Eukaryotic species diverge after the point P′. The process from protoarchaebacteria (P) to protoeukaryotes (P′) will be discussed in the second half of the chapter.

20.2 The last common ancestor commonote

Based on the tree topology of Figure 20.1 and parsimony, it is possible to speculate on the characteristics of the last common ancestor. Organisms on both branches of the tree have the same genetic system: the chromosome is replicated by DNA polymerase, transcribed by RNA polymerase, translated by machinery consisting of ribosome, tRNAs, and other factors, using universal codons. These characteristics are likely to have originated from the common ancestor unless these systems have been developed independently.

Based on the principle of parsimony, the characteristics shared by eubacteria and one of the other groups, archaebacteria or eukaryotes, is likely to be that of the common

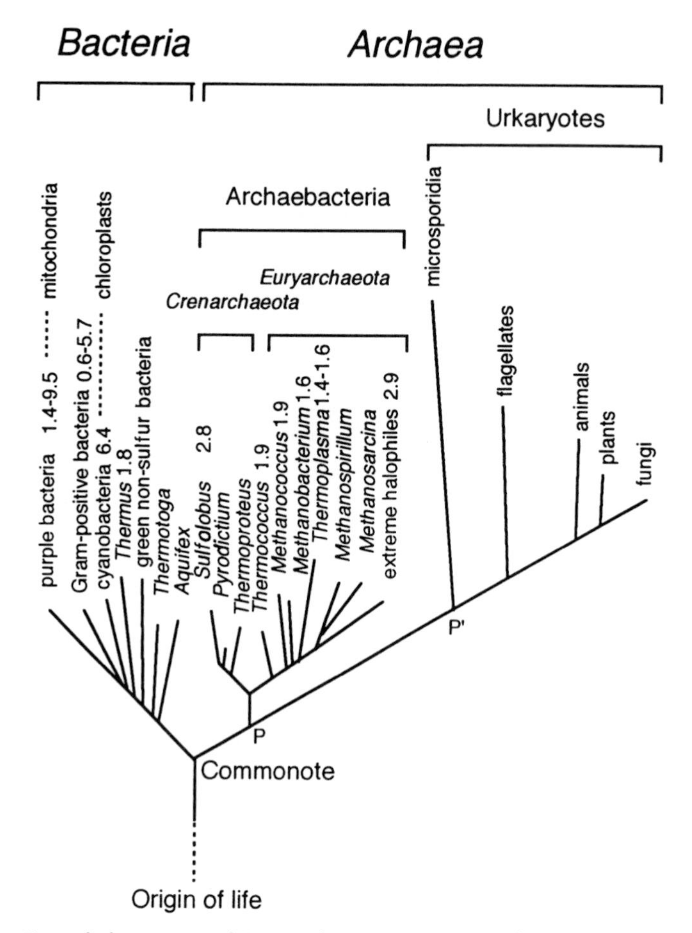

Figure 20.1 Size of chromosomal DNA of microorganisms. The general phylogenetic tree was constructed based on Woese *et al.* (1990) and classified according to Yamagishi and Oshima (1995). P and P′ represent protoarchaebacteria and protoeukaryotes. Data were collected from a review by Krawiec and Riley (1995) and other references (Charlebois *et al.*, 1991; Sitzmann and Klein, 1991; Stetter and Leisinger, 1992; Cohen *et al.*, 1992; Borges and Bergquist, 1993; Kondo *et al.*, 1993; Lopez-Garcia *et al.*, 1992; Tabata *et al.*, 1993).

ancestor. Accordingly, the common ancestor seems to have been an organism of size about 1 μm surrounded by a single lipid membrane. The last common ancestor seems to have had eukaryotic α-type DNA polymerase, because these are found also in archaebacterial and eubacterial branches (Ito and Braithwaite, 1991; Iwasaki *et al.*, 1991; Pisani *et al.*, 1992; Forterre, 1992; Uemori *et al.*, 1993).

Figure 20.1 also summarises the size of chromosomal DNAs of microorganisms. The size varies from several hundreds of kilobases (kb) to several megabases (Mb). The range is rather similar between eubacteria and archaebacteria. It is noted that organisms that branched at points close to the root tend to have genome size of 1.5–2.0 Mb. This size may be that of the chromosome of the common ancestor. These characteristics suggest

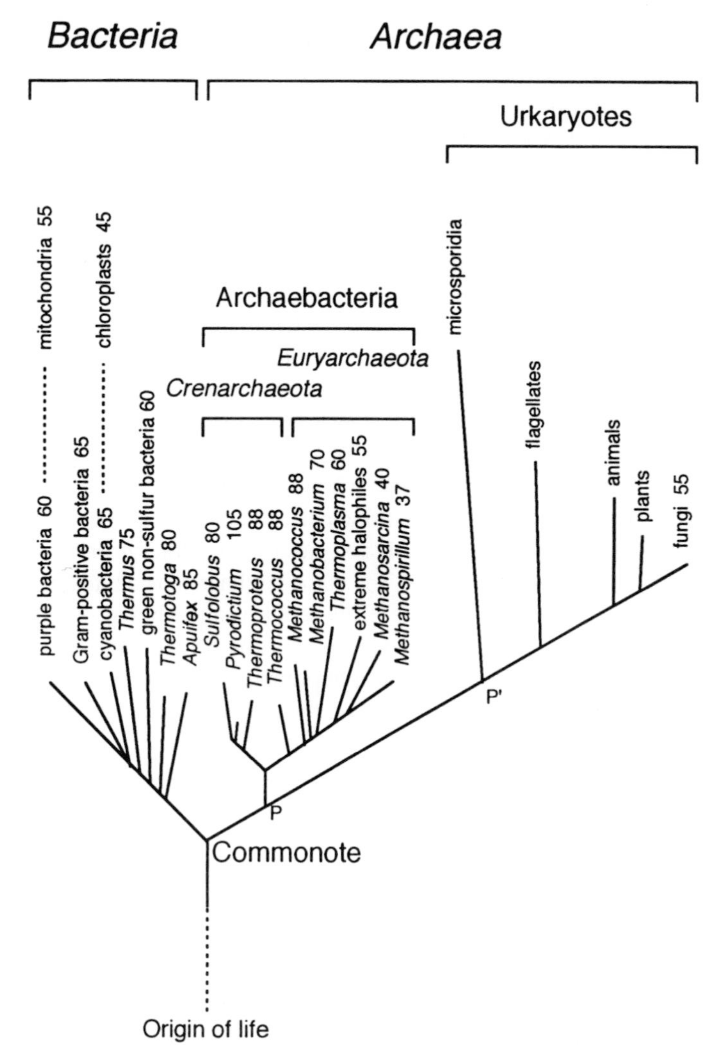

Figure 20.2 Optimum growth temperatures of microorganisms. The optimum growth temperature of each species or the optimum temperature of the most thermophilic species in each group is shown. The data are collected from the work of Brock (1978), *Bergey's Manual of Systematic Bacteriology* (1989) and other references (Huber *et al.*, 1986; Huber *et al.*, 1992). Only data obtained from laboratory culture experiments are included. The general phylogenetic tree was constructed based on Woese *et al.* (1990) and classified according to Yamagishi and Oshima (1995). P and P′ represent protoarchaebacteria and protoeukaryotes, respectively.

that the last common ancestor commonote was a organism very similar to the contemporary eubacteria and archaebacteria.

Figure 20.2 summarises the optimum growth temperatures of microorganisms. The antiquity of thermophilic microorganisms has been proposed by several authors (Pace *et al.*, 1986; Achenbach-Richter *et al.*, 1987; Pace, 1991; Burggraf *et al.*, 1992). The proposal was based on the findings that the many eubacterial groups contain thermophilic microorganisms, and the ultrathermophilic microorganisms can be found at a position close to the root in the general phylogenetic tree.

However, Gogarten-Boekels *et al.* (1995) have discussed the possible effects of heavy meteorite bombardment on the early evolution of life on the earth. Gogarten-Boekels *et al.* (1995) and Forterre (1996) discussed the possibility that the selection of thermophilic phenotype can explain the distribution of thermophilic organisms in the general philogenetic tree. Thus the common ancestor need not necessarily have been an ultrathermophile. Nevertheless, these authors also agree that the ancestors of life on the earth were thermophilic at some stage in the early history of the evolution of life. Figure 20.2 gives further support to the antiquity of thermophilic microorganisms on the earth. An organism close to the root of the tree tends to have higher growth temperature. An organism that branched far from the root tends to have lower optimum growth temperature. This result can be interpreted as meaning that the organism with lower optimum growth temperature branched when the ambient temperature had decreased to that temperature. The temperature at which the protoeukaryotic organism separated from the protoarchaebacteria may be 50–60 °C: eukaryotes can grow at the highest temperature. It is also likely that both the ancestors of eubacteria and of archaebacteria were ultrathermophilic microorganisms.

Figures 20.1 and 20.2 also show the principal classification of the living organisms. We have proposed including archaebacteria and eukaryotes (Urkaryotes) in the same taxon Archaea (Yamagishi and Oshima, 1995). The taxon Archaea thus defined is monophyletic whatever the branching within the group. This proposition is based on the fact that the most important branching point of the living organisms on the earth is point C in Figures 20.1 and 20.2. Accordingly, we think it appropriate to make a principal division at this point. We also think it appropriate to retain the common name archaebacteria for the group consisting of Crenarchaeota and Euryarchaeota. We have previously proposed to handle the eukaryotes as the composition or chimera of prokaryotes (Yamagishi and Oshima, 1995).

20.3 The process from the protoarchaebacteria to protoeukaryotic cells

Figure 20.1 suggest that the ancestor of the eukaryotes separated from the ancestor of archaebacteria somewhere near the point P. If there is a distinct line between the point P and the branching point of Euryarchaeota and Crenarchaeota, then the protoeukaryotes evolved from the protoarchaebacteria. Accordingly, there is no special archaebacterial species which is especially closely related to the protoeukaryotes as a whole. Nevertheless, it is still possible that some of the contemporary archaebacteria retain some of the characteristics of the protoarchaebacteria.

There are large differences between eukaryotic cells and archaebacterial cells. The largest difference is the presence of nucleus and other organelles such as mitochondria and chloroplasts in eukaryotic cells. The endosymbiotic origin of mitochondria and chloroplasts is well established. The mitochondrion originated from a microorganism related to proteobacteria (Yang *et al.*, 1985). The chloroplast originated from cyanobacteria (Palenik and Heselkon, 1992).

In addition to the symbiotic nature of eukaryotic cells, there are several other differences in characteristics between archaebacteria and eukaryotes. The sizes of the cells are different: eukaryotic cells are about ten times larger in diameter than prokaryotic cells. The genome size of eukaryotes is also much larger than that of prokaryotic cells in general. There are also extremely complex membrane systems in eukaryotic cells: endoplasmic reticulum, lysosome, vacuole, Golgi body, etc.

20.3.1 *The structure of macrocells of* Thermoplasma

We have analysed some characteristics of the cell wall-free archaebacterium *Thermoplasma acidophilum*. *Thermoplasma* is an acidothermophilic archaebacterium with optimum growth temperature of 50–60 °C. The archaebacterium was found by Darland *et al.* (1970) and was analysed extensively by Searcy and his colleagues (Searcy *et al.*, 1981; Hixon and Searcy, 1993). This archaebacterium is proposed to be the candidate for the archaebacterium which derived the nuclear and cytoplasmic moiety of eukaryotic cells (Searcy *et al.*, 1981; Margulis, 1993).

We have isolated several new strains of *Thermoplasma* from hot springs near Tokyo (Yasuda *et al.*, 1995) and have analysed microbiological characteristics of the strains. The characteristics are indistinguishable from those of *Thermoplasma acidophilum* (Yasuda *et al.*, 1995). The cells of the type strain are irregular (Hixon and Searcy, 1993). We have analysed the cellular structure by scanning electron microscopy (SEM). Strain HO-121 showed irregular cell shape similar to that reported for the type strain of *Thermoplasma acidophilum*, although one of the strains, HO-51, showed spherical cells with smooth surface. Another strain, HO-12, had a spherical cell surface with knobs.

The cells of these strains formed cotton-like aggregates which could be seen with the naked eye after the prolonged incubation of the culture, and the structure of this cotton-like aggregate was investigated by SEM. Individual spherical cells were recognisable in the aggregate of the cells of strain HO-51; the structure is shown schematically in Figure 20.3A. The aggregated cells of strain HO-54 showed significantly different structure; the cells are connected each other to form a continuous lamellar structure and individual cells could not be distinguished. The structure is shown schematically in Figure 20.3B. It appeared that strain HO-54 formed multinucleate cells.

The lamellar structure may be the form of the intermediate stage between the proto-archaebacteria and protoeukaryotic cells. Figures 20.3 A and B show the cellular structures we have observed in the *Thermoplasma* strains HO-51 and HO-54, respectively. If we assume that the extralamellar structure is decreased in volume, and the cytoplasmic compartment of the lamellar structure is increased in volume, the expected structure can be represented as shown in Figure 20.3C. The extralamellar structure is connected to the outside and the topology is the same as the inside of the endoplasmic reticulum of eukaryotic cells. Thus the continuous transition from Figure 20.3A through 20.3B to 20.3C can explain the process from protoarchaebacteria to protoeukaryotic cells. The increase in cell volume and the genome size and formation of endoplasmic reticulum can be explained by this process.

The model presented here does not deny the endosymbiotic theory of the origin of organelles surrounded by double membranes, such as mitochondria and chloroplasts in eukaryotic cells. Instead, the larger size of the protoeukaryotic cell rather enables the endosymbiosis of other prokaryotic microorganisms. The model also explains the origin of the single-membrane-bounded structures in eukaryotic cells, i.e. vacuole, lysosome, Golgi body, etc., as well as endoplasmic reticulum. These single-membrane-bounded structures may originate from cytoplasmic membrane of multinucleate cells of protoarchaebacteria.

Increase in the size of a cell provides space for larger chromosomal DNA. Multinucleate cells contain many copies of genomic DNA. The increase in genome size is initially just that of the total number of copies. Multiple copying of the genomic DNA is expected to accelerate mutation of genes. Although a gene may lose its original function through mutations, other copies of the gene can support the function if it is necessary. Gene duplications are often found in many organisms and they are considered to play a role in the creation

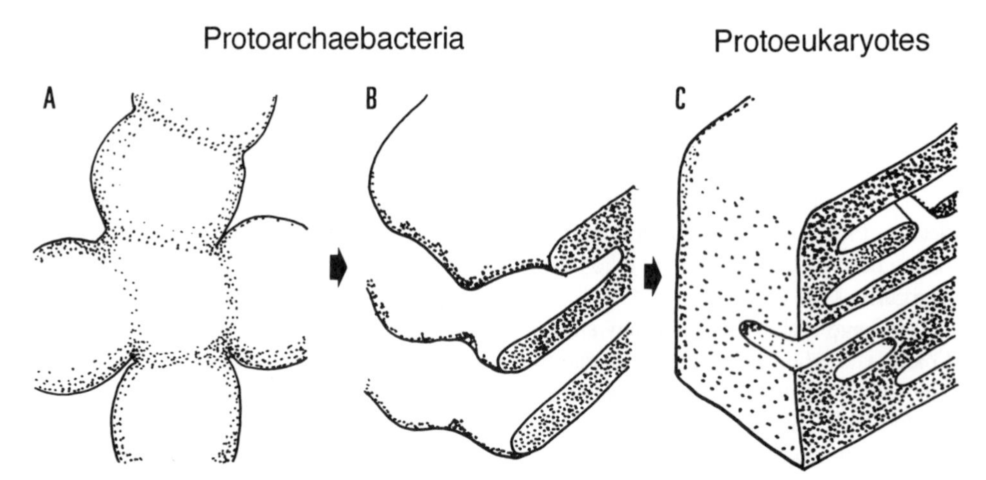

Figure 20.3 Model of the process from the protoarchaebacteria to protoeukaryotes. (A) and (B) represent the structures of cells observed in strains of *Thermoplasma*. Expansion of the cytoplasmic space of the structure from (B) to (C) can explain the formation of protoeukaryotic cells. The process can explain the increase in size of the cells and in genome size, and the origin of endoplasmic reticulum and other single-membrane-surrounded structures in eukaryotic cells.

of new functional genes. Multiple copies of chromosomal DNA must facilitate the creation of new genes.

In general, cells have to incorporate substrates from outside to inside and export waste matter from inside to outside. Such transportation becomes difficult in larger cells. The multinucleate cells found in the strain HO-54 retained the extracellular space as the channel system through the cytoplasmic space. These channel system must support the transport of compounds between internal cytoplasmic space and the external medium.

20.3.2 *Cytoskeleton in* Thermoplasma *cells*

The significant differences in the cellular structures of the cells of different strains of *Thermoplasma* suggest the presence of mechanical structure to maintain the respective shape of the cells. Because there is no cell wall around the cell membrane, there must be some structural component which supports the membrane from the cytoplasmic space. Cells of strain HO-51 were treated with Triton X-100, DNAase and RNAase. The solution was centrifuged and the pellet was investigated by atomic force microscopy (AFM). AFM revealed membrane structure of size about 1 μm (S. Kasas, K. Ito, G. Takhashi, A. Ikai, T. Oshima and A. Yamagishi, unpublished). The membrane structure is overlapped with mesh-like structure. The average hole size of the mesh was 30 nm × 18 nm, and the average height of the mesh was 1.6 nm. It is not clear whether the mesh is flat and attached to the back side of the cytoplasmic membrane in the cell. Alternatively, the mesh-like structure may be formed by flattening of three-dimensional mesh or network during the drying process of preparation of the structure for AFM observation.

Nevertheless, the structure was retained after treatment with neutral detergent Triton X-100 and recovered by centrifugation. The mesh portion resisted the pressure of the tip of an atomic force microscope during observation, suggesting the mechanical strength of the mesh-like structure. The structure seems to represent cytoskeleton in *Thermoplasma* cells. Hixon and Searcy (1993) reported the formation of mesh-like structure from the cell extract of *Thermoplasma* in the presence of Ca^{2+} and ATP. The relation between the structure reported by Hixon and Searcy (1993) and the mesh-like structure observed in Triton-treated cell in our experiments is not yet clear.

20.3.3 *Tetraether lipid biosynthesis in* Thermoplasma

The membrane lipids in archaebacteria are different from those in eubacteria and eukaryotes. Cytoplasmic membrane of *Thermoplasma* consists of tetraether lipids with some diether lipids. The ether lipid biosynthetic pathway is essentially the same as that of isoprenoid biosynthesis. Tetraether lipids have a structure which is made by connecting the adjacent diether lipid molecules at the heads of their isopranyl alcohol moiety.

We have analysed the effect of squalene epoxidase inhibitors on tetraether lipid biosynthesis (T. Kon, A. Yamagishi and T. Oshima, unpublished). The squalene epoxidase inhibitor terbinafine inhibited the growth of *Thermoplasma*, while the growth of *Escherichia coli* and *Halobacterium* was not inhibited by the inhibitor.

We have analysed the effect of terbinafine on the biosynthesis of tetraether lipids in *Thermoplasma*. *Thermoplasma* cells were cultured with [^{13}C]mevalonic acid. The cells were harvested and lipids were extracted from the cells. The lipids were analysed by thin-layer chromatography (TLC) after acid methanolysis. Significant incorporation of ^{13}C was observed in the tetraether lipid fraction on TLC. Terbinafine significantly affected the ether lipid biosynthesis. The radioactivity recovered in the tetraether lipid fraction was reduced to about 10% of the control by the addition of 0.1 mg/ml terbinafine. The radioactivity of the diether lipid fraction was increased simultaneously. [^{13}C]Mevalonic acid and the inhibitor were removed by centrifugation and the cell pellet was resuspended in fresh culture medium. After incubation of the cells in the fresh medium, cells were harvested and lipids were analysed. The radioactivity in the diether fraction was decreased and that of tetraether fraction was increased. These results clearly indicate that the tetraether lipids are synthesised from the diether lipids. These results also indicate that the biosynthesis from diether lipids to tetraether lipids is inhibited by terbinafine.

It is interesting to note that terbinafine is a specific inhibitor of squalene epoxidase, which catalyses the reaction in the biosynthesis of steroids in eukaryotes. These results suggest that the structure of the enzyme catalysing the tetraether lipid biosynthesis and that of the squalene epoxidase may be similar at least at the site of the inhibition by terbinafine. It is also important to note that the biosynthesis of squalene in eukaryotic cells is localised in microsomes. The location is compatible with the scheme presented in Figure 20.3: the endoplasmic reticulum of eukaryotic cells originated from the cytoplasmic membrane of protoarchaebacteria.

20.4 Conclusions

Although the question whether the common ancestor commonote was thermophilic or not is still a matter of debate, it is likely that the ancestors both of eubacteria and archaebacteria were ultrathermophilic microorganisms.

The multinucleate lamellar structure of macrocells was found in a strain of the acidothermophilic cell wall-less archaebacterium *Thermoplasma*. We propose that the formation of multinucleate cells of protoarchaebacteria participated in the process of forming the protoeukaryotic cells before the endosymbiosis of mitochondria and chloroplasts. The multinucleate cells must have contributed to the increase in cell size and genome size, and to the formation of single-membrane-surrounded systems such as endoplasmic reticulum and vacuoles in eukaryotic cells.

20.5 Summary

The early history of the evolution of living organisms on the earth is discussed based on the analysis of thermophilic archaebacteria. We propose that the formation of multinucleate macrocells is the important step in the process from protoarchaebacteria to protoeukaryotic cells. The increase in both cell volume and size of genome and formation of a single membrane surrounding intracellular structures such as endoplasmic reticulum in eukaryotic cells can be explained by the formation of multinucleate cells with lamellar structure.

References

Bergey's Manual of Systematic Bacteriology (1989) vol. 3 (Baltimore: Williams and Wilkins).

ACHENBACH-RICHTER, L., GUPTA, R., STETTER, K. O. and WOESE, C. R. (1987) Were the original eubacteria thermophiles? *System. Appl. Microbiol.*, **9**, 34–39.

BORGES, K. M. and BERGQUIST, P. L. (1993) Genomic restriction map of the extremely thermophilic bacterium *Thermus thermophilus* HB8. *J. Bacteriol.*, **175**, 103–110.

BROCK, T. D. (1978) *Thermophilic Microorganisms and Life at High Temperatures* (New York: Springer-Verlag).

BURGGRAF, S. X., OLSEN, G. J., STETTER, K. O. and WOESE, C. R. (1992) A phylogenetic analysis of *Aquifex pyrophilus*. *System. Appl. Microbiol.*, **15**, 352–356.

CHARLEBOIS, R. L., SCHALKWYK, L. C., HOFMAN, J. D. and DOOLITTLE, W. F. (1991) Detailed physical map and set of overlapping clones covering the genome of the archaebacterium *Haloferax volcanii* DS2. *J. Mol. Biol*, **222**, 509–524.

COHEN, A., LAM, W. L., CHARLEBOIS, R. L., DOOLITTLE, W. F. and SCHALKWYK, L. C. (1992) Localising genes on the map of the genome of *Haloferax volcanii* one of the Archaea. *Proc. Natl Acad. Sci. USA*, **89**, 1602–1606.

DARLAND, G., BROCK, T. D., SAMSONOFF, W. and CONTI, S. F. (1970) A thermophilic, acidophilic mycoplasma isolated from a coal refuse pile. *Science*, **170**, 1416–1418.

FORTERRE, P. (1992) The DNA polymerase from the archaebacterium *Pyrococcus furiosus* does not testify for a specific relationship between archaebacteria and eukaryotes. *Nucleic Acids Res.*, **20**, 1181.

FORTERRE, P. (1996) A hot topic: the origin of hyperthermophiles. *Cell*, **85**, 789–792.

GOGARTEN-BOEKELS, M., HILARIO, E. and GOGARTEN, J. P. (1995) The effects of heavy meteorite bombardment on the early evolution – the emergence of the three domains of life. *Origins of Life and Evolution of the Biosphere*, **25**, 251–264.

GOGARTEN, J. P., KIBAK, H., DITTRICH, P. *et al.* (1989) Evolution of the vacuolar H^+-ATPase: Implication for the origin of eukaryotes. *Proc. Natl Acad. Sci. USA*, **86**, 6661–6665.

HIXON, W. and SEARCY, D. G. (1993) Cytoskeleton in the archaebacterium *Thermoplasma acidophilum*? Viscosity increase in soluble extracts. *BioSystems*, **29**, 151–160.

HUBER, R., LANGWORTHY, T. A., KOENIG, H. *et al.* (1986) *Thermotoga maritima* sp. nov. represents a new genus of unique extremely thermophilic eubacteria growing up to 90 °C. *Arch. Microbiol.*, **144**, 324–333.

HUBER, R., WILHARM, T., HUBER, D. (1992) *Aquifex pyrophilus* gen. nov. sp. nov., represents a novel group of marine hyperthermophilic hydrogen-oxidising bacteria. *System. Appl. Microbiol.*, **15**, 340–351.

ITO, J. and BRAITHWAITE, D. K. (1991) Compilation and alignment of DNA polymerase sequences. *Nucleic Acids Res.*, **19**, 4045–4057.

IWABE, N., KUMA, K., HASEGAWA, M., OSAWA, S. and MIYATA, T. (1989) Evolutionary relationship of archaebacteria, eubacteria, and eukaryotes inferred from pylogenetic trees of duplicated genes. *Proc. Natl Acad. Sci. USA*, **86**, 9355–9359.

IWASAKI, H., ISHINO, Y., TOH, H., NAKATA, A. and SHINAGAWA, H. (1991) *Escherichia coli* DNA polymerase II is homologous to alpha-like DNA polymerases. *Mol. Gen. Genet.*, **226**, 24–33.

KONDO, S., YAMAGISHI, A. and OSHIMA, T. (1993) A physical map of the sulfur-dependent archaebacterium *Sulfolobus acidocaldarius* 7 chromosome. *J. Bacteriol.*, **175**, 1532–1536.

KRAWIEC, S. and RILEY, M. (1995) Organisation of bacterial chromosome. *Microbiol. Rev.*, **54**, 502–539.

LOPEZ-GARCIA, P., PASCUAL, J., SMITH, C. and AMILS, R. (1992) Genomic organisation of the halophilic archaeon *Haloferax mediterranei*: physical map of the chromosome. *Nucleic Acids Res.*, **20**, 2459–2464.

MARGULIS, L. (1993) *Symbiosis in Cell Evolution*, 2nd edn (New York: W. H. Freeman).

PACE, N. R. (1991) Origin of life – facing up to the physical setting. *Cell*, **65**, 531–533.

PACE, N. R., OLSEN, G. J. and WOESE, C. R. (1986) Ribosomal RNA phylogeny and the primary lines of evolutionary descent. *Cell*, **45**, 325–326.

PALENIK, B. and HASELKON, R. (1992) Multiple evolutionary origins of prochlorophytes, the chlorophyll b-containing prokaryotes. *Nature*, **355**, 265–267.

PISANI, F. M., DEMARTINO, C. and ROSSI, M. (1992) A DNA polymerase from the archaeon *Sulfolobus solfataricus* shows sequence similarity to family B DNA polymerases, *Nucleic Acids Res.*, **20**, 2711–2716.

SEARCY, D. G., STEIN, D. B. and SEARCY, K. B. (1981) A mycoplasma-like archaebacterium possibly related to the nucleus and cytoplasm of eukaryotic cells. *Ann. NY. Acad. Sci.*, **361**, 312–324.

SITZMANN, J. and KLEIN, A. (1991) Physical and genetic map of the *Methanococcus voltae* chromosome. *Mol. Microbiol.*, **5**, 505–513.

STETTER, R. and LEISINGER, T. (1992) Physical map of the *Methanobacterium thermoautotrophicum* Marburg chromosome. *J. Bacteriol.*, **174**, 7227–7234.

TABATA, K., KOSUGE, T., NAKAHARA, T. and HOSHINO, T. (1993) Physical map of the extremely thermophilic bacterium *Thermus thermophilus* HB27 chromosome. *FEBS Lett.*, **331**, 81–85.

UEMORI, T., ISHINO, Y., TOH, H., ASADA, K. and KATO, I. (1993) Organisation and nucleotide sequence of the DNA polymerase gene from the archaeon *Pyrococcus furiosus. Nucleic Acids Res*, **21**, 259–265.

WOESE, C. R., KANDLER, O. and WHEELIS, M. L. (1990) Towards a natural system of organisms: proposal for the domains Archaea, Bacteria, and Eukarya. *Proc. Natl Acad. Sci. USA*, **87**, 4576–4579.

YAMAGISHI, A. and OSHIMA, T. (1995) Return to dichotomy: *Bacteria* and *Archaea*. In *Chemical Evolution: Self-organisation of the Macromolecules of Life*, ed. J. Chela-Flores, M. Chadha, A. Negron-Mendoza and T. Oshima (Hampton, VA: Deepak Publishing).

YANG, D., OYAIZU, Y., OYAIZU, H., OLSEN, G. J. and WOESE, C. R. (1985) Mitochondrial origins. *Proc. Natl Acad. Sci. USA*, **82**, 4443–4447.

YASUDA, M., OYAIZU, H., YAMAGISHI, A. and OSHIMA, T. (1995) Morphological variation of new *Thermoplasma acidophilum* isolates from Japanese hot springs. *Appl. Environ. Microbiol.*, **61**, 3482–3485.

PART EIGHT

Life at High Temperature

21

Primitive Coenzymes and Metabolites in Archaeal/Thermophilic Metabolic Pathways

R. M. DANIEL

Thermophile Research Unit, Department of Biological Sciences, The University of Waikato, Hamilton, New Zealand

21.1 Introduction

There is now evidence that life arose at high temperature (Woese, 1987) and therefore that all the metabolic pathways we now observe are evolutionary successors of pathways that operated at high temperatures – possibly above 100 °C. 'Modern', mesophilic, pathways will have adapted to overcome the difficulties presented by low-temperature metabolism; but those successors of the ancestral organism which still operate at high temperature are presumably closest in nature to it, and so an examination of thermophilic (and probably archaeal) metabolic pathways is the best place to seek clues to the nature of early metabolism. In particular, we may be helped by a study of the constraints which may have been imposed on these early metabolic pathways by high temperature. The first step must be a judgement as to the temperature at which the ancestral organism(s) lived. Overall, the evidence in favour of any particular temperature is not strong, but the best indications probably come from those taxonomic trees which show that, in general, thermophiles are more deeply rooted than mesophiles (Figure 21.1), and the more deeply rooted a taxonomic branch, the higher the growth temperature of the organism. Thus, among the prokaryotes *Aquifex* (t_{opt} ~ 85 °C) is more deeply rooted than *Thermotoga* (t_{opt} ~ 80–85 °C), which is more deeply rooted than *Thermus* (t_{opt} ~ 65–75 °C) (Woese, 1987; Achenbach Richter *et al.*, 1987; Olsen *et al.*, 1994). The highest known temperature optimum for growth is 105 °C for *Pyrodictium* (Stetter, 1982) and we must therefore consider the possibility that the ancestral organism grew at a higher temperature; but a growth temperature in the range 90–125 °C seems likely.

21.2 Macromolecular stability

What factors limit metabolism at these temperatures? The four main candidates are proteins, lipids, DNA/RNA, and low-molecular-mass compounds. I think that protein instability can be safely ruled out as a factor limiting metabolism up to 125 °C. A number of proteins have half-lives above 100 °C which are comparable with the doubling times of their host organisms (Table 21.1). Some proteins have been shown to be relatively stable

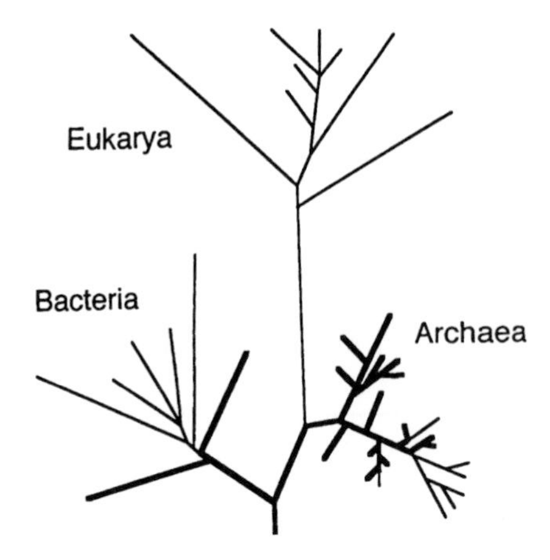

Figure 21.1 Location of hyperthermophiles within a taxonomic tree of the three groups of living organisms (Stetter, 1996). (Hyperthermophile lineage in bold.)

Table 21.1 Some enzymes stable above 100 °C

Enzyme	Source	Half-life (h)	Temperature (°C)
Glyceraldehyde-3-phosphate dehydrogenase	*Thermotoga maritima*	>2	100
Hydrogenase	*Pyrococcus furiosus*	2	100
Amylase	*P. woesei*	6	100
DNA-RNA polymerase	*Thermoproteus tenax*	2	100
Glutamate dehydrogenase	*P. furiosus*	10	100
Cellobiohydrolase	*Thermotoga* sp.	1.1	108
Amylase	*P. furiosus*	2	120

Data from Coolbear *et al.* (1992) and Adams (1993).

at 130 °C, with and without stabilising treatments and conditions (Figure 21.2). In the case of the xylanase and α-glucosidase we have demonstrated activity at 130 °C consistent with the stability data (Simpson *et al.*, 1991; Piller *et al.*, 1996). There are no theoretical reasons why proteins cannot be conformationally stable up to 125 °C, and work by Hensel *et al.* (1992), confirmed by us (Daniel *et al.*, 1996), indicates that (at least at 100 °C) irreversible degradative reactions do not take place, or are greatly slowed, in conformationally intact proteins. The evidence is stronger that the upper limit of protein stability is set by the upper temperature limit of living organisms than vice versa (for a fuller discussion, see Daniel *et al.*, 1996). To date there is no evidence that biological lipids are unstable up to 125 °C, although there is speculation as to the extent to which the lipids found in archaeal membranes are a product of taxonomic position or a requirement for high temperature life (Gambacorta *et al.*, 1994). Potential mechanisms for stabilisation of DNA and RNA are well established. DNA (and possibly RNA) can be stabilised by binding to protein, and such very stable proteins have been found in hyperthermophilic archaea (Grayling *et al.*, 1994; Ronimus and Musgrave, 1996). A variety of RNA modifications are known, some of which apparently have a stabilising function (Kowalak, 1994).

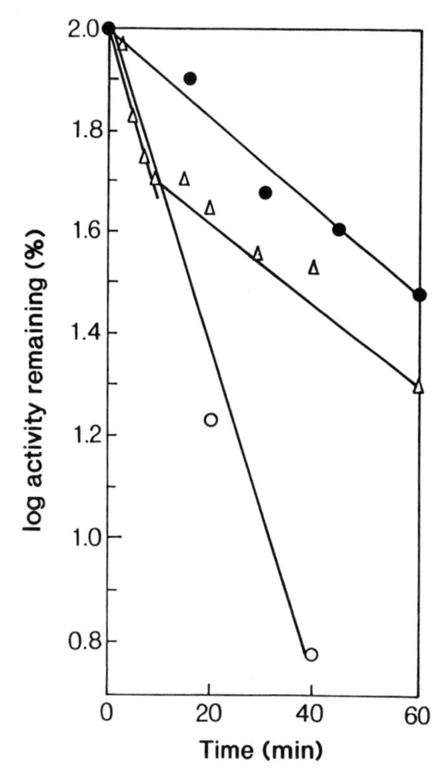

Figure 21.2 Stability of three enzymes at 130 °C. The plot shows the loss of activity with time of the unpurified amylase from *Pyrococcus furiosus* (○), the α-glucosidase from *Thermococcus* strain AN1 in 90% sorbitol (●), and the immobilised xylanase from *Thermotoga maritima* sp. FjSS3B1 in molten sorbitol (Δ). (Data from Koch *et al.*, 1990; Piller *et al.*, 1996; Simpson *et al.*, 1991.)

21.3 Coenzyme and metabolite stability: general considerations

Biological macromolecules all seem to have the potential to be stable up to 125 °C. The case is not so clear cut for low-molecular-mass metabolites and coenzymes, some of which have quite short half-lives even at 105 °C (see, for example, Table 21.2). However, in many cases, stability is very dependent upon conditions. For example, ATP stability is greatly affected by pH and metal ions (Tetas and Lowenstein, 1963; Ramirez *et al.*, 1980), so microenvironment protection is possible. In dilute buffer at around pH 7, the half-life of NADH is only a few minutes at 95 °C (Lowry *et al.*, 1961; Walsh *et al.*, 1983; Hudson *et al.*, 1993), and we have failed to stabilise NADP using various additives, including a stable dehydrogenase with a low K_m for NADP/H (Hudson *et al.*, 1993); but the end products of degradation have not been identified and rapid re-synthesis is a possibility. In any event, since it is used as a coenzyme in modern archaea growing at 105 °C, such organisms obviously have a way of circumventing this instability.

There are a number of other mechanisms by which metabolite/coenzyme thermal instability may be overcome at high growth temperatures.

Table 21.2 Metabolite and coenzyme stabilities

Metabolite/coenzyme	Percentage remaining after	
	1 h/95 °C	3 h/105 °C
NAD	<5	n.d.
FAD	100	85
FMN	75	65
Pyridoxal phosphate	40	0
Glucose	100	100
Glucose 6-phosphate	100	70
Glucose 1,6-diphosphate	90	50
Gluconate	100	100
6-Phosphogluconate	100	90
Glycerate	100	100
3-Phosphoglycerate	100	100
Acetate	100	100
Acetyl phosphate	<10	n.d.
CoASH	100	45
Acetyl-CoA	100	75
ATP	40	0
ADP	50	0
AMP	95	60

10 mmol/l solutions in distilled water containing 1 mmol/l KI were heated in sealed glass tubes, and degradation was assessed by changes to the electrospray mass spectrum.

1 **Metabolic channelling**. Van de Casteele *et al.* (1990) have suggested that the thermal instability of carbamoyl phosphate can be circumvented at high temperatures by the physical juxtaposition of the enzyme producing the metabolite (carbamoyl-phosphate synthetase) and the next enzyme in the metabolic pathway (ornithine carbamoyltransferase), which consumes the metabolite. Evidence for such a juxtaposition has been obtained for *Pyrococcus furiosus* (Legrain *et al.*, 1995).

2 **Catalytic efficiency**. Sterner *et al.* (1995) have presented evidence that the thermal instability of phosphoribosyl anthranilate (an intermediate in the tryptophan biosynthesis pathway; $t_{1/2}$ = 39 s at 80 °C) is overcome in *Thermotoga maritima* by the very high efficiency of the phosphoribosyl anthranilate isomerase enzyme, which has a very low K_m and a high K_{cat}.

3 **Modification of local conditions**. While there are no specific examples of this, the thermal stability of many metabolites is highly dependent on local conditions. The half-life of ATP is shortened by an order of magnitude in the presence of some bivalent metals, and the stability of both ATP and NADH is highly pH dependent (Tetas and Lowenstein, 1963; Lowry *et al.*, 1961).

4 **Substitution**. Perhaps the most interesting mechanism is substitution, where the thermal instability of a metabolite has been dealt with by using a more stable substitute. Of course, this is putting the situation back-to-front. What has presumably occurred is that evolution into lower-temperature environments has eventually allowed the use of less-stable (but more effective) metabolites/coenzymes.

21.4 NAD(P) stability and the role of non-haem iron proteins

NAD/P is a metabolite which I believe we can identify as being a low-temperature substitution, a relatively late evolutionary development that succeeded the use of non-haem iron proteins (Daniel and Danson, 1995), for the following reasons.

1. NAD and NADP are unstable at 95 °C, ($t_{1/2} \sim 2$ min)
2. It is widely accepted that non-haem iron proteins occurred early in evolution. Various researchers have proposed an involvement of iron sulfide in the origin of life (Russel *et al.*, 1988, 1990, 1993, 1994; Wächterhäuser, 1988, 1990, 1992; Cairns-Smith *et al.*, 1992), and Hall *et al.* (1971) have proposed that a ferredoxin may have been among the earliest proteins formed. Others have supported this view (e.g. Wächtershäuser, 1992). Some non-haem iron redox proteins are stable and functional at 100 °C (Adams, 1993).
3. There are a number of enzymes in the central metabolic pathways of thermophilic archaea that use non-haem iron protein for oxidation–reduction reactions, where NAD and NADP are used in their non-archaeal (and less primitive) counterparts (Danson, 1988). Figure 21.3 summarises some of these findings. For example, in eukaryotic and the majority of bacterial organisms, pyruvate is oxidatively decarboxylated to acetyl-CoA via the pyruvate dehydrogenase multienzyme complex (Perham, 1991). This enzyme system uses NAD as the oxidant: a similar oxidative decarboxylation of pyruvate is carried out by the archaea, but the enzyme catalyst is an iron–sulfur pyruvate oxidoreductase, having no dependency on NAD/P (Kerscher and Oesterhelt, 1982). The pyruvate oxidoreductase is found in all archaeal phenotypes (thermophiles, halophiles, and methanogens) and also in anaerobic bacteria; Kerscher and Oesterhelt (1982) have proposed that this enzyme existed before the divergence of the three domains and that the NAD-linked dehydrogenase complex evolved after the development of oxidative phosphorylation. Much of the evidence on the non-haem iron-linked enzymes in archaea comes from the group of Adams. Makund and Adams (1991) have shown that *P. furiosus* oxidises glucose to pyruvate via a nonphosphorylated Entner–Doudoroff pathway in which the redox reactions are catalysed by ferredoxin-linked oxidoreductases (Figure 21.3). In the less-primitive and the less-thermophilic archaea *Sulfolobus* and *Thermoplasma* (Kjems *et al.*, 1992), glucose is also catabolised via this nonphosphorylated pathway, but the equivalent redox enzymes (glucose dehydrogenase and glyceraldehyde dehydrogenase) are NAD/P linked (De Rosa *et al.*, 1984; Budgen and Danson, 1986).

   Other examples of the tendency for non-haem iron protein to replace NAD/P in more primitive/more thermophilic organisms include: the finding (Iwasaki *et al.*, 1994, 1995) that in *Sulfolobus* the 2-oxoglutarate oxidoreductase is ferredoxin linked, rather than NAD/P linked; the non-haem iron protein linkage of the aldehyde oxidoreductase from ES-4 (Johnson *et al.*, 1993) (already known in *P. furiosus*); the 2-ketoisovalerate: ferredoxin oxidoreductase from several hyperthermophilic archaea (Heider *et al.*, 1996); and the indole:pyruvate oxidoreductase in *P. furiosus* (Mai and Adams, 1994). An increased dependence upon non-haem iron protein rather than NAD has also now been shown in one of the most primitive and thermophilic bacteria, *Thermotoga*, at the pyruvate:ferredoxin oxidoreductase step, although the enzyme mechanism is different from that found in Archaea (Smith *et al.*, 1994).
4. A *P. furiosus* non-haem iron protein transfers electrons to a metal electrode without the need for a mediator (Park *et al.*, 1991). If this is due to a reduced distance between

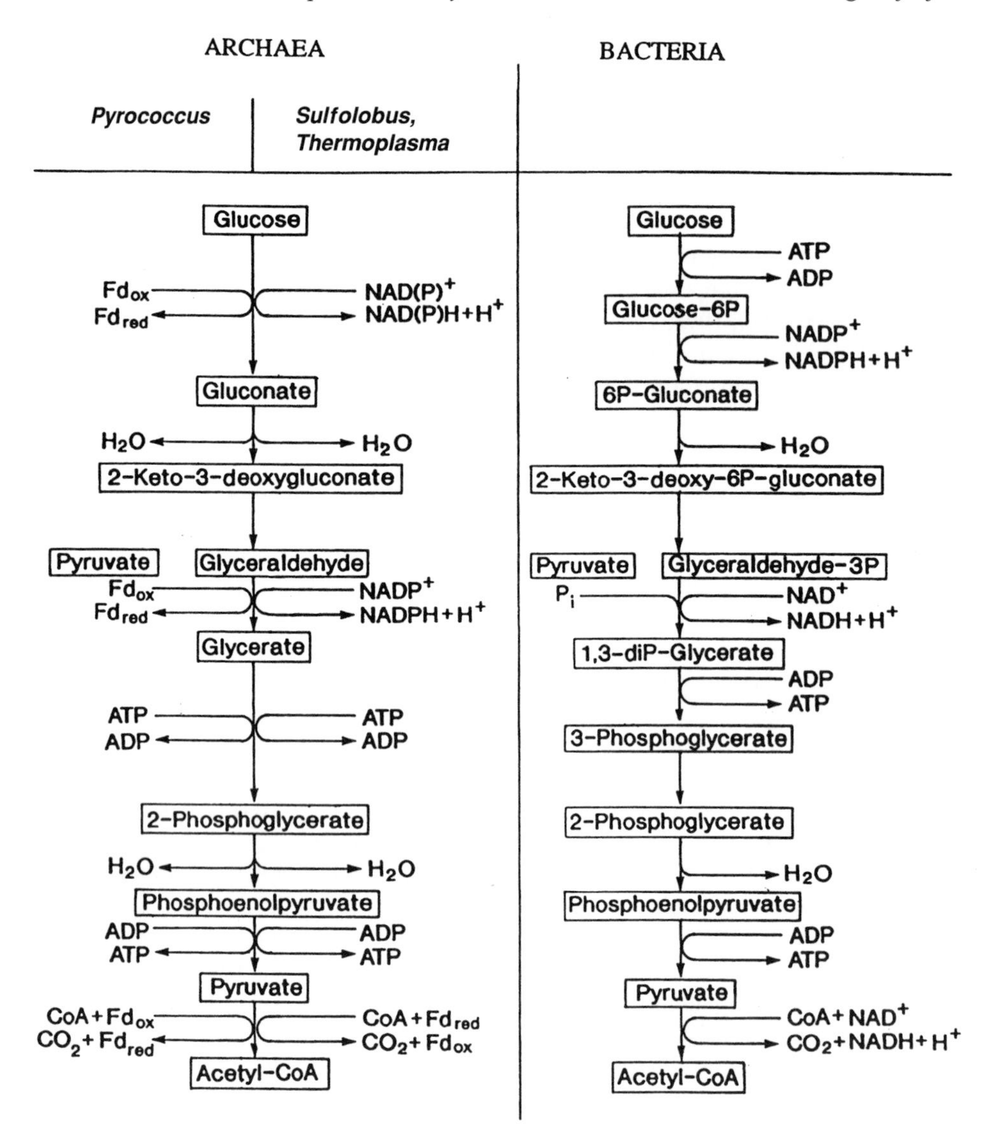

Figure 21.3 The Entner–Doudoroff pathways of glucose oxidation in thermophilic archaea and in bacteria. (From Daniel and Danson, 1995.)

the electrode and the redox centre of the enzyme, it implies a relatively exposed site for the iron–sulfur cluster, possibly facilitating the exchange of electrons between different non-haem iron proteins, and thus a function as a 'redox currency' in the same way that NAD/P(H) couples function in modern organisms. In any event, the wider redox potential range over which non-haem iron proteins act compared with NAD/P(H) may have favoured their early use as redox coenzymes.

5 Some properties of the NAD/P(H)-linked dehydrogenases in archaeal enzymes may be due to a less well developed enzyme–coenzyme interaction than in the bacterial and eukaryotic equivalents. The dehydrogenases from thermophilic archaea can often use both NAD and NADP, in some cases with equal efficiency, whereas most dehydrogenases

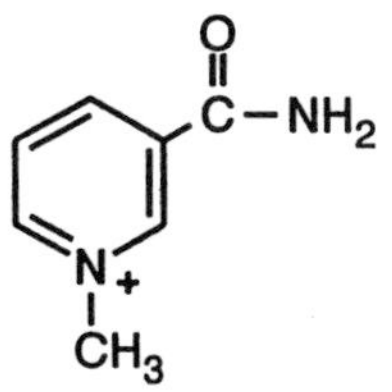

Figure 21.4 1-Methylnicotinamide (*N*-methylnicotinic acid amide).

from bacteria and eukaryotes are specific for one or the other (Perham, 1991). Dual-cofactor specificity is much more common in thermophilic archaea (Daniel and Danson, 1995).

The evidence for an evolutionary progression from the use of non-haem iron protein to NAD/P as a redox currency is fairly compelling. So much so that a search for possible intermediates in this progression may be worthwhile. One possible candidate is 1-methylnicotinamide (*N*-methylnicotinic acid amide) (Figure 21.4). This compound could obviously be a precursor of NAD/P, and it is an effective redox agent in at least one biological system (Daniel and Gray, 1976). Its redox potential ($E^{o\prime} = -419$ mV at 30 °C) (Loach, 1970) is very similar to that of most hyperthermophilic iron–sulfur proteins ($E^{o\prime}$ around -400 mV at 25 °C; Smith *et al.*, 1995) and close to those of NAD and NADP ($E^{o\prime} = -320$ and -324 mV, respectively). Variants with other groups replacing methyl are also possibilities.

21.5 The stability of phosphorylated coenzymes and metabolites

There are a number of other low-molecular-mass compounds which, on the basis of their instability, may have been later evolutionary developments (Table 21.2). Pyridoxal phosphate is somewhat unstable and, as already indicated above, NAD(P) is quite unstable. Both of these do occur in thermophilic archaea, but we may find that pyridoxal phosphate is somewhat less common than in mesophiles. Given the presence of an Entner–Doudoroff pathway based on nonphosphorylated carbohydrates in thermophilic archaea (Danson, 1988), these are another interesting possibility for substitution. However, the differences in stability are small (Table 21.2), even though the nonphosphorylated forms are generally more stable. Nevertheless, although the relative importance of the phosphorylated and nonphosphorylated pathways is not clear within the thermophilic archaea, the nonphosphorylated pathway is confined to this group.

Acetyl phosphate is particularly unstable relative to acetate and acetyl-CoA (Table 21.2), and its use seems likely to be a later development (as suggested by Schäfer *et al.*, 1993), although the direct conversion of acetyl-CoA to acetate (Figure 21.5) is an archaeal characteristic rather than a thermophilic one.

The most important phosphorylated derivatives are of course ATP and ADP. The stability of this class of compounds is in the order pyrosphosphate/AMP > ADP > ATP, and ATP is relatively unstable at 95 °C. There is some evidence for the 'substitution' of the more stable compounds (PP and ADP) for ATP in thermophilic archaea. For example, Kengan and colleagues (1994) have found that in *Pyrococcus* both hexokinase and phosphofructokinase are, uniquely, ADP linked. In the case of hexokinase, glucose is normally phosphorylated by ATP, which is less stable than ADP. In the case of phosphofructokinase fructose 6-phosphate is also normally phosphorylated by ATP; Siebers and Hensel (1993)

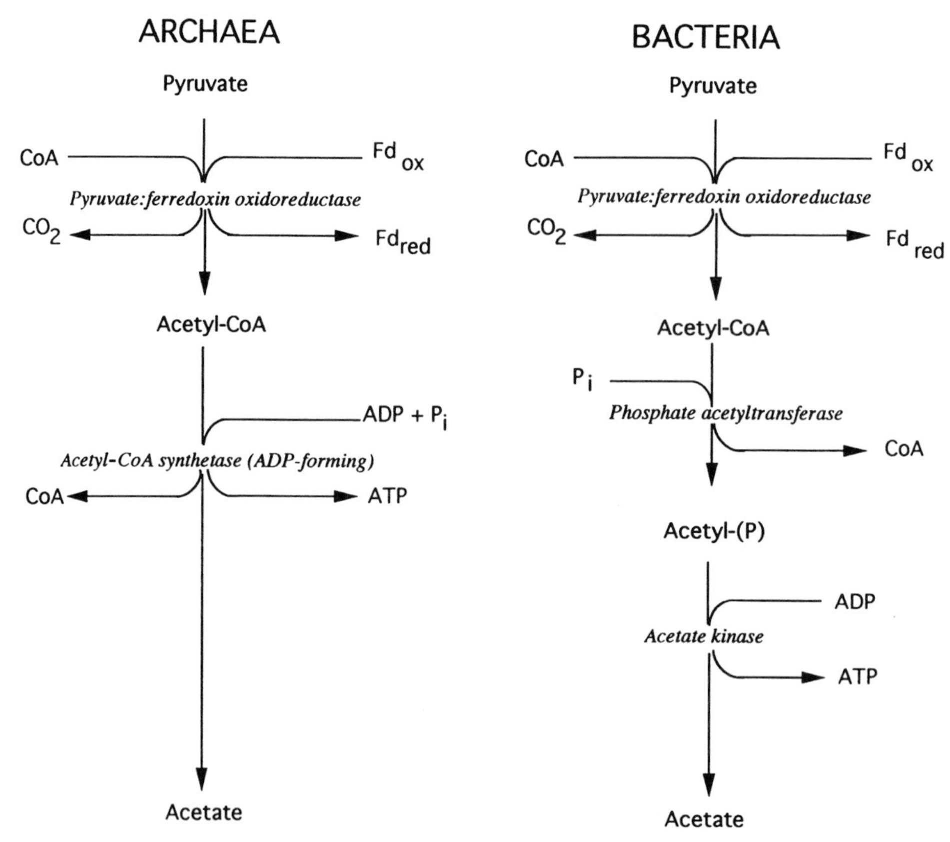

Figure 21.5 Mechanisms of acetate formation and ATP synthesis from pyruvate in archaea and anaerobic bacteria (Schafer *et al.*, 1993).

have shown that in *Thermoproteus tenax*, as in some primitive eukaryotic parasites, pyrophosphate is used.

Irrespective of stability considerations, it would not of course be surprising if a simple compound like pyrophosphate preceded the use of the adenine nucleotides as a cellular energy currency. But the lower thermal stability of ATP may be giving us opportunities to observe some preserved residual primitive use of pyrophosphate in modern organisms growing at high temperatures.

21.6 Conclusions

While the existence of more stable coenzymes and metabolites in more primitive organisms can reasonably be taken to support a high-temperature origin for life, this support is not unequivocal. It could be argued that pyrophosphate is just a simpler compound than ATP, and that non-haem iron proteins were likely to be conveniently available before NAD(P) (providing one is content to accept that enzymes preceded DNA and RNA). A lot more research is needed, particularly into the mechanism(s) by which NAD(P), for example, survives/is stabilised in hyperthermophilic archaea growing above 100 °C. If simple mechanisms are found enabling stabilisation to significantly higher temperatures, this would somewhat weaken the support currently given by this work to a high-temperature origin for life, unless such mechanisms could themselves be shown to be late evolutionary developments.

Research in this general field can hardly be regarded as scientific unless it can lead to testable predictions and/or suggest new lines of enquiry. Some of the predictions based on the suggestion that non-haem iron proteins preceded the use of NAD/P (Daniel and Danson, 1995) have already been fulfilled, and I suggest that searches for redox molecules which may have been evolutionary intermediates (e.g. 1-methylnicotinamide) may be fruitful. The evidence is less compelling in the case of other relatively unstable metabolites/coenzymes, but the more frequent use of pyrophosphate as an energy currency, and an absence of acetyl phosphate, may reasonably be expected in more deeply rooted (primitive) microorganisms.

21.7 Summary

There is little evidence that the instability of biological macromolecules limits the upper temperature for life to its current maximum of about 110 °C. However, some key coenzymes and metabolites, including NAD(P), pyridoxal phosphate, acetyl phosphate, ATP, and ADP are quite unstable above 100 °C. While the presence of some of these compounds in archaea growing above 100 °C indicates the existence of mechanisms for circumventing this instability in 'modern' organisms, there is also evidence that in primitive organisms more stable compounds (e.g. non-haem iron protein) replaced some or all of the functions of these (e.g. NAD(P)). The less-stable compounds may therefore have been a later evolutionary development, supporting a high-temperature origin for life.

The evidence is most compelling for the later replacement of non-haem iron proteins by NAD(P), but there is some evidence supporting the case for a later (cooler) evolutionary origin of acetyl phosphate, ATP, and ADP. In the case of NAD(P) it is suggested that since a number of archaea use a mixture of non-haem iron proteins and NAD(P), redox coenzymes intermediate in the progressive evolutionary 'takeover' by NAD(P) may exist, and that 1-methylnicotinamide (or a related compound) is a possible candidate.

References

ACHENBACH-RICHTER, L., GUPTA, R., STETTER, K. O. and WOESE, C. R. (1987) Were the original eubacteria thermophiles? *System. Appl. Microbiol.*, **9**, 34–39.

ADAMS, M. W. W. (1993) Enzymes and proteins from organisms that grow near and above 100 °C. *Anna. Rev. Microbiol.*, **47**, 627–658.

BUDGEN, N. and DANSON, M. J. (1986) Metabolism of glucose via a modified Entner–Doudoroff pathway in the thermoacidophilic arachaebacterium *Thermoplasma acidophilum*. *FEBS Lett.*, **196**, 207–210.

CAIRNS-SMITH, A. G., HALL, A. J. and RUSSELL, M. J. (1992) Mineral theories of the origin of life and an iron sulfide example. *Origins of Life and Evolution of the Biosphere*, **22**, 161–180.

COOLBEAR, T., DANIEL, R. M. and MORGAN, H. W. (1992) The enzymes from extreme thermophiles. *Adv. Biochem. Eng. Biotechnol.*, **45**, 57–98.

DANIEL, R. M. and DANSON, M. J. (1995) Did primitive microrganisms use nonhaem iron proteins in place of NAD/P? *J. Mol. Evol.*, **40**, 559–563.

DANIEL, R. M. and GRAY, J. (1976) Nitrate reductase from anaerobically grown *Rhizobium japonicum*. *J. Gen. Microbiol.*, **96**, 247–251.

DANIEL, R. M., DINES, M. and PETACH, H. H. (1996) The denaturation and degradation of stable enzymes at high temperatures. *Biochem. J.*, **317**, 1–11.

DANSON, M. J. (1988) Archaebacteria, the comparative enzymology of their central metabolic pathways. *Adv. Microb. Physiol.*, **29**, 165–231.

DEROSA, M., GAMBACORTA, A., NICHOLAUS, B., GIARDINA, P., PAERIO, E. and BUONOCORE, V. (1984) Glucose metabolism in the extreme thermoacidophilic archaebacterium *Sulfolobus solfataricus, J. Biochem.*, **224**, 407–414.

GAMBACORTA, A., TRINCONE, A., NICHOLAUS, B., LAMA, L. and DE ROSA, M. (1994) Unique features of lipids of archaea. In *Molecular Biology of Archaea*, ed. F. Pfeifer *et al.*, pp. 18–27 (Stuttgart: Gustav Fischer Verlag).

GRAYLING, R. A., SANDERSON, K. and REEVE, J. A. (1994) Archaeal DNA – binding proteins and chromosome structure. *System. Appl. Microbiol.*, **16**, 582–590.

HALL, D. O., CAMMACK, R. and RAO, K. K. (1971) Role for ferredoxins in the origin of life and biological evolution. *Nature*, **233**, 136–138.

HEIDER, J., MAI, X. and ADAMS, M. W. W. (1996) Characterisation of 2-ketoisovalerate ferredoxin oxidoreducatase, a new and reversible coenzyme A-dependant enzyme involved in peptide fermentation by hyperthermophilic archaea. *J. Bacteriol.*, **178**, 780–787.

HENSEL, R., JAKOB, I., SCHEER, H. and LOTTSPEICH, R. (1992) Proteins from hyperthermophilic archaea: stability towards covalent modification of the peptide chain. *Biochem. Soc. Symp.*, **58**, 127–133.

HUDSON, R. D., RUTTERSMITH, L. D. and DANIEL, R. M. (1993) Glutamate dehydrogenase from the extremely thermophilic achaebacterial isolate AN1. *Biochim. Biophys. Acta*, **1202**, 244–250.

IWASAKI, T., WAKAGI, T., ISOGAI, Y., TANAKA, K., IIZUKA, T. and OSHIMA, T. (1994) Functional and evolutionary implications of a [3Fe-4s] cluster of the dicluster-type ferredoxin from the thermoacidophilic archeon, *Sulfolobus* sp. strain 7. *J. Biol. Chem.*, **269**, 29444–29450.

IWASAKI, T., WAKAGI, T. and OSHIMA, T. (1995) Ferredoxin-dependent redox system of a thermoacidophilic archeon, *Sulfolobus* sp. strain 7. *J. Biol. Chem.*, **270**, 17878–17883.

JOHNSON, J. L., RAJAGOPALAN, K. V., MAKUND, S. and ADAMS, M. W. W. (1993) Identification of molybdopterin as the organic component of the tungsten cofactor in four enzymes from hyperthermophilic archaea. *J. Biol. Chem.*, **268**, 4848–4853.

KENGAN, S. W. M., DE BOK, F. A. M., VAN LOO, N. D., DIJKEMA, C., STAMS, A. J. M. and DE VOS, W. M. (1994) Evidence for the operation of a novel Emden–Meyerhof pathway that involves ADP-dependent kinases during sugar fermentation by *Pyrococcus furiosus*. *J. Biol. Chem.*, **269**, 17537–17541.

KERSCHER, L. and OESTERHELT, D. (1982) Pyruvate ferridoxin oxidoreductase – new findings on an ancient enzyme. *Trends Biochem. Sci.*, **7**, 371–374.

KJEMS, J., LARSEN, N., DALGAARD, J. Z., GARRETT, R. A. and STETTER, K. O. (1992) Phylogenetic relationships among the hyperthermophilic Archaea determined from partial 23S rRNA gene sequences. *System. Appl. Microbiol.*, **15**, 203–208.

KOCH, R., ZABLOWSKI, P., SPREINAT, A. and ANTRANIKIAN, G. (1990) Extremely thermostable amylolytic enzyme from the archaebacterium *Pyrococcus furiousus*. *FEMS Microbiol. Lett.*, **71**, 21–26.

KOWALAK, J. A., DALLUGE, J. J., MCCLOSKEY, J. A. and STETTER, K. O. (1994) The role of post-transcriptional modifications in stabilisation of transfer RNA from hyperthermophiles. *Biochemistry*, **33**, 7869–7876.

LEGRAIN, C., DEMAREZ, M., GLANSDORFF, N. and PIERARD, A. (1995) Ammonia-dependent synthesis and metabolic channelling of carbamoyl phosphate in the hyperthermophilic archeon *Pyrococcus furiosus*. *Microbiology*, **141**, 1093–1099.

LOACH, P. A. (1970) Oxidation reduction potentials, absorbance bands and molar absorbance of compounds used in biochemical studies. In *Handbook of Biochemistry*, 2nd edn, ed. H. A. Sober, pp. J-33–J-40 (Cleveland, OH. Chemical Rubber Company).

LOWRY, O. H., PASSONNEAU, J. V. and ROCK, M. K. (1961) The stability of pyridine nucleotides. *J. Biol. Chem.*, **236**, 2756–2759.

MAI, X. and ADAMS, M. W. W. (1994) Indolepyruvate ferredoxin oxidoreductase from the hyperthermophilic archaeon, *Pyroccus furiosus*, a new enzyme involved in peptide fermentation. *J. Biol. Chem.*, **269**, 16726–16732.

MUKUND, S. and ADAMS, M. W. W. (1991) The novel tungsten–iron–sulfur protein of the hyperthermophilic archaebacterium, *Pyrococcus furiousus*, is an aldehyde ferredoxin oxidoreductase, evidence for its participation in a unique glycolytic pathway. *J. Biol. Chem.*, **266**, 14208–14216.

OLSEN, G. J., WOESE, C. R. and OVERBEEK, R. (1994) The winds of (evolutionary) change: breathing new life into microbiology. *J. Bacteriol.*, **176**, 1–6.

PARK, J-B., FAN, C., HOFFMAN, B. M. and ADAMS, M. W. W. (1991) Potentiometric and electron nuclear double resonance properties of the two spin forms of the $[4Fe\text{-}4S]^{1+}$ cluster in the novel ferredoxin from the hyperthermophilic archaebacterium, *Pyrococcus furiosus*. *J. Biol. Chem.*, **266**, 19351–19356.

PERHAM, R. N. (1991) Domains, motifs and linkers in 2-oxo acid dehydrogenase multienzyme complexes. A paradigm in the design of a multifunctional protein. *Biochemistry*, **30**, 8501–8512.

PILLER, K., DANIEL, R. M. and PETACH, H. (1996) Properties and stabilisation of an extracellular α-glucosidase from the extremely thermophilic archaebacteria *Thermococcus* strain AN1: enzyme activity at 130 °C. *Biochim. Biophys. Acta*, **1292**, 197–205.

RAMIREZ, F., MARECEK, J. F. and SZAMOSI, J. (1980) Magnesium and calcium ion effects on hydrolysis rates of adenosine-5′-triphosphate. *J. Org. Chem.*, **45**, 4748–4752.

RONIMUS, R. and MUSGRAVE, D. R. (1996) Purification and characterisation of a histone-like protein from the archaeal isolate AN1, a member of the thermococcales. *Mol. Microbiol.*, **20**, 77–86.

RUSSELL, M. J., DANIEL, R. M. and HALL, M. J. (1993) On the emergence of life via catalytic iron–sulfide membranes. *Terra Nova*, 343–347.

RUSSELL, M. J., HALL, A. J. and GIZE, A. P. (1990) Pyrite and the origin of life. *Nature*, **344**, 387.

RUSSELL, M. J., HALL, A. J., CAIRNS-SMITH, A. G. and BRATEMAN, P. S. (1988) Submarine hot springs and the origin of life. *Nature*, **336**, 117.

RUSSELL, M. J., DANIEL, R. M., HALL, A. J. and SHERRINGHAM, J. A. (1994) A hydrothermally precipitated catalytic iron sulfide membrane as a first step tewards life. *J. Mol. Evol.*, **39**, 231–243.

SCHAFER, T., SELIG, M. and SCHONHEIT, P. (1993) Acetyl-CoA synthetase (ADP-forming) in archaea, a novel enzyme involved in acetate formation and ATP synthesis. *Arch. Microbiol.*, **159**, 72–83.

SIEBERS, B. and HENSEL, R. (1993) Glucose catabolism of the hyperthermophilic Archaeum *Thermoproteus tenax*. *FEMS Microbiol. Lett.*, **111**, 1–8.

SIMPSON, H., HAUFLER, U. R. and DANIEL, R. M. (1991) An extremely thermostable xylanase from the thermophilic eubacterium *Thermotoga*. *J. Biochem.*, **277**, 413–427.

SMITH, E. T., BLAMEY, J. M. and ADAMS, M. W. W. (1994) Pyruvate ferredoxin oxidoreductases of the hyperthermophilic archaeon *Pyrococcus furiosus* and the hyperthermophilic bacterium *Thermotoga maritima* have different catalytic mechanisms. *Biochemistry*, **33**, 1008–1016.

SMITH, E. T., BLAMEY, J. M., ZHOU, Z. H. and ADAMS, M. W. W. (1995) A variable-temperature direct electrochemical study of metalloproteins from hyperthermophilic microorganisms involved in hydrogen production from pyruvate. *Biochemistry*, **34**, 7161–7169.

STERNER, R., DAHM, A., DARIMONT, B., IVENS, A., LEIBL, W. and KIRSCHNER, K. (1995) 8-Barrel proteins of tryptophan biosynthesis in the hyperthermophilic *Thermotoga maritima*. *EMBO J.*, **14**, 4395–4402.

STETTER, K. O. (1982) Ultrathin mycelia forming organisms from submarine volcano areas having an optimum growth temperature of 105 °C. *Nature*, **300**, 258–260.

STETTER, K. O. (1996) Hyperthermophilic prokaryotes. *FEMS Microbiol. Rev.*, **18**, 149–158.

TETAS, M. and LOWENSTEIN, J. M. (1963) The effect of bivalent metal ions on the hydrolysis of adenosine di- and triphosphate. *Biochemistry*, **2**, 350–357.

VAN DE CASTEELE, M., DEMAREZ, M., LEGRAIN, C., GLANSDORFF, N. and PIERARD, A. (1990) Pathways of arginine biosynthesis in extremely thermophilic archaeo-and eubacteria. *J. Gen. Microbiol.*, **136**, 1177–1183.

WÄCHTERSHÄUSER, G. (1988) Pyrite formation, the first energy source for life; a hypothesis. *System. Appl. Microbiol.*, **10**, 207–210.

WÄCHTERSHÄUSER, G. (1990) The case for the chemoautotrophic origin of life in an iron–sulfur world. *Origins of Life and Evolution of the Biosphere*, **20**, 173–176.

WÄCHTERSHÄUSER, G. (1992) Ground works for an evolutionary biochemistry: the iron–sulfur world. *Biophys. Mol. Biol.*, **58**, 85–202.

WALSH, K. A. J., DANIEL, R. M. and MORGAN, H. W. (1983) A soluble NADH dehydrogenase from *Thermus aquaticus* strain T351. *J. Biochem.*, **201**, 427–433.

WOESE, C. R. (1987) Bacterial evolution. *Microbiol. Rev.*, **51**, 221–271.

22

3-Phosphoglycerate Kinase and Triosephosphate Isomerase from Hyperthermophilic Archaea: Features of Biochemical Thermoadaptation

REINHARD HENSEL[1], ALEXANDER SCHRAMM[1], DANIEL HESS[1] AND RUPERT J. M. RUSSELL[2]

[1] *FB 9 Mikrobiologie Universität GH Essen, Essen, Germany*
[2] *School of Biology and Biochemistry, University of Bath, Bath, UK*

22.1 Introduction

We compare structural properties of the glycolytic enzymes 3-phosphoglycerate kinase (PGK; EC 2.7.2.3) and triosephosphate isomerase (TIM; EC 5.3.1.1) from mesophilic and hyperthermophilic Archaea to deduce common traits in protein thermoadaptation. PGKs and TIMs from Bacteria and Eucarya have been thoroughly investigated with respect to structure and function. For both enzymes, more than 30 amino acid sequences are known; the three-dimensional structure could be determined in four (PGK) or seven (TIM) cases, respectively. The PGKs of eucaryal and bacterial sources represent monomeric enzymes and exhibit a bilobal structure (Blake and Evans, 1974; Watson *et al.*, 1982; Harlos *et al.*, 1992; Davies *et al.*, 1993). For its catalytic activity, an induced-fit mechanism with excessive hinge bending has been proposed (Sinev *et al.*, 1989; Haran *et al.*, 1992), which has recently been confirmed by crystal analysis of the closed conformation (Bernstein *et al.*, 1997).

All bacterial and eucaryal triosephosphate isomerases are usually homomeric dimers. A remarkable exception is represented by the bifunctional PGK/TIM fusion protein in *Thermotoga maritima*, which forms a homomeric tetramer (Schurig *et al.*, 1995). The catalytic mechanism of TIM has been studied in great detail by numerous investigations (Knowles, 1991; cf. Delboni *et al.*, 1995, and literature cited therein). The tertiary structure of TIM is characterised by a $(\beta/\alpha)_8$ barrel topology with the active-site residues located at the top of the barrel (Wierenga *et al.*,1992; cf. Delboni *et al.*, 1995). Comparisons of the three-dimensional structures of the enzymes from the moderately thermophilic *Bacillus stearothermophilus* with those of mesophilic sources have been conducted to provide insights into the determinants of protein thermostability (Davies *et al.*, 1993;

Delboni *et al.*, 1995). As main thermophilic features, increased internal hydrophobicity, reduction of the entropy of unfolding by preferring bulky residues, stabilisation of secondary structure elements, reduction of cavities, protection against heat-induced modification of the peptide chain, and strengthening of intersubunit contacts have been proposed.

To obtain insight into the determinants of protein thermoadaptation allowing enzymes to be stable and active at the upper borderline of temperature, we have focus on the structural properties of the PGK and TIM from the hyperthermophilic archaea *Methanothermus fervidus* (growth optimum 83 °C; Stetter *et al.*, 1981) and *Pyrococcus woesei* (growth optimum 100 °C; Zillig *et al.*, 1987) and compared them with those of the enzyme homologues from the mesophilic methanogen *Methanobacterium bryantii*. Based on the observation that the enzymes from the hyperthermophilic Archaea show a higher subunit association state than their mesophilic counterparts, a correlation between oligomerisation and thermoadaptation has been suggested. Sequence comparisons of TIMs from mesophilic and thermophilic sources revealed putative thermophilic features of TIM from the mesophile *M. bryantii,* which were interpreted as reminiscent of the thermophilic past of this organism. The use of such molecular relicts is proposed for reconstructing the direction of thermoadaptation.

22.2 Quarternary structure of the PGK and TIM from mesophilic and thermophilic Archaea

Whereas the subunit molecular masses of the archaeal PGK and TIM were of similar size to those usually found for the enzyme homologues from Bacteria and Eucarya, molecular mass determinations under nondenaturing conditions using different methods (gel filtration, sucrose density centrifugation, sedimentation equilibrium centrifugation) yielded significantly higher values for the enzymes from the hyperthermophiles *M. fervidus* and *P. woesei* (Table 22.1): The size of the native PGK and TIM from both hyperthermophiles was twice as high as found for the protein species of Bacteria and Eucarya (PGK ~90–100 kDa; TIM~100 kDa). Considering the subunit masses of the enzymes, we must conclude that the PGK and TIM of *M. fervidus* and *P. woesei* represent dimers or tetramers, respectively. In the case of PGK, the dimeric association could be confirmed by chemical

Table 22.1 Molecular mass of archaeal triosephosphate isomerases and phosphoglycerate kinases

	Triosephosphate isomerase		3-Phosphoglycerate kinase	
	$M_{r,subunit}$	$M_{r,native}$	$M_{r,subunit}$	$M_{r,native}$
M. bryantii	24 000[a]	57 000[c]	44 000[a]	42 000[c]
M. fervidus	23 500[b]	100 000[c]	46 000[b]	96 000[c]
P. woesei	24 000[b]	100 000[c, d]	46 000[b]	95 000[c]

[a] Deduced from the coding gene.
[b] Deduced from the coding gene and SDS-PAGE.
[c] Deduced from gel filtration.
[d] Confirmed by sucrose density centrifugation and sedimentation equilibrium centrifugation.
Data from Kohlhoff *et al.* (1996), Hess *et al.* (1995), Fabry *et al.* (1990) and this study.

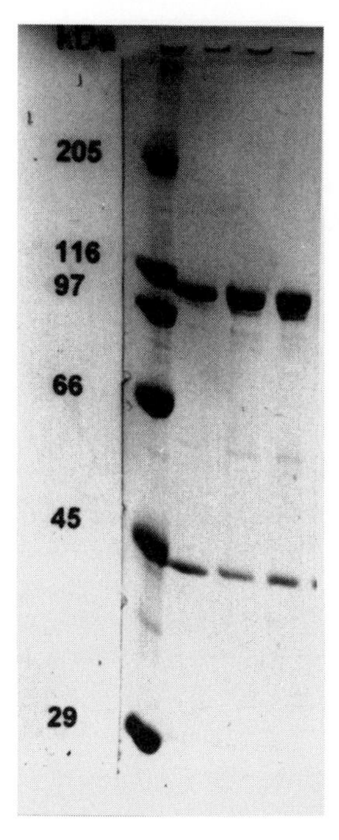

Figure 22.1 Cross-linking of PGK from *Methanothermus fervidus* by bis(maleimido) hexane (BMH). Reaction conditions were: 0.1 mol/l Hepes/KOH, pH 7.0; 0.6 mol/l KCl; 50 μmol/l PGK; 0.5 mmol/l BMH. The reaction was performed at room temperature for 2 to 15 min. Analysis was performed by SDS-PAGE. Lanes from left to right: molecular mass standard, and after incubation for 2, 5 and 15 min.

cross-linking (Figure 22.1). Contrary to this finding, the apparent molecular masses of the enzyme homologues from the mesophilic archaeum *M. bryantii* showed the common size (PGK 40 kDa; TIM 57 kDa) indicative of monomers (PGK) or dimers (TIM), respectively. From this we must conclude that the higher aggregation state of the enzymes from the hyperthermophilic archaea does not represent a domain-specific feature but rather correlates with adaptation to higher temperature.

22.3 Homology-based models for the subunit assembly of PGK and TIM from the hyperthermophilic *M. fervidus* and *P. woesei*

22.3.1 *Dimeric PGK*

To identify the subunit association in dimeric PGK, cross-linking experiments with the PGK from *M. fervidus* were performed. High recoveries (80–90%) of cross-linked dimers could be obtained with the cysteine specific maleimido esters *N,N′*-bis(maleimido)-1,6-hexane (BMH) and N,N′-bis(3-maleimidopropyl)-2-hydroxy-1,3-propandiamine (MPHPD), differing in hydrophobicity and spacer length (16.1 Å for BMH; 24.1 Å for MPHPD). The recovery in cross-linked dimers, however, depends strongly upon the oxidation state of the protein: preparations performed without protection against oxygen yielded only a 10–20% recovery of the cross-linked product, indicating that the reactive cysteines are susceptible to oxygen, being thus highly exposed.

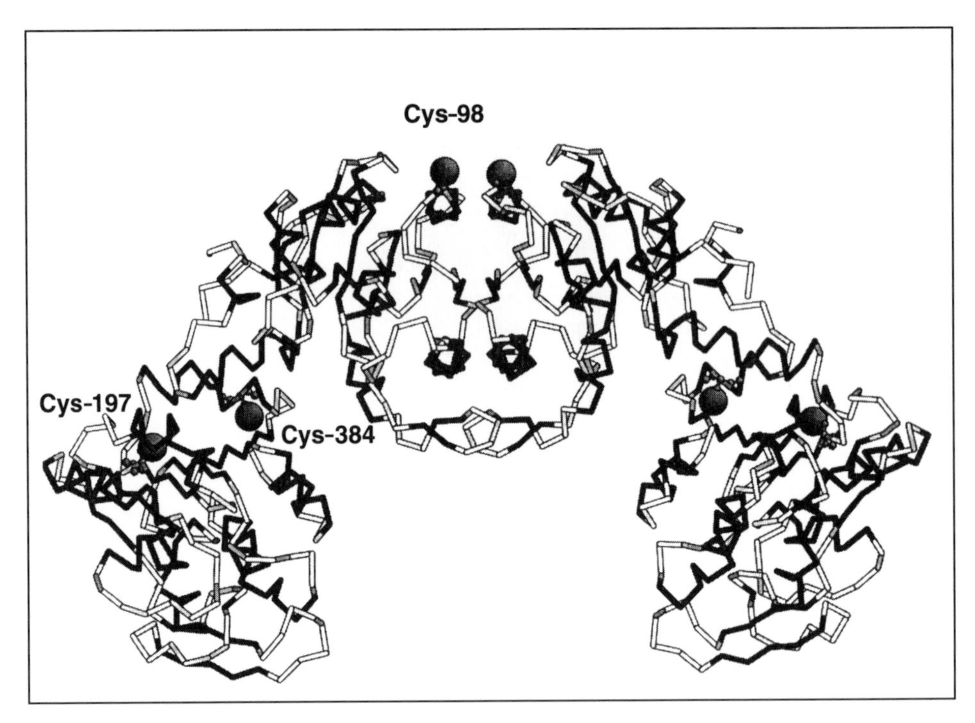

Figure 22.2 Model of the subunit association in dimeric PGK from *M. fervidus*. The single subunit was modelled on the structure of *B. stearothermophilus* PGK using the program package *What if* (Vriend, 1990). The picture was generated using the Rasmol program (Sayle and Milner-White, 1995). α-Helices are depicted in dark grey, β-strands in black, turns and loops in light grey. Cysteine residues are numbered as indicated according to the *M. fervidus* sequence and shown as ball-and-stick models. Sulfur atoms of cysteines are shown as 1.2 Å spheres.

Molecular modelling of the three-dimensional structure of the *M. fervidus* PGK monomer on the basis of the spatial structure of the *B. stearothermophilus* PGK (Davies *et al.*, 1993) favours Cys-98 (*M. fervidus* numbering; Fabry *et al.*, 1990) as docking residue for chemical cross-linking (Figure 22.2): the modelling experiments account for a solvent-exposed position of this residue within the N-terminal domain of the polypeptide, thus explaining its reactivity towards oxygen and cross-linking reagents. The other two Cys residues present in the *M. fervidus* PGK polypeptide (Cys-198, Cys-384) were predicted to be located in a more buried position close to the active site. As indicated by cross-linking experiments, both residues are accessible only by the hydrophobic reagent BMH but not by the hydrophilic MPHPD: after blocking the exposed Cys-98 with *N*-ethylmaleimide (NEM), incubation with BMH resulted in a protein species exhibiting a higher electrophoretic mobility on SDS-PAGE, probably owing to a more compact conformation caused by the covalent cross-link. Respective modification of these residues using MPHPD failed, presumably owing to the poor accessibility by the more hydrophilic thiol reagent. Compatible with their assigned position near the active site, their covalent modification can be hindered by adding the substrates ATP and 3-phosphoglycerate (Figure 22.3). These modification experiments confirm the validity of the deduced model and support the correct positioning of Cys-98 within the spatial structure of the PGK monomer

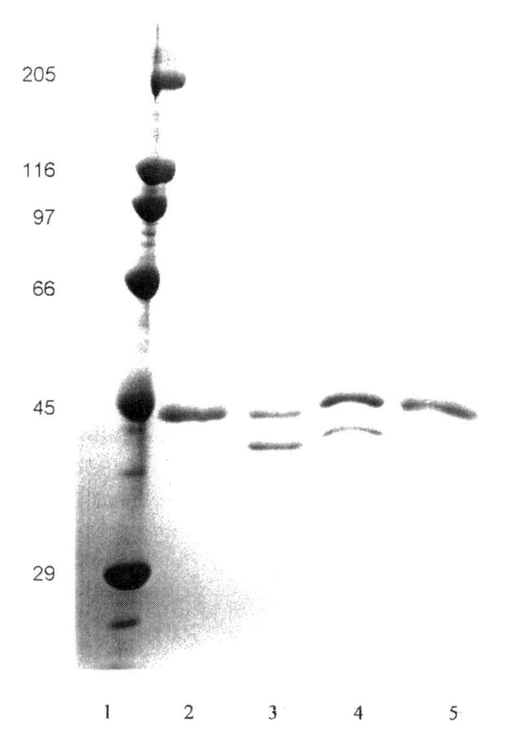

Figure 22.3 Cross-linking of PGK from *M. fervidus* by BMH after blocking of solvent-exposed cysteines with *N*-ethylmaleimide (NEM). Blocking of solvent exposed cysteines of the *M. fervidus* PGK with NEM was performed for 30 min at 60 °C. Reaction conditions were: 0.1 mol/l Hepes, pH 7.0; 0.6 mol/l KCl; 50 μg/ml PGK; 0.5 μmol/l NEM. After dialysis against 0.1 mol/l Hepes/KOH, pH 7.0, cross-linking was performed at 60 °C for 30 min in the presence of 0.5 mmol/l BMH. The reaction was stopped by adding cysteine to 50 mmol/l. Analysis was performed by SDS-PAGE. Lanes from left to right: 1, molecular mass standard; 2, control (NEM-treated PGK, no cross-linker added); 3, NEM-treated PGK cross-linked by BMH; 4, NEM-treated PGK cross-linked by BMH in the presence of ATP (10.5 mmol/l); 5, NEM-treated PGK cross-linked by BMH in the presence of ATP (10.5 mmol/l) and 3-phosphoglycerate (13.5 mmol/l).

of *M. fervidus* (Hess, 1995). We therefore assume that the monomers are associated by their N-terminal domains, thus bringing the Cys-98 residues of both subunits to an appropriate distance (< 16 Å) to be cross-linked. At the moment we cannot define the dimer contacts more precisely. As possible dimer interface (Figure 22.2) we discuss the helices α4 and α6 and the intermediate β strands (corresponding to positions 70 to 108 in the *B. stearothermophilus* PGK structure). This assembly would allow the formation of an extended 12-stranded antiparallel β-sheet across the subunit contacts. A corresponding arrangement in the mesophilic structure of the *M. bryantii* PGK seems to be less favoured because of the presence of a charged arginyl residue (position 88), which replaces the hydrophobic methionyl residue in the structure of the hyperthermophilic *M. fervidus* (position 87) and *P. woesei* PGK (position 82; Hess *et al.*, 1995). The proposed association would not restrict the hinge movement of the domains and would therefore not interfere with the enzyme's function. Crystallographic analyses will show whether the proposed subunit assembly really applies.

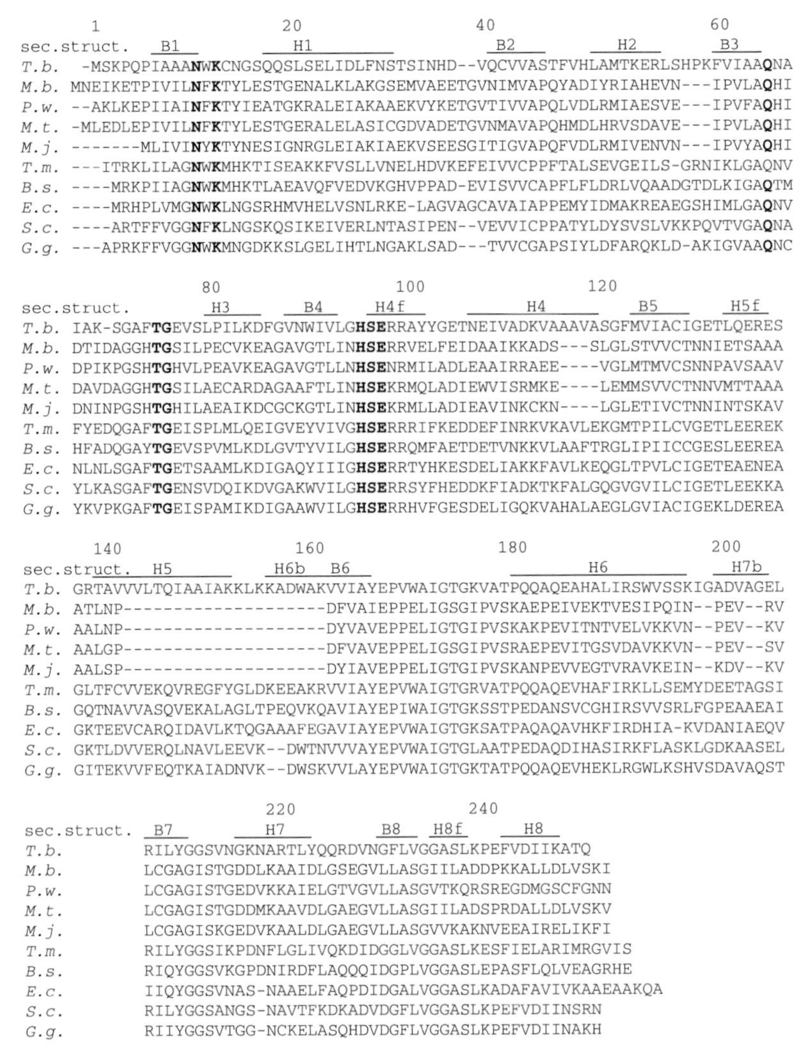

Figure 22.4 Alignment of TIM sequences from organisms of the three domains of life. The secondary structural elements of the three-dimensional structure of *T. brucei* TIM (Wierenga *et al.*, 1992) are given above the sequences. Amino acids involved in interface binding (Wierenga *et al.*, 1992) are indicated in bold type. Gaps introduced for optimal alignment are marked by hyphens (-). Abbreviations used are: sec.struct., secondary structure; H, α-helix; B, β-strand; T.b., *Trypanosoma brucei* (Swinkel *et al.*, 1986); M.b., *M. bryantii* (this study); P.w., *P. woesei* (Kohlhoff *et al.*, 1996); M.t., *Methanobacterium thermoautotrophicum* (Smith *et al.*, 1997); M.j., *Methanococcus jannaschii* (Bult *et al.*, 1996); T.m., *T. maritima* (Schurig *et al.*, 1995); B.s., *B. stearothermopilus* (Artavanis and Harris, 1980); E.c., *Escherichia coli* (Pichersky *et al.*, 1984); S.c., *Saccharomyces cerevisiae* (Alber and Kawasaki, 1982); G.g., *Gallus gallus* (Straus and Gilbert, 1985); Z.m., *Zea mays* (Marchionni and Gilbert, 1986). The *M. bryantii* TIM sequence was deduced from the *tpi* gene sequence. The fragment of the genomic DNA of *M. bryantii* harbouring the *tpi* gene was identified by hybridizing Southern blots with a PCR product derived from the adjacent *pgk* gene (Fabry *et al.*, 1990). A 4.2 kb-*Eco*RI/*Hin*dIII-fragment giving positive signals was cloned into pBluescript KS+ (Stratagene, La Jolla, CA, USA) using standard procedures (Sambrook *et al.*, 1989). Sequencing revealed that this genomic fragment contained both the *tpi* and the *pgk* gene.

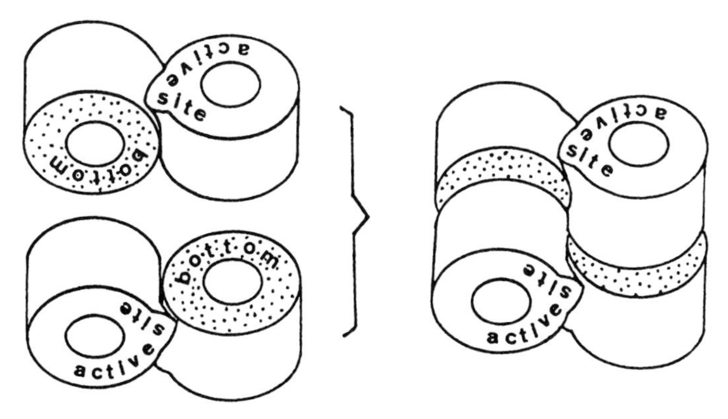

Figure 22.5 Model of the monomer arrangement in the *P. woesei* TIM tetramer (from Kohlhoff *et al.*, 1996; with permission of the publisher).

22.3.2 Tetrameric TIM

In contrast to the dimeric PGK, no experimental data are yet available for the subunit assembly of the tetrameric TIM. As shown by sequence comparisons, the archaeal TIM is – like the archaeal PGK (Hess *et al.*, 1995) – homologous to the bacterial and eucaryal counterparts (Kohlhoff *et al.*, 1996). From the observation that out of the 14 highly conserved residues involved in the dimer interface of bacterial and eucaryal TIMs (Wierenga *et al.*, 1992; Delboni *et al.*, 1995) 8 are present in the *P. woesei* and *M. bryantii* sequence (Figure 22.4), we assume similar subunit contacts also in the archaeal enzymes. The proposed model for the tetrameric association, however, is based mainly on striking sequence differences between the archaeal TIMs and their obligately dimeric enzyme homologues from Bacteria and Eucarya. As first noticed for the *P. woesei* structure, the archaeal sequence shows more or less extended gaps, which have to be introduced for optimal alignment (Figure 22.4). Assuming a similar secondary structure arrangement as determined for the bacterial and eucaryal TIM, secondary structure elements (Figure 22.4; loop between α-helix H2 and β-strand B3, α-helices H4, H5, H6) are affected by these changes, which are located at the barrel bottom and at the barrel flank without obvious functional importance. These changes were suggested to be correlated with the formation of new subunit binding sites in the tetrameric TIM favouring dimer–dimer contacts through the barrel bottom and the functionally inert flank of the barrel (Figure 22.5). A similar mode of association could be found for the dimeric TIM barrel protein phosphoribosyl anthranilate isomerase (PRAI) of *Thermotoga maritima*, which forms strong subunit contacts via the barrel bottom (Hennig *et al.*, 1997). For steric reasons, the slender barrel of the archaeal TIM with its shortened helices (especially helix H5) but without additional secondary structure elements at its periphery would also allow intimate contacts via the functionally inert flank of the barrel. In contrast, the tetrameric TIM barrel protein β-glycosidase of *Sulfolobus solfataricus* seems to be a less suited model for subunit assembly of the archaeal TIMs because of its considerable extensions at the outside of the β-barrel, which may require a specific subunit association (Aguilar *et al.*, 1997).

Unexpectedly, these characteristic sequence features could be found not only in the TIM sequences of the hyperthermophilic Archaea: the amino acid sequence of the dimeric TIM of the mesophilic *M. bryantii* showed equivalent gaps of the same size and at the same positions (Figure 22.4). Thus, if the striking gaps are correlated with the subunit association, they obviously represent only a necessary but not a sufficient prerequisite for tetramer formation.

22.4 Thermoadaption through oligomerisation – a common strategy?

The trend to higher state of oligomerisation does not seem to be restricted to archaeal PGK and TIM. As an additional, rather fancy, example, the tetrameric TIM/PGK fusion protein from the hyperthermophilic bacterium *T. maritima* (Schurig *et al.*, 1995) confirms the thermophilic trend to higher aggregation state. Preference for higher subunit assembly could also be found in other enzyme proteins from hyperthermophiles. Thus, the PRAI from *T. maritima* represents a homomeric dimer, whereas the enzyme homologues from mesophilic bacteria are monomers (Hennig *et al.*, 1997). A similar tendency could be observed within the glycosyl hydrolase family 1: the β-glucosidase/β-glycosidase from the hyperthermophilic archaea *P. furiosus* and *S. solfataricus* (Kengen *et al.*, 1993; Aguilar *et al.*, 1997) are tetramers; the β-glucosidase from the hyperthermophilic *T. maritima* represents a dimer; whereas the β-glucosidases of mesophilic or moderately thermophilic bacteria are mainly monomers – with the only known exception of *Agrobacterium* sp., which forms a dimeric β-glucosidase (Day and Withers, 1986; Wakarchuk *et al.*, 1988; Paavilainen *et al.*, 1993). Although the relationship between oligomerisation and thermo-adaptation has been hotly discussed, the presented examples support such a correlation. From experimental and comparative investigations (Schultes and Jaenicke, 1991; Facchiano *et al.*, 1995; Beernink and Tolan, 1994) that demonstrate the crucial role of intersubunit contacts for conformational stability of the whole protein molecule, and from the recent findings that the subunit contacts especially in proteins from hyperthermophiles are intensified by enhancement of electrostatic (mainly ionic) or hydrophobic interactions (Kirino *et al.*, 1994; Yip *et al.*, 1995; Korndörfer *et al.*, 1995; Hennig *et al.*, 1997; Russell *et al.*, 1997), we conclude that the higher order of oligomerisation serves mainly protein thermo-stabilisation by additional intersubunit contacts. These additional interactions may be of special importance at the upper temperature limit of life for compensating the decreasing contributions of intrasubunit electrostatic and hydrophobic interactions. We must, however, assume that the stabilisation potential of the additional contacts can only be used by proteins, whose topology allows subunit association without essential restriction of the catalytic properties, i.e. without reducing the conformational flexibility of the active site and without hindering its accessibility by substrates. Despite these restrictions, we expect that, with more profound knowledge about the structure of proteins from hyperthermophiles, the number of enzyme proteins following the trend to higher oligomerisation state will increase.

22.5 Following the direction of thermoadaptation – a hypothetical approach

Although the question of the direction of thermoadaptation (from hot to cold or from cold to hot) is highly relevant not only for the mechanism of thermoadaptation but also for our understanding of the evolution of life, no convincing strategies for determining the direction can yet be proposed. To stimulate further ideas, we present a rather hypothetical approach to addressing this intriguing question by comparative and experimental studies on proteins.

Whether this question can be answered on the protein level depends mainly on the availability of features in proteins reminiscent of the mesophilic or thermophilic past and on how to recognise them. In principle, we must focus on properties that confer a 'thermophilic' or 'mesophilic' phenotype on the protein (i.e. allowing the protein to be 'sufficiently' stable and active at a given temperature) but are also compatible (and not counteractive) with the structural and functional requirements under changed temperature conditions. Since these properties do not counteract with the requirements dictated by the new temperature (because of their nonexclusive character), they will be tolerated and thus could serve as molecular relicts of the temperature past.

Certainly, at the present stage of knowledge it seems premature to propose an approach for analysing the direction of thermoadaptation, since clear ideas do not exist even about typically mesophilic or thermophilic properties of proteins nor are hints available for the existence of such reminiscent features. Nevertheless, the following suggestions for interpreting structural features of TIMs from mesophilic and thermophilic sources with respect to the direction of thermoadaptation should motivate further activities in this field.

As shown above, TIMs – like other proteins – showed a striking parallelism between oligomerisation state and the temperature to which the proteins are adapted. This parallelism has been interpreted in terms of thermoadaptation: the additional intersubunit contacts in the higher order oligomers of the enzyme homologues from the hyperthermophiles would lead to an increase of the overall stability of the protein in line with the assumption that higher thermostability is mainly caused by an increase of noncovalent interactions (electrostatic and hydrophobic interactions) and bulky residues resulting in a more rigid conformation (Vihinen, 1987; cf. Jaenicke *et al.*, 1996). Since, however, for the adaptation to lower temperatures more flexibility is required for functional reasons, structural elements governing the rigidity/flexibility act as exclusive features: a rigidified conformation will hardly be accepted by an enzyme protein which has to function at lower temperature; and a more flexible conformation (as a prerequisite for enzyme function at lower temperature) seems to be a poor basis for thermostability. As such, structural elements directly determining rigidity/flexibility are probably not suitable candidates for molecular relicts reflecting the temperature past.

As outlined above, not only the tetrameric TIMs from the hyperthermophilic *M. fervidus* and *P. woesei* but also the dimeric TIM from the mesophilic *M. bryantii* show the same deletions in the primary structure, which we assume to be correlated with the formation of the new subunit binding sites. Although the correlation between the structural deviations and the formation of the new subunit binding sites in TIMs of hyperthermophiles remains to be proved by three-dimensional structure analysis and directed mutagenesis, the occurrence of the equivalent deviations in the dimeric *M. bryantii* TIM may be interpreted as a feature reminiscent of the thermophilic past of *M. bryantii*. In contrast to the lower oligomerisation state found for the *M. bryantii* TIM, obviously dictated by the requirement for higher flexibility at lower temperature, the shortenings – or even eliminations – of secondary structural elements, presumably serving a tighter subunit association in the thermophilic structure, could be conserved in the mesophilic structure because they do not interfere with the demand for higher conformational flexibility.

Further indications for molecular relicts reminiscent of the thermophilic past of *M. bryantii* TIM can be deduced from sequence comparisons including mesophilic and thermophilic enzyme homologues of all three domains (Figure 22.4). Obviously specific for the TIMs from mesophilic Bacteria and Eucarya, a remarkable conservation of an Asn residue can be observed at position 13 (for numbering see Figure 22.4). This residue possesses importance in the spatial arrangement of the active site (Wierenga *et al.*, 1992; cf. Delboni *et al.*, 1995). As shown by Ahern *et al.* (1987) for the yeast enzyme, this residue is susceptible to heat-induced deamidation, thus governing the heat stability of the enzyme.

The observation that in thermophiles – independent of their bacterial or archaeal origin – Asn is replaced either by His or Tyr residues may therefore be interpreted as protection against heat-induced chemical modification of the peptide chain. Strikingly, the TIM of the mesophilic archaeon *M. bryantii* shows the same Asn replacements as found for the TIMs from archaeal hyperthermophiles. Since there are no obvious reasons for a respective protection of the Asn residue in the mesophile, we might suggest that the observed avoidance of Asn is a relict of the thermophilic past. However, we cannot exclude that this Asn/Tyr exchange merely represents a domain-specific feature. For a profound conclusion, the thermophily-specific character of the reminiscent feature in question must be confirmed by experimental studies. Also, analyses of other proteins are necessary to support the suggestion that *M. bryantii* stems from a thermophilic ancestor.

22.6 Conclusions

Protein thermoadaptation turned out to be extremely complex and no general rules for thermal adaptation can be defined yet. Nature's strategies to adapt proteins to different environmental temperatures seem to be highly variable and may depend not only on the inherent structural and functional properties of the protein under consideration but also on the temperature range, to which the protein has been adapted. Thus, as argued by Russell *et al.* (1997), the strikingly high number of ion pairs recently observed in intra- and intersubunit interactions of proteins from hyperthermophiles may be interpreted as a specific strategy for guaranteeing stability at the extreme temperatures at which these organisms live. Also, the higher oligomerisation states observed especially with proteins from hyperthermophiles may be interpreted as a specific adaptation strategy for the upper temperature limit of growth: possibly, the additional intersubunit contacts compensate the decreasing individual contributions of noncovalent (especially hydrophobic) interactions to the overall stability at these extreme temperatures.

Since, under these extreme conditions, the peptide chain runs the risk of being chemically modified, protection mechanisms against heat-induced chemical modifications also seem to be essential for the integrity for the protein structure at the upper temperature range. Protection against heat-induced chemical modifications (such as oxidation, deamidation, hydrolysis of the peptide bond, formation of isoaspartyl bonds) can be achieved by rigidifying the conformation to avoid the formation of critical transition states of the undesired reactions, by shielding the labile residue from the solvent or just by avoidance of the labile residue at critical sites (Delboni *et al.*, 1995; Hess *et al.*, Russell *et al.*, 1997).

As discussed above, at least the avoidance of heat-labile residues at critical sites may fulfil additional, diagnostic purposes: since the reduction in thermolabile residues does not necessarily interfere with the demands of an enzyme protein working under mesophilic conditions (i.e. higher flexibility) this feature could serve as molecular relict reminiscent of the thermophilic past. For determining such reminiscent features, there should be available not only a detailed structural analysis of homologous proteins from thermophiles and mesophiles but also experimental data that confirm the heat lability of the avoided residues in the mesophilic structure but exclude alternative functions of the replacing residue. Presumably, the most reliable determinations are to be expected from studying closely related proteins, thus allowing insights only into more recent adaptation events. But, with accumulation of data, it might also be possible to trace the evolution back to the origin of life.

22.7 Summary

3-Phosphoglycerate kinase (PGK) and the triose-phosphate isomerase (TIM) from the hyperthermophilic archaea *Methanothermus fervidus* and *Pyrococcus woesei* were analysed with respect to quarternary structure. The enzymes from these hyperthermophiles were homomeric dimers (PGK) or tetramers (TIM), respectively, in contrast to the their mesophilic counterparts of archaeal, bacterial and eucaryal sources representing monomers (PGK) or dimers (TIM), respectively. Cross-linking experiments performed with the *M. fervidus* PGK indicate a dimerisation of the monomers through the N-terminal domain. For the tetrameric TIM, a dimerisation of dimers through new subunit contacts at the bottom and the functionally inactive flank of the $(\beta/\alpha)_8$ barrel was postulated. Supported by several studies on various oligomeric proteins emphasising the importance of the inter-subunit contacts for thermostabilisation, we expect that the higher association state primarily serves thermostabilisation. In a second part, we address the question whether from the protein structure indications about the direction of thermoadaptation can be obtained. On the basis of sequence comparisons of TIMs from the hyperthermophilic *M. fervidus* and *P. woesei* with the enzyme from the mesophilic archaeon *Methanobacterium bryantii*, we discuss the existence of molecular relicts reminiscent of past temperatures.

Acknowledgements

The work was supported by grants of the Deutsche Forschungsgemeinschaft and the Fonds der Chemischen Industrie. Thanks are due to Professor K. Kirschner and Dr R. Sterner (Biozentrum der Universität Basel) for stimulating discussions. We are also indebted to Mr A. Lustig (Biozentrum der Universität Basel) for performing the sedimentation equilibrium centrifugation experiments.

References

Aguilar, C. F., Sanderson, I., Moracci, M. *et al.* (1997) Crystal structure of the beta-glycosidase from the hyperthermophilic archaeon *Sulfolobus solphataricus*: resilience as a key factor in thermostability. *J. Mol. Biol.*, **271**, 789–802.

Ahern, T. J., Casal, J. I., Petsko, G. A. and Klibanov, A. M. (1987) Control of oligomeric enzyme thermostability by protein engineering. *Proc. Natl Acad. Sci. USA*, **84**, 675–679.

Alber, T. and Kawasaki, G. (1982) Nucleotide sequence of the triose phosphate isomerase gene of *Saccharomyces cerevisiae*. *J. Mol. Appl. Genet.*, **1**, 419–434.

Artavanis-Tsakonas, S. and Harris, J. I. (1980) Primary structure of triosephosphate isomerase from *Bacillus stearothermophilus*. *Eur. J. Biochem.*, **108**, 599–611.

Beernink, P. T. and Tolan, D. R. (1994) Subunit interface mutants of rabbit muscle aldolase form active dimers. *Protein Sci.*, **3**, 1383–1391.

Bernstein, B. E., Michels, P. A. and Hol, W. G. (1997) Synergistic effects of substrate-induced conformational changes in phosphoglycerate kinase activation. *Nature*, **385**, 275–278.

Blake, C. C. F. and Evans, P. R. (1974) Structure of horse muscle phosphoglycerate kinase, some results on the chain conformation, substrate binding and evolution of the molecule from a 3 Å Fourier map. *J. Mol. Biol.*, **84**, 585–601.

Bult, C. J., White, O., Olsen, G. J. *et al.* (1996) Complete genome sequence of the methanogenic archaeon, *Methanococcus jannaschii*. *Science*, **273**, 1058–1073.

Davies, G. J., Gamblin, S. J., Littlechild, J. A. and Watson, H. C. (1993) The structure of a thermally stable 3-phosphoglycerate kinase and a comparison with its mesophilic equivalent. *Proteins*, **15**, 283–289.

DAY, A. G. and WITHERS, S. G. (1986) The purification and characterisation of a beta-glucosidase from *Alcaligenes faecalis*. *Can. J. Biochem.*, **64**, 914–922.

DELBONI, L. F., MANDE, S. C., RENTIER-DELRUE, F. *et al.* (1995) Crystal structure of recombinant triosephosphate isomerase from *Bacillus stearothermophilus*. An analysis of potential thermostability factors in six isomerases with known three-dimensional structures points to the importance of hydrophobic interactions. *Protein Sci.*, **4**, 2594–2604.

FABRY, S., HEPPNER, P., DIETMAIER, W. and HENSEL, R. (1990) Cloning and sequencing the gene encoding 3-phosphoglycerate kinase from mesophilic *Methanobacterium bryantii* and thermophilic *Methanothermus fervidus*. *Gene*, **91**, 19–25.

FACCHIANO, A., RAGONE, R., CONSALVI, V., SCANDURRA, R., DE ROSA, M. and COLONNA, G. (1995) Molecular properties of glutamate dehydrogenase from the extreme thermophilic archaebacterium *Sulfolobus solfataricus*. *Biochim. Biophys. Acta*, **1251**, 170–176.

FURTH, A. J., MILMAN, J. D., PRIDDLE, J. D. and OFFORD, R. E. (1974) Studies on the subunit structure and amino acid sequence of triosephosphate isomerase from chicken breast muscle. *Biochem. J.*, **139**, 11–22.

GABELSBERGER, J. LIEBL, W. and SCHLEIFER, K. H. (1993) Purification and properties of recombinant beta-glucosidase of the hyperthermophilic bacterium *Thermotoga maritima*. *Appl. Microbiol. Biotechnol.*, **40**, 44–52.

HARAN, G., HAAS, E., SZPIKOWSKA, B. K. and MAS, M. T. (1992) Domain motions in phosphoglycerate kinase – determination of interdomain distance distribution by site-specific-labeling and time-resolved fluorescence energy transfer. *Proc. Natl Acad. Sci. USA*, **89**, 11764–11768.

HARLO, K., VAS, M. and BLAKE, C. F. (1992) Crystal structucture of pig muscle phosphoglycerate kinase and its substrate 3-phosphoglycerate. *Proteins*, **12**, 133–144.

HENNIG, M., STERNER, R., KIRSCHNER, K. and JANSONIUS, J. N. (1997) Crystal structure at 2.0 Å resolution of phosphoribosyl anthranilate isomerase from the hyperthermophile *Thermotoga maritima*: possible determinants of protein stability. *Biochemistry*, **36**, 6009–6016.

HESS, D. (1995) Strukturelle und funktionelle Charakterisierung von 3-Phosphoglyceratkinasen aus hyperthermophilen Archaea (Ph.D. thesis, University of Essen).

HESS, D., KRÜGER, K., KNAPPIG, A., PALM, P. and HENSEL, R. (1995) Dimeric 3-phosphoglycerate kinase from hyperthermophilic Archaea: cloning, sequencing and expression of the 3-phosphoglycerate kinase gene of *Pyrococcus woesei* in *Escherichia coli* and characterisation of the protein. Structural and functional comparison with the 3-phosphoglycerate kinase of *Methanothermus fervidus*. *Eur. J. Biochem.*, **233**, 227–237.

JAENICKE, R., SCHURIG, H., BEAUCAMP, N. and OSTENDORP, R. (1996) Structure and stability of hyperstable protein: glycolytic enzymes from hyperthermophilic bacterium *Thermotoga matitima*. *Adv. Protein Chem.*, **48**, 181–269.

KENGEN, S. W., LUESINK, E. J., STAMS, A. J. and ZEHNDER, A. J. (1993) Purification and characterisation of an extremely thermostable beta-glucosidase from the hyperthermophilic archaeon *Pyrococcus furiosus*. *Eur. J. Biochem.*, **213**, 305–312.

KIRINO, H., AOKI, M., AOSHIMA, M. (1994) Hydrophobic interaction at the subunit interface contributes to the thermostability of 3-isopropylmalate dehydrogenase from the extreme thermophile, *Thermus thermophilus*. *Eur. J. Biochem.*, **220**, 275–281.

KNOWLES, J. R. (1991) Enzyme catalysis: not different, just better. *Nature*, **350**, 121–124.

KOHLHOFF, M., DAHM, A. and HENSEL, R. (1996) Tetrameric triosephosphate isomerase from hyperthermophilic Archaea. *FEBS Lett.*, **383**, 245–250.

KORNDÖRFER, I., STEIPE, B., HUBER, R., TOMSCHY, A. and JAENICKE, R. (1995) The crystal structure of holo-glyceraldehhyde-3-phosphate dehydrogenase from the hyperthermophilic bacterium *Thermotoga maritima* at 2.5 Å resolution, *J. Mol. Biol.*, **146**, 511–521.

MARCHIONNI, M. and GILBERT, W. (1986) The triosephosphate isomerase gene from maize: introns antedate the plant–animal divergence. *Cell*, **46**(1), 133–141.

PAAVILAINEN, S., HELLMAN, J. and KORPELA, T. (1993) Purification, characterisation, gene cloning, and sequencing of a new beta-glucosidase from *Bacillus circulans* subsp. *alkaliphilus*. *Appl. Environ. Microbiol.*, **59**, 927–932.

PICHERSKY, E., GOTTLIEB, L. D. and HESS, J. F. (1984) Nucleotide sequence of the triose phosphate isomerase gene of *Escherichia coli*. *Mol. Gen. Genet.*, **195**, 314–320.

RUSSELL, R. J. M., FERGUSON, J. M. C., HOUGH, D. W., DANSON, M. J. and TAYLOR, G. L. (1997) The crystal structure of citrate synthase from the hyperthermophilic Archaeon *Pyrococcus furiosus* at 1.9 Å resolution. *Biochemistry*, **36**, 9983–9994.

SAMBROOK, J., FRITSCH, E. F. and MANIATIS, T. (1989) *Molecular Cloning: A Laboratory Manual*, 2nd edn (Cold Spring Harbour NY: Cold Spring Harbor Laboratory Press).

SAYLE, R. and MILNER-WHITE,-E. J. (1995) Biomolecular graphics for all. *Trends Biochem. Sci.*, **20**, 374.

SCHULTES, V. and JAENICKE, R. (1991) Folding intermediates of hyperthermophilic D-glyceraldehyde-3-phosphate dehydrogenase of *Thermotoga maritima* are trapped at low temperature. *FEBS Lett.*, **290**, 235–238.

SCHURIG, H., BEAUCAMP, N., OSTENDORP, R., JAENICKE, R., ADLER, E. and KNOWLES, J. R. (1995) Phosphoglycerate kinase and triosephosphate isomerase from the hyperthermophilic bacterium *Thermotoga maritima* from a covalent bifunctional enzyme complex. *EMBO J.*, **14**, 442–451.

SINEV, M. A., RAZULYAEV, O. I., VAS, M., TIMCHEMKO, A. A. and PTITSYN, O. B. (1989) Correlation between activity and hinge-bending domain displacement in 3-phosphoglycerate kinase. *Eur. J. Biochem.*, **180**, 61–66.

SMITH, D. R., DOUCETTE-STAMM, L. A., DELROUGHERY, C. *et al.* (1997) Complete genome sequence of *Methanobacterium thermoautotrophicum* delta H: functional analysis and comparative genomics. *J. Bacteriol.*, **179**, 7135–7155.

STETTER, K. O., THOMM, M., WINTER, J. *et al.* (1981) *Methanothermus fervidus*, sp. nov., a novel extremely thermophilic methanogen islated from an Icelandic hot spring. *Zentralblatt der Bakteriologie Mikrobiologie und Hygiene, I Abteilung Originale C2*, 166–178.

SWINKEL, B. W., GIBSON, W. C., OSINGA, K. A., KRAMER, R., VEENEMAN, G. H., VAN BOOM, J. H. and BORST, P. (1986) Characterization of the gene for microbody (glycosomal) triosephosphate isomerase of *Trypanosoma brucei*. *EMBO J.* **5**, 1291–1298.

VIHINEN, M. (1987) Relationship of protein flexibility to thermostability. *Protein Eng.*, **1**, 477–480.

VRIEND, G. (1990) WHAT IF: a molecular modeling and drug design program. *J. Mol. Graphics*, **8**, 52–56.

WAKARCHUK, W. W., GREENBERG, N. M., KILBURN, D. G., MILLER, R. C. JR. and WARREN, R. A. (1988) Structure and transcription analysis of the gene encoding a cellobiase from *Agrobacterium* sp. Strain ATCC 21400. *J. Bacteriol.*, **170**, 301–307.

WATSON, H. C., WALKER, N. P. C., SHAW, P. J. *et al.* (1982) Sequence and structure of yeast phosphoglycerate kinase. *EMBO J.*, **1**, 1635–1640.

WIERENGA, R. K., NOBLE, M. E. M. and DAVENPORT, R. C. (1992) Comparison of the refined crystal structures of liganded and unliganded chicken, yeast and trypanosomal triosephosphate isomerase. *J. Mol. Biol.*, **224**, 1115–1126.

YIP, K. S. P., STILLMAN, T. J., BRITTON, K. L. *et al.* (1995) The structure of *Pyrococcus furiosus* glutamate dehydrogenase reveals a key role for ion-pair networks in maintaining enzyme stability at extreme temperatures. *Structure*, **3**, 1147–1158.

ZILLIG, W., KLENK, H.-P., TRANT, J. *et al.* (1987) *Pyrococcus woesei*, sp. nov., an ultrathermophilic marine archaebacterium, representing a novel order, *Thermococcales*. *System. Appl. Microbiol.*, **9**, 62–70.

23

The Evolutionary Significance of the Metabolism of Tungsten by Microorganisms Growing at 100 °C

MICHAEL W. W. ADAMS

Department of Biochemistry and Molecular Biology and Center for Metalloenzyme Studies, University of Georgia, Athens, Georgia, USA

23.1 Introduction

A fundamental question is whether microorganisms that thrive near the upper temperature limit for life (currently 113 °C; Blöchl *et al.*, 1997) require unusual growth substrates, substances not typically needed by more conventional life forms. If so, what are such substances used for, and do they have any evolutionary significance? In this chapter we focus on the utilisation of the metals tungsten (W, atomic number 74) and molybdenum (Mo, 42) by microorganisms that grow at 100 °C. These two elements are chemically analogous but they have very different biological functions, and microorganisms growing at extreme temperatures have been central in our understanding of why biological systems differentiate between tungsten and molybdenum. The reader is referred to the recent reviews on the role of tungsten (Kletzin and Adams, 1996; Johnson *et al.*, 1996) and molybdenum (Hille, 1996; Howard and Rees, 1996) in biological systems and the properties of enzymes that contain these elements.

Several types of microorganism are now known that can grow at temperatures of 100 °C and above. These include species of the genera *Thermofilum*, *Pyrodictium*, *Pyrococcus*, *Hyperthermus*, *Stetteria*, *Pyrobaculum*, *Aerophilum*, *Pyrolobus* and *Methanopyrus* (Stetter, 1996; Sako *et al.*, 1996; Blöchl *et al.*, 1997; Jochimsen *et al.*, 1997). Of these, the best-studied from a physiological and biochemical perspective is *Pyrococcus furiosus* (Fiala and Stetter, 1986). This organism grows optimally at 100 °C by the fermentation of carbohydrates and peptides, with organic acids, H_2 and CO_2 as the main products. It also reduces elemental sulfur (S^0) to H_2S if S^0 is added to the growth medium. An indication that *P. furiosus* might utilise unusual fermentative pathways came with the finding that it requires tungsten for optimal growth (Bryant and Adams, 1989). This element is rarely used in biological systems; indeed, at that time, only one tungstoenzyme had been characterised (from a moderately thermophilic bacterium; Yamamoto *et al.*, 1983). In contrast, the essential role of molybdenum in biology had been known for decades and a large number of molybdoenzymes have been extensively characterised (Hille, 1996). They play intimate roles in the global cycles of nitrogen, carbon and sulfur, with nitrogenase, nitrate reductase, formate dehydrogenase and xanthine oxidase being well-studied examples.

Indeed, virtually all life forms that have been examined have been found to harbour molybdoenzymes; for example, humans contain three distinct types and *Escherichia coli* contains six varieties (Hille, 1996). In fact, tungsten is generally regarded as a molybdenum-antagonist, since most microorganisms when grown in the presence of tungsten produce either inactive, metal-free molybdoenzymes or tungsten-substituted enzymes with little or no catalytic activity (Johnson *et al.*, 1996). Why then does *P. furiosus* utilise tungsten? Is this a characteristic of microorganisms growing near the upper temperature limit? And what are the evolutionary implications?

23.2 Role of tungsten in the metabolism of *Pyrococcus furiosus*

To date, three different tungstoenzymes have been purified from *P. furiosus* and related species. They all catalyse the oxidation of various aldehydes to the corresponding acid (equation 23.1), a reaction of very low potential (acetaldehyde/acetate, E°′ = –580 mV).

$$RCHO + H_2O \quad RCOO^- + 3\,H^+ + 2\,e^- \qquad (23.1)$$

All three enzymes also use the redox protein ferredoxin as the electron carrier, but they differ in their substrate specificities. One of them, aldehyde:ferredoxin oxidoreductase or AOR, oxidises a broad range of both aliphatic and aromatic aldehydes and shows the highest catalytic efficiency with the aldehyde derivatives of the common amino acids, such as acetaldehyde (from alanine), isovalerylaldehyde (from valine) and phenylacetaldehyde (from phenylalanine; Mukund and Adams, 1991; Heider *et al.*, 1995). These compounds are generated by the transamination of amino acids and their subsequent decarboxylation by 2-ketoacid oxidoreductase which also produces the corresponding coenzyme A derivatives (Adams and Kletzin, 1996). Thus, as shown in Figure 23.1, AOR is thought to play a key

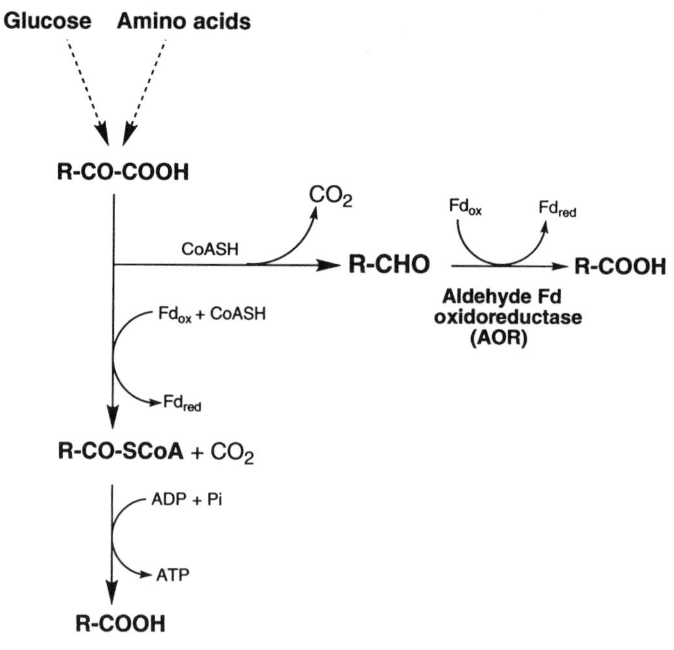

Figure 23.1 Role of aldehyde ferredoxin oxidoreductase (AOR) in the peptide metabolism of *P. furiosus*. Fd_{ox} and Fd_{red} represent the oxidised and reduced forms, respectively, of ferredoxin. Modified from Kletzin and Adams (1996).

role in peptide fermentation by *P. furiosus* by removing aldehydes generated during amino acid oxidation (Ma *et al.*, 1997). The second tungstoenzyme in this organism is termed glyceraldehyde-3-phosphate:ferredoxin oxidoreductase, or GAPOR. This enzyme is absolutely specific for glyceraldehyde-3-phosphate, a substrate for which it has a very high affinity (Mukund and Adams, 1995). GAPOR plays a central role in the unusual glycolytic pathway of *P. furiosus* in which it functions in place of the expected NAD(P)-dependent glyceraldehyde-3-phosphate dehydrogenase, the enzyme found in more conventional organisms (Figure 23.2). The third tungstoenzyme in *P. furiosus* is formaldehyde: ferredoxin oxidoreductase or FOR (Mukund and Adams, 1993, 1996). In contrast to AOR and GAPOR, the true substrate for FOR is unclear. The enzyme oxidises only C_1–C_3 unsubstituted aldehydes but has only very low affinity for them (K_m values > 20 mmol/l), suggesting that their oxidation is not of physiological significance. Nevertheless, like AOR and GAPOR, FOR is present at a relatively high concentration in *P. furiosus* cells and it is assumed to be part of an unusual metabolic route probably involved with peptide metabolism (Mukund and Adams, 1993).

23.3 Properties of the tungstoenzymes

23.3.1 *Tungstoenzymes from* P. furiosus

As indicated in Table 23.1, although AOR, FOR and GAPOR from *P. furiosus* have different quaternary structures, the sizes and cofactor contents of their subunits are very similar, with each containing one W atom and approximately four Fe atoms. The crystal structure of AOR, obtained to 2.3 Å resolution (Chan *et al.*, 1995), showed that the W atom in its subunit is coordinated in part by four sulfur atoms originating from the dithiolene groups of two organic pterin cofactors (Figure 23.3). Hence, the AOR subunit contains a mononuclear tungsto-bispterin site. The W atom is not directly coordinated to the protein; rather, spectroscopic analyses indicate that the W coordination sphere is completed by a terminal oxo and a hydroxy group, as shown in Figure 23.4 (Koehler *et al.*, 1996). Such analyses have also established that, like the Mo site in all known molybdoenzymes, the W site in AOR undergoes a two-electron redox reaction, consistent with aldehyde oxidation (equation 23.1), in which the tungsten cycles between the IV, V and VI oxidation states. The crystal structure of AOR also showed that a [4Fe–4S] cluster is situated about 8 Å from the W atom, the presumed site of aldehyde oxidation, and the 4Fe cluster is thought to mediate electron transfer to ferredoxin, the external electron carrier.

The close relationship between the subunits of AOR, FOR and GAPOR was first suggested by the similarity in their N-terminal amino acid sequences (Johnson *et al.*, 1996), and this was subsequently confirmed by the complete amino acid sequences of all three enzymes which show a high degree of similarity (G. Schut and W. De Vos, unpublished data; Kletzin *et al.*, 1995). This sequence similarity suggested a similarity in the structure and function of these three enzymes, and this has recently been confirmed with the determination of the crystal structure of FOR (Y. Hu, R. Roy, M. W. W. Adams and D. C. Rees, unpublished data). Like AOR, the subunit of FOR contains a mononuclear tungsto-bispterin site with an adjacent [4Fe–4S] cluster, although unlike AOR (or GAPOR), FOR contains a Ca^{2+} ion which is situated next to one of the pterin molecules. It is thought that the calcium ion plays a role in substrate binding. As yet there is no structural information on GAPOR, but the similarity in amino acid sequence and cofactor content between it and AOR and FOR suggests that it too contains a tungsto-bispterin site with an

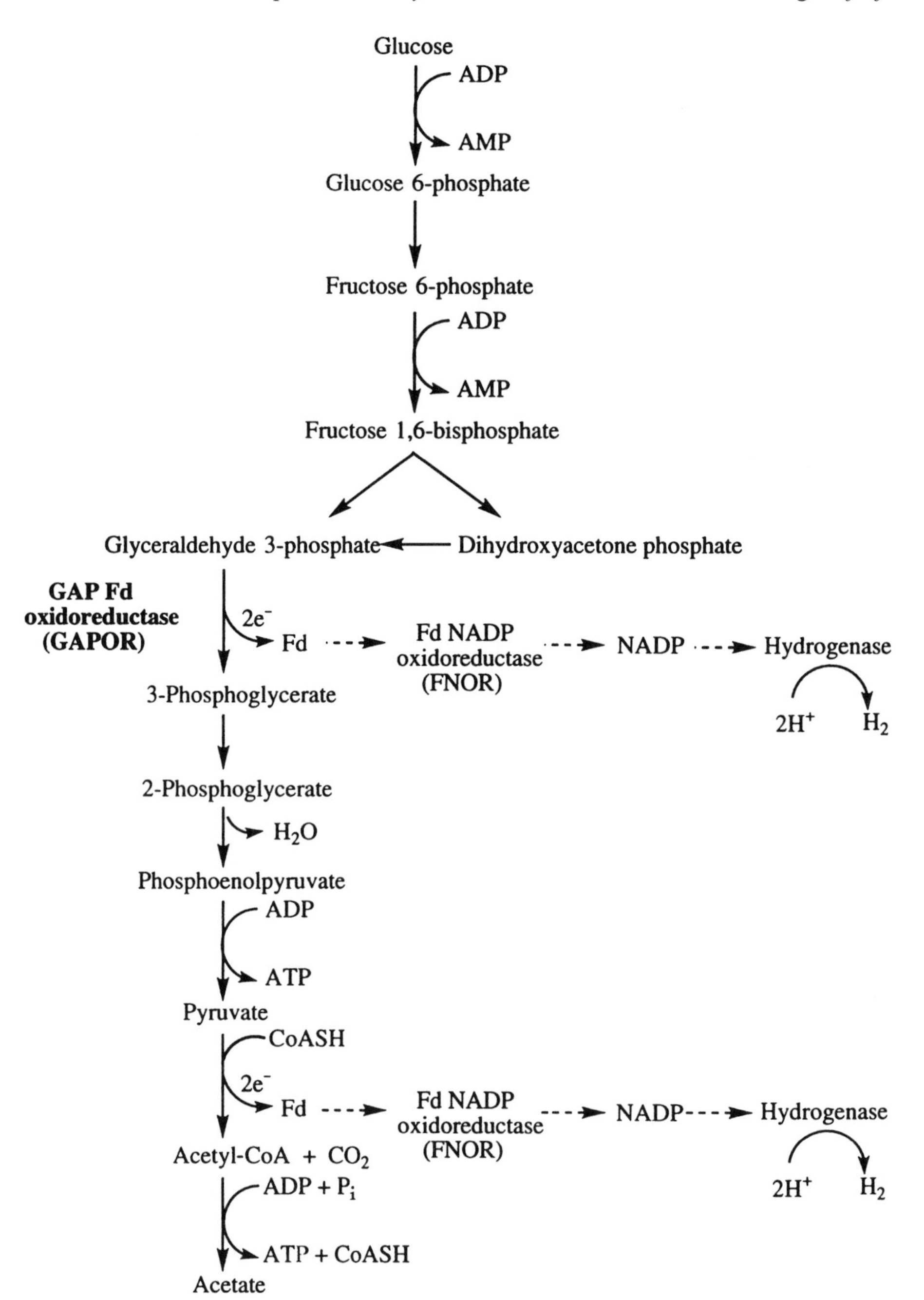

Figure 23.2 Role of glyceraldehyde-3-phosphate:ferredoxin oxidoreductase (GAPOR) in the glycolytic pathway of *P. furiosus*. Fd represents ferredoxin. Modified from Mukund and Adams (1995).

Table 23.1 Molecular properties of tungstoenzymes

Organism and enzyme[a]	Holoenzyme M_r	Subunits	Subunit M_r	W content[b]	FeS or cluster content[c]
I. AOR-type (aldehyde-oxidising)					
P. furiosus AOR	136	α_2	67	2 W	2 [Fe_4S_4] + 1 Fe
P. furiosus FOR	280	α_4	69	4 W	4 [Fe_4S_4]
P. furiosus GAPOR	63	α	63	1 W	~6 Fe
C. formicoaceticum CAR	134	α_2	67	2 W	~11 Fe, ~16 S
D. gigas ADH	132	α_2	65	2 W	~10 Fe
II. F(M)DH-type (CO_2-reducing)					
C. themoaceticum FDH	340	$\alpha_2\beta_2$	96,76	2 W	2 Se, ~40 Fe
M. wolfei FMDH II	130	$\alpha\beta_\gamma$	64,51,35	1 W	2–5 Fe
M. thermoautotrophicum FMDH II	160	$\alpha_\beta\gamma_\delta$	65,53,31,15	1 W	~8 Fe
III. AH-type (C_2H_2-hydrating)					
Pr. acetylenicus AH	73	α	73	1 W	4–5 Fe

[a] The sources of the enzymes are given in the text. Data taken from Adams and Kletzin (1996) Kletzin and Adams (1996) and Johnson *et al.* (1996).
[b] Expressed as an integer value per mole of holoenzyme.
[c] Cluster contents are expressed per mole of holoenzyme and are based on EPR spectroscopy or crystallography.

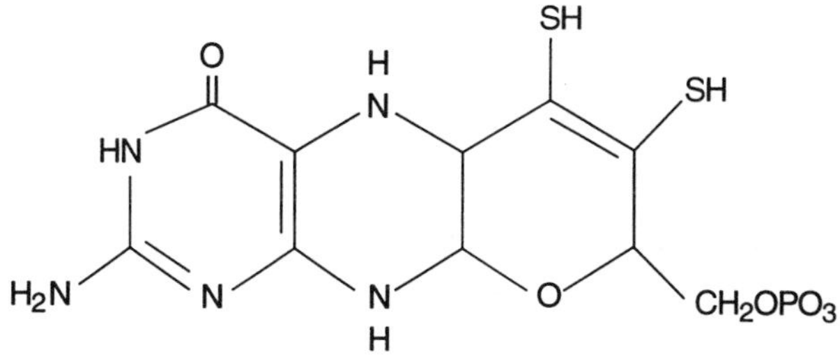

Figure 23.3 Structure of the pterin cofactor that coordinates the tungsten atom in *P. furiosus* AOR. The tungsten site is coordinated in part by four sulfur atoms from the dithiolene group of two pterin cofactors. Modified from Chan *et al.* (1995).

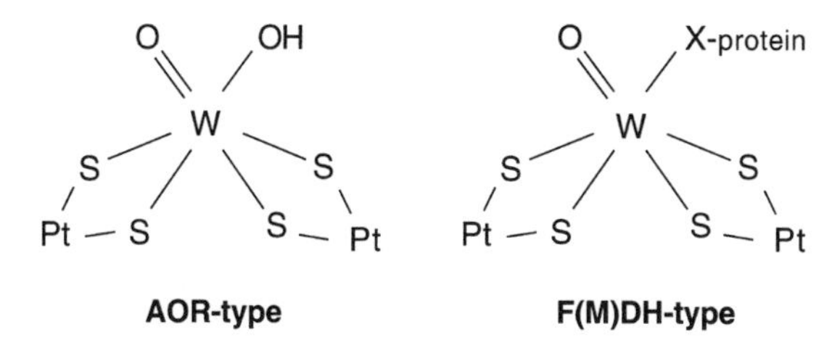

Figure 23.4 Proposed structures of the tungsten sites in the AOR and F(M)DH types of tungstoenzyme; pt represents the pterin cofactor. For the F(M)DH type, X is sulfur (from cysteine) or selenium (from selenocysteine) for FMDH and FDH, respectively. Modified from Johnson *et al.* (1996).

adjacent 4Fe cluster. GAPOR also contains two zinc atoms per subunit, an element not found in AOR and FOR, and these positively charged ions likely play a role in binding the negatively charged phosphate group of its substrate, glyceraldehyde 3-phosphate. In any event, it is clear that AOR, FOR and GAPOR are part of the same enzyme family.

Although the three tungstoenzymes of *P. furiosus* show similarities in their amino acid sequences, they show no similarity whatsoever to any of the more than 30 amino acid sequences available for molybdoenzymes (Kletzin *et al.*, 1995). This is very surprising, since all molybdoenzymes known, with the notable exception of nitrogenase, also contain a pterin cofactor, and most also contain one or more iron–sulfur clusters. In fact, crystallographic analyses have been carried out with three different molybdoenzymes, and the structure of their pterin cofactor is the same as that in AOR. For two of the enzymes, DMSO reductase (from a photosynthetic bacterium) and formate dehydrogenase (from *Escherichia coli*; Schindelin *et al.*, 1996; Schneider *et al.*, 1996; Boyington *et al.*, 1997), the Mo atom is coordinated by four sulfur atoms from two pterin molecules, as in *P. furiosus* AOR, while in the other, aldehyde oxidoreductase (from a sulfate-reducing bacterium; Romao *et al.*, 1995), the Mo atom is ligated by two sulfur atoms from a single pterin. The Mo sites in both DMSO reductase and formate dehydrogenase differ from the W site in AOR, however, in that the molybdenum is also coordinated by the protein (via a serinate and a cysteinate residue, respectively), as well as by two pterins. It therefore appears that, while the structures of the catalytic sites in tungsten and molybdenum enzymes are similar, the two enzyme types must have diverged very early on the evolutionary time scale, such that present-day versions show no detectable sequence similarity.

23.3.2 ***Tungstoenzymes from other microorganisms***

In addition to the three tungstoenzymes of *P. furiosus*, five other types are known and their properties are summarised in Table 23.1. Carboxylic acid reductase (CAR; White *et al.*, 1989) has been purified from two species of acetogenic *Clostridia*, from the moderate thermophile *C. thermoaceticum* and from the mesophile *C. formicoaceticum*, while aldehyde dehydrogenase (ADH; Hensgens *et al.*, 1995) is from the mesophilic sulfate reducer, *Desulfovibrio gigas*. Like the *P. furiosus* enzymes, CAR and ADH catalyse the oxidation of aldehydes, although their physiological roles and their electron carriers have not been established. As indicated in Table 23.1, the sizes of the subunits of CAR and ADH and their cofactor contents are similar to those of the *P. furiosus* tungstoenzymes, and a relationship between all of these enzymes has been confirmed by the similarity in their N-terminal amino acid sequences (Kletzin and Adams, 1996). Thus, while the complete amino acid sequences of CAR and ADH have yet to be reported, it seems that both enzymes also belong to the aldehyde-oxidising family of tungstoenzymes, of which *P. furiosus* AOR is the prototypical example. Indeed, from a comparison of the spectroscopic properties reported for CAR and ADH, it appears that all of the AOR family of tungstoenzymes have a common active-site structure (Figure 23.4; Koehler *et al.*, 1996; Johnson *et al.*, 1996).

Of the three other types of tungstoenzyme known, one is termed acetylene hydratase (AH) and this has been obtained from the anaerobic acetylene-utiliser, *Pelobacter acetylenicus* (Rosner and Schink, 1995). AH hydrates C_2H_2 to acetaldehyde. Thus, in contrast to the other tungstoenzymes known, this enzyme does not formally catalyse a redox reaction. The catalytic role played by tungsten in AH is not known, nor is the relationship between this enzyme and the other tungstoenzymes.

The other two tungstoenzymes known do not catalyse aldehyde oxidation (or acetylene hydration); rather, they catalyse the activation of CO_2. One type is found in certain species of methanogen. All methanogens contain the enzyme formylmethanofuran dehydrogenase (FMDH) which reduces CO_2 to a formyl group using methanofuran (MFR) as the C-1 carrier. This is the first step in methanogenesis and, like the AOR-type reactions, is also a reaction of low potential ($E_m = -497$ mV; equation 23.2).

$$CO_2 + MFR^+ + H^+ + 2\ e^- \quad CHO\text{-}MFR + H_2O \tag{23.2}$$

In virtually all methanogens, FMDH is a molybdoenzyme, but two moderately thermophilic *Methanobacterium* species, *M. thermoautotrophicum* and *M. wolfei*, when grown in the presence of tungsten, produce a second type of FMDH which contains tungsten (Schmitz *et al.*, 1992; Bertram *et al.*, 1994).

The other example in the CO_2-activating family of tungstoenzymes is formate dehydrogenase (FDH) from the acetogen *C. thermoaceticum*. This was the first tungstoenzyme to be characterised and is also unusual in that it contains selenium as selenocysteine (Yamamoto *et al.*, 1993). Using NADPH as the electron donor, FDH reduces CO_2 to formate as the first step in acetogenesis, which is also a very low-potential reaction ($E_m = -432$ mV; equation 23.3).

$$CO_2 + H^+ + 2\ e^- \quad HCOO^- \tag{23.3}$$

As shown in Table 23.1, compared to the members of the AOR family of tungstoenzymes, the FDH and FMDH type are much more complex enzymes with at least two different types of subunit, and these show no N-terminal amino acid sequence similarity with the subunits of the AOR family (Kletzin and Adams, 1996). That these two types of tungstoenzyme are unrelated is confirmed by analyses of the complete amino acid sequences of both FDH and FMDH (D. Gollin and L. G. Ljungdahl, unpublished data; Hochheimer *et al.*, 1995), which show no similarity to sequence of *P. furiosus* AOR (see Johnson *et al.*, 1996). However, the sequences of one of the subunits of both FDH and FMDH do show high similarity to the sequences of some molybdoenzymes, in particular those represented by DMSO reductase and formate dehydrogenase. In fact, sequence alignments predict that, like the Mo site in these two molybdoenzymes, the W site in both FDH and FMDH is likely to be coordinated by an amino acid residue of the protein (by selenocysteine and cysteine, respectively), in contrast to the case with AOR. Hence, the F(M)DH type of tungstoenzyme, representing FMDH and FDH, falls within the much larger class of molybdoenzymes, while the AOR group (see Table 23.1) appear to be 'true' tungstoenzymes in that they are phylogenetically distinct and, in spite of common cofactors, are unrelated to the enzymes that contain molybdenum.

23.4 Tungsten- and molybdenum-containing isoenzymes

A complicating factor in evaluating why certain organisms have chosen to use tungsten rather than molybdenum is that, with one exception, all of the known tungsten-utilising organisms can use molybdenum in addition to tungsten. That is, the specific species of methanogen, clostridia, sulfate-reducer and acetylene-utiliser mentioned above that are known to produce tungstoenzymes also synthesise molybdenum-containing isoenzymes when they are grown in the presence of molybdenum rather than tungsten (see Kletzin and Adams, 1996). These molybdenum-containing enzymes are distinct from their tungsten-containing counterparts and encoded by separate genes but they appear to serve the same physiological role as their W-containing analogue. In the case of tungsten-containing FMDH, CAR and ADH, their corresponding Mo-isoenzymes have been well characterised as members of the large family of molybdoenzymes. However, in contrast to the situation with the mesophilic and moderate thermophiles, growth of *P. furiosus* in the presence of molybdenum does not lead to the production of molybdenum-containing, aldehyde-oxidising isoenzymes of AOR, FOR or GAPOR (Mukund and Adams, 1996). Remarkably, *P. furiosus* is able to scavenge the tungsten that contaminates the medium (~15 nmol/l when additional tungsten is not added), even in the presence of a 6500-fold excess of molybdenum, to produce active tungsten-containing forms of AOR and FOR (GAPOR activity was ~1% of that present in tungsten-sufficient cells). Thus, the evidence at present suggests that *P. furiosus* is obligately dependent upon tungsten for growth.

It should be noted that of all microorganisms that are capable of growing near 100 °C, so far it is only fermentative species, such as *P. furiosus* and its relatives (*Pyrococcus* strain ES4, *Thermococcus litoralis* and *Thermococcus* strain ES-1; Heider *et al.*, 1995; Kletzin and Adams, 1996; Koehler *et al.*, 1996) whose ability to metabolise tungsten and molybdenum has been thoroughly investigated and from which tungstoenzymes have been purified. The growth of the methanogen *Methanopyrus kandleri*, which grows up to 110 °C, was reported to be stimulated by tungsten but not by molybdenum (Vorholt *et al.*, 1997), suggesting that it contains the W-isoenzyme of FMDH, although the enzyme has yet to be purified. Clearly, further studies with other organisms that grow near 100 °C need to be carried out to determine whether all of them are indeed dependent upon tungsten and incapable of utilising molybdenum.

In any event, there are clearly two types of tungstoenzyme known at present, the evolutionarily unique AOR type, and the molybdoenzyme-related F(M)DH type. Within both families, molybdenum-containing isoenzymes are known, and these are part of the much larger class of molybdoenzyme. The only exceptions are the *P. furiosus* AOR-type enzymes which have no 'Mo-equivalent'.

23.5 Why tungsten and not molybdenum?

Why have microorganisms such as *P. furiosus* that thrive near the upper temperature limit of life chosen to use tungsten rather than molybdenum at the active sites of key enzymes, when most life forms use exclusively molybdenum? Some insight into this was recently provided by considering the properties of synthetic tungsten and molybdenum complexes (Johnson *et al.*, 1996). Specifically, one can ask what are the structural requirements of a W site that enable it to cycle between the IV, V and VI redox states, and to do so within the biological range of reduction potential (–500 to +800 mV) and temperature (≤113 °C)? Unfortunately, the chemistry of mononuclear tungsten complexes is much less developed than that of molybdenum, which is in part due to the thermodynamic instability of oxo-W(IV) complexes compared to the equivalent oxo-Mo(IV) complex. This obviously suggests that W sites of biological relevance would not have predominant oxo coordination, although this could be the case with molybdenum. Thus, sulfur donor ligands may be required to tune the W(IV/V/VI) redox couples within the biological range. Indeed, monooxo and dioxo W(VI/V) complexes with four sulfur ligands (from bis-dithiolate and bis-dithiolene groups) have been characterised and, in contrast to the oxo-W complexes, these can be reduced to the W(IV) state. However, such complexes are very O_2-sensitive compared to the equivalent molybdenum complexes, indicating that such a W site within an enzyme would likely function only under anaerobic conditions (or would need to be protected from O_2 if used in aerobic organisms). The bisthiolate/thiolene W(VI) complexes also exhibit strongly enhanced thermal stability compared to the corresponding Mo(VI) complex, suggesting that only the former would be stable enough to be utilised at temperatures near the normal boiling point. This enhanced bond strength of W(VI) complexes may also account for the observation that such complexes are generally kinetically slower than the equivalent Mo(VI) complexes. Thus, one would expect that higher temperatures might be a requirement for W centres to catalyse reactions at high rates, reactions that could be performed at similar rates by Mo centres at much lower temperatures. The relevant tungsten complexes also have much lower redox potentials than the analogous molybdenum complex (typically 300–400 mV more negative), indicating that they are better suited to catalyse lower-potential redox processes.

Hence, from the properties of synthetic complexes, one would predict that if tungsten were utilised in biology, it would only catalyse low-potential reactions under anaerobic conditions, and then significant catalytic rates would be observed only at high temperatures. Conversely, the instability of relevant molybdenum complexes at high temperatures might preclude their utilisation, otherwise such complexes should be catalytically competent over the whole biological range of potentials under both aerobic and anaerobic conditions. Remarkably, these conclusions from the properties of synthetic W/Mo complexes fit very well with our current understanding of the properties of tungstoenzymes. So far, all have been found in anaerobic organisms (Table 23.1) and they all catalyse reactions of low potential (equations 23.1–23.3). Moreover, the tungstoenzymes in the mesophiles and moderate thermophiles can be replaced by analogous molybdoenzymes, while the three tungstoenzymes in *P. furiosus* appear to be obligately dependent upon tungsten. Apparently, the latter enzymes are carrying out reactions near the limits of biological systems – very low potential reactions at extreme temperatures. It seems that tungsten, but not molybdenum, is capable of remaining in the (VI) oxidation state under these conditions, a state that is essential for catalytic activity. The analyses of the synthetic complexes also suggests that the W site in tungstoenzymes would be coordinated by at least four sulfur atoms. We know that this is the case with *P. furiosus* AOR (and also FOR), and it has been suggested (Johnson *et al.*, 1996) that this is also true for the FM(D)H family of tungstoenzymes (Figure 23.4).

23.6 Evolutionary considerations

What are the evolutionary ramifications of life forms thriving at 100 °C and above requiring tungsten, while virtually the rest of biology utilises molybdenum? A rather appealing scenario can be formulated if one assumes that the original life forms on this planet did indeed first evolve at temperatures near 100 °C. Under such conditions, it seems plausible that the ancestor to the pterin cofactor was present as a tungsten-containing bispterin species (Figure 23.5). Such a cofactor would be highly thermostable (with four coordinating sulfur atoms) and would be able to catalyse two-electron redox chemistry at high temperatures under anaerobic conditions. This would not be true if such a cofactor contained molybdenum, or if it were coordinated by only a single pterin (or by two sulfur atoms). Thus, as depicted in Figure 23.5, the ancestral tungsto-bispterin cofactor during evolution would have been readily incorporated into enzymes to give specificity to the two-electron redox reaction, but such enzymes must have rapidly diverged into the 'AOR'

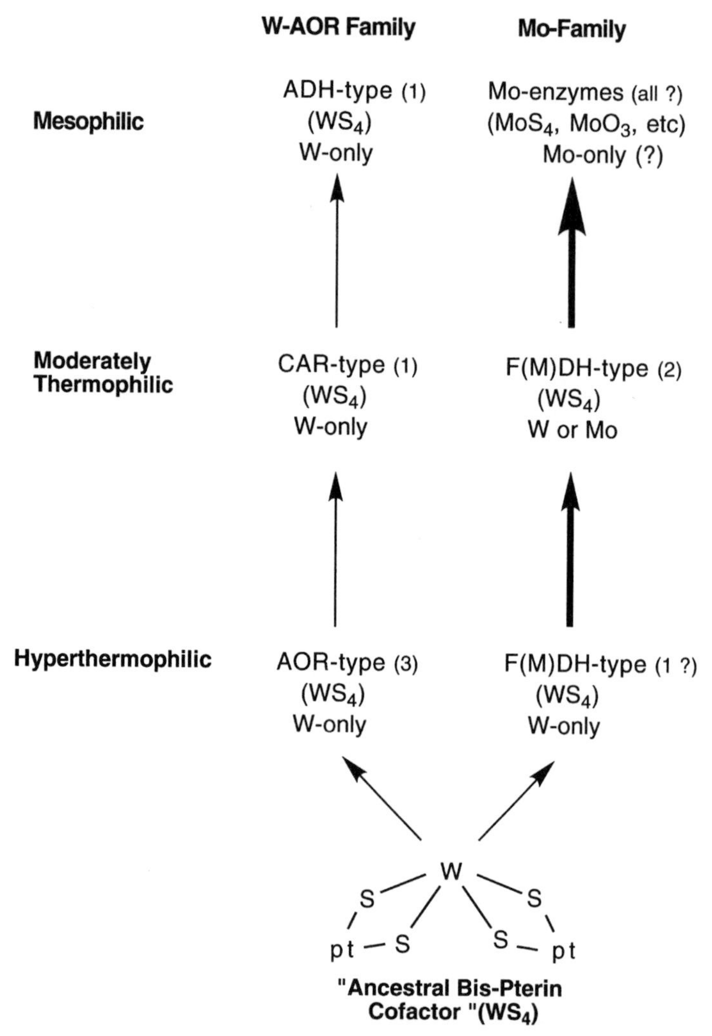

Figure 23.5 Proposed evolution of tungsten- and molybdenum-containing enzymes from an 'ancestral tungsten-bispterin cofactor'; pt indicates pterin. The currently known number of examples of each enzyme type is indicated, together with whether they use tungsten or molybdenum or both. See text for details.

type, of which there are three present-day examples (AOR, FOR and GAPOR of *P. furiosus*), and the F(M)DH-type, perhaps represented currently in the methanogen *M. kandleri*. As evolution progressed and life became adapted to moderately thermophilic and eventually mesophilic temperatures, the AOR lineage barely survived, such that this 'true' family of tungstoenzymes is now present in only a very limited number of micro-organisms (CAR in certain clostridia and ADH in certain sulfate-reducing bacteria). These enzymes still only utilise tungsten and appear to also maintain a bispterin site. In contrast, the F(M)DH lineage flourished as the ambient temperature decreased, and in moderately thermophilic organisms molybdenum could be substituted, e.g. in FDH and FMDH, although the metal was still coordinated by the bispterin site. Eventually, this line of descent became completely dominated by molybdenum, and rapidly diverged to give the large array of molybdoenzymes that are currently present in mesophilic organisms.

Hence, present-day molybdoenzymes catalyse a range of two-electron redox reactions, in both anaerobic and aerobic organisms, and some contain mono-pterin sites (Figure 23.5).

Of course, while the scenario depicted in Figure 23.5 is a useful hypothesis, there are several caveats that must be mentioned. For example, while it seems unlikely that a molybdenum-containing pterin cofactor could survive at temperatures near 100 °C, this scenario obviously does not prove that life must have had a hyperthermophilic origin, it is merely consistent with it. Thus, we cannot rule out that the ancestral pterin form was a molybdenum-containing mono- or bispterin cofactor in an F(M)DH type enzyme under moderately thermophilic conditions. This form would then have diverged to give rise to all molybdopterin-containing enzymes at lower (mesophilic) temperatures, and to the tungsten-containing FMDH type at high temperatures. This same cofactor must also have been 'assimilated' by a second type of enzyme, the AOR type, to give rise to the high-temperature tungstoenzymes now found in *P. furiosus* and related species. Thus, while this evolutionary pathway is not as straightforward as that depicted in Figure 23.5, it is equally valid. The notion of a pterin cofactor, whether coordinated by one or two pterins and whether tungsten- or molybdenum-associated, being 'utilised' during evolution by unrelated enzymes is consistent with our current knowledge of how a wide range of enzymes that contain either haem or flavin cofactor must have evolved. Thus, we cannot rule out parallel evolution of tungsten- and molybdenum-containing enzymes once a metal-pterin cofactor was 'invented'.

A final comment concerns the abundance of tungsten and molybdenum in natural environments. Both are relatively scarce metals, ranking 54th and 53th, respectively, in the abundance of elements on this planet (see Kletzin and Adams, 1996). However, while most rock formations contain similar concentrations of tungsten and molybdenum, the concentration of molybdenum in freshwater environments is typically at least two orders of magnitude higher than that of tungsten, owing to differences in the solubility of the relevant complexes. Similarly, molybdenum (110 nmol/l) is far more abundant than tungsten (1 pmol/l) in aerobic marine environments. On the other hand, the situation is very different in deep-sea hydrothermal vents where, compared to the open ocean, tungsten is present at dramatically higher concentrations. For example, hydrothermal vent fluids that have been examined contain ~2 μmol/l tungsten, while the tungsten concentration (~500 mg/kg) in the chimney walls of a 'black smoker' was about tenfold greater than that of molybdenum (see Kletzin and Adams, 1996). Thus, it may not be coincidental that life has been proposed to have originated at extreme temperatures in deep-sea hydrothermal systems, and that at least some of the present-day marine hyperthermophiles, such as *P. furiosus* and related species, appear to be obligately tungsten-dependent. Consequently, a potential factor in the present-day dominance of molybdenum-containing over tungsten-containing enzymes in the mesophilic world may well be the result of utilisation of the more readily available of two elements during the course of the evolution.

23.7 Conclusions

A survey of the distribution and properties of tungsten-containing enzymes shows that only microorganisms growing at extreme temperatures, or at least those represented by the heterotrophic anaerobe *P. furiosus*, absolutely require this element for growth. In stark contrast, the few species of moderately thermophilic and mesophilic microorganisms that are able to utilise tungsten also produce molybdenum-containing isoenzymes to catalyse the same reaction. These data suggest that tungsten was preferred over molybdenum by ancestral microorganisms so that tungsten-containing enzymes are the evolutionary precursors of all molybdenum-containing enzymes, which themselves are virtually ubiquitous in present-day biological systems. However, it cannot be ruled out that high-temperature microorganisms evolved to utilise tungsten to catalyse very specific reactions in unique tungsten-rich environments such as hydrothermal vents. Hence, we cannot distinguish between (a) the simultaneous evolution of tungsten- and molybdenum-containing enzymes at moderately thermophilic temperatures and (b) a tungsto-bispterin site of the type found in *P. furiosus* AOR perhaps having been the precursor to all present-day molybdenum-containing enzymes. The latter scenario, depicted in Figure 23.5, is consistent with the known properties of tungsten and molybdenum synthetic complexes, but assumes rather than predicts a hyperthermophilic origin. Clearly, further insight into the evolutionary significance of tungsten and molybdenum will require additional studies of microorganisms growing at 100 °C and above to determine whether their key enzymes are, indeed, dependent upon tungsten rather than molybdenum.

23.8 Summary

The archaeon *Pyrococcus furiosus*, grows optimally at 100 °C by the fermentation of peptides and carbohydrates. Three aldehyde-oxidising enzymes, known as AOR, FOR and GAPOR, play key roles in the fermentation pathways of this organism. All three enzymes are unusual in that they contain tungsten, an element rarely utilised in biological systems. In contrast, the chemically analogous element molybdenum plays an essential role in virtually all other life forms, and molybdoenzymes are ubiquitous. However, *P. furiosus* does not utilise molybdenum. The tungsten sites of AOR, FOR and GAPOR are very similar. All are mononuclear and coordinated in part by four sulfur atoms from the dithiolene groups of two organic pterin molecules. Although the same pterin moiety also coordinates the molybdenum atom in molybdoenzymes, these enzymes show no amino acid sequence similarity with the sequences of AOR, FOR and GAPOR. Therefore, the latter enzymes must have diverged from all molybdoenzymes very early on the evolutionary time scale. A second type of tungstoenzyme, termed F(M)DH, is found in certain moderately thermophilic and mesophilic microorganisms (represented by the CO_2 – reducing enzymes, furanate dehydrogerane (FDH) formye methanofuran dehydrogerane (FMDH)) this class does show sequence similarity to the large class of molybdoenzymes. A comparison of the properties of synthetic tungsten and molybdenum complexes indicates that tungsten, but not molybdenum, is chemically suited to catalyse low-potential reactions (such as aldehyde oxidation) at extreme temperatures, but only if the tungsten site has predominant sulfur coordination. It is therefore hypothesised that an ancestral tungsto-bispterin cofactor was the precursor to two lineages of tungstoenzyme, the AOR and the F(M)DH type. The AOR type is now present in *P. furiosus* and related high-temperature organisms, but otherwise has a very limited distribution. On the other hand, the F(M)DH type of tungstoenzyme is postulated to be the precursor of the large family of molybdoenzymes which are today distributed throughout the biological world.

Acknowledgements

Research in the author's laboratory is supported by the US Department of Energy, the US National Science Foundation and the US National Institutes of Health.

References

ADAMS, M. W. W. and KLETZIN, A. (1996) Oxidoreductase-type enzymes and redox proteins involved in the fermentative metabolisms of hyperthermophilic archaea. *Adv. Protein Chem.*, **48**, 101–180.

BERTRAM, P. A., SCHMITZ, R. A., LINDER, D. and THAUER, R. K. (1994) Tungstate can substitute for molybdate in sustaining growth of *Methanobacterium thermoautotrophicum* – identification and characterisation of a tungsten isoenzyme of formylmethanofuran dehydrogenase. *Arch. Microbiol.*, **161**, 220–228.

BLÖCHL, RACHEL, R., BURGGRAF, S., HAFENBRANDL, D., JANNASCH, H. W. and STETTER, K. O. (1997) *Pyrolobus fumarii*, gen. and sp. nov., represents a novel group of archaea, extending the upper temperature limit for life to 113 °C. *Extremophiles*, **1**, 14–21.

BOYINGTON, J. C., GLADYSHEV, V. N., KHANGULOV, S. V., STADTMAN, T. C. and SUN, P. D. (1997) Crystal structure of formate dehydrogenase H: catalysis involving Mo, molybdopterin, selenocysteine, and an Fe4S4 cluster. *Science*, **275**, 1305–1308.

BRYANT, F. O. and ADAMS, M. W. W. (1989) Characterisation of hydrogenase from the hyperthermophilic archaebacterium, *Pyrococcus furiosus. J. Biol. Chem.*, **264**, 5070–5079.

CHAN, M. K., MUKUND, S., KLETZIN, A., ADAMS, M. W. W. and REES, D. C. (1995) Structure of the hyperthermophilic tungstoprotein enzyme aldehyde ferredoxin oxidoreductase. *Science*, **267**, 1463–1469.

FIALA, G. and STETTER, K. O. (1986) *Pyrococcus furiosus* sp. nov. represents a novel genus of marine heterotrophic archaebacteria growing optimally at 100 °C. *Arch. Microbiol.*, **145**, 56–61.

HEIDER, J., MA, K. and ADAMS, M. W. W. (1995) Purification, characterization and metabolic function of aldehyde ferredoxin oxidoreductase from the hyperthermophilic and proteolytic archaeon, *Thermococcus* strain ES-1, *J. Bacteriol.*, **177**, 4757–4764.

HENSGENS, C. M. H., HAGEN, W. R. and HANSEN, T. H. (1995) Purification and characterisation of a benzyl viologen-linked, tungsten-containing aldehyde oxidoreductase from *Desulfovibrio gigas. J. Bacteriol.*, **177**, 6195–6200.

HILLE, R. (1996) The mononuclear molybdenum enzymes. *Chem. Rev.*, **96**, 2757–2816.

HOCHHEIMER, A., SCHMITZ, R. A., THAUER, R. K. and HEDDERICH, R. (1995) The tungsten formylmethanofuran dehydrogenase from *Methanobacterium thermoautotrophicum* contains sequence motifs characteristic for enzymes containing molybdopterin dinucleotide. *Eur. J. Biochem.*, **234**, 910–920.

HOWARD, J. B. and REES, D. C. (1996) Structural basis for nitrogen fixation. *Chem. Rev.*, **96**, 2965–2982.

JOCHIMSEN, B., PEINEMANN-SIMON, S., VÖLKER, H. *et al.* (1997) *Stetteria hydrogenophila*, gen. nov. and sp. nov., a novel mixotrophic sulfur-dependent *crenarchaeote* isolated from Milos, Greece. *Extremophiles*, **2**, 67–73.

JOHNSON, M. K., REES, D. C. and ADAMS, M. W. W. (1996) Tungstoenzymes. *Chem. Rev.*, **96**, 2817–2839.

KLETZIN, A. and ADAMS, M. W. W. (1996) Tungsten in biology. *FEMS Microbiol. Rev.*, **18**, 5–64.

KLETZIN, A., MUKUND, S., KELLEY-CROUSE, T. L., CHAN, M. K., REES, D. C. and ADAMS, M. W. W. (1995) Molecular characterisation of tungsten-containing enzymes from hyperthermophilic archaea: aldehyde ferredoxin oxidoreductase from *Pyrococcus furiosus* and formaldehyde ferredoxin oxidoreductase from *Thermococcus litoralis. J. Bacteriol.*, **177**, 4817–4819.

KOEHLER, B. P., MUKUND, S., CONOVER, R. C. *et al.* (1996) Spectroscopic characterisation of the tungsten and iron-sulfur centres in aldehyde ferredoxin oxidoreductases from two hyperthermophilic archaea. *J. Am. Chem. Soc.*, **118**, 12391–12405.

MA, K., HUTCHINS, A., SUNG, S.-H. S. and ADAMS, M. W. W. (1997) Pyruvate ferredoxin oxidoreductase from the hyperthermophilic archaeon, *Pyrococcus furiosus*, functions as a coenzyme A-dependent pyruvate decarboxylase, *Proc. Natl. Acad. Sci.*, **94**, 9608–9613.

MUKUND, S. and ADAMS, M. W. W. (1991) The novel tungsten-iron-sulfur protein of the hyperthermophilic archaebacterium, *Pyrococcus furiosus*, is an aldehyde ferredoxin oxidoreductase: evidence for its participation in a unique glycolytic pathway, *J. Biol. Chem.*, **266**, 14208–14216.

MUKUND, S. and ADAMS, M. W. W. (1993) Characterization of a novel tungsten-containing formaldehyde ferredoxin oxidoreductase from the extremely thermophilic archaeon, *Thermococcus litoralis*. a role for tungsten in peptide catabolism, *J. Biol. Chem.*, **268**, 13592–13600.

MUKUND, S. and ADAMS, M. W. W. (1995) Glyceraldehyde-3-phosphate ferredoxin oxidoreductase, a novel tungsten-containing enzyme with a potential glycolytic role in the hyperthermophilic archaeon, *Pyrococcus furiosus, J. Biol. Chem.*, **270**, 8389–8392.

MUKUND, S. and ADAMS, M. W. W. (1996) Molybdenum and vanadium do not replace tungsten in the three tungstoenzymes of the hyperthermophilic archaeon *Pyrococcus furiosus, J. Bacteriol.*, **178**, 163–167.

ROMAO, M. J., ARCHER, M., MOURA, I. (1995) Crystal structure of the xanthine oxidase-related aldehyde oxido-reductase from *D. gigas*. *Science*, **270**, 1170–1176.

ROSNER, B. and SCHINK, B. (1995) Purification and characterisation of acetylene hydratase of *Pelobacter acetylenicus*, a tungsten iron–sulfur protein. *J. Bacteriol.*, **177**, 5767–5772.

SAKO, Y., NOMURA, N., UCHIDA, A. *et al.* (1996) *Aeropyrum pernix* gen. nov., sp. nov., a novel aerobic hyperthermophilic archaeon growing at temperatures up to 100 °C. *Int. J. System. Bacteriol.*, **46**, 1070–1077.

SCHINDELIN, H., KISKER, C., HILTON, J., RAJAGOPALAN, K. V. and REES, D. C. (1996) Crystal structure of DMSO reductase: redox-linked changes in molybdopterin coordination. *Science*, **272**, 5268–5271.

SCHMITZ, R. A., ALBRACHT, S. P. J. and THAUER, R. K. (1992) Properties of the tungsten-substituted molybdenum formylmethanofuran dehydrogenase in *Methanosarcina wolfei*. *FEBS Lett.*, **309**, 78–81.

SCHNEIDER, F., LOWE, J., HUBER, R., SCHINDELIN, H., KISKER, C. and KNABLEIN, J. (1996) Crystal structure of dimethyl sulfoxide reductase from *Rhodobacter capsulatus* at 1.88 angstrom resolution. *J. Mol. Biol.*, **263**, 53–69.

STETTER, K. O. (1996) Hyperthermophilic prokaryotes. *FEMS Microbiol. Rev.*, **18**, 149–158.

VORHOLT, J. A., VAUPEL, M. and THAUER, R. K. (1997) A selenium-dependent and a selenium-independent formylmethanofuran dehydrogenase and their transcriptional regulation in the hyperthermophilic *Methanopyrus kandleri*. *Mol. Microbiol.*, **23**, 1033–1042.

WHITE, H., STROBL, G., FEICHT, R. and SIMON, H. (1989) Carboxylic acid reductase: a new tungsten enzyme which catalyses the reduction of non-activated carboxylic acids to aldehydes. *Eur. J. Biochem.*, **184**, 89–96.

YAMAMOTO, I., SAIKI, T., LIU, S.-M. and LJUNGDAHL, L. G. (1983) Purification and properties of NADP-dependent formate dehydrogenase from *Clostridium thermoaceticum*, a tungsten-selenium-iron protein. *J. Biol. Chem.*, **258**, 1826–1832.

Index